Staphylococcus aureus

This book presents a comprehensive overview of the intricate mechanisms of virulence, antimicrobial resistance, immune evasion, and biofilm formation of *Staphylococcus aureus*. It provides insights into mechanisms by which *S. aureus* acquires and sustains drug resistance, including gene mutation, efflux pumps, and horizontal gene transfer. The book examines the latest advancements in pathogen detection, application of artificial intelligence, and systems biology approaches in *S. aureus* research. In addition, the book examines host immune responses to persistent *S. aureus* infections, focusing on immune modulation, chronic inflammation, and evasion mechanisms. Further, the chapters review novel therapeutic interventions, including quorum-sensing inhibitors, plant-derived antimicrobials, and innovative vaccine development. By integrating genomic research, computational biology, and emerging drug discovery techniques, this book outlines the next generation of treatment options for *S. aureus*-related infections.

Key Features:

- Explores the molecular and genetic mechanisms underlying *S. aureus* virulence, drug resistance, and host adaptation
- Examines antimicrobial resistance mechanisms and the growing challenge of multidrug-resistant *S. aureus* strains
- Discusses the role of biofilm formation and quorum sensing in disease persistence and treatment resistance
- Presents state-of-the-art diagnostic technologies, including AI-driven detection and molecular profiling techniques
- Reviews novel therapeutic approaches, including plant-based antimicrobials, quorum-sensing inhibitors, and the development of vaccines

This book is an important resource for students and researchers of microbiology and pharmaceutical sciences.

Staphylococcus aureus
Pathogenesis, Drug Resistance, and Innovative Therapeutics

Edited by
Anjaneyulu Musini, Narasaiah Kolliputi,
and Vinod Kumar Yata

CRC Press
Taylor & Francis Group
Boca Raton London New York

CRC Press is an imprint of the
Taylor & Francis Group, an **informa** business

Designed cover image: Shutterstock

First edition published 2026
by CRC Press
2385 NW Executive Center Drive, Suite 320, Boca Raton FL 33431

and by CRC Press
4 Park Square, Milton Park, Abingdon, Oxon, OX14 4RN

CRC Press is an imprint of Taylor & Francis Group, LLC

ISBN: 978-1-041-08813-4 (hbk)
ISBN: 978-1-041-08814-1 (pbk)
ISBN: 978-1-003-64707-2 (ebk)

DOI: 10.1201/9781003647072

Typeset in Times
by Apex CoVantage, LLC

Contents

Foreword

Infectious diseases have long shaped the course of human health and medicine, but few pathogens have proven to be as persistent, adaptable, and clinically challenging as *Staphylococcus aureus*. Once considered a relatively manageable bacterium, *S. aureus* has evolved into a global health concern primarily due to its ability to resist antibiotics, evade immune responses, and thrive in both hospital and community settings. This book presents a timely and comprehensive examination of *S. aureus* from its basic biology and mechanisms of disease to the latest breakthroughs in detection, treatment, and prevention. It is a testament to the multidisciplinary nature of modern microbiology, blending microbiological research, clinical insight, genomics, artificial intelligence, systems biology, and therapeutic innovation.

What sets this work apart is its depth and breadth. Readers are taken on a journey through the fundamental science of *S. aureus*, including its role in serious infections, its remarkable resistance mechanisms, and its impact on public health systems. At the same time, the book explores critical forward-looking themes such as vaccine development, biofilm disruption strategies, phage therapy, and the use of AI in both diagnostics and drug discovery. Importantly, the authors do not shy away from the complexity of the problem. They approach *S. aureus* not simply as a microbial enemy but as a sophisticated organism that interacts dynamically with its environment, hosts, and other microbes. This perspective is vital if we are to develop truly effective and sustainable strategies to combat it.

Whether you are a student just beginning to explore microbiology, a researcher investigating antibiotic resistance, a healthcare professional battling infections on the front lines, or a policy-maker concerned with antimicrobial stewardship, this book offers critical insights and practical knowledge. It reflects not only scientific progress but also a call to action, urging us to continue innovating, collaborating, and responding with urgency and precision.

I commend the authors and contributors for creating a resource that is both academically rigorous and accessible. This book will no doubt become a valuable reference in the ongoing battle against *S. aureus* and antimicrobial resistance more broadly.

Dr. Naga Raju Maddela
Professor
Departamento de Ciencias Biológicas,
Facultad de Ciencias de la Salud
Universidad Técnica de Manabí,
Portoviejo, 130105, Ecuador

Preface

Staphylococcus aureus is a formidable pathogen that has challenged clinicians, researchers, and public health systems for decades. Once considered a mere skin colonizer, it has now emerged as one of the most adaptive and persistent bacteria responsible for a wide spectrum of diseases, from superficial infections to severe systemic conditions like pneumonia, sepsis, endocarditis, and even cancer-associated complications. Its ability to rapidly evolve, acquire resistance to multiple antibiotics, form biofilms, and evade host immune responses has made it a major threat in both community and healthcare settings. This book is a comprehensive compilation of current knowledge, emerging research, and future directions in understanding, detecting, managing, and preventing *S. aureus* infections. Spanning 16 detailed chapters, it presents a multidisciplinary overview of the pathogen's biological behavior, genetic complexity, host interactions, resistance mechanisms, and the global implications of its rise as a drug-resistant superbug.

Beginning with an introduction to the bacterium's general microbiological and clinical features, the text delves into the molecular mechanisms of its pathogenesis, the evolution of methicillin-resistant strains, and the intricacies of biofilm development and quorum sensing. Subsequent chapters explore cutting-edge topics such as host immune responses, zoonotic aspects like bovine mastitis, and the intersection between chronic infection and carcinogenesis. The book further investigates technological advancements in diagnostics, AI-based research models, systems biology, and novel therapeutic strategies, including vaccine development, quorum-sensing inhibitors, bacteriophage therapy, and nanotechnology. By emphasizing both foundational principles and novel innovations, it equips students, researchers, clinicians, microbiologists, and biotechnologists with the insights necessary to confront this ever-adapting pathogen.

We acknowledge the collaborative contributions of scientists and healthcare professionals worldwide who continue to push the boundaries of our understanding of *S. aureus*. As the global burden of antimicrobial resistance escalates, this work underscores the urgent need for integrated, interdisciplinary approaches to surveillance, treatment, and prevention.

It is our hope that this book will serve as a valuable resource in academic, clinical, and research settings, fostering informed action and inspiring future innovations to control and ultimately conquer the threat of *S. aureus*.

Anjaneyulu Musini, PhD
Assistant Professor
Jawaharlal Technological University Hyderabad
Kukatpally, Hyderabad, India

Narasaiah Kolliputi, PhD
Associate Professor
University of South Florida
Tampa, Florida, USA

Vinod Kumar Yata, PhD
Associate Professor
Malla Reddy University, Maisammaguda
Hyderabad, Telangana, India

Acknowledgments

We would like to express our sincere gratitude to all the contributors who have shared their valuable insights, cutting-edge research, and clinical perspectives in this volume. Their expertise has been instrumental in shaping this comprehensive resource on *Staphylococcus aureus*, encompassing its pathogenesis, mechanisms of drug resistance, and emerging therapeutic strategies. We are deeply thankful to CRC Press (Taylor & Francis Group) for its continued support and for providing the platform to disseminate this important work to the global scientific community. Special thanks to the editorial and production teams for their guidance and professionalism throughout the publishing process. We also acknowledge the academic and research institutions that supported the authors and editors who have enabled the pursuit of scientific advancements in infectious diseases and antimicrobial resistance. We dedicate this book to the countless healthcare professionals, microbiologists, and researchers worldwide who continue to battle against antibiotic-resistant pathogens and strive to improve public health outcomes through innovation and collaboration. Last but not least, we are profoundly thankful to our families and friends for their unwavering support, patience, and encouragement. Their understanding and sacrifices have been the foundation that allowed us to devote the time and energy required for this endeavor.

Dr. Anjaneyulu Musini
Dr. Narasaiah Kolliputi
Dr. Vinod Kumar Yata

About the Editors

Anjaneyulu Musini is Assistant Professor at the Centre for Biotechnology, University College of Engineering, Science and Technology, Jawaharlal Nehru Technological University (JNTU), Hyderabad, India. He earned his Ph.D. in Biotechnology in 2016 from the Institute of Science and Technology, JNTUH, and has over 15 years of academic and research experience in the areas of Biotechnology, Microbiology, and Natural Product Pharmacology. Dr. Musini completed his M.Tech. research at the Department of Biochemistry, Indian Institute of Science (IISc), Bangalore, where he developed a strong foundation in anticancer and antimicrobial research. His current research interests include the discovery and characterization of bioactive compounds from medicinal plants, drug-resistant pathogens, fermentation-based production of biomolecules such as L-Arginine, and natural product pharmacology. He has also contributed to interdisciplinary research in the fields of bioplastics, biofilms, nanoparticles, and quorum-sensing inhibitors. He has successfully completed a TEQIP-III-funded R&D project investigating the antibacterial properties of *Salacia oblonga* against drug-resistant *Staphylococcus aureus*. His doctoral work focused on the extensive phytochemical and biological characterization of *Salacia oblonga*, resulting in several high-impact publications. Dr. Musini has published over 16 peer-reviewed journal articles, 10 book chapters, and 1 book. He holds one published patent and has filed a design patent. His research has been presented at numerous national and international conferences. In addition to his research, Dr. Musini actively contributes to academic development through curriculum design, participation in the Internal Quality Assurance Cell (IQAC) and involvement in NBA accreditation processes at JNTUH. He has guided M.Sc. and M.Tech. students in Biotechnology and Microbiology and is skilled in diverse technical areas ranging from molecular biology to fermentation technology. His qualifications and recognitions include the GATE (2007), DST-TIFAC Senior Research Fellowship (2010), Canadian Commonwealth Fellowship (2012), and UGC-SET qualification (2018). He has also completed several NPTEL-FDP certifications. With nine years of post-doctoral teaching and research experience, Dr. Musini was recently appointed as an editorial board member of Scientific Reports (Springer Nature).

Narasaiah Kolliputi is distinguished Associate Professor at the University of South Florida (USF), Morsani College of Medicine, with interdisciplinary appointments in the Departments of Pathology & Cell Biology, Pharmaceutical Sciences, Internal Medicine, Pediatrics, and Molecular Medicine. He joined USF in 2010 as Assistant Professor in the Division of Allergy and Immunology and was tenured and promoted to Associate Professor in 2013. Dr. Kolliputi completed his postdoctoral training in Physiology at Oklahoma State University (2004–2006), followed by advanced research in Pulmonary and Critical Care Medicine at Massachusetts General Hospital and Harvard Medical School (2006–2008), where he subsequently served as a Research Scientist (2008–2009) before joining USF. A recognized expert in lung injury, inflammasome biology, oxidative stress, and redox signaling, Dr. Kolliputi has built a translational research program that bridges basic molecular mechanisms with clinical implications in acute lung injury, pulmonary fibrosis, asthma, and inflammatory cell death (PANoptosis). His research has been continuously supported by the NIH, including R01, R21, and R56 awards totaling nearly $2.91 million (2013–2020), an AHA National Scientist Development Grant ($308K), and multiple USF institutional awards (Internal, PEG, G, and GME). In total, his PI-level portfolio exceeds $4.49M, and his cumulative research portfolio, including Co-I and trainee-linked funding, exceeds $5.73M.

Dr. Kolliputi's research productivity is reflected in over 100 peer-reviewed publications, including 27 papers from the past five years in high-impact journals such as *Scientific Reports*, *Frontiers in Pharmacology*, *FASEB Journal*, *Aging Cell*, and *the American Journal of Physiology*. His laboratory has been awarded two U.S. patents, and he was elected to the USF Chapter of the National Academy of Inventors (NAI), with a dedicated brick at the Inventors Commons in the USF Research Park. A committed mentor and educator, Dr. Kolliputi has trained over 100 undergraduate, graduate, and medical students, many of whom have achieved national recognition through AHA postdoctoral and predoctoral fellowships, GME Research Awards, an AΩA Student Research Fellowship, and USF Health Research Day honors. In recognition of his contributions to science and education, he has received numerous institutional awards, including the USF Outstanding Research Achievement Award (2017), the Internal Medicine Outstanding Research Award (2017), and the Theodore and Venette Askounes-Ashford Distinguished Scholar Award (2018). He also received the Young Investigator Award from the Society for Experimental Biology and Medicine (SEBM).

Dr. Kolliputi serves as Chairman of NIH Study Sections and Special Emphasis Panels (NHLBI) and as Associate Editor for *Frontiers in Pharmacology*, *Frontiers in Physiology*, and *Frontiers in Genetics*, while also serving as a reviewer for the NIH, DoD, AHA, and leading scientific journals. Through his innovative research, mentorship, and leadership, Dr. Kolliputi continues to advance USF's translational mission, expanding the national and international visibility of the Division of Allergy and Immunology.

Vinod Kumar Yata is an Associate Professor at the School of Allied and Healthcare Sciences, Malla Reddy University, Hyderabad, Telangana, India. Previously, he served as Assistant Professor at the Department of Molecular Biology, Central University of Andhra Pradesh, India, prior to which he worked as a Research Scientist at the Center of Excellence for Aging and Brain Repair at the University of South Florida (USF) in Tampa, Florida, USA. He also worked as a Research Associate at the National Dairy Research Institute in Karnal, India. Previously, he also worked as Assistant Professor at the Department of Biotechnology, Dr. B R Ambedkar National Institute of Technology, Jalandhar, Punjab, India. He received his Ph.D. in Biotechnology from the Indian Institute of Technology Guwahati. He specializes in interdisciplinary research, which includes Animal Biotechnology, Microbiome in Aging, Cancer Biology, Nanotechnology, Microfluidics, and Bioinformatics. He has developed a microfluidic device for the separation of live and motile spermatozoa from cattle semen samples. He has published several research articles in peer-reviewed international journals and presented papers at several international conferences. He published eight books as an editor and two books as an author.

Section I

Pathogenesis and Molecular Foundations of Staphylococcus aureus

1 Introduction to *Staphylococcus aureus*

General Overview

Anjaneyulu Musini and P. Gnana Deepu
Centre for Biotechnology, University of College Engineering Science and Technology, Jawaharlal Nehru Technological University, Hyderabad, India

ABBREVIATIONS

DNA	deoxyribose nucleic acid
IgG	immunoglobulin G
LAMP	loop-mediated isothermal amplification
MIRA	multienzyme isothermal rapid amplification
MRSA	methicillin-resistant *Staphylococcus aureus*
PBP2A	penicillin-binding protein
PCR	polymerase chain reaction
PVL	panton–valentine leukocidin
S. aureus	*Staphylococcus aureus*
S. epidermidis	*Staphylococcus epidermidis*
SCCmec	staphylococcal chromosomal cassette mec
SSSS	staphylococcal scalded skin syndrome
SSTIs	skin and soft tissue infections
TSS	toxic shock syndrome
VISA	vancomycin-intermediate *Staphylococcus aureus*
VRSA	vancomycin-resistant *Staphylococcus aureus*

1.1 OVERVIEW OF *STAPHYLOCOCCUS AUREUS*

Staphylococcus aureus, often known as *S. aureus* or staph, is a gram-positive bacterium that causes a wide range of diseases, from mild skin infections to serious, life-threatening issues (Tong et al., 2015). It belongs to the Staphylococcaceae family and is a harmless bacterium that can be found on the skin and mucous membranes of people and animals. It was first found in the 1880s by Scottish doctor Sir Alexander Ogston and has been linked to wound infections thereafter. The bacterium's name is derived from the Greek words "staphyle," meaning bunch of grapes, and "kokkos," meaning berry (Kazmi, 2022), alluding to its clustered appearance under a microscope. The species name "aureus" relates to its golden-yellow hue, which is commonly observed in colonies (Mustapha et al., 2014).

S. aureus is a spherical, cocci-shaped bacterium with grape-like clusters that may be seen under a microscope. The purple stain suggests a strong peptidoglycan coating in the cell wall, which is typical of gram-positive bacteria (Masalha et al., 2001). Growing in high-salt conditions, they look golden or yellow on growth media (Missiakas & Schneewind, 2013). It is a prevalent bacterium in both the environment and human flora, frequently found on the skin and mucous membranes, particularly in the nasal area. While colonization does not necessarily result in disease, it can be troublesome if germs enter the body through skin breaches (Mustapha et al., 2014).

DOI: 10.1201/9781003647072-2

S. aureus is a pathogenic bacterium renowned for producing various extracellular proteins and polysaccharides, such as enzymes and toxins, allowing it to infiltrate host tissues and elude the immune system (Tam & Torres, 2019). This complex pathophysiology makes it a difficult pathogen to control, and a better understanding of its features, infection processes, and interactions with the host is needed. It generates diverse toxins and enzymes, including hemolysins, leukocidins, exfoliative toxins, enterotoxins, and toxic shock syndrome toxin-1 (Abril et al., 2020). These toxins increase the bacterium's pathogenicity by allowing it to cling to host tissues, avoid the immune system, gain nutrition, and destroy host cells. Its ability to form biofilms on surfaces like medical devices or tissues is a significant virulence trait, protecting the bacteria from antibiotics and immune responses. Enzymes like coagulase and staphylokinase help evade the immune system, allowing the bacteria to colonize, invade, and spread in the host, resulting in various infections (Tam & Torres, 2019). It is essential to comprehend the virulence factors of *S. aureus* for effectively managing and preventing diseases caused by this bacterium.

S. aureus, a widespread bacterium, colonizes around 30% of the human population asymptomatically, particularly in the anterior nares (nostrils). The frequency of colonization varies with age, geography, and healthcare exposure (Gehrke et al., 2023). *S. aureus* is a leading source of hospital-acquired and community-acquired infections, and methicillin-resistant *S. aureus* (MRSA) poses considerable public health risks. MRSA strains are especially dangerous in hospital settings because they are resistant to many medications and can cause serious infections. *S. aureus* may cause everything from minor skin irritations to fatal illnesses, contributing to major public health problems (Guo et al., 2020). It is a significant bacterium that causes a wide range of illnesses, from simple skin problems to more serious disorders such as blood poisoning, lung inflammation, and heart valve infection. Its pathogenicity stems from a mix of virulence factors, such as toxins, enzymes, and immune evasion mechanisms (Cheung et al., 2021). The bacterium's antibiotic resistance, particularly in MRSA, complicates treatment choices and increases morbidity and mortality (Guo et al., 2020). The economic impact of *S. aureus*-related infections is significant, including the expenditures of extended hospital stays, costly treatments, and public health measures. It is a major cause of community-acquired and healthcare-associated infections, such as skin and soft tissue infections (SSTIs), bacteremia, endocarditis, osteomyelitis, and pneumonia (Prestinaci et al., 2015).

S. aureus is a type of bacterium having round cells about 0.5 to 1.5 micrometers in size. When examined under a microscope, these cells cluster together like grape clusters, a distinguishing feature of this variety of bacteria (Mirbagheri et al., 2023). The cell wall is an important component that contributes to both structural integrity and pathogenicity. A thick peptidoglycan layer, which accounts for almost half of the cell wall's dry weight, offers stiffness and resistance to osmotic pressure (Beeby et al., 2013). Teichoic acids and lipoteichoic acids are also found in the cell wall, where they serve important roles in cell wall maintenance, ion homeostasis, and host tissue adhesion. The peptidoglycan layer's thickness and presence of teichoic acids help preserve the crystal violet stain during Gram staining, giving *S. aureus* its distinctive purple color under the microscope (Silhavy et al., 2010) (Figure 1.1).

It can generate a protective outer coating of carbohydrates known as a capsule, which encases the cell wall and plays an important role in its ability to cause damage. Bacteria with capsules are typically more dangerous than those without capsules. Protein A binds to the Fc region of immunoglobulin G (IgG), which inhibits opsonization and subsequent phagocytosis (Gao et al., 2024; Santos-López et al., 2021).

S. aureus' genetic material is arranged into a single circular chromosome with approximately 2.8 million base pairs in length. This chromosome encodes the bacterium's critical capabilities, including protein synthesis, cell division, and crucial virulence proteins that contribute to its pathogenicity (Zubair et al., 2015). In addition to chromosomal DNA, *S. aureus* typically carries plasmids. These plasmids are tiny, circular, double-stranded DNA molecules that may replicate

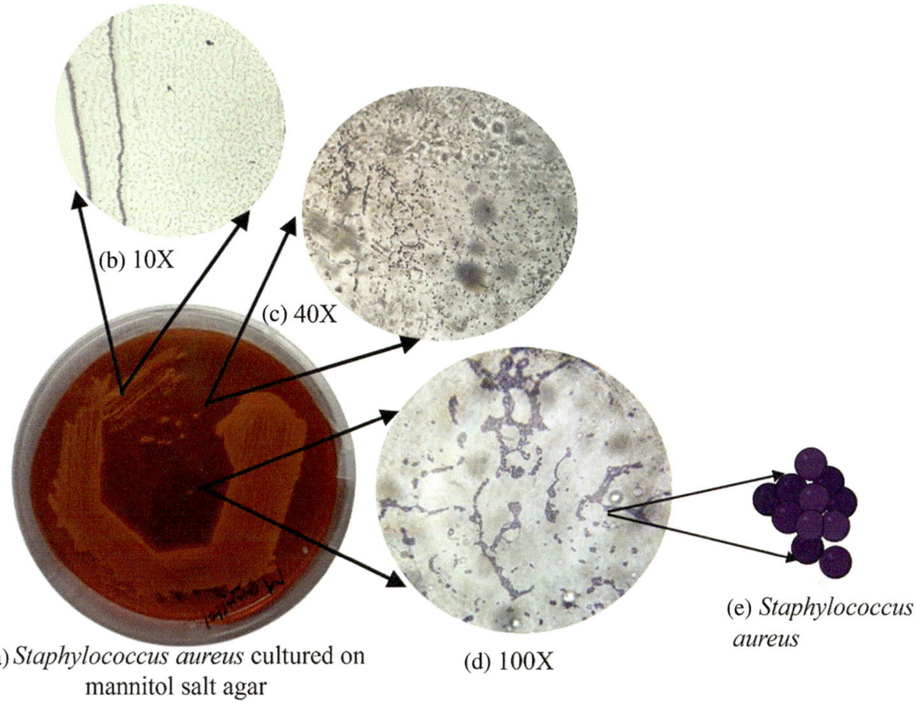

FIGURE 1.1 *Staphylococcus aureus* cultured on mannitol salt agar, a selective medium. The bacterial growth is visible as yellow colonies on the agar plate, indicating mannitol fermentation. Microscopic images of *S. aureus* at different magnifications 10X, 40X, and 100X, which appears as clusters of cocci. The characteristic grape-like clusters of *S. aureus* observed under higher magnification.

independently and contain additional genes for a variety of tasks. Plasmids can carry genes that confer antibiotic resistance, encode toxins, and promote other characteristics that enable the bacteria to thrive in a variety of conditions and cause infections (Malachowa& DeLeo, 2010; Kwong et al., 2017).

1.2 HABITAT AND NATURAL RESERVOIRS OF *S. AUREUS*

S. aureus prefers to live on human skin and mucous membranes. Secondary habitats for *S. aureus* include clinical and community settings (Linz et al., 2023; Tong et al., 2015). In healthcare environments such as hospitals and long-term care institutions, *S. aureus* can contaminate numerous surfaces, medical equipment, and healthcare workers' hands, allowing it to spread. Furthermore, *S. aureus* may be found in community settings such as homes, schools, and public transportation, where it can be transmitted by direct contact with sick people or contaminated surfaces (Jaradat et al., 2020; Jabłońska-Trypuć et al., 2022).

1.2.1 Habitat of *S. aureus*

1.2.1.1 Nasal Passage

S. aureus is known to live mostly in the nasal tube, namely, the anterior nares. *S. aureus* lives permanently in the nasal passages of around 20% of people and is only present occasionally in the remaining 60%. These bacteria flourish in the moist squamous epithelium of the nasal septum near the nasal ostium, which contains a high number of *S. aureus* cells (Danelli et al., 2020). The nasal epithelium has different structures. The anterior nares have a keratinized, stratified squamous epithelium, whereas the posterior nares have pseudostratified, columnar ciliated epithelial cells (Sakr et al., 2018).

1.2.1.2 Skin

S. aureus is commonly found on human skin, especially in the axillae, groin, and perineum. It can temporarily colonize healthy human skin, particularly in wet regions. Atopic dermatitis patients frequently have persistent *S. aureus* colonization on both inflamed and non-inflamed skin areas (Kim et al., 2019). Coagulase-negative staphylococci, such as *S. epidermidis*, thrive in moist environments like the axillae and are also frequent in the anterior nares. Although not as diverse as the gut microbiome, the skin's collection of bacteria is critical for *S. aureus* colonization and interaction with the body (Gehrke et al., 2023).

1.2.1.3 Other Body Parts

S. aureus can be detected in several other bodily areas, including the rectum, vagina, gastrointestinal system, and axilla (Figure 1.2). The anterior nares are a key reservoir from which *S. aureus* can spread to various cutaneous and mucosal locations, interacting with epithelial cell ligands such as loricrin and cytokeratin 10 (K10) (Sakr et al., 2018). *S. aureus*' capacity to travel widely demonstrates how it may live in many different regions of the body, operating as both a typical resident and a disease-causing agent (Sakr et al., 2018; Tong et al., 2015).

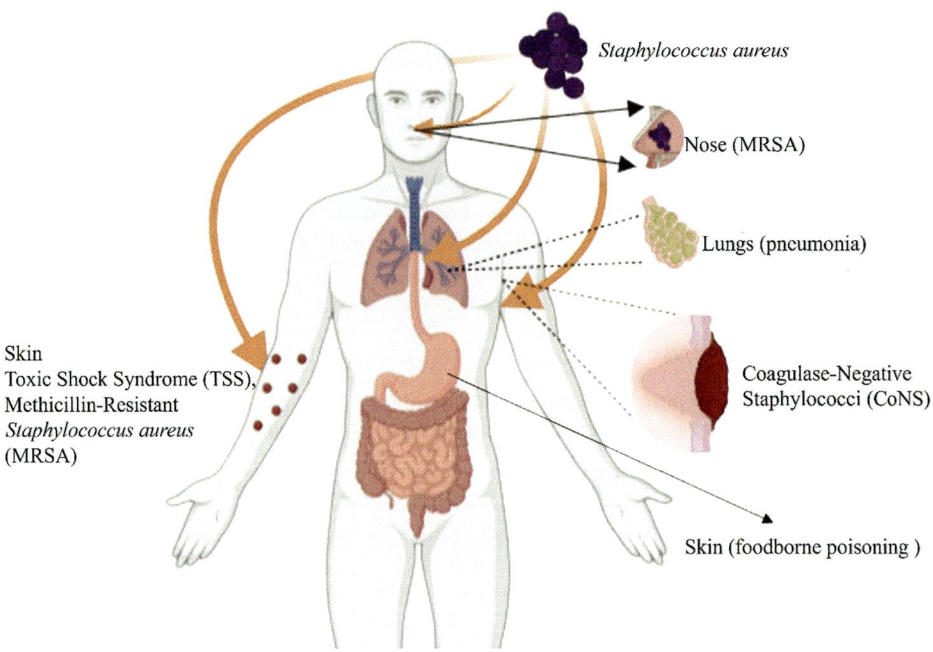

FIGURE 1.2 Several anatomical sites where *Staphylococcus aureus* can grow and cause infection. The organism is often detected in the nose, where methicillin-resistant *S. aureus* (MRSA) can live. It can also infect the lungs, resulting in pneumonia, and the skin, causing toxic shock syndrome (TSS) and MRSA infections, and *S. aureus* is linked to food poisoning and skin diseases.

1.2.2 Factors Influencing Its Survival and Growth in Different Environments

The survival and growth of *S. aureus* are heavily influenced by the environmental conditions it finds itself in. These ecological factors are essential in determining an *S. aureus*'s ability to survive, reproduce, and ultimately thrive in a particular habitat. From temperature and humidity to the availability of resources such as nutrients and oxygen, various factors come into play and interact with each other, shaping the adaptations and strategies employed by *S. aureus* (Balasubramanian et al., 2017). Understanding these factors and their impact on survival and growth is essential for studying and conserving biodiversity, as well as predicting the response of organisms to changes in their environment, such as climate change or habitat destruction (Weiskopf et al., 2020)

Factors like Temperature and Humidity: *S. aureus* can thrive in a variety of temperatures but likes moderate humidity. Its capacity to withstand dry conditions for lengthy periods helps it stick to surfaces (Kozajda et al., 2019). Nutrient Availability: *S. aureus* needs nutrients to develop and reproduce. It thrives in nutrient-rich settings like skin and mucous membranes, where it has access to proteins, lipids, and other organic substances. In hospital settings, it may also use nutrients contained in biological fluids and contaminated medical equipment (Krismer et al., 2014). pH levels: *S. aureus* favors neutral to slightly alkaline pH values; however, it may survive in a variety of pH conditions. This adaptability enables it to colonize multiple bodily places and live in a variety of environmental situations (Zhou & Fey, 2020). Oxygen Availability: Because *S. aureus* can develop in both oxygen-rich and oxygen-depleted situations, it can survive in both aerobic and anaerobic settings. This adaptability allows it to colonize both oxygen-rich surfaces like the skin and oxygen-deficient habitats like deep tissue infections (Yagüe et al., 2021); it greatly influences the survivability and growth of the *S. aureus*.

1.2.3 Prevalence of *S. aureus* in Various Ecological Niches

S. aureus is widely dispersed in various kinds of ecological environments, demonstrating its flexibility and tenacity. Humans and animals serve as the principal natural reservoirs for *S. aureus*. In humans, the anterior nares are the major reservoirs, but colonization can spread to other skin locations and mucous membranes (Balasubramanian et al., 2017; Tong et al., 2015). *S. aureus* colonizes the skin, nasal passages, and gastrointestinal tracts of cattle, pigs, and poultry, as well as pets like dogs, cats, and wild animals, forming a reservoir for zoonotic transmission. It can contaminate goods and surfaces in food production facilities, especially those that handle dairy and meat products, creating a risk of foodborne disease (Kwiecinski & Horswill, 2020).

Mechanisms like Adhesion: *S. aureus* has surface proteins that aid in attachment to host tissues, such as clumping factors, fibronectin-binding proteins, and collagen-binding proteins. These adhesins allow the bacteria to bind to epithelial cells in the nares, skin, and mucous membranes (Paharik & Horswill, 2016). Immunological Evasion: *S. aureus* generates a variety of components that aid in evading the host immunological response, including protein A, which binds to the Fc region of IgG and inhibits opsonization and phagocytosis. The polysaccharide capsule also prevents phagocytosis (De Jong et al., 2019). Biofilm Formation: *S. aureus* may develop biofilms on various surfaces, including medical equipment and host tissues. Biofilms create a protective habitat that improves bacterial survival, promotes chronic colonization, and increases antibiotic resistance and immunological responses (Peng et al., 2022). Toxin Production: *S. aureus* produces and sustains infections by producing toxins such as hemolysins, leukocidins, and exfoliative toxins, which contribute to tissue damage and immune evasion (Cheung et al., 2021); it helps understand how *S. aureus* survives in various ecological niches.

Understanding the habitat and common locations is critical for understanding the disease's epidemiology, transmission dynamics, and pathogenic components. *S. aureus* is a versatile and opportunistic organism that causes a wide range of infections, including simple skin infections and serious, life-threatening diseases, including sepsis, pneumonia, and endocarditis (Tong et al., 2015). Researchers can develop better strategies for preventing and controlling the spread of *S. aureus* by studying its habitat and common locations, such as implementing targeted sanitation protocols and surveillance measures, particularly in healthcare settings where it is a leading cause of hospital-acquired infections (Haque et al., 2020; Piewngam & Otto, 2024). Furthermore, understanding where it is usually found might help create targeted treatments, such as decontamination protocols and immunization techniques, to lessen the burden of *S. aureus* infections in both human and animal populations.

1.3 VIRULENCE FACTORS AND PATHOGENESIS

1.3.1 Virulence Factors

S. aureus develops various types of virulence factors that contribute to its pathogenesis. The host environment influences the synthesis of these virulence factors, with iron availability, for example, having a substantial impact on their expression. Global regulators such as RNA III govern the production of virulence factors, demonstrating the bacteria's flexibility to the host environment (Tam & Torres, 2019).

These include the following. Surface Proteins: These help cells cling to host tissues and evade the immune system. For example, protein A attaches to the Fc region of immunoglobulins, whereas clumping factor binds to fibrinogen (Foster, 2019). Exoenzymes: *S. aureus* creates enzymes such as coagulase, which clogs blood plasma to avoid the immune system, and hyaluronidase, which degrades hyaluronic acid in connective tissue, allowing infection to spread (Tam & Torres, 2019). Toxins: The bacteria produce toxins such as hemolysins, α-toxin and β-toxin, which lyse red and white blood cells, causing tissue injury and immune evasion (Otto, 2014). Leukocidins: Panton–valentine leukocidin (PVL) kills leukocytes, resulting in severe skin infections and necrotizing pneumonia (Adler et al., 2006). Exfoliative Toxins: These toxins cause the outer skin layer to slough, which can result in disorders like staphylococcal scalded skin syndrome (Adler et al., 2006). Enterotoxins: These are heat-stable toxins that cause staphylococcal food poisoning through vomiting and diarrhea (Hennekinne et al., 2012). Immune Evasion Mechanisms: *S. aureus* may survive inside host cells and form biofilms, which protect the bacterial population from the host's immune system and antibiotics (Cheung et al., 2021; Peng et al., 2022).

1.3.2 Pathogenesis

S. aureus is a widespread skin and mucous membrane colonizer that affects around 30% of the population. It adheres to host tissues thanks to cell wall-associated components such as clumping factors A and B, which bind to extracellular matrix proteins and fibrin (Linz et al., 2023). Once colonized, *S. aureus* can infiltrate local tissues, causing infections ranging from simple skin irritations to more serious diseases. These infections often present as SSTIs, but they can also lead to more serious results, such as osteomyelitis and infective endocarditis, when bacteria adhere to damaged heart valves or bone tissue (González-García et al., 2023).

1.3.2.1 Systemic Infection

The ability of *S. aureus* to generate systemic infections is important to its pathophysiology. If the germs enter the circulation, they can spread throughout the body, causing potentially fatal illnesses, including septicemia and septic shock (Kwiecinski & Horswill, 2020). The bacteria's resilience, as seen by its capacity to withstand severe environments such as drying, high temperatures, and stomach acid, increases its virulence and adds to the intricacy of systemic infections. Furthermore, *S. aureus*'s capacity to attach to and grow on the surfaces of implanted devices, such as artificial joints and cardiac pacemakers, complicates treatment plans and increases the risk of persistent infection (Linz et al., 2023; Yamazaki et al., 2024).

1.3.2.2 Pathogenic Mechanisms

S. aureus employs various kinds of mechanisms to cause infection, avoid the immune system, and penetrate host tissues. These pathways are essential to the bacterium's pathogenicity and capacity to cause illness (Cheung et al., 2021).

1.3.2.3 Host Colonization

S. aureus is highly proficient in colonizing the human body, particularly the nasal passages and skin. Cell wall-associated components such as clumping factors A and B help the bacteria cling to host tissues, enhancing colonization, particularly in the cardiac endothelium (Bhattacharya & Horswill, 2024). Persistent *S. aureus* carriers with a high bacterial load are more likely to get infected. This is owing to the bacterium's capacity to avoid host defenses and establish itself within host tissues (Tong et al., 2015).

1.3.2.4 Immune Evasion

One of the most formidable characteristics of *S. aureus* is its ability to escape the immune system. It secretes a variety of proteins that inhibit neutrophil chemotaxis, complement activation, opsonization, and phagocytic death of bacteria (Thammavongsa et al., 2015). Furthermore, the bacteria create coagulases, which combine with prothrombin to form fibrin clots, successfully protecting themselves from host phagocytes. This process is critical for abscess development in contaminated tissues (Cheung et al., 2021). Moreover, *S. aureus* generates protein A, which binds to immunoglobulin G (IgG) and inhibits opsonization and phagocytosis, limiting the host's capacity to mount an efficient immune response (Falugi et al., 2013).

1.3.2.5 Tissue Invasion

S. aureus can enter host tissues through a variety of methods. FnBPA, an invasive protein, connects fibronectin to α5β1 integrins on endothelial cells, allowing bacteria to invade them (Shinji et al., 2011). This connection causes the bacteria to internalize into host cells, which is critical for the development of diseases, including infective endocarditis and metastatic abscess formation. *S. aureus*'s propensity to infiltrate and remain within host cells makes therapy difficult and raises the risk of chronic infection (Cheung et al., 2021).

1.3.3 Modes of Transmission

S. aureus spreads by direct contact with contaminated surfaces, person-to-person contact, and, sometimes, via airborne pathways (Kozajda et al., 2019). Colonization usually happens on the skin and mucosal surfaces, especially in the anterior nares, axillae, groin, and perineum. Carriers of *S. aureus*, particularly those infected with MRSA, can act as reservoirs for the bacterium's spread in both community and clinical settings (Tong et al., 2015).

1.3.3.1 Types of *S. aureus* Infections

The most frequent *S. aureus* infections are SSTIs, such as impetigo, folliculitis, furuncles (boils), carbuncles, cellulitis, and abscesses. SSTIs can range from medium to severe, occasionally developing

TABLE 1.1
Types of *Staphylococcus aureus* Infections

S. No.	Types of *Staphylococcus aureus* Infections	Examples/Descriptions	References
1	Skin and soft tissue infections	Boils, abscesses, and cellulitis	Knox et al. (2015); Ki and Rotstein (2008)
2	Bloodstream infections (bacteremia)	Can lead to sepsis	Kwiecinski and Horswill (2020)
3	Pneumonia	Often hospital-acquired	Nakou et al. (2009)
4	Endocarditis	Infection of heart valves	Grapsa et al. (2022)
5	Osteomyelitis	Bone infection	Colston and Atkins (2018)

into necrotizing fasciitis (Knox et al., 2015; Ki & Rotstein, 2008). *S. aureus* bloodstream infections, also known as bacteremia, can cause serious consequences such as sepsis, septic shock, and metastatic infections, including endocarditis and osteomyelitis (Kwiecinski & Horswill, 2020). *S. aureus* can cause pneumonia, both in the community and in the hospital. MRSA is a common cause of severe necrotizing pneumonia, particularly in healthy young people (Nakou et al., 2009). *S. aureus* infection of heart valves can cause acute bacterial endocarditis with a fast onset and severe symptoms. It is associated with a significant morbidity and death rate (Grapsa et al., 2022). Osteomyelitis and Septic Arthritis: These diseases affect bones and joints and are commonly caused by hematogenous spread or direct injection. *S. aureus* is a leading cause of osteomyelitis and septic arthritis, especially in youngsters and those with artificial joints (Colston & Atkins, 2018). *S. aureus* causes toxin-mediated illnesses, such as Staphylococcal scalded skin syndrome (SSSS), TSS, and food poisoning (Table 1.1).

1.4 CLINICAL MANIFESTATIONS OF *S. AUREUS* INFECTIONS

1.4.1 Skin and Soft Tissue Infections

S. aureus is the most common cause of SSTIs, which can be superficial or profound. Folliculitis, a frequent symptom, is characterized by clusters of little red papules or pustules that may cure without leaving scars (Knox et al., 2015; Ki & Rotstein, 2008). More severe types, such as furunculosis and carbuncles, affect deeper layers of the skin and may lead to more serious consequences. Impetigo, another SSTI caused by *S. aureus*, is distinguished by a golden crust growing over red lesions and primarily affects youngsters (Knox et al., 2015).

1.4.2 Respiratory Infections

Staphylococcal pneumonia, a serious respiratory infection, frequently accompanies a viral infection like influenza and can result in consequences like bacteremia or sepsis (Kwiecinski & Horswill, 2020). It can be caused by either direct inhalation of germs or hematogenous dissemination from other infected locations. Because of their susceptibility to fast illness progression, the infection is especially harmful in children and the elderly (Archibald & Quisling, 2013).

1.4.3 Food Poisoning

Staphylococcal food poisoning is caused by the intake of enterotoxins generated by *S. aureus* in inadequately kept or handled foods. Symptoms such as nausea, vomiting, and diarrhea normally appear quickly after consuming the contaminated food and are self-limiting, usually disappearing within 24 hours (Argudín et al., 2010). Creamy sweets, processed meats, and dairy products are all regularly implicated foods. Proper food handling and storage techniques play an important role in prevention (Lorenzo et al., 2018).

1.4.4 Nosocomial Infections

S. aureus is a well-known source of nosocomial infections, especially MRSA, which presents considerable issues in hospital settings due to its resistance to several medications (Tong et al., 2015). These infections can cause longer hospital stays and higher healthcare expenses. MRSA is frequently linked with serious and sometimes fatal diseases such as septicemia and endocarditis, particularly in patients with invasive devices or weakened immune systems (Guo et al., 2020).

1.4.5 Symptoms and Clinical Presentation

The clinical appearance of *S. aureus* infections varies greatly depending on the location and intensity of the infection. SSTIs are characterized by regional redness, swelling, discomfort, and sometimes purulent discharge. Severe instances may appear with systemic symptoms such as fever and malaise. Bacteremia and sepsis can cause high fever, chills, hypotension, and organ failure (Knox et al., 2015). Pneumonia symptoms include coughing, chest discomfort, dyspnea, and purulent sputum production, which may be followed by systemic symptoms. Endocarditis causes fever, heart murmurs, and embolic events, which can progress to heart failure if left untreated. Complications might include valve damage, septic emboli, and the need for surgical intervention (Yamaguchi et al., 2009). Osteomyelitis and septic arthritis are distinguished by localized pain, edema, redness, and limited mobility of the afflicted region, as well as systemic signs (Colston & Atkins, 2018). Toxin-mediated disorders, such as SSSS, are characterized by extensive skin blistering and desquamation; TSS by a quick onset of high fever, rash, hypotension, and multi-organ involvement; and food poisoning by vomiting and diarrhea (Gaither, 2011).

1.4.6 Antibiotic Resistance

1.4.6.1 Methicillin-Resistant *S. aureus*

MRSA is a significant clinical challenge due to its resistance to multiple antibiotics, including methicillin and other beta-lactams. This resistance is principally provided by the mecA gene on the staphylococcal chromosomal cassette mec (SCCmec), which encodes the penicillin-binding protein 2A (PBP2A) (Dhungel et al., 2021). PBP2A has a poor affinity for beta-lactams, which allows the bacterium to resist antibiotic treatment. The widespread use of antibiotics, frequently for diseases that do not require them, has accelerated MRSA's development (Fishovitz et al., 2014). This bacterium is capable of causing serious infections such as pneumonia, sepsis, and surgical-site infections, especially in hospital settings where antibiotic usage is common (Alghamdi et al., 2023).

1.4.6.2 Vancomycin-Intermediate and Vancomycin-Resistant *S. aureus*

Initially, vancomycin was the drug of choice for MRSA infections. However, strains of *S. aureus* showing intermediate resistance (VISA) and even full resistance (VRSA) to vancomycin have emerged (Cong et al., 2020). These strains carry the vanA gene cluster, which changes the peptidoglycan cell wall precursor targets of vancomycin, dramatically lowering the drug's efficacy (Van Der Aart et al., 2016). The vanA gene cluster is frequently found on mobile genetic elements, allowing it to be transferred across bacteria, such as from *Enterococcus* species to *S. aureus*, potentially spreading resistance to other bacterial species. VRSA viruses are particularly problematic since they have resistance to both methicillin and vancomycin, leaving fewer treatment choices (Loomba et al., 2010).

1.4.6.3 Resistance Mechanisms

The resistance mechanisms in MRSA involve several strategies beyond the mecA gene. The SCCmec element, which contains mecA, is categorized into many categories, each with its own set of genes, such as the ccr and mec complex classes (Paterson et al., 2014). These components allow bacteria to not only survive methicillin but also adapt to various antimicrobial environments, hence increasing their survival and propagation. MRSA strains can also express a modified penicillin-binding protein, PBP2A, which takes over the function of other inactivated PBPs, leading to resistance (Hanssen & Sollid, 2006).

1.4.6.4 Biofilm Formation and Drug Resistance

One of the key factors contributing to the pathogenicity and drug resistance of *S. aureus* is its ability to form biofilms (Otto 2018). Biofilms are structured communities of bacteria encased in a self-produced extracellular matrix that adheres to surfaces. In the context of *S. aureus*, biofilm formation is a significant virulence factor for the following reasons: protection from host immune response, adherence to surfaces, intercellular communication, and heterogeneous environment (Musini et al., 2022). Biofilm formation in *Staphylococcus aureus* is strongly linked to enhanced antibiotic resistance, driven by multiple synergistic mechanisms, including limited antibiotic penetration, altered metabolic states of biofilm-embedded cells, and the upregulation of efflux pumps that actively expel antibiotics from bacterial cells. Horizontal Gene Transfer: The close proximity of bacteria within a biofilm facilitates the transfer of resistance genes between cells, contributing to the spread of antibiotic resistance. The presence of persister cells, which are a subpopulation of bacteria within the biofilm that are highly tolerant to antibiotics, plays a crucial role in chronic and recurrent infections (Costerton et al., 1999). The biofilm mode of growth significantly complicates the treatment of *S. aureus* infections, particularly in the case of MRSA. Conventional antibiotic therapy often fails to eradicate biofilms, leading to chronic infections and the need for surgical intervention, particularly in infections associated with medical devices (Kiedrowski & Horswill, 2011).

1.5 DIAGNOSIS AND TRANSMISSION

1.5.1 Diagnosis

S. aureus may be detected using a variety of laboratory tests. The technique begins by isolating the bacteria from the afflicted regions, such as skin infections or nasal discharge. Microscopic inspection reveals *S. aureus*'s typical blue or purple hue, which aids in identification. For a definite diagnosis, collected samples are grown on media such as blood agar or mannitol salt agar, which promote *S. aureus* growth while inhibiting other bacteria (Gnanamani et al., 2017). Confirmatory assays, such as coagulase or catalase, are then used to clearly identify *S. aureus*. Swabs from the injection site, blood cultures, respiratory secretions, tissue biopsies, and bodily fluids such as synovial or cerebrospinal fluid are all needed for an accurate diagnosis of *S. aureus* infection (Tong et al., 2015). Proper aseptic methods are critical for avoiding contamination. Culture, biochemical testing, molecular procedures, and mass spectrometry are examples of microbiological techniques used for isolation and identification (Karmakar et al., 2016). Catalase and coagulase assays are critical for identifying *S. aureus*, whereas polymerase chain reaction (PCR) can detect *S. aureus*-associated genes. MALDI-TOF mass spectrometry is a useful method for quickly identifying bacterial species, including *S. aureus* (Manukumar & Umesha, 2017).

Antimicrobial susceptibility testing is critical for determining suitable treatment. Methods include disk diffusion, broth dilution, e-testing, and the use of automated systems such as VITEK and BD Phoenix. *S. aureus* has developed a variety of antibiotic resistance mechanisms, including beta-lactamase synthesis, altered penicillin-binding proteins (PBPs), efflux pumps, and target site alterations (Mlynarczyk-Bonikowska et al., 2022). The rise of antibiotic-resistant forms of *S. aureus*, particularly MRSA, presents substantial public health issues, resulting in increased morbidity, death, and healthcare expenditures due to restricted treatment choices and the need for more expensive and toxic medications (Gurung et al., 2020). Combating antibiotic resistance in *S. aureus* requires efforts such as antibiotic stewardship programs, the discovery of novel antimicrobial drugs, and continued research into alternative treatment techniques such as bacteriophage therapy, immunotherapy, and vaccines (Abebe & Birhanu, 2023).

1.5.1.1 Rapid Diagnostic Tools

Advances in diagnostic technologies have resulted in quick tests that greatly shorten the time necessary to detect *S. aureus*. Real-time PCR and quantitative PCR are two noteworthy examples, providing rapid results within hours and becoming widely used in clinical settings due to their accuracy and efficiency in identifying *S. aureus* (Heng et al., 2022). Isothermal amplification methods, such as multienzyme isothermal rapid amplification (MIRA), provide another quick diagnostic option. MIRA can identify *S. aureus* directly from clinical samples in around 25 minutes with a high level of specificity and efficiency, making it ideal for regular diagnostic usage (Heng et al., 2022).

When MRSA is suspected, specialized MRSA tests are performed, such as PBP2a testing or cefoxitin disk diffusion (Heng et al., 2022). These tests may detect MRSA from nose swabs or wound samples in a couple of hours, utilizing either culture-based methods or molecular assays, which are speedier and can produce findings in as little as five hours. Loop-mediated isothermal amplification (LAMP) and PCR are used to identify *S. aureus* in food samples because of their quickness and sensitivity. These approaches are quicker and more sensitive than conventional culturing procedures. LAMP, in particular, may be combined with a smartphone-based detection system for point-of-care testing, resulting in fast and reliable findings (Zhao et al., 2013). These rapid diagnostic methods are critical for the timely and precise identification of *S. aureus*, allowing for the successful management and treatment of infections caused by this versatile bacterium.

1.5.2 Transmission and Dissemination

Modes of Transmission of *S. aureus* from Natural Reservoirs to Human and Animal Hosts: *S. aureus* is transmitted through several routes. Direct Contact: Skin-to-skin contact is the main mechanism of transmission. This can happen among individuals or between humans and animals (Kwiecinski & Horswill, 2020). Direct contact between healthcare personnel and patients can spread *S. aureus*, especially if adequate hand hygiene is not followed. Indirect Contact: *S. aureus* may persist on surfaces and objects (fomites) for long periods of time. Indirect transmission happens when people come into contact with infected surfaces and then touch their nose, mouth, or skin, which promotes colonization (Jaradat et al., 2020). Airborne Transmission: Although less frequent, *S. aureus* can spread by respiratory droplets when an infected person coughs or sneezes. This technique of communication is very useful in congested environments (Dhand & Li, 2020). Foodborne Transmission: Contaminated food products, especially those handled by carriers or manufactured in unclean settings, might be a source of *S. aureus*. Ingestion of food contaminated with the bacterium's enterotoxins can cause food poisoning (Kadariya et al., 2014).

1.5.3 Factors Contributing to the Spread and Transmission of *S. aureus*

Several factors contribute to the spread and transmission of *S. aureus*:

- **Colonization Density**: People with high colonization densities are more likely to spread *S. aureus* to others. Healthcare personnel and patients with high bacterial loads have a higher risk of transmitting the infection (Tong et al., 2015).
- **Environmental Contamination**: Contaminated surfaces and items in healthcare facilities, households, and public spaces promote indirect transfer. High-touch surfaces, such as doorknobs, bedrails, and medical equipment, are typical sources of contamination (Suleyman et al., 2018).

- **Human Behavior**: Poor hand hygiene, insufficient cleaning and disinfection, and incorrect food handling all contribute to the spread of *S. aureus*. Close contact in crowded places, such as schools, jails, and sports facilities, raises the chance of transmission.
- **Animal–Human Interactions**: Close contact with colonized animals, especially pets and livestock, can result in zoonotic transmission. Veterinary practitioners and those dealing with animals are more likely to contract *S. aureus* (Khairullah et al., 2023).

1.6 PREVENTION AND CONTROL STRATEGIES

1.6.1 Preventive Strategies

To prevent the spread of *S. aureus*, especially MRSA, in hospital settings, it is important to maintain strict hand hygiene standards and utilize alcohol-based hand sanitizers. Contact Precautions: Use gloves and gowns and isolate known MRSA carriers or infected patients. Environmental Cleaning: Surfaces and equipment should be cleaned and disinfected regularly to prevent contamination. Screening Programs: Active monitoring and screening for MRSA colonization, particularly in high-risk settings such as critical care units (Samuel et al., 2023).

To manage *S. aureus*, public health strategies include surveillance, which involves monitoring and reporting infections and resistance patterns through national and international networks; antibiotic stewardship, which are programs aimed at optimizing antibiotic usage to decrease resistance development (Walia et al., 2019); education – increasing awareness among healthcare professionals and the general public on the significance of infection control measures and appropriate antibiotic usage (Khadse et al., 2023).

Infection control systems are critical for managing and preventing infections in healthcare facilities. These programs should involve continuing infection control training and competence evaluations in areas such as environmental cleaning, disinfection, and the use of personal protective equipment. Monitoring compliance with these principles and providing feedback on performance are essential components. Implementing the CDC Core Elements of Hospital Antibiotic Stewardship Programs is also advised for managing and monitoring antibiotic usage, hence avoiding the development of resistance. Hospitals should use a multidisciplinary strategy that includes nurses for patient care, environmental services for cleanliness, and infection prevention teams for monitoring and executing control measures. Regular audits and feedback systems should be implemented to ensure that the techniques used are compliant and effective (Haque et al., 2020).

1.6.2 Treatment and Management

1.6.2.1 Antibiotic Therapy for *S. aureus* Infections

Antibiotic therapy is determined by the nature and severity of the infection, as well as the *S. aureus* strain's susceptibility profile. Common antibiotics used are beta-lactams: For methicillin-susceptible *S. aureus*, penicillinase-resistant penicillins like nafcillin and oxacillin, as well as first-generation cephalosporins like cefazolin, are preferable (Akanbi et al., 2017; Esposito et al., 2023). Glycopeptides: For MRSA infections, vancomycin is frequently the first-line therapy. However, vancomycin-intermediate *S. aureus* (VISA) and vancomycin-resistant *S. aureus* (VRSA) strains require different treatments (Pardo et al., 2024). Other drugs: Linezolid, daptomycin, tigecycline, and newer drugs such as ceftaroline and delafloxacin are available to treat MRSA and complex *S. aureus* infections (Guo et al., 2020).

1.6.2.2 Challenges and Considerations in Antibiotic Resistance

Antibiotic resistance in *S. aureus*, particularly MRSA, presents substantial problems. Resistance mechanisms include the following. Beta-lactamase Production: *S. aureus* generates beta-lactamase, which hydrolyzes penicillin's beta-lactam ring and renders it useless (Bush & Bradford, 2016). mecA Gene: This gene produces a PBP2a that has a poor affinity for beta-lactams, resulting in methicillin resistance (Lade & Kim, 2023). Efflux Pumps and Mutations: Resistance to other antibiotic classes, such as macrolides, tetracyclines, and fluoroquinolones, can be caused via efflux pumps or target site mutations (Alghamdi et al., 2023).

1.6.3 Alternative Treatment Approaches

In addition to antibiotic therapy, *S. aureus* infections may require surgical intervention, such as draining abscesses, debridement of diseased tissues, and removal of contaminated prosthetic devices (Tong et al., 2015); supportive care – to treat sepsis and other severe infections, supportive treatments such as fluid resuscitation, hemodynamic support, and organ dysfunction control are required (Mph, n.d. 2024); immunotherapy – research on vaccines and monoclonal antibodies that target *S. aureus* virulence factors is ongoing, but no effective vaccine is presently available for clinical use (Clegg et al., 2021).

1.7 CONCEPTUAL FRAMEWORK AND EMERGING PERSPECTIVES IN *S. AUREUS* RESEARCH

S. aureus remains a formidable human pathogen, causing diseases from benign skin abscesses to life-threatening conditions (bacteremia, endocarditis, toxic shock syndrome) thanks to its sophisticated arsenal of virulence factors (Touaitia et al., 2025). These include surface adhesins (MSCRAMMs) and secreted toxins that enable tissue colonization, immune evasion, and host cell damage. For example, MSCRAMMs promote biofilm formation on medical devices, a key contributor to chronic infections, while toxins such as hemolysins, exfoliative toxins, and superantigens (e.g., TSST-1, PVL) directly disrupt host barriers and dysregulate immunity. In parallel, *S. aureus* has evolved extensive antibiotic resistance via gene acquisition and mutation. Hospital-associated MRSA strains carry large SCCmec elements (I–III) encoding the PBP2a β-lactamase that renders most β-lactams ineffective. Community MRSA lineages utilize smaller, more mobile SCCmec (type IV) elements and often co-acquire virulence loci (e.g., PVL genes) for rapid spread outside hospitals. High-throughput genomics has revealed that mobile genetic elements (plasmids, phages, SCCs) shuffle both resistance and toxin genes across strains. This co-evolution of virulence and resistance makes understanding the molecular pathogenesis of *S. aureus* critical for new therapies (Touaitia et al., 2025).

Host–pathogen interaction studies highlight the many strategies by which *S. aureus* persists and evades immune clearance. Besides toxin-mediated damage, *S. aureus* uses factors like protein A and capsule to bind antibodies and avoid opsonization (not shown). The agr quorum-sensing system further modulates this by upregulating toxins while repressing adhesins once infection is established. Chronic infections (e.g., endocarditis, osteomyelitis, device-associated infection) result when bacteria survive in biofilms or small-colony variants, often leading to relapse. Notably, *S. aureus* is a major cause of mastitis: it is "a leading cause of bovine mastitis worldwide" and frequently responsible for subclinical mammary infections that incur large economic losses (Campos et al., 2022). Recent epidemiological studies even link *S. aureus* infection with cancer. For instance, survivors of *S. aureus* bacteremia were found to have significantly higher incidence of subsequent malignancies (multiple myeloma, leukemia, sarcoma, cervical, liver, pancreatic, and urinary tract cancers) compared to matched controls (Wei et al., 2022). Likewise, genomic surveys of tumor microbiomes have detected *Staphylococcus* as a common genus – for example, one study found *Staphylococcus* DNA in 22 of 23 breast-cancer tissue samples (Wei et al., 2022). These observations suggest complex, bidirectional interactions between chronic *S. aureus* infection and host pathology, warranting further investigation of immune evasion and inflammation in oncogenesis.

Biofilm formation and quorum sensing are central to *S. aureus* persistence. Attachment via MSCRAMMs (ClfA, ClfB, FnBPs, etc.) not only mediates colonization but also seeds biofilms – structured communities encased in polysaccharide or protein matrices (Touaitia et al., 2025). In biofilms, metabolic dormancy and limited drug penetration render cells highly tolerant to antibiotics and immunity (Peng et al., 2022). The agr quorum-sensing circuit coordinates this lifestyle: Agr activation triggers RNAIII expression, which upregulates secreted toxins and degradative enzymes while downregulating surface adhesins, thereby promoting dissemination from biofilm to acute infection. Inhibition of this system has become a therapeutic goal: recent studies show that targeting agr (with small-molecule inhibitors, peptide analogs of the autoinducing peptide, etc.) can disrupt biofilm maturation and restore antibiotic efficacy (Le & Otto, 2015). Accordingly, many control strategies focus on preventing biofilm establishment or breaking it down – for example, surface coatings (e.g., anti-adhesive polymers, dPNAG inhibitors) and enzymes (DNAases, dispersin B) have been explored to block biofilm formation (Touaitia et al., 2025).

Advances in diagnostics and systems-level analyses are transforming *S. aureus* tracking. Traditional culture and PCR methods have been supplemented by proteomic and genomic tools: MALDI-TOF MS rapidly identifies *S. aureus* at the species level, and multiplex PCR arrays screen for resistance and virulence genes. Notably, CRISPR-Cas-based assays are emerging as highly sensitive nucleic acid tests for MRSA and virulence markers (Wang et al., 2024). Strikingly, natural CRISPR loci in *S. aureus* are inversely correlated with resistance gene carriage (Wang et al., 2024), hinting that CRISPR elements may limit horizontal spread of resistance. At the epidemiological level, whole genome sequencing (WGS) and multi-locus sequence typing are now standard for outbreak investigation and surveillance of clones. These large genomic datasets feed into computational "systems biology" models of *S. aureus*: transcriptomic and proteomic profiling under different conditions is revealing how gene networks respond to host environments, antibiotics, or vaccines. Such omics approaches have already uncovered novel toxin variants and regulatory circuits (Touaitia et al., 2025), and they underpin machine-learning efforts to predict resistance phenotypes from sequence. Cutting-edge artificial intelligence (AI) methods are also being applied: for example, deep-learning algorithms have been used to design new antibiotic scaffolds that are active against MRSA. These AI-driven tools promise faster diagnostics and rational drug design by learning patterns in large clinical and experimental datasets (Branda & Scarpa, 2024).

Emerging therapies reflect these multifaceted insights. Despite many trials, no vaccine against *S. aureus* is yet approved, in part due to the pathogen's diverse immune evasion. Current vaccine candidates are multivalent formulations targeting surface proteins (ClfA, FnBPs, IsdB), capsular polysaccharides (CP5/CP8), or toxins (Hla, PVL, TSST-1) to elicit opsonophagocytic antibodies and Th17/T cell responses. Passive immunization with monoclonal antibodies (e.g., anti-ClfA, anti-SpA) is also being tested, as well as adjunctive use of intravenous immunoglobulin. In parallel, novel antimicrobials are under development: next-generation β-lactams (ceftobiprole, ceftaroline), lipoglycopeptides (dalbavancin, oritavancin), and oxazolidinones (tedizolid) aim to overcome current resistance (Clegg et al., 2021). Non-traditional agents are equally promising. Bacteriophage therapy – both natural phages and engineered phages – can directly lyse *S. aureus*; for instance, phage StAP1 lyses nearly half of clinical MRSA strains, and phage-derived endolysins (e.g., LysK, Exebacase) have shown potent synergy with vancomycin to eliminate biofilms (Touaitia et al., 2025). At the molecular level, CRISPR-Cas9 systems delivered via phagemids have been used in the lab to selectively cut resistance genes (mecA, blaZ) or biofilm genes (icaA), re-sensitizing MRSA to antibiotics or disassembling biofilms (Touaitia et al., 2025). Nanotechnology is also being applied: multifunctional nanoparticles (metallic or polymeric) can carry high local drug loads to infection sites, improving efficacy and reducing host toxicity (Andrade et al., 2023). Finally, the antibiotic pipeline is being reinvigorated by AI-driven screens (as above) and by bioprospecting (e.g., teixobactin analogs, antimicrobial peptides). Nevertheless, the rapid evolution of *S. aureus* and shrinking interest in antibiotic R&D pose ongoing challenges, reinforcing the need for diversified approaches.

In summary, modern *S. aureus* research is highly interdisciplinary. It integrates detailed molecular microbiology of virulence and resistance with cutting-edge technologies across disciplines. Emerging approaches include the following: AI/ML and Computational Design: Machine-learning models and deep learning are accelerating the discovery of new antibiotics and enhancing diagnostics, such as AI-assisted analysis of MALDI-TOF spectra or WGS data (With AI, Researchers Identify a New Class of Antibiotic Candidates, 2023). CRISPR-Cas Tools: CRISPR systems enable precise gene editing for research (dissecting gene function) and are being developed as sequence-specific antimicrobials (targeting resistance plasmids or toxin genes) (Wang et al., 2024; Touaitia et al., 2025). Phage Therapy and Lysins: Lytic phages and phage-derived enzymes offer targeted killing of *S. aureus*, especially for drug-resistant and biofilm-associated infections (Touaitia et al., 2025). Nanotechnology: Engineered nanoparticles serve as drug delivery vehicles (e.g., antibiotic-loaded liposomes or functionalized silica NPs) that concentrate treatment at infection sites and enhance penetration into biofilms (Andrade et al., 2023). Omics and Systems Biology: High-throughput genomics, transcriptomics, proteomics, and metabolomics elucidate global regulatory networks and host–pathogen interactions. These omics datasets, combined with computational modeling, are uncovering novel drug/vaccine targets and informing precision therapies (Touaitia et al., 2025).

1.8 CONCLUSION

This chapter presents a thorough examination of *S. aureus*, a highly adaptable and diverse organism that causes a variety of diseases due to its virulence factors and drug resistance mechanisms. *S. aureus*'s complexity and adaptability make it a formidable pathogen in both community and clinical settings. Understanding its virulence factors and the environment that promotes its proliferation and transmission is critical for establishing effective diagnostic, preventive, and control measures for the diseases it causes.

These findings have major implications for public health strategy, clinical practices, and future research areas. Emphasizing preventive measures and implementing strong infection control methods is critical to limiting the spread of this bacterium, particularly methicillin-resistant forms. The debate underlines the importance of continued research into innovative treatment strategies and vaccines for successful *S. aureus* infection control. It is a common and versatile bacterium that can colonize a variety of environments and natural reservoirs, the principal ones being humans and animals. Understanding the parameters that influence *S. aureus* survival, colonization, and transmission is critical for creating effective infection prevention and control measures. Integrating knowledge from microbiology, epidemiology, and public health can help us better handle the difficulties faced by this tough virus.

REFERENCES

Abebe, A., & Birhanu, A. (2023). Methicillin resistant *Staphylococcus aureus*: Molecular mechanisms underlying drug resistance development and novel strategies to combat. *Infection and Drug Resistance*, *16*, 7641–7662. https://doi.org/10.2147/idr.s428103

Abril, A. G., Villa, T. G., Barros-Velázquez, J., Cañas, B., Sánchez-Pérez, A., Calo-Mata, P., & Carrera, M. (2020). *Staphylococcus aureus* exotoxins and their detection in the dairy industry and mastitis. *Toxins*, *12*(9), 537. https://doi.org/10.3390/toxins12090537

Adler, A., Temper, V., Block, C., Abramson, N., & Moses, A. (2006). Panton-Valentine leukocidin–producing *Staphylococcus aureus*. *Emerging Infectious Diseases*, *12*(11), 1789–1790. https://doi.org/10.3201/eid1211.060726

Akanbi, O. E., Njom, H. A., Fri, J., Otigbu, A. C., & Clarke, A. M. (2017). Antimicrobial susceptibility of *Staphylococcus aureus* isolated from recreational waters and beach sand in eastern cape province of South Africa. *International Journal of Environmental Research and Public Health*, *14*(9), 1001. https://doi.org/10.3390/ijerph14091001

Alghamdi, B. A., Al-Johani, I., Al-Shamrani, J. M., Alshamrani, H. M., Al-Otaibi, B. G., Almazmomi, K., & Yusof, N. Y. (2023). Antimicrobial resistance in methicillin-resistant *Staphylococcus aureus. Al-Mi'galaẗ Al-sa'udiyaẗLī-ulum Al-ḥayaẗ, 30*(4), 103604. https://doi.org/10.1016/j.sjbs.2023.103604

Andrade, S., Ramalho, M. J., Santos, S. B., Melo, L. D. R., Santos, R. S., Guimarães, N., . . . Pereira, M. C. (2023). Fighting methicillin-resistant *Staphylococcus aureus* with targeted nanoparticles. *International Journal of Molecular Sciences, 24*(10), 9030. https://doi.org/10.3390/ijms24109030

Archibald, L. K., & Quisling, R. G. (2013). Central nervous system infections. In *Springer eBooks* (pp. 427–517). https://doi.org/10.1007/978-1-4471-5226-2_22

Argudín, M. N., Mendoza, M. C., & Rodicio, M. R. (2010). Food poisoning and *Staphylococcus aureus* enterotoxins. *Toxins, 2*(7), 1751–1773. https://doi.org/10.3390/toxins2071751

Balasubramanian, D., Harper, L., Shopsin, B., & Torres, V. J. (2017). *Staphylococcus aureus* pathogenesis in diverse host environments. *Pathogens and Disease*, ftx005. https://doi.org/10.1093/femspd/ftx005

Beeby, M., Gumbart, J. C., Roux, B., & Jensen, G. J. (2013). Architecture and assembly of the gram-positive cell wall. *Molecular Microbiology, 88*(4), 664–672. https://doi.org/10.1111/mmi.12203

Bhattacharya, M., & Horswill, A. R. (2024). The role of human extracellular matrix proteins in defining *Staphylococcus aureus* biofilm infections. *FEMS Microbiology Reviews, 48*(1). https://doi.org/10.1093/femsre/fuae002

Branda, F., & Scarpa, F. (2024). Implications of artificial intelligence in addressing antimicrobial resistance: Innovations, global challenges, and healthcare's future. *Antibiotics, 13*(6), 502. https://doi.org/10.3390/antibiotics13060502

Bush, K., & Bradford, P. A. (2016). β-lactams and β-lactamase inhibitors: An overview. *Cold Spring Harbor Perspectives in Medicine, 6*(8), a025247. https://doi.org/10.1101/cshperspect.a025247

Campos, B., Pickering, A. C., Rocha, L. S., Aguilar, A. P., Fabres-Klein, M. H., De Oliveira Mendes, T. A., . . . De Oliveira Barros Ribon, A. (2022). Diversity and pathogenesis of *Staphylococcus aureus* from bovine mastitis: Current understanding and future perspectives. *BMC Veterinary Research, 18*(1). https://doi.org/10.1186/s12917-022-03197-5

Cheung, G. Y. C., Bae, J. S., & Otto, M. (2021). Pathogenicity and virulence of *Staphylococcus aureus. Virulence, 12*(1), 547–569. https://doi.org/10.1080/21505594.2021.1878688

Clegg, J., Soldaini, E., McLoughlin, R. M., Rittenhouse, S., Bagnoli, F., & Phogat, S. (2021). *Staphylococcus aureus* vaccine research and development: The past, present and future, including novel therapeutic strategies. *Frontiers in Immunology, 12*. https://doi.org/10.3389/fimmu.2021.705360

Colston, J., & Atkins, B. (2018). Bone and joint infection. *Clinical Medicine, 18*(2), 150–154. https://doi.org/10.7861/clinmedicine.18-2-150

Cong, Y., Yang, S., & Rao, X. (2020). Vancomycin resistant *Staphylococcus aureus* infections: A review of case updating and clinical features. *Journal of Advanced Research, 21*, 169–176. https://doi.org/10.1016/j.jare.2019.10.005

Costerton, J. W., Stewart, P. S., & Greenberg, E. P. (1999). Bacterial biofilms: A common cause of persistent infections. *Science, 284*(5418), 1318–1322.

Danelli, T., Duarte, F. C., De Oliveira, T. A., Da Silva, R. S., Alfieri, D. F., Gonçalves, G. B., . . . Yamada-Ogatta, S. F. (2020). Nasal carriage by *Staphylococcus aureus* among healthcare workers and students attending a University Hospital in Southern Brazil: Prevalence, phenotypic, and molecular characteristics. *Interdisciplinary Perspectives on Infectious Diseases, 2020*, 1–11. https://doi.org/10.1155/2020/3808036

De Jong, N. W. M., Van Kessel, K. P. M., & Van Strijp, J. a. G. (2019). Immune evasion by *Staphylococcus aureus. Microbiology Spectrum, 7*(2). https://doi.org/10.1128/microbiolspec.gpp3-0061-2019

Dhand, R., & Li, J. (2020). Coughs and sneezes: Their role in transmission of respiratory viral infections, including SARS-CoV–2. *American Journal of Respiratory and Critical Care Medicine, 202*(5), 651–659. https://doi.org/10.1164/rccm.202004-1263pp

Dhungel, S., Rijal, K. R., Yadav, B., Dhungel, B., Adhikari, N., Shrestha, U. T., Adhikari, B., Banjara, M. R., & Ghimire, P. (2021). Methicillin-resistant *Staphylococcus aureus* (MRSA): Prevalence, antimicrobial susceptibility pattern, and detection of mecA gene among cardiac patients from a Tertiary Care Heart Center in Kathmandu, Nepal. *Infectious Diseases Research and Treatment, 14*, 117863372110373. https://doi.org/10.1177/11786337211037355

Esposito, S., Blasi, F., Curtis, N., Kaplan, S., Lazzarotto, T., Meschiari, M., . . . Bassetti, M. (2023). New antibiotics for *Staphylococcus aureus* infection: An update from the world association of infectious diseases and immunological disorders (WAidid) and the Italian society of anti-infective therapy (SITA). *Antibiotics, 12*(4), 742. https://doi.org/10.3390/antibiotics12040742

Falugi, F., Kim, H. K., Missiakas, D. M., & Schneewind, O. (2013). Role of protein A in the evasion of host adaptive immune responses by *Staphylococcus aureus. mBio, 4*(5). https://doi.org/10.1128/mbio.00575-13

Fishovitz, J., Hermoso, J. A., Chang, M., & Mobashery, S. (2014). Penicillin-binding protein 2a of methicillin-resistant *Staphylococcus aureus*. *IUBMB Life*, *66*(8), 572–577. https://doi.org/10.1002/iub.1289

Foster, T. J. (2019). Surface Proteins of *Staphylococcus aureus*. *Microbiology Spectrum*, *7*(4). https://doi.org/10.1128/microbiolspec.gpp3-0046-2018

Gaither, J. B. (2011). Toxic shock syndrome. In *Elsevier eBooks* (pp. 348–351). https://doi.org/10.1016/b978-0-323-07167-3.00059-0

Gao, S., Jin, W., Quan, Y., Li, Y., Shen, Y., Yuan, S., . . . Wang, Y. (2024). Bacterial capsules: Occurrence, mechanism, and function. *Npj Biofilms and Microbiomes*, *10*(1). https://doi.org/10.1038/s41522-024-00497-6

Gehrke, A. K. E., Giai, C., & Gómez, M. I. (2023). *Staphylococcus aureus* adaptation to the skin in health and persistent/recurrent infections. *Antibiotics*, *12*(10), 1520. https://doi.org/10.3390/antibiotics12101520

Gnanamani, A., Hariharan, P., & Paul-Satyaseela, M. (2017). *Staphylococcus aureus*: Overview of bacteriology, clinical diseases, epidemiology, antibiotic resistance and therapeutic approach. In *InTech eBooks*. https://doi.org/10.5772/67338

González-García, S., Hamdan-Partida, A., Valdez-Alarcón, J. J., Bustos-Hamdan, A., & Bustos-Martínez, J. (2023). Main factors of *Staphylococcus aureus* associated with the interaction to the cells for their colonization and persistence. In *Infectious diseases*. https://doi.org/10.5772/intechopen.107974

Grapsa, J., Blauth, C., Chandrashekhar, Y., Prendergast, B., Erb, B., Mack, M., & Fuster, V. (2022). *Staphylococcus Aureus* infective endocarditis. *Journal of the American College of Cardiology*, *79*(1), 88–99. https://doi.org/10.1016/j.jacc.2021.10.015

Guo, Y., Song, G., Sun, M., Wang, J., & Wang, Y. (2020). Prevalence and therapies of antibiotic-resistance in *Staphylococcus aureus*. *Frontiers in Cellular and Infection Microbiology*, *10*. https://doi.org/10.3389/fcimb.2020.00107

Gurung, R. R., Maharjan, P., & Chhetri, G. G. (2020). Antibiotic resistance pattern of *Staphylococcus aureus* with reference to MRSA isolates from pediatric patients. *Future Science OA*, FSO464. https://doi.org/10.2144/fsoa-2019-0122

Hanssen, A. M., & Sollid, J. U. E. (2006). SCCmecin staphylococci: Genes on the move. *FEMS Immunology and Medical Microbiology*, *46*(1), 8–20. https://doi.org/10.1111/j.1574-695x.2005.00009.x.

Haque, M., McKimm, J., Sartelli, M., Dhingra, S., Labricciosa, F. M., Islam, S., . . . Charan, J. (2020). Strategies to prevent healthcare-associated infections: A narrative overview. *Risk Management and Healthcare Policy*, *13*, 1765–1780. https://doi.org/10.2147/rmhp.s269315

Heng, P., Liu, J., Song, Z., Wu, C., Yu, X., & He, Y. (2022). Rapid detection of *Staphylococcus aureus* using a novel multienzyme isothermal rapid amplification technique. *Frontiers in Microbiology*, *13*. https://doi.org/10.3389/fmicb.2022.1027785

Hennekinne, J. A., De Buyser, M. L., & Dragacci, S. (2012). *Staphylococcus aureus*and its food poisoning toxins: Characterization and outbreak investigation. *FEMS Microbiology Reviews*, *36*(4), 815–836. https://doi.org/10.1111/j.1574-6976.2011.00311.x/

Jabłońska-Trypuć, A., Makuła, M., Włodarczyk-Makuła, M., Wołejko, E., Wydro, U., Serra-Majem, L., & Wiater, J. (2022). Inanimate surfaces as a source of hospital infections caused by fungi, bacteria and viruses with particular emphasis on SARS-CoV–2. *International Journal of Environmental Research and Public Health*, *19*(13), 8121. https://doi.org/10.3390/ijerph19138121

Jaradat, Z. W., Ababneh, Q. O., Sha'aban, S. T., Alkofahi, A. A., Assaleh, D., & Shara, A. A. (2020). Methicillin Resistant *Staphylococcus aureus* and public fomites: A review. *Pathogens and Global Health*, *114*(8), 426–450. https://doi.org/10.1080/20477724.2020.1824112

Kadariya, J., Smith, T. C., & Thapaliya, D. (2014). *Staphylococcus aureus*and staphylococcal food-borne disease: An ongoing challenge in public health. *BioMed Research International, 2014*, 1–9. https://doi.org/10.1155/2014/827965

Karmakar, A., Dua, P., & Ghosh, C. (2016). Biochemical and molecular analysis of *Staphylococcus aureus* clinical isolates from hospitalized patients. *˜the œCanadian Journal of Infectious Diseases & Medical Microbiology/Canadian Journal of Infectious Diseases & Medical Microbiology, 2016*, 1–7. https://doi.org/10.1155/2016/9041636

Kazmi, S. Y. (2022). The etymology of microbial nomenclature and the diseases these cause in a historical perspective. *Saudi Journal of Biological Sciences*, *29*(11), 103454. https://doi.org/10.1016/j.sjbs.2022.103454

Khadse, S. N., Ugemuge, S., & Singh, C. (2023). Impact of antimicrobial stewardship on reducing antimicrobial resistance. *Curēus*. https://doi.org/10.7759/cureus.49935

Khairullah, A. R., Sudjarwo, S. A., Effendi, M. H., Ramandinianto, S. C., Gelolodo, M. A., Widodo, A., . . . Kurniawati, D. A. (2023). Pet animals as reservoirs for spreading methicillin-resistant *Staphylococcus aureus* to human health. *Journal of Advanced Veterinary and Animal Research*, *10*(1), 1. https://doi.org/10.5455/javar.2023.j641

Ki, V., & Rotstein, C. (2008). Bacterial skin and soft tissue infections in adults: A review of their epidemiology, pathogenesis, diagnosis, treatment and site of care. *Canadian Journal of Infectious Diseases and Medical Microbiology*, *19*(2), 173–184. https://doi.org/10.1155/2008/846453

Kiedrowski, M. R., & Horswill, A. R. (2011). New approaches for treating staphylococcal biofilm infections. *Annals of the New York Academy of Sciences*, *1241*(1), 104–121.

Kim, J., Kim, B. E., Ahn, K., & Leung, D. Y. M. (2019). Interactions between atopic dermatitis and *Staphylococcus aureus* infection: Clinical implications. *Allergy Asthma and Immunology Research*, *11*(5), 593. https://doi.org/10.4168/aair.2019.11.5.593

Knox, J., Uhlemann, A. C., & Lowy, F. D. (2015). *Staphylococcus aureus* infections: Transmission within households and the community. *Trends in Microbiology*, *23*(7), 437–444. https://doi.org/10.1016/j.tim.2015.03.007

Kozajda, A., Jeżak, K., & Kapsa, A. (2019). Airborne *Staphylococcus aureus* in different environments – A review. *Environmental Science and Pollution Research International*, *26*(34), 34741–34753. https://doi.org/10.1007/s11356-019-06557-1

Krismer, B., Liebeke, M., Janek, D., Nega, M., Rautenberg, M., Hornig, G., . . . Peschel, A. (2014). Nutrient limitation governs *Staphylococcus aureus* metabolism and niche adaptation in the human nose. *PLOS Pathogens*, *10*(1), e1003862. https://doi.org/10.1371/journal.ppat.1003862

Kwiecinski, J. M., & Horswill, A. R. (2020). *Staphylococcus aureus* bloodstream infections: Pathogenesis and regulatory mechanisms. *Current Opinion in Microbiology*, *53*, 51–60. https://doi.org/10.1016/j.mib.2020.02.005

Kwong, S. M., Ramsay, J. P., Jensen, S. O., & Firth, N. (2017). Replication of staphylococcal resistance plasmids. *Frontiers in Microbiology*, *8*. https://doi.org/10.3389/fmicb.2017.02279

Lade, H., & Kim, J. S. (2023). Molecular determinants of β-lactam resistance in methicillin-resistant *Staphylococcus aureus* (MRSA): An updated review. *Antibiotics*, *12*(9), 1362. https://doi.org/10.3390/antibiotics12091362

Le, K. Y., & Otto, M. (2015). Quorum-sensing regulation in staphylococci – An overview. *Frontiers in Microbiology*, *6*. https://doi.org/10.3389/fmicb.2015.01174

Linz, M. S., Mattappallil, A., Finkel, D., & Parker, D. (2023). Clinical impact of *Staphylococcus aureus* skin and soft tissue infections. *Antibiotics*, *12*(3), 557. https://doi.org/10.3390/antibiotics12030557

Loomba, P. S., Taneja, J., & Mishra, B. (2010). Methicillin and vancomycin-resistant *S. aureus* in hospitalized patients. *Journal of Global Infectious Diseases*, *2*(3), 275. https://doi.org/10.4103/0974-777x.68535

Lorenzo, J. M., Munekata, P. E., Dominguez, R., Pateiro, M., Saraiva, J. A., & Franco, D. (2018). Main groups of microorganisms of relevance for food safety and stability. In *Elsevier eBooks* (pp. 53–107). https://doi.org/10.1016/b978-0-12-811031-7.00003-0

Malachowa, N., & DeLeo, F. R. (2010). Mobile genetic elements of *Staphylococcus aureus*. *Cellular and Molecular Life Sciences*, *67*(18), 3057–3071. https://doi.org/10.1007/s00018-010-0389-4

Manukumar, H. M., & Umesha, S. (2017). MALDI-TOF-MS based identification and molecular characterization of food associated methicillin-resistant *Staphylococcus aureus*. *Scientific Reports*, *7*(1). https://doi.org/10.1038/s41598-017-11597-z

Masalha, M., Borovok, I., Schreiber, R., Aharonowitz, Y., & Cohen, G. (2001). Analysis of transcription of the *Staphylococcus aureus* aerobic Class Ib and anaerobic Class III ribonucleotide reductase genes in response to oxygen. *Journal of Bacteriology*, *183*(24), 7260–7272. https://doi.org/10.1128/jb.183.24.7260-7272.2001

Mirbagheri, V. S., Alishahi, A., Petroudi, S. H. R. H., & Ojagh, S. M. (2023). Transcriptome analysis of some genes of *Staphylococcus aureus* exposed to alkylated, nano- and native chitosans. *Carbohydrate Polymer Technologies and Applications*, *6*, 100351. https://doi.org/10.1016/j.carpta.2023.100351

Missiakas, D. M., & Schneewind, O. (2013). Growth and laboratory maintenance of *Staphylococcus aureus*. *Current Protocols in Microbiology*, *28*(1). https://doi.org/10.1002/9780471729259.mc09c01s28

Mlynarczyk-Bonikowska, B., Kowalewski, C., Krolak-Ulinska, A., & Marusza, W. (2022). Molecular mechanisms of drug resistance in *Staphylococcus aureus*. *International Journal of Molecular Sciences*, *23*(15), 8088. https://doi.org/10.3390/ijms23158088

Mph, A. K. M. (n.d. Jun 26, 2024). Septic shock treatment & management: Approach considerations, goals of hemodynamic support. *Fluid Resuscitation*. https://emedicine.medscape.com/article/168402-treatment?form=fpf

Musini, A., Chilumoju, S. P., & Giri, A. (2022). Biofilm formation in drug-resistant pathogen *Staphylococcus aureus*. In *Microbial Biofilms* (pp. 23–46). CRC Press.

Mustapha, M., Bukar-Kolo, Y. M., Geidam, Y. A., & Gulani, I. A. (2014). Review on methicillin-resistant *Staphylococcus aureus* (MRSA) in dogs and cats. *International Journal of Animal and Veterinary Advances*, *6*(2), 61–73. https://doi.org/10.19026/ijava.6.5619

Nakou, A., Woodhead, M., & Torres, A. (2009). MRSA as a cause of community-acquired pneumonia. *European Respiratory Journal*, *34*(5), 1013–1014. https://doi.org/10.1183/09031936.00120009
Otto, M. (2014). *Staphylococcus aureus* toxins. *Current Opinion in Microbiology*, *17*, 32–37. https://doi.org/10.1016/j.mib.2013.11.004
Otto, M. (2018). Staphylococcal biofilms. *Microbiology Spectrum*, *6*(4), 10–1128.
Paharik, A. E., & Horswill, A. R. (2016). The staphylococcal biofilm: Adhesins, regulation, and host response. *Microbiology Spectrum*, *4*(2). https://doi.org/10.1128/microbiolspec.vmbf-0022-2015
Pardo, L., Mota, M. I., Parnizari, A., Varela, A., Algorta, G., & Varela, G. (2024). Detection of vancomycin resistance among methicillin-resistant *Staphylococcus aureus* strains recovered from children with invasive diseases in a reference pediatric hospital. *Antibiotics*, *13*(4), 298. https://doi.org/10.3390/antibiotics13040298
Paterson, G. K., Harrison, E. M., & Holmes, M. A. (2014). The emergence of mecC methicillin-resistant *Staphylococcus aureus*. *Trends in Microbiology*, *22*(1), 42–47. https://doi.org/10.1016/j.tim.2013.11.003
Peng, Q., Tang, X., Dong, W., Sun, N., & Yuan, W. (2022). A review of biofilm formation of *Staphylococcus aureus* and its regulation mechanism. *Antibiotics*, *12*(1), 12. https://doi.org/10.3390/antibiotics12010012
Piewngam, P., & Otto, M. (2024). *Staphylococcus aureus* colonisation and strategies for decolonisation. *The Lancet Microbe*, *5*(6), e606–e618. https://doi.org/10.1016/s2666-5247(24)00040-5
Prestinaci, F., Pezzotti, P., & Pantosti, A. (2015). Antimicrobial resistance: A global multifaceted phenomenon. *Pathogens and Global Health*, *109*(7), 309–318. https://doi.org/10.1179/2047773215y.0000000030
Sakr, A., Brégeon, F., Mège, J. L., Rolain, J. M., & Blin, O. (2018). *Staphylococcus aureus* nasal colonization: An update on mechanisms, epidemiology, risk factors, and subsequent infections. *Frontiers in Microbiology*, *9*. https://doi.org/10.3389/fmicb.2018.02419
Samuel, P., Kumar, Y. S., Suthakar, B. J., Karawita, J., Kumar, D. S., Vedha, V., . . .Thakkar, K. (2023). Methicillin-resistant *Staphylococcus aureus* colonization in intensive care and burn units: A narrative review. *PubMed*, *15*(10), e47139. https://doi.org/10.7759/cureus.47139
Santos-López, A., Rodríguez-Beltrán, J., & Millán, L. S. (2021). The bacterial capsule is a gatekeeper for mobile DNA. *PLoS Biology*, *19*(7), e3001308. https://doi.org/10.1371/journal.pbio.3001308
Shinji, H., Yosizawa, Y., Tajima, A., Iwase, T., Sugimoto, S., Seki, K., & Mizunoe, Y. (2011). Role of fibronectin-binding proteins A and B in invitro cellular infections and invivo septic infections by *Staphylococcus aureus*. *Infection and Immunity*, *79*(6), 2215–2223. https://doi.org/10.1128/iai.00133-11
Silhavy, T. J., Kahne, D., & Walker, S. (2010). The bacterial cell envelope. *Cold Spring Harbor Perspectives in Biology*, *2*(5), a000414. https://doi.org/10.1101/cshperspect.a000414
Suleyman, G., Alangaden, G., & Bardossy, A. C. (2018). The role of environmental contamination in the transmission of nosocomial pathogens and healthcare-associated infections. *Current Infectious Disease Reports*, *20*(6). https://doi.org/10.1007/s11908-018-0620-2
Tam, K., & Torres, V. J. (2019). *Staphylococcus aureus* secretes toxins and extracellular enzymes. *Microbiology Spectrum*, *7*(2). https://doi.org/10.1128/microbiolspec.gpp3-0039-2018
Thammavongsa, V., Kim, H. K., Missiakas, D., & Schneewind, O. (2015). Staphylococcal manipulation of host immune responses. *Nature Reviews Microbiology*, *13*(9), 529–543. https://doi.org/10.1038/nrmicro3521
Tong, S. Y. C., Davis, J. S., Eichenberger, E., Holland, T. L., & Fowler, V. G. (2015). *Staphylococcus aureus* infections: Epidemiology, pathophysiology, clinical manifestations, and management. *Clinical Microbiology Reviews*, *28*(3), 603–661. https://doi.org/10.1128/cmr.00134-14
Touaitia, R., Mairi, A., Ibrahim, N. A., Basher, N. S., Idres, T., & Touati, A. (2025). *Staphylococcus aureus*: A review of the pathogenesis and virulence mechanisms. *Antibiotics*, *14*(5), 470. https://doi.org/10.3390/antibiotics14050470
Van Der Aart, L. T., Lemmens, N., Van Wamel, W. J., & Van Wezel, G. P. (2016). Substrate inhibition of VanA by d-alanine reduces vancomycin resistance in a VanX-dependent manner. *Antimicrobial Agents and Chemotherapy*, *60*(8), 4930–4939. https://doi.org/10.1128/aac.00276-16
Walia, K., Ohri, V. C., Madhumathi, J., & Ramasubramanian, V. (2019). Policy document on antimicrobial stewardship practices in India. *Indian Journal of Medical Research*, *149*(2), 180. https://doi.org/10.4103/ijmr.ijmr_147_18
Wang, J., Liu, F., Long, J., Jin, Y., Chen, S., Duan, G., & Yang, H. (2024). The application of CRISPR-Cas system in *Staphylococcus aureus* infection. *Heliyon*, *10*(14), e34383. https://doi.org/10.1016/j.heliyon.2024.e34383
Weiskopf, S. R., Rubenstein, M. A., Crozier, L. G., Gaichas, S., Griffis, R., Halofsky, J. E., . . . Whyte, K. P. (2020). Climate change effects on biodiversity, ecosystems, ecosystem services, and natural resource management in the United States. The Science of the Total Environment, 733, 137782. https://doi.org/10.1016/j.scitotenv.2020.137782

Wei, Y., Sandhu, E., Yang, X., Yang, J., Ren, Y., & Gao, X. (2022). Bidirectional functional effects of *Staphylococcus* on carcinogenesis. *Microorganisms*, *10*(12), 2353. https://doi.org/10.3390/microorganisms10122353

With AI, researchers identify a new class of antibiotic candidates. (2023, December 20). @Broadinstitute. https://www.broadinstitute.org/news/researchers-use-ai-identify-new-class-antibiotic-candidates

Yagüe, D. P., Mihaljevic, J., Mbegbu, M., Wood, C., Hepp, C., Kyman, S., . . . Pearson, T. (2021). Survival of *Staphylococcus aureus* on sampling swabs stored at different temperatures. *Journal of Applied Microbiology*, *131*(3), 1030–1038. https://doi.org/10.1111/jam.15023

Yamaguchi, M., Minamide, Y., Terao, Y., Isoda, R., Ogawa, T., Yokota, S. I., . . . Kawabata, S. (2009). Nrc of *Streptococcus pneumoniae* suppresses capsule expression and enhances anti-phagocytosis. *Biochemical and Biophysical Research Communications*, *390*(1), 155–160. https://doi.org/10.1016/j.bbrc.2009.09.099

Yamazaki, Y., Ito, T., Tamai, M., Nakagawa, S., & Nakamura, Y. (2024). The role of *Staphylococcus aureus* quorum sensing in cutaneous and systemic infections. *Inflammation and Regeneration*, *44*(1). https://doi.org/10.1186/s41232-024-00323-8

Zhao, X., Li, Y., Park, M., Wang, J., Zhang, Y., He, X., . . . Oh, D. H. (2013). Loop-mediated isothermal amplification assay targeting the femA gene for rapid detection of *Staphylococcus aureus* from clinical and food samples. *Journal of Microbiology and Biotechnology*, *23*(2), 246–250. https://doi.org/10.4014/jmb.1207.07022

Zhou, C., & Fey, P. D. (2020). The acid response network of *Staphylococcus aureus*. *Current Opinion in Microbiology*, *55*, 67–73. https://doi.org/10.1016/j.mib.2020.03.006

Zubair, S., Fischer, A., Liljander, A., Meens, J., Hegerman, J., Gourlé, H., . . . Jores, J. (2015). Complete genome sequence of *Staphylococcus aureus*, strain ILRI_Eymole1/1, isolated from a Kenyan dromedary camel. *Standards in Genomic Sciences*, *10*(1). https://doi.org/10.1186/s40793-015-0098-6

2 Unraveling the Pathogenesis of *Staphylococcus aureus*

Erugu Ankitha[1], *Rama Prasanna*[1], *Vinjamuri Nishakhya*[2], *and Brandi C. Miller*[3,4,5]

[1]Centre for Biotechnology, University College of Engineering, Science and Technology, JNTUH, Hyderabad, India

[2]Government Degree College for Women, Begumpet, Hyderabad, India

[3]USF Center for Microbiome Research, Microbiomes Institute, University of South Florida Morsani College of Medicine, Tampa, FL, USA

[4]Center for Excellence in Aging and Brain Repair, University of South Florida Morsani College of Medicine, Tampa, FL, USA

[5]Department of Neurosurgery and Brain Repair, University of South Florida Morsani College of Medicine, Tampa, FL, USA

ABBREVIATIONS

agr	accessory gene regulator
AIP	autoinducing peptide
APC	antigen-presenting cell
ArlRS	autolysis-related locus regulatory system
CA-MRSA	community-associated methicillin-resistant *Staphylococcus aureus*
CHIPs	chemotaxis inhibitory protein of *Staphylococcus aureus*
CnBP	collagen-binding protein
ClfA/B	clumping factor A/B
EAP	extracellular adherence protein
EBP	extracellular binding protein
ECM	extracellular matrix
EPS	extracellular polymeric substances
ET	exfoliative toxin
FFA	free fatty acid
FgBP	fibrinogen-binding protein
FLIPr	formyl peptide receptor-like 1 inhibitor protein
FMN	flavin mononucleotide riboswitch
FnBP	fibronectin-binding protein
FPR	formylated peptides receptor
HK	histidine kinase
HysA	hyaluronidase A
ICAM	intercellular adhesion molecule-1
IE	infective carditis
IFN-γ	interferon-γ
IgG	immunoglobulin G
IL	interleukin
Isd	iron-regulated surface determinant system

DOI: 10.1201/9781003647072-3

LA-MRSA	livestock-associated methicillin-resistant *Staphylococcus aureus*
MDR	multidrug resistance
MHC-II	major histocompatibility complex II
MLS	macrolides–lincosamide–streptogramin antibiotics
MRSA	methicillin-resistant *Staphylococcus aureus*
MSCRAMMs	microbial surface components recognizing adhesive matrix molecules
PBP2a	penicillin-binding protein 2a
PSGL-1	P-selectin glycoprotein ligand-1
PSM	phenol-soluble modulins
PVL	panton–valentine leukocidin
RBC	red blood cell
ROS	reactive oxygen species
ROT	repressor of toxins
RR	response regulator
rRNA	ribosomal RNA
SaeRS	staphylococcal exoprotein expression regulatory system
Sak	staphylokinase
SAL	*Staphylococcus aureus* lipase
SAM	S-adenosylmethionine riboswitch
sarA	staphylococcal accessory regulator
S. aureus	*Staphylococcus aureus*
SCIN	staphylococcal complement inhibitor
SE	staphylococcal enterotoxin
sRNA	small regulatory RNA
SSL	staphylococcal superantigen-like
SSSS	staphylococcal scalded skin syndrome
SSTI	soft skin and tissue infections
TCR	T cell receptor
TCS	two-component system
TNF-α	tumor necrosis factor-α
TPP	thiamine pyrophosphate
TSS	toxic shock syndrome
TSST-1	toxic shock syndrome toxin-1
UTI	urinary tract infection
VISA	vancomycin-intermediate *Staphylococcus aureus*
VraSR	vancomycin resistance-associated regulatory system
VRSA	vancomycin-resistant *Staphylococcus aureus*

2.1 INTRODUCTION

Staphylococcus aureus is a commensal bacterium of the human microbiome, most prominently colonizing in mucous membranes and on the skin in approximately 30% of the global population. Under favorable conditions, it can become pathogenic and is the causative agent of many infections. Pathogenic *S. aureus* colonizes human tissues inconspicuously and produces toxic compounds, enhancing its virulence and antibiotic resistance. It is a gram-positive, round-shaped (cocci), facultative anaerobe with a diameter of 0.5–1.0 µm, and it is non-motile and non-spore forming. Cell division occurs in three planes, causing it to form clusters and grape-like bunches (Ramos-León et al., 2024). *S. aureus* possesses a polysaccharide capsule, which is a crucial virulence factor which inhibits phagocytosis, causing invasive infections (Lowy, 1998; O'Riordan & Lee, 2004). Contact with infected people or fomites spreads *S. aureus*; it can also propagate via airborne particles or through medical devices in healthcare settings (Adams & Dancer, 2020). This includes antibiotic-resistant strains, such as methicillin-resistant *S. aureus* (MRSA). Epidemiology involves adhesion, colonization, virulence factors, reservoirs, transmission, and antibiotic resistance. Around 20% of individuals are recurring nasal carriers of *S. aureus*, while 60% are sporadic carriers, and approximately 20% are non-carriers (Chmielowiec-Korzeniowska et al., 2020). The nasal cavity is a significant reservoir for *S. aureus*, increasing invasion of the bacterium and increasing the risk of infection (Chmielowiec-Korzeniowska et al., 2020).

S. aureus can tolerate many standard antibiotics, making antibiotic resistance an ongoing and urgent public health issue. Methicillin, cephalosporin, and penicillin are common β-lactam antibiotics against which *S. aureus* has acquired resistance. Overuse of antibiotics is one primary factor contributing to the rapid development of antibiotic-resistant *S. aureus* (Congdon et al., 2023). Various combinations of antibiotics can enter the body through animal-based foods or through frequent antibiotic use (Van Boeckel et al., 2015). Mutations in virulence genes and horizontal gene transfer also play a crucial role (Edet et al., 2025). Additionally, global transportation contributes to the spread of *S. aureus* and antibiotic resistance traits, thus increasing the risk. In the next section, we introduce the most common antibiotics, their mechanisms, and how *S. aureus* confers resistance against them.

Macrolide-Lincosamide-Streptogramin (MLS) Antibiotics. The first MLS antibiotic, erythromycin, was introduced in 1952 (Ungureanu, 2010). They are bacteriostatic and target the 50s subunit of the bacterial ribosome to effectively inhibit protein synthesis (Yao et al., 2019). MLS antibiotics are one of the best classes of antibiotics used to combat staphylococcal infections as well as penicillin-related allergies. Due to the prevalence of resistance genes and increased administration, *S. aureus* isolates quickly acquire resistance toward MLS antibiotics. There are two mechanisms of MLS resistance in *S. aureus*. The first is ribosomal modification by 23s ribosomal RNA (rRNA) methylases, which are regulated by *erm* genes (i.e., *ermA*, *ermB*, and *ermC*) and prevent the antibiotics from binding to the ribosome (Yao et al., 2019). The other mechanism, active efflux, is regulated by *msr* genes (such as *msrA*) and pumps antibiotics out via ATP-dependent mechanisms, which reduces intracellular antibiotic concentrations and prevents ribosomal binding (Ojo et al., 2006).

Aminoglycosides. Aminoglycosides are broad-spectrum antibiotics which enter the bacterial cell through energy-independent and energy-dependent mechanisms, including active transport across the cytoplasmic membrane. They interfere with mRNA transcription and protein translation by binding to the 16s rRNA. Despite their development in 1944, *S. aureus* became resistant to them by the 1950s (Gootz, 1990). The most common resistance mechanisms in *S. aureus* include (1) chromosomal mutations which alter the ribosomal site, binding affinity, or membrane permeability and (2) production of enzymes (i.e., acetyltransferases and phosphotransferases), which inactivate aminoglycoside activity (Woodford, 2005).

Fluoroquinolones. In the 1980s, fluoroquinolones were developed primarily for gram-negative bacteria, but later demonstrated activity against gram-positive bacteria, including pneumococci and streptococci. Quinolones target bacterial enzymes, including DNA gyrase and topoisomerase, by direct binding, which impairs or inhibits the replication of bacterial DNA, eventually leading to cell death. Quinolone resistance is developed by *S. aureus*, particularly MRSA, via mutations in topoisomerase and DNA gyrase (Alseqely et al., 2021).

β-Lactam Antibiotics. Penicillins, including penicillin-G, amoxicillin, and ampicillin, are β-lactam antibiotics originally isolated from *Penicillium chrysogenum* and were reportedly active against a wide range of gram-negative and gram-positive cocci since their discovery in 1928 (Lobanovska & Pilla, 2017). Other β-lactams include cephalosporins, monobactams, and carbapenems. β-lactams inhibit peptidoglycan cross-linking, which impairs bacterial cell wall synthesis. However, growing resistance to penicillin has become an urgent public health issue in recent decades. In 1940, 85% of *S. aureus* strains were susceptible to penicillin-G; however, *S. aureus* developed resistance within 3 years of its widespread use (Harkins et al., 2017). Approximately 90% of *Staphylococcus* strains produce β-lactamase, an enzyme which hydrolyzes the β-lactam ring and makes *S. aureus* resistant to β-lactams (Lade & Kim, 2023). *S. aureus* produces four β-lactamase subtypes (A, B, C, and D), which enhance its resistance to penicillin and other β-lactams, including methicillin (Lade & Kim, 2023).

Glycopeptides. Discovered in 1957, vancomycin is a prominent glycopeptide antibiotic effective against gram-positive bacteria due to its inhibition of bacterial cell wall synthesis (Levine, 2006). *Enterococcus* strains were the first to acquire vancomycin resistance, and *S. aureus* acquired resistance in 2002, with increased resistance over the past few decades (Levine, 2006). The *vanA* gene is carried by vancomycin-resistant *S. aureus* (VRSA) and is associated with the *Tn1546* transposon (Périchon & Courvalin, 2009). The major mechanism of VRSA is modification of bacterial cell walls by *vanA* and other *van* genes (i.e., *vanH* and *vanX*), making vancomycin impotent or inefficient in binding to the cell wall. Vancomycin-intermediate *S. aureus* (VISA) has intermediate resistance and is the precursor to VRSA.

2.2 VIRULENCE FACTORS

S. aureus is commonly found on skin and in nasal passages and possesses a remarkable capacity to transition from a commensal microbe to a formidable pathogen. This transformation is promoted by its extensive arsenal of virulence factors, which enable its colonization and invasion of host tissues and immune system evasion, which contribute to disease progression. Studying these virulence factors is crucial for the development of effective strategies to combat *S. aureus* infections, ranging from minor skin irritations to life-threatening systemic illnesses, such as pneumonia, bacteremia, and endocarditis (Linz et al., 2023; Lowy, 1998). Table 2.1 shows the different categories of virulence factors possessed by *S. aureus* and their mechanisms of action.

- **Cell Wall Components**: The cell wall of *S. aureus* serves as a dynamic interface and is comprised of several virulence factors. Peptidoglycan is a major structural component of its cell wall and is composed of amino sugars and tetrapeptides, including *N*-acetylglucosamine and *N*-acetylmuramic acid disaccharides (Sharif et al., 2009). Furthermore, the peptidoglycan layer is characterized by a pentapeptide stem of L-alanine-D-iso-glutamine-L-lysine-D-alanine-D-alanine and a pentaglycyl bridge (Sharif et al., 2009). Together, these biomolecules create a rigid meshwork, which can stimulate the innate immune system and trigger the release of cytokines that contribute to inflammation. Excessive or chronic inflammation can result in long-term tissue damage and exacerbate disease severity (Matsuo & Iwasa, 2025). Anionic polymers embedded within the peptidoglycan layer (wall teichoic acids) or cell membrane (lipoteichoic acids), are also essential for cell wall integrity (Wanner et al., 2017). They mediate the attachment of *S. aureus* to host cells, particularly epithelial cells, which activate complement, increase inflammation, and facilitate immune evasion (Cheung et al., 2021; Wanner et al., 2017).
- **Protein A**: Protein A is a prominent surface protein and is covalently anchored to the peptidoglycan layer in *S. aureus* via sortases (Ton-That et al., 1999). N-terminal immunoglobulin-binding domains on Protein A facilitate binding to the Fc region of immunoglobulin G (IgG) antibodies, blocking opsonization of *S. aureus* (Atkins et al., 2008; Deisenhofer, 1981). Protein A also blocks the cell proliferation of B cells, further impairing host defenses. These mechanisms allow *S. aureus* to cloak itself from the immune defenses of the host.
- **Polysaccharide Capsule**: The capsule of *S. aureus* is a polysaccharide layer surrounding the bacterial cell; serotypes 5 and 8 are two common subtypes (O'Riordan & Lee, 2004). The polysaccharide capsule masks surface proteins (i.e., opsonins), effectively blocking phagocytosis by innate immune cells (i.e., neutrophils). This allows *S. aureus* to persist longer in the host and establish a robust infection.
- **Adhesins**: Adhesins are surface proteins that mediate the binding of bacteria to host tissues and implanted medical devices. They exhibit remarkable specificity, targeting various components of the extracellular matrix (ECM) and plasma proteins, including fibronectin, fibrinogen, and collagen (Foster et al., 2014; Josse et al., 2017). Fibronectin-binding

TABLE 2.1
Functions of Major Virulence Factors in *Staphylococcus aureus*

Category	Examples and Major Function(s) in Virulence	References
Cell wall components	• *Peptidoglycan:* Facilitates bacterial adhesion • *Teichoic acids:* Mediate bacterial adhesion • *Polysaccharide capsule:* Masks the bacterial surface and prevents phagocytosis • *Protein A:* Binds the Fc region of antibodies	Atkins et al. (2008); Kuipers et al. (2016); Sharif et al. (2009); Wanner et al. (2017)
Adhesins	• *FnBPs:* Bind fibronectin • *FgBPs:* Bind fibrinogen • *CnBPs:* Bind collagen • *EBPs:* Bind elastin • *ClfA/B:* Bind fibrinogen	Downer et al. (2002); Elasri et al. (2002); Herman-Bausier et al. (2018); Josse et al. (2017); Zhang et al. (2022)
Toxins	• *Hemolysins:* Form pores • *Leukocidins:* Form pores • *Superantigens:* Overstimulate T cells, leading to "cytokine storm" • *ETs:* Cleave desmoglein-1 • *Enterotoxins:* Overstimulate T cells ("cytokine storm")	Argudín et al. (2010); Colin et al., (1994); Ladhani et al., (1999); Otto (2014); Tuffs et al. (2019)
Enzymes	• *Coagulase:* Forms fibrin clots around bacteria • *Hyaluronidase:* Degrades hyaluronic acid • *Staphylokinase:* Activates plasminogen to plasmin and dissolves fibrin clots • *Proteases:* Cleave host proteins • *Lipases:* Hydrolyze host lipids • *DNases*: Degrade extracellular DNA	Cheng et al. (2010); Ibberson et al. (2014); Kengmo Tchoupa et al. (2024); Ohbayashi et al. (2011); Peetermans et al. (2014); Tanaka et al. (2018)
Immune evasion factors	• *SCIN:* Blocks complement • *CHIPs:* Inhibits chemotaxis (immune cell recruitment) • *EAP:* Facilitates adhesion and inhibits chemotaxis	Cheung et al. (2021); Van Wamel et al. (2006)
Regulatory factors	• *agr:* Switches virulence gene expression from adherence/biofilm genes to toxin production and dissemination • *sarA:* Regulates biofilm formation, adhesion, and toxin production	Beenken et al. (2010); Boles and Horswill (2008); Queck et al. (2008)

Abbreviations: agr: Accessory Gene Regulator, CHIPs: Chemotaxis inhibitory protein of *S. aureus*, ClfA/B: Clumping factor A/B, CnBPs: Collagen-binding proteins, EAP: Extracellular adherence protein, EBPs: Elastin-binding proteins, FgBPs: Fibrinogen-binding proteins, FnBPs: Fibronectin-binding proteins, sarA: Staphylococcal accessory regulator A, SCIN: Staphylococcal complement inhibitor.

proteins (FnBPs, i.e., FnBPA and FnBPB) on *S. aureus* recognize adhesive matrix molecules (MSCRAMMs) which bind to the glycoprotein fibronectin in tissues, on fibronectin-coated devices, or on endothelial cells lining blood vessels (Josse et al., 2017; Speziale & Pietrocola, 2020). Additionally, extracellular fibrinogen-binding proteins (FgBPs) promote biofilm formation and immune evasion of *S. aureus* by interfering with fibrinogen, a key plasma protein involved in platelet aggregation and wound healing (Zhang et al., 2022). Biofilms are comprised of bacterial communities encapsulated within a self-assembled matrix and are highly resistant to therapeutic interventions, leading to sustained *S. aureus* colonization and infection (Periasamy et al., 2012). Collagen-binding proteins (CnBPs) facilitate the adhesion of *S. aureus* to collagen, a primary structural component of connective tissues and the ECM; CnBPs are highly relevant in osteomyelitis and septic arthritis. Cna binds to collagen and also to the C1q component of the complement system, blocking phagocytic clearance of

S. aureus and promoting biofilm formation (Herman-Bausier et al., 2016). Elastin-binding proteins (EBPs) mediate adherence between *S. aureus* and elastin, an ECM protein linked with enhanced elasticity and flexibility in multiple tissues (Downer et al., 2002). Thus, binding of EBPs, such as EbpS, contributes to the development of infections in elastin-rich tissues, including the skin, lungs, and vasculature. Clumping Factors A (ClfA) and B (ClfB) are cell wall components of *S. aureus* which facilitate its aggregation ("clumping") and adhesion to fibrinogen and fibrin, respectively (Herman-Bausier et al., 2018). This clumping can exacerbate tissue damage due to the development of microthrombi and formation of biofilms while promoting the evasion of the host immune system (Herman-Bausier et al., 2018).

- **Toxins**: The α, β, γ, and δ-toxins of *S. aureus* are hemolysins which facilitate cell lysis and death. Pore-forming α-toxin can lyse erythrocytes, leucocytes, and endothelial cells and was one of the first identified pore-forming toxins, and increased expression has been linked with increased *S. aureus* virulence (Montgomery et al., 2008; Tavares et al., 2014). Another notable pore-forming toxin is panton–valentine leukocidin (PVL), which leads to severe pneumonia and skin infections due to its extensive damage to tissues and leukocytes (Cheung et al., 2021; Labandeira-Rey et al., 2007). The β-toxin mediates membrane disruption by hydrolyzing sphingomyelin, a key sphingolipid comprising cell membranes (Huseby et al., 2007). Furthermore, γ-toxins like HlgAB and HlgCB are composed of two subunits that assemble on the surface of leukocytes and erythrocytes, forming pores that lead to cell lysis. They can also bind to chemokine receptors on neutrophils and monocytes or promote the production of pro-inflammatory cytokines, promoting cell damage and death (Gillman et al., 2017). Exfoliative toxins (ETs; i.e., ETA and ETB) induce blistering of the epidermis, primarily in infants and young children, leading to aggressive skin conditions like staphylococcal scalded skin syndrome (SSSS) (Ladhani et al., 1999). Another class of toxins, called superantigens, is associated with a massive and destructive production of pro-inflammatory cytokines, such as tumor necrosis factor-α (TNF-α) and interleukins (ILs), via uncontrollable activation of T cells and T cell receptors (TCR) (Kaempfer et al., 2017). Toxic Shock Syndrome Toxin-1 (TSST-1) is one prominent superantigen produced by *S. aureus* and leads to the development of toxic shock syndrome (TSS), a life-threatening condition characterized by elevated fever, rash, hypotension, cognitive deficits/confusion, and multi-organ dysfunction (Tuffs et al., 2019). Collectively, these toxins contribute to the mortality and morbidity associated with highly virulent *S. aureus* strains.
- **Enzymes**: *S. aureus* secretes a diverse array of enzymes that degrade host tissue, facilitate bacterial dissemination, and provide essential nutrients, all contributing to its pathogenesis. Coagulase converts fibrinogen to fibrin and aids in the formation of fibrin clots around the bacteria, which protect them from immune cells and phagocytosis (Cheng et al., 2010). It is also a key biomarker used to diagnose *S. aureus* infections, including those of the skin, blood, and lungs (Sperber & Tatini, 1975). Hyaluronidase (HysA) degrades hyaluronic acid (an ECM component), facilitating bacterial spread and detachment of hyaluronic acid from biofilms (Ibberson et al., 2014). *S. aureus* produces four classes of proteases, including (1) staphylococcal serine proteases, (2) cysteine proteases, (3) metalloproteases, and (4) staphopain (Karlsson & Arvidson, 2002). These proteases degrade many host proteins, including antimicrobial peptides, heavy chain proteins of immunoglobulins, and elastin; proteolysis ultimately aids in immune evasion and tissue destruction. Lipases, such as *S. aureus* lipase (SAL), break down lipids, allowing *S. aureus* to colonize sebaceous areas of the skin and obtain nutrients (for energy metabolism or membrane integrity) through the release of fatty acids or their modification (Kengmo Tchoupa et al., 2024). Staphylokinase (Sak) aids in dissemination of *S. aureus* (due to its thrombolytic activities) and neutralizes antimicrobial peptides. Furthermore, deoxyribonucleases degrade DNA, leaving nutrients for *S. aureus* and serving as a potent diagnostic marker for identifying *S. aureus* (Tanaka et al., 2018).

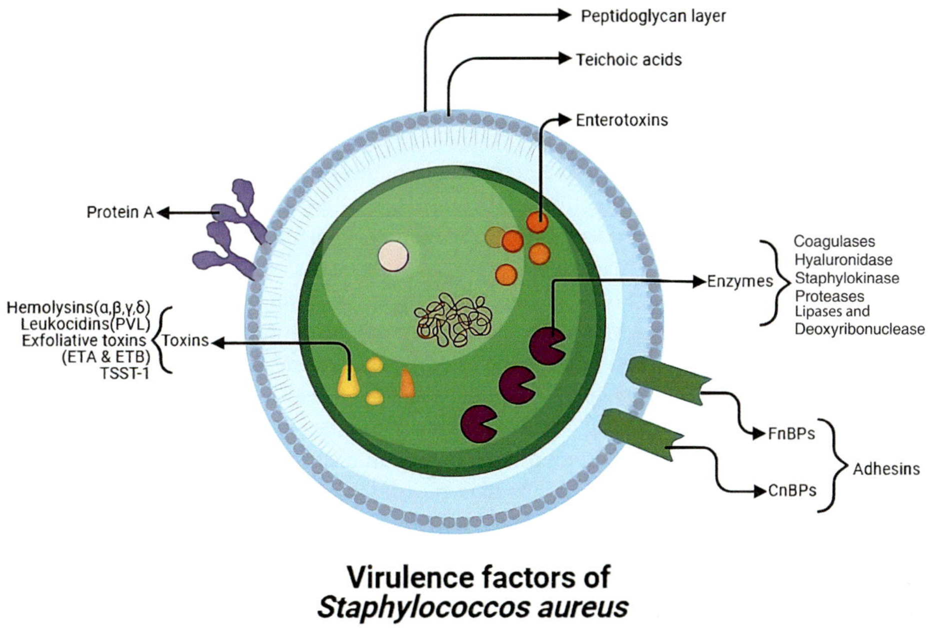

FIGURE 2.1 Virulence factors of *Staphylococcus aureus*. The major virulence mechanisms include cell surface structures, toxins, and secreted enzymes. These factors are regulated in response to environmental cues, enhancing the pathogenicity of *S. aureus* in both community-acquired and nosocomial infections. Abbreviations: CnBPs: collagen-binding proteins, ETA/B: exfoliative toxin A/B, FnBPs: fibronectin-binding proteins, PVL: panton–valentine leucocidin, *S. aureus*: *Staphylococcus aureus*, TSST-1: toxic shock syndrome toxin-1.

- **Immune Evasion and Regulatory Factors**: *S. aureus* evades the host immune system via factors like staphylococcal complement inhibitor (SCIN), chemotaxis inhibitory protein of *S. aureus* (CHIPs), and extracellular adherence protein (EAP), which disrupt complement activation, neutrophil recruitment, and immune factor functions (Cheung et al., 2021). The expression of virulence genes in *S. aureus* is orchestrated by multiple regulatory elements, including the accessory gene regulator (agr) system and staphylococcal accessory regulator (sarA) family of transcriptional regulators (Beenken et al., 2010). These systems allow *S. aureus* to adapt to diverse environmental conditions and optimize pathogenicity based on bacterial density (Table 2.1). The staphylococcal exoprotein expression (SaeRS) and autolysis-related locus (ArlRS) two-component systems (TCS) in *S. aureus* regulate virulence gene expression in response to the environment (Figure 2.1). SaeRS regulates genes for secreted or surface proteins, while ArlRS regulates expression of cell wall- and biofilm-related genes (Crosby et al., 2020; Voyich et al., 2009).

2.3 PATHOGENICITY OF *S. AUREUS*

S. aureus infections can arise from systemic or asymptomatic colonization. Asymptomatic colonization often occurs in hospital settings, where nosocomial infections may be transmitted from infected individuals and other patients or via contaminated surfaces (fomites). Spread of the bacteria to other body sites (through behaviors like nose picking) can lead to invasive infections (Foster et al., 2014; Räz et al., 2023). *S. aureus* can also be transmitted through contact with animals, such as pigs and cattle, leading to the emergence of livestock-associated MRSA (LA-MRSA), which is also a significant global public health concern (Anjum et al., 2019; Campos et al., 2022). Understanding the intricate mechanisms employed by *S. aureus* to cause infection is crucial for the development of effective strategies. The next section will delve deeper into the step-by-step mechanism of pathogenic *S. aureus* within the host.

2.3.1 Adhesion

A crucial factor which makes *S. aureus* a remarkable pathogen is its ability to adhere to the host tissue. Bacterial penetration of protective epithelium is required for the development of systemic infection. This is an energy-dependent cascade reaction which includes the encoding of surface proteins. MSCRAMMS are one major class of adhesion proteins, which function like "molecular velcro," allowing firm attachment of *S. aureus* to host tissue (Foster et al., 2014; Ghasemian et al., 2015). Several flavors of MSCRAMMS are anchored to peptidoglycan and extend outward to bind ECM proteins. FnBPs are regulated by the *fnbA* and *fnbB* genes and are key MSCRAMMs utilized by *S. aureus*, with slightly different genes depending on the strain; this allows *S. aureus* to colonize different tissues (Schwarz-Linek et al., 2003; Speziale & Pietrocola, 2020).

Collagen is one of the most abundant proteins in human tissues, accounting for ~30% of protein mass. *S. aureus* expresses the CnBP Cna, which is encoded by the *cna* gene and allows the bacteria to adhere to collagen, facilitating their colonization in collagen-rich tissues like skin, bone, and cartilage (Foster et al., 2014; Josse et al., 2017; Saei, 2012). Consequently, Cna plays an important role in the development of infections in bones (osteomyelitis) and joints (septic arthritis) (Elasri et al., 2002). As *S. aureus* moves into the epithelial tissue, it binds to ClfA and ClfB, which bind to fibrinogen (blood clotting protein) (Foster et al., 2014; Herman-Bausier et al., 2018). Encoded by the *clfA* gene, ClfA plays a crucial role in *S. aureus* aggregation in the presence of fibrinogen, resulting in clumping and contributing to the formation of biofilms (Herman-Bausier et al., 2018). ClfA and ClfB both contribute to the ability of *S. aureus* to adhere to endothelial cells and the lining of blood vessels, resulting in severe conditions like endocarditis or bacteremia (Josse et al., 2017; O´ Brien et al., 2002). In the following text, we describe the precise mechanisms of adhesins and how they regulate infection by *S. aureus*.

Mechanisms of Adhesins in *S. aureus*

Step 1: Adhesin Synthesis and Trafficking. Adhesins are typically large and complex molecules encoded by specific genes, including *clfA*, *clfB*, *fnbA*, *fnbB*, *cna*, and *spa* (Cheung et al., 2021; Herman-Bausier et al., 2018; Herman-Bausier et al., 2016; Speziale & Pietrocola, 2020). Following translation to proteins, many adhesins possess an N-terminal signal peptide, a short amino acid sequence (commonly containing a YSIRK/GXXS motif) indicating that the protein will be secreted (Mazmanian et al., 2000). During translocation across the cytoplasmic membrane, the signal peptide is cleaved, and the protein undergoes folding and post-translational modifications, such as glycosylation, to achieve its functional conformation.

Step 2: Functional Display-Surface Penetration. Other adhesins have a sorting signal at their C-terminal end, indicating they will be displayed on the cell surface (Mazmanian et al., 2000). The most prominent mechanism of adhesin anchoring to the cell wall is sortase-mediated anchoring, where sortase enzymes covalently link the adhesin to peptidoglycan (Deisenhofer, 1981). Many crucial *S. aureus* adhesins contain the LPXTG (leucine–proline–X-threonine–glycine) motif at their C-terminus, which is cleaved by Sortase A between its threonine (T) and glycine (G) residues (Mazmanian et al., 2000). The threonine carbonyl is then covalently linked to the peptidoglycan cross-bridge, leaving the binding domain exposed to the extracellular environment (Mazmanian et al., 2000; Otto, 2014). Although less common, other mechanisms involve direct integration of the adhesin into the host membrane through ionic or hydrophobic interactions or direct binding to host receptors (Feuillie et al., 2017; Foster et al., 2014). The polysaccharide capsule has a notable role in blocking adhesin interactions with the host, but some adhesins like

ClfA are able to extend outwards to bind host receptors which provides an extended reach for interaction with host cells (Dinges et al., 2000; Risley et al., 2007).

Step 3: Host Cell Interaction – The Molecular Handshake. MSCRAMMs bind to various ECM components, such as fibronectin, collagen, laminin, and elastin. As described above, *S. aureus* utilizes specialized adhesins for different ECM components – FnBPs bind to fibronectin to facilitate attachment to damaged tissues and promote its internalization (Schwarz-Linek et al., 2003). Furthermore, Cna binds to collagen, aiding in *S. aureus* colonization in bones and cartilage (Saei, 2012), while clumping factors (i.e., ClfA and ClfB) bind to fibrinogen and cytokeratin 10, allowing *S. aureus* to adhere to blood clots (Herman-Bausier et al., 2018; O´ Brien et al., 2002). Some *S. aureus* strains can also interact with epithelial cells, either through adhesins like EAP or indirectly through mucins, initiating internalization and invasion (Shuter et al., 1996).

Step 4: Downstream Effects – Triggering Cascade. The binding of adhesins to host cell ligands triggers a cascade which leads to colonization at the site of infection and enhances the virulence and persistence of *S. aureus*. One downstream effect is biofilm formation, which creates a protective environment for *S. aureus* and enhances its antibiotic resistance and evasion (Xu et al., 2024). These bacteria form irreversible bonds with the host tissue surfaces, aggregate to form microcolonies, and secrete extracellular polymeric substances (EPSs), creating a matrix (Periasamy et al., 2012). Mature biofilms can disperse to other sites, spreading the bacterial infection to other tissues (Boles & Horswill, 2008). Adhesins can modulate the host immune response, suppressing inflammation, evading host defenses, or triggering cytokine storms (Pietrocola et al., 2017).

2.3.2 Invasion by *S. aureus* Virulence Factors

The pathogenic success of *S. aureus* relies heavily on its ability to invade host tissues and produce potent toxins that damage cells and disrupt immune defenses. Although primarily an extracellular pathogen, *S. aureus* can enter host endothelial and epithelial cells via surface protein-mediated interactions (Shuter et al., 1996). After integration into host cells, several enzymes exacerbate *S. aureus* invasion and further aid its dissemination. Hyaluronidase degrades hyaluronic acid in connective tissues, while Sak breaks down fibrin clots (Ibberson et al., 2014). Furthermore, lipases and proteases break down host lipids and proteins, respectively, to promote their spread. Biofilms protect the bacteria from immune cells and antibiotics, enabling their persistence in tissues and on devices (Periasamy et al., 2012). Significant invasion and dissemination of *S. aureus* also lead to the production of several toxins, which constitute a major arsenal by which it causes severe infection and tissue damage. There are several flavors of *S. aureus* toxins, including superantigens, hemolysins, ETs, leukocidins, and enterotoxins.

Mechanisms of toxins in *S. aureus*

- **Superantigens**. Superantigens, such as TSST-1, overstimulate the immune system and produce a massive cytokine response by bridge formation and direct binding to Major Histocompatibility Complex II (MHC-II) molecules on antigen-presenting cells (APCs) and TCRs (Dinges et al., 2000; Moza et al., 2007). Furthermore, nonspecific T cell activation occurs when superantigens bind to the TCR outside of the antigen-binding groove, allowing activation of a large T cell population (Otto, 2014). This massive T cell activation leads to the release of excessive amounts of cytokines, including interleukin IL-1, IL-2, TNF-α, and interferon-γ (IFN-γ) (Kaempfer et al., 2017; Moza et al., 2007). This "cytokine storm" leads to widespread inflammation, a hallmark of many diseases, including TSS, which is characterized by fever, rash, hypotension, and multi-organ failure (Cheung et al., 2021). Similarly, staphylococcal enterotoxins can lead to food poisoning, where cytokine release is localized to the gut, resulting in severe nausea, vomiting, and diarrhea (Argudín et al., 2010).

- **Hemolysins**. *S. aureus* produces several cytolytic toxins to cause cell death. Hemolysins are a specific type of toxin which are specific to erythrocytes (red blood cells; RBCs), and α-, β-, γ-, and δ-toxins are key examples. RBCs are lysed by either pore formation and damage to the membrane (α-, γ-, and δ-toxins) or via lipid degradation (β-toxin), which disrupts osmotic balance and induces swelling (Seilie & Wardenburg, 2017). During RBC lysis, the heme portion of hemoglobin is released. Heme is an iron-rich molecule used by *S. aureus* as a nutrient to enhance colonization and virulence. *S. aureus* uses the Iron-regulated surface determinant system (Isd), which is semi-conserved in bacteria, to acquire iron from heme (Grigg et al., 2010). Iron (and other metals) are critical for the growth of *S. aureus* and are major cofactors because they supply nutrients to the bacteria in nutrient-deprived tissues, promote the expression of other virulence factors, fuel key metabolic processes, and aid in immune evasion by quenching toxic compounds like reactive oxygen species (ROS) (Corbin et al., 2008). Although less common, *S. aureus* also obtains iron through siderophores (i.e., staphyloferrin A and B), which quench inorganic iron from other protein sources (Corbin et al., 2008). Thus, the destruction of RBCs and acquisition of iron are significant virulence factors utilized by *S. aureus* and both play a key role in its pathogenicity. Below, we describe four major hemolysins and their biological relevance in promoting *S. aureus* infection.
    a. α-toxin (also called Hla) is a pore-forming toxin. It binds to the cell membrane via the ADAM10 receptor, which is expressed on multiple cell types, including epithelium, endothelium, immune cells, and RBCs (Tavares et al., 2014; Wilke & Wardenburg, 2010). On the cell membrane, α-toxin oligomerizes to form a heptameric β-barrel ring (Wilke & Wardenburg, 2010). This ring creates a β-barrel pore, which allows ions and small molecules to freely pass into the cell, disrupting osmotic balance and leading to cell lysis. Specific mechanisms include (1) efflux of K+ and influx of water into RBCs, leading to swelling and lysis (Wilke & Wardenburg, 2010), (2) calcium influx in nucleated cells (Krones et al., 2021), and (3) disruption of tight junction proteins (i.e., E-cadherin) (Becker et al., 2017), making host epithelial/endothelial barriers "leakier" and enhancing *S. aureus* dissemination.
    b. Furthermore, β-toxin (also called sphingomyelinase C) hydrolyzes sphingomyelin, a major component of cell membranes, particularly in neurons (endothelial cells), RBCs, and leukocytes (Huseby et al., 2007). Cleavage of sphingomyelin produces ceramide and phosphorylcholine and causes major disruption of membrane structures, also leading to cell lysis (Cheung et al., 2021; Huseby et al., 2007). In RBCs, β-toxin can also function as a hemolysin, performing a unique phenomenon called "hot-cold" lysis, where lysis only occurs at cold temperatures (Huseby et al., 2007; Li et al., 2025). The toxin also demonstrates magnesium-dependent cytotoxicity on leukocytes (Huseby et al., 2007). As a result, β-toxin is commonly associated with lower immune cell and leukocyte viability, increased inflammation, and RBC and vasculature damage.
    c. Bicomponent toxins, such as γ-toxin, consist of two proteins that work together to form β-barrel pores in cell membranes. Membrane disruption allows the influx of ions and small molecules, resulting in cell lysis. The structure of γ-toxin (also called Hlg) includes two S-components (HlgA and HlgC), which bind to the F-subunit HlgB to form two leukocidins, HlgAB and HlgCB (Yamashita et al., 2011). HlgAB binds to chemokine receptors (i.e., CXCR1 and CXCR2) on RBCs, monocytes, and macrophages in humans and animals (Spaan et al., 2014). HlgCB is specific to humans and targets leukocytes (Spaan et al., 2014).
    d. δ-toxin (also known as Hld) is a small, amphipathic peptide, and it belongs to a class of toxins called phenol-soluble modulins (PSMs). This toxin is not pore-forming; instead, it disrupts cell membranes by inserting itself into the cell surface or phospholipid bilayer via detergent (surfactant)-like mechanisms (Cheung et al., 2015). It is involved in the formation of membrane vesicles, which transport bacterial components including DNA and proteins, and the development of amyloid aggregates and fibrils, enhancing

S. aureus virulence (Cheung et al., 2015; Wang et al., 2023). Oligomerization of the toxin on the membrane of RBCs enhances its hemolytic capacity (Cheung et al., 2015). Perturbations in lipid organization in the membrane result in increased membrane permeability and free ion exchange, both of which culminate in cell lysis. δ-toxin also prominently alters immune system function by activating neutrophils and other granulocytes, leading to TNF-α production (Kasimir et al., 1990).

Exfoliative Toxins. ETs, including ETA, ETB, and ETE, are serine proteases highly specific for desmosomes, which are structures that hold skin cells (keratinocytes) together in animals and humans. These toxins cleave desmoglein-1 (encoded by *dsg1*), a cadherin that is highly expressed in the epidermis and is critical for desmosome structure (Ladhani et al., 1999). Its cleavage at the extracellular domain 3 disrupts desmosome integrity, resulting in characteristic blistering and peeling of the skin associated with SSSS, a condition primarily affecting infants and young children (Dinges et al., 2000; Ladhani et al., 1999; Melish & Glasgow, 1970).

Leukocidins. Leukocidins have similar structures to hemolysins and form pores in cell membranes. The key difference is that hemolysins target RBCs or epi/endothelial cells to acquire nutrients (i.e., iron) for bacterial growth and persistence (Seilie & Wardenburg, 2017), while leukocidins lyse white blood cells (WBCs; i.e., neutrophils and macrophages) to facilitate immune evasion (Cheung et al., 2021). PVL, LukSF-PV, LukGH, and LukAB are major leukocidins produced by *S. aureus*. They are bicomponent toxins like the γ-toxin, but the two are functionally distinct. The "S" subunit of leukocidins binds to host cell receptors, which recruit the "F" subunit and allow the formation of an octameric pore on the membrane (Colin et al., 1994). In a cohort of immunocompromised individuals and in animal studies, PVL has been associated with severe necrotizing pneumonia, which is an aggressive lung condition characterized by rapid necrosis of lung tissue and can lead to sepsis or respiratory failure (Labandeira-Rey et al., 2007). However, the role of PVL in *S. aureus* (including MRSA) virulence remains controversial, especially in animal studies (Bubeck Wardenburg et al., 2008). PVL, lukS-PV, and lukF-PV most commonly form pores in neutrophils, releasing ROS and proteases, which contribute to functional impairments in the cell and tissue necrosis (Labandeira-Rey et al., 2007). Different leukocidins target different WBCs; while PVL and LukED primarily target neutrophils and macrophages, LukAB is more specific for dendritic cells (Gordon & Lowy, 2008). Nonetheless, major mechanisms of leukocidins involve K^+ and Ca^{2+} flux in macrophages, which is linked to the activation of NLRP3 inflammasomes and caspase-1, leading to the production of pro-inflammatory cytokines like IL-6 and IL-1β as well as chemokines (Holzinger et al., 2012). By killing immune cells, leukocidins allow *S. aureus* to evade the host defenses and establish infection.

Mechanisms of enzymes in *S. aureus*

Enzymes produced by *S. aureus* act as molecular "invasion tools," facilitating everything from adhesion to tissue degradation and significantly contributing to the severity and spread of infection. These enzymes work in concert to effectively loosen and break down host tissue structures, dismantling protective barriers, facilitating the rapid invasion and spread of *S. aureus*, and promoting the risk of severe infection.

- **Breaking Down Barriers**: The Role of Hyaluronidase and Collagenase. Hyaluronic acid, a glycosaminoglycan, is a major constituent of the ECM and pericellular matrix, providing structure to the cell and connective tissues, acting as a barrier against microbes, and contributing to immune regulation. However, *S. aureus* produces hyaluronidase (encoded by the *hysA* gene on the *CodY* global regulator), a major spreading factor which degrades hyaluronic acid in the ECM via hydrolysis of its glycosidic bonds (Ibberson et al., 2014).

Degradation of hyaluronic acid loosens the connection between cells and allows *S. aureus* and its biofilms to disperse and penetrate deeper into the tissues, promoting rapid tissue damage and heightening inflammation. Mutations in *hysA* are linked with lower *S. aureus* load and reduced disease severity in lung infections and abscesses (Hart et al., 2013; Ibberson et al., 2014). *S. aureus* uses a similar mechanism to target collagen, which is the most abundant structural protein comprising the ECM. Collagenase hydrolyzes several peptide bonds present in collagen, thus destabilizing its rigid structure (Urso et al., 2024). Transcriptomics analyses show that many clinical isolates exhibit upregulation of collagenase, which contributes to chronic infection (Urso et al., 2024). Breakdown of collagen promotes bacterial spread into deeper tissue layers like the dermis and fascia or even blood vessels, facilitating the development of abscesses or deep tissue infections (Cheung et al., 2021; Urso et al., 2024).

- **Evading Host Defenses**: The Crucial Roles of Sak and Coagulase. Following the disruption of ECM components, Sak targets fibrin clots in the host. Plasmin, a serine protease, in complex with Sak facilitates the digestion of fibrin clots; Sak also activates plasminogen (inactive), converting it into plasmin and allowing further breakdown of clots (Ibberson et al., 2014; Peetermans et al., 2014). Fibrinolysis allows *S. aureus* to escape from confined sites, such as abscesses, and spread to new areas of the body, undermining the host defense system. Coagulase is another *S. aureus*-secreted enzyme which works in concert with Sak. It binds prothrombin, a major clotting factor, to form a staphylothrombin complex, which converts fibrinogen to fibrin and produces a fibrin clot to protect *S. aureus* from phagocytosis and antibiotics (Cheng et al., 2010). These fibrin clots act as camouflage. Via cycles of clotting and fibrinolysis, Sak and coagulase allow *S. aureus* to remain protected during migration to different tissues and then facilitate invasion in these tissues (Peetermans et al., 2014). These cloaking mechanisms allow *S. aureus* to establish a protected niche for persistent colonization and replication.
- **Damaging Host Cells**. *S. aureus* produces multiple lipases and proteases, which degrade host lipids and proteins, respectively. Lipases, such as glycerol ester hydrolase and lipase 2, hydrolyze triglycerides and other lipids into free fatty acids (FFAs) and glycerol (Foster et al., 2014). Lipolysis has a strong impact on host cell membranes, sebaceous glands, and skin due to the high concentration of lipids in these regions, promoting colonization and the formation of abscesses. Furthermore, *S. aureus* can use FFAs as a carbon source, via the fatty acid kinase system, which enhances its growth and proliferation (Foster et al., 2014; Frank et al., 2020). Other FFAs, such as linoleic acid, are antimicrobial; however, *S. aureus* can modify them, making them ineffective against bacterial infection (Kengmo Tchoupa et al., 2024). Staphopains (i.e., ScpA and ScpB) are cysteine proteases which target structural and cytoskeletal proteins like fibronectin and collagen and cause damage to immunoglobulins and neutrophils (Ohbayashi et al., 2011). V8 Protease has similar functions, targeting immunoglobulins and aiding in tissue disintegration (Mustor et al., 2025). On the other hand, aureolysin activity most notably prevents opsonization and phagocytosis (as part of the complement system) and inactivates antimicrobial peptides (Laarman et al., 2011). Thus, these proteases contribute to tissue destruction and invasion and inactivate host immune factors, collectively enhancing *S. aureus* pathogenesis and infection.

2.3.3 Evasion of the Immune System

Following the invasion of host tissues, *S. aureus* implements mechanisms to evade the host immune system. Next, we describe these processes utilized by *S. aureus* to evade host defenses.

Step 1: Inhibiting Chemotaxis and Leukocyte Recruitment. Several proteins produced by *S. aureus* block immune cell chemotaxis and recruitment into infected tissue. CHIPs potently inhibit chemotaxis of neutrophils and monocytes by targeting formylated peptide receptors (FPRs) and the C5aR receptor of the complement system (Cheung et al., 2021; Van Wamel et al., 2006). FPR-like 1 Inhibitory Protein (FLIPr) is very similar in that it inhibits neutrophil chemotaxis and the inflammatory response; however, FLIPr mainly targets the FPR-like 1 receptor (Prat et al., 2009). FLIPr also has purported roles in inhibiting opsonophagocytosis and antibody-mediated cell death. Furthermore, SCIN targets C3 convertases and blocks the complement cascade, which limits neutrophil chemotaxis and phagocyte recruitment (Gordon & Lowy, 2008; Van Wamel et al., 2006). Another way by which *S. aureus* impairs host immune cell recruitment is by targeting adhesion molecules. EAP and its homologs bind to intercellular adhesion molecule-1 (ICAM-1) on endothelial cells, which hinders the adherence of leukocytes to the walls of blood vessels and their ability to transmigrate into infected tissues (Cheung et al., 2021). Staphylococcal superantigen-like proteins (SSLs), such as SSL5, SSL10, and SSL11, also interfere with immune cell trafficking. SSLs bind to P-selectin glycoprotein ligand-1 (PSGL-1), impairing leukocyte recruitment and rolling (Bestebroer et al., 2007). The combination of these key proteins together impairs chemotaxis, blocks adhesion, and inhibits effective recruitment of immune cells, allowing *S. aureus* to increase infectivity before the host immune system elicits an immune response.

Step 2: Neutralizing Phagocytosis. Another mechanism of *S. aureus* involves blocking key steps of phagocytosis, including recognition of phagocytes and bacterial opsonization and engulfment. Protein A binds to the Fc region of IgG antibodies in an altered antibody orientation, preventing receptor recognition and proper opsonization and hindering phagocytosis by neutrophils and macrophages (Atkins et al., 2008). Additionally, Protein A triggers B cell activation and promotes inflammation, potentially leading to autoimmune-like responses (Atkins et al., 2008; Deisenhofer, 1981). Staphylococcal binder of IgG utilizes a similar mechanism, targeting IgG and C3 (part of the complement system), thereby blocking opsonization and complement-mediated phagocytosis (Atkins et al., 2008; Koch et al., 2012). The negatively charged and hydrophilic polysaccharide capsule impairs the uptake of complement proteins and antibodies and also reduces the susceptibility of *S. aureus* to antimicrobials (O'Riordan & Lee, 2004). Another key mechanism involves the secretion of inhibitory molecules, which block the complement cascade. SCINs, extracellular fibrinogen-binding protein, and extracellular complement-binding protein bind to the C3 component of complement and impair opsonization (Koch et al., 2012; Van Wamel et al., 2006). Furthermore, as described previously, leukocidins and hemolysins directly lyse cells, including immune cells/WBCs and RBCs, by forming pores or breaking down sphingomyelin. These toxins also directly target phagocytes, neutralizing or killing them, which prevents bacterial engulfment and opsonophagocytosis. Sak, through its fibrinolytic properties, produces plasmin, which can lower efficient opsonization through the proteolysis of opsonins like IgG and C3b (Ibberson et al., 2014).

Step 3: Suppressing Intracellular Killing within Phagocytes. If *S. aureus* is unable to block immune cells extracellularly, another approach involves targeting phagocytosis. Neutrophils and other phagocytes produce a massive amount of ROS, such as superoxides, hydrogen peroxide, and free radicals, as well as nitric oxide species, in a process called "respiratory burst," which is directly involved in cell death during phagocytosis (Gaupp et al., 2012). *S. aureus* produces antioxidants like catalase, superoxide dismutase, and staphyloxanthin, which scavenge ROS and nitric species, thereby reducing phagocytosis (Gaupp et al., 2012). By detoxifying potent antimicrobial compounds, *S. aureus* can survive ROS-mediated damage, which contributes to its virulence. During phagosome-lysosome fusion, phagosomes containing *S. aureus* fuse with lysosomes, creating an acidic environment rich in proteases and antimicrobial peptides that facilitate cell lysis (Lee et al., 2020). However, *S. aureus* proteins, such as PSMs and SSL-3, can impair the formation and/or maturation of phagolysosomes via complement system inhibition or binding immunoglobulins (Surewaard et al., 2013).

Step 4: Modulation of Adaptive Immunity by Superantigens. In sustained *S. aureus* infections, specific immune responses are elicited. Adaptive immunity involves antigen presentation, T cell activation, production of antibodies, and development of memory B and T cells. *S. aureus* produces several superantigens, such as TSST-1 and staphylococcal enterotoxins (SEs), which alter the adaptive immune response by bypassing normal antigen presentation and directly and non-specifically activating a large proportion of T cells, leading to cytokine storm (Tuffs et al., 2019). While initially triggering a strong immune response, this uncontrolled activation can ultimately lead to immune suppression and tissue damage, such as in sepsis and other long-term infections (Dinges et al., 2000; Hotchkiss et al., 2013).

2.3.4 Regulation of *S. aureus* Virulence Factors

S. aureus uses multiple conductors to activate or repress gene expression that aid in its survival and virulence. They regulate expression in response to environmental cues. They are clinically relevant and, because their mutations drastically impair the virulence and pathogenesis of *S. aureus*, are potential therapeutic targets. In this section, we describe some of the most critical regulatory factors of *S. aureus*.

Two-Component Systems. TCSs act as primary environmental sensors and are comprised of two parts – a sensor histidine kinase (HK) and a response regulator (RR) (Bleul et al., 2022). The HK is membrane-bound and detects a specific signal (i.e., pH, osmolarity, antibiotic concentration, or stress) and phosphorylates the RR (Bleul et al., 2022). Once phosphorylated, cytoplasmic RR binds to DNA, which either activates or represses transcription of the target gene, allowing *S. aureus* to alter expression of its virulence factors. *S. aureus* possesses numerous TCS, including the agr, SaeRS, and vancomycin resistance-associated (VraSR) systems (Josse et al., 2017). Agr is a quorum-sensing system and includes the AgrC sensor and AgrA RR. It senses the accumulation of autoinducing peptides (AIPs), allowing detection of bacterial density, and is critical for the transition from colonization to invasive infection (Beenken et al., 2010). At low densities, surface proteins and other adhesins are upregulated to enhance colonization, while toxins and proteases are produced at high densities to promote evasion, dissemination, and tissue damage (Periasamy et al., 2012). SaeRS comprises the SaeS HK and an SaeR regulator and responds to ROS and antimicrobial peptides in the host (Voyich et al., 2009). It also regulates several genes involved in immune evasion, such as leukocidins and hemolysins, as well as factors central to adaptive immunity, especially in the blood (Voyich et al., 2009). VraSR, comprising the VraS sensor and VraR regulator, plays a major role in cell wall homeostasis and resistance to cell wall-targeting antibiotics like vancomycin (Yin et al., 2006). It identifies cell wall stress and damage from common antibiotics like glycopeptides and β-lactams and activates genes involved in cell wall repair (Howden et al., 2023). The VraSR regulon has clinical relevance as it is highly associated with VRSA, which is difficult to treat due to vancomycin resistance.

Transcription Factors. *S. aureus* transcription factors coordinate with TCS regulators or each other to modulate the expression of genes related to adhesion, invasion, biofilm formation, and metabolism. SarA binds DNA to activate or repress other regulators like agr and SaeRS to influence biofilm formation- and toxin-related genes (Beenken et al., 2010). *S. aureus* strains with mutations in the *sarA* gene have demonstrated significantly lower virulence compared to non-mutant strains (Beenken et al., 2025). CodY also binds DNA and acts as a metabolic sensor by activating or repressing gene transcription based on nutrient levels (Ibberson et al., 2014; Roux et al., 2014). In environments rich in nutrients such as GTP and amino acids (i.e., isoleucine and valine), CodY is activated and regulates genes related to amino acid transport and catabolism. Repressor of toxins (ROT) acts as a counter-regulator to agr; ROT is activated in early infection or when bacterial numbers are low, while agr is activated with higher bacterial density (Roux et al., 2014). It represses the expression of several toxins and proteases to ensure virulence factors are kept low during the initial stages of infection (Cheung et al., 2021). Small regulatory RNAs (sRNAs) are short and non-coding and regulate gene expression by binding to mRNA or interacting with proteins to promote or inhibit translation. There have been more than 50 sRNAs (such as RNAIII) identified in *S. aureus*, and these are involved in virulence, metabolism, and stress responses. RNAIII regulates virulence factor expression as a downstream effector of agr by either inhibiting or activating translation, which facilitates the shift from adhesion to invasion during infection (Queck et al., 2008). Nutrient acquisition is a critical process in *S. aureus* pathogenesis, and a class of regulators, riboswitches, coordinate this process. Riboswitches are located at the 5′ untranslated region of mRNA and are specific for small metabolites. S-adenosylmethionine (SAM) riboswitches are involved in methionine and sulfur metabolisms (Traykovska & Penchovsky, 2022), while flavin mononucleotide (FMN) riboswitches regulate riboflavin, based on FMN levels (Wang et al., 2017). Vitamin biosynthesis (i.e., thiamine) is also tightly regulated by thiamine pyrophosphate (TPP) levels via its binding to the TPP riboswitch (Müller et al., 2009). Riboswitches are highly conserved in bacteria, including *S. aureus*, and do not exist in mammalian cells, making them an attractive therapeutic target, especially for antibiotic-resistant strains. Inhibitors such as TPP and FMN analogs could restrict access to nutrients, limiting the viability of *S. aureus* and reducing its threat.

2.3.5 Proliferation of *S. aureus*

Under favorable conditions (37°C and in the presence of oxygen), *S. aureus* takes up nutrients and rapidly proliferates via binary fission, with a doubling time of 20–30 minutes. Its growth is impacted by temperature, pH, and oxygen levels. Proliferation continues until regulatory systems (i.e., CodY, agr, VraSR) trigger the bacteria to enter survival or virulence mode. It is tightly regulated by a complex network of genes involved in nutrient acquisition, stress responses, and cell wall integrity (i.e., *vraSR*), biofilm formation (i.e., *fnbA*, *fnbB*, *clfA*, *clfB*, and *sarA*), and balancing virulence factors (i.e., *agr*, *rot*, *sarA*, and *codY*). Quorum sensing coordinates proliferation based on bacterial density and cell-to-cell communication, triggering the production of virulence factors and influencing growth (Otto, 2014; Roux et al., 2014).

Sufficient nutrient levels are vital for the growth and survivability of *S. aureus*. It has high flexibility and can utilize a wide range of nutrients and adapt to nutrient-deprived and oxygen-depleted (anaerobic) conditions (Gordon & Lowy, 2008; Voyich et al., 2009). This is clinically relevant as it allows *S. aureus* to proliferate and survive in various host regions, including skin, blood and vasculature, and bones. *S. aureus* prefers glucose as its primary carbon source but can metabolize other sugars like mannose and galactose. It also has adopted several mechanisms for adapting to amino acid and nitrogen levels; these include CodY regulation and arginine deiminase (Roux et al., 2014; Voyich et al., 2009). Furthermore, *S. aureus* has several pathways to utilize various nutrients and adapt to suboptimal conditions, including an anaerobic environment, where *S. aureus* can use fermentation pathways or nitrate to facilitate growth. *S. aureus* can also readily adapt to changes in iron acquisition, redox requirements, pH and stress responses, and biofilm metabolism. Collectively, these pathways remodel the metabolism of *S. aureus* so it can proliferate and survive in nutrient-depleted conditions or under stress.

2.3.6 Cell Lysis

S. aureus also controls its own cell lysis (autolysis) to enhance its virulence and facilitate its dispersal. This process involves the breakdown of the bacterial cell wall and membrane, releasing cellular contents into the surrounding environment. Autolysins, including Atl and Sle1, are key peptidoglycan hydrolases which carry out cell lysis (Porayath et al., 2018). Their activity is tightly regulated to prevent uncontrolled lysis, which is detrimental for *S. aureus* as it may enhance susceptibility to antibiotics and lower its persistence. TCSs (i.e., VraSR), agr, sarA, and environmental factors (i.e., β-lactam antibiotics/glycopeptides, high concentrations of sodium or other cations, and detergents or oxidative species) are all critical factors regulating autolysin functions. Atl also facilitates processes such as cell separation after division (Fernández et al., 2017). Furthermore, some *S. aureus* strains have prophages (i.e., ϕ11, ϕ12, and ϕ13) in their genomes or encounter lytic bacteriophages, which have direct impacts on autolysis or can indirectly influence it by modulating bacterial regulatory systems (Fernández et al., 2017). Prophages can exacerbate *S. aureus* infection by increasing inflammation and immune evasion; on the other hand, phage therapy is a viable approach for reducing *S. aureus* load (Liu et al., 2024). Interestingly, several virulence factors can also contribute to or promote autolysis. PSMs exhibit surfactant-like properties and insert themselves in bacterial membranes, leading to their disruption and lysis (Cheung et al., 2021; Surewaard et al., 2013). CidA (holin-like protein) and LrgAB (antiholin-like protein) work together to balance autolysis and cell survival in *S. aureus* and have reported roles in transporting carbohydrate metabolites (Endres et al., 2022). CidA promotes lysis, while LgrAB inhibits it (Endres et al., 2022). The consequences of *S. aureus* cell lysis are multifaceted, especially in the host (Figure 2.2).

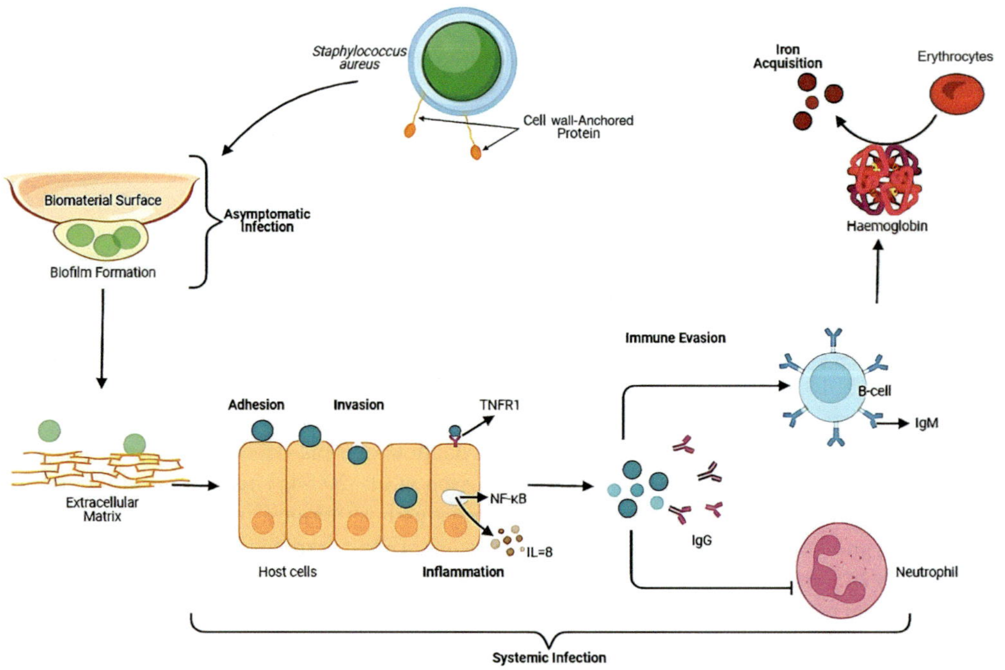

FIGURE 2.2 *Staphylococcus aureus* pathogenesis and immune evasion mechanisms. *S. aureus* has a complex pathogenesis and employs multiple mechanisms to evade the host immune system. The process begins with asymptomatic colonization of biomaterial surfaces, where *S. aureus* adheres and initiates biofilm formation, creating an extracellular matrix that facilitates persistence and confers antibiotic resistance. Cell wall-anchored and surface proteins promote adhesion to host tissues and biomedical devices. Following adhesion, *S. aureus* produces various exotoxins and enzymes, such as hemolysins and leukocidins, which lyse erythrocytes, immune cells, and endothelial cells and ultimately contribute to tissue destruction and inflammation. The inflammatory response is driven by the release of cytokines (i.e., IL-6, IL-8, and TNF-α), neutrophil recruitment, and the generation of ROS and nitrogen species. Iron, which is essential for bacterial growth and systemic dissemination, is acquired from heme during RBC lysis. Immune evasion includes inhibiting opsonization and phagocytosis by binding to the Fc region of immunoglobulins (facilitated by Protein A), degradation of antibodies, and interference with complement activation. Ultimately, these interactions can lead to systemic infections such as sepsis, particularly in vulnerable or immunocompromised individuals. *Abbreviations*: IgG: immunoglobulin G, IgM: immunoglobulin M, IL: interleukin, NF-κB: nuclear factor kappa B, RBCs: red blood cells, ROS: reactive oxygen species, TNF-α: tumor necrosis factor-α.

Lysis releases intracellular virulence factors, which increase inflammation and contribute to tissue damage and immune evasion. The release of extracellular DNA during lysis also contributes to the dispersal of biofilms, allowing colonization at new sites; nutrient deprivation is one major environmental trigger that may lead to programmed cell lysis and biofilm dispersal (Foster et al., 2014; Periasamy et al., 2012). The DNA released after lysis is taken up by other bacteria through horizontal gene transfer, which leads to the spread of antibiotic-resistance genes and virulence factors within the *S. aureus* population (Campos et al., 2022; Woodford, 2005). However, excessive and uncontrolled lysis can cause bacterial cell death, and a strong immune response could eliminate the infection.

2.4 DISEASE DIVERSITY OF *S. AUREUS*

S. aureus is notable for its remarkable adaptability to host tissues and ability to cause a wide range of infections in both humans and animals. A significant global public health concern is the emergence of MRSA, which is characterized by its resistance to β-lactam antibiotics and regulation of the *mecA* resistance gene (Lade & Kim, 2023). *mecA* encodes a penicillin-binding protein 2a (PBP2a) that has a reduced affinity for β-lactam antibiotics, enhancing peptidoglycan cross-linking and survival even at high antibiotic concentrations (Lade & Kim, 2023). Community-associated MRSA (CA-MRSA) is particularly alarming due to its increased virulence and rapid transmission among healthy individuals, with a prevalence reaching nearly 40% in young kids and around 10–20% in the public (Wong et al., 2018). Furthermore, LA-MRSA presents a zoonotic risk, largely due to overuse of antibiotics in agricultural practices (Van Boeckel et al., 2015). Additionally, *S. aureus* has shown resistance to vancomycin (VRSA/VISA) via enhanced cell wall thickening and has exhibited multidrug resistance, making treatment increasingly challenging (Congdon et al., 2023; Périchon & Courvalin, 2009). The inherent disease diversity of *S. aureus* infections is determined based on a range of factors, including strain variation, host immunity, virulence factors, site of infection, and antibiotic resistance.

2.4.1 Diseases in Humans

S. aureus can infect virtually any organ system within the human body, resulting in a diverse array of pathological conditions (Figure 2.3). Skin and soft tissue infections (SSTIs) are the most prevalent manifestation of pathogenic *S. aureus*, are highly recurring (even after treatment), and include diseases ranging from mild skin infections to severe and life-threatening conditions like necrotizing fasciitis (Ladhani et al., 1999; Linz et al., 2023). Folliculitis is a common SSTI characterized by small, red papules and pus due to infection of fair follicles (Linz et al., 2023). Impetigo is similar, but it also impacts the epidermis, frequently affects children and is characterized by honey-colored crusts, which arise due to the bursting of pus-filled blisters (Linz et al., 2023). Similarly, atopic dermatitis causes dry, itchy, and inflamed skin, which often forms a rash, and is also very common in children. A recent study suggests that *S. aureus* extracellular vesicles are involved in the pathogenesis of atopic dermatitis by transporting toxins and other effectors and disrupting skin barriers (Lee et al., 2009). Unfortunately, *S. aureus* can also cause infections in deeper dermal and tissue layers. Furuncles (commonly known as boils) are deeper tissue infections which impact hair follicles as well as dermal and subcutaneous tissues. They are characterized by painful, pus-filled nodules and can interconnect to form clusters, called carbuncles, which often extend deeper and are more aggressive (Del Giudice, 2020; Ladhani et al., 1999). Cellulitis is an infection of the deeper layers of the skin and subcutaneous tissue, causing redness, swelling, pain, and warmth, with no distinct pustule (Del Giudice, 2020). This differs from abscesses, which are localized collections of pus. A very severe and invasive SSTI related to *S. aureus* is necrotizing fasciitis (also known as "flesh-eating disease"). It is a rapidly progressing, life-threatening infection that destroys the soft tissues, including the fascia and subcutaneous tissue. Initial infection is characterized by mild

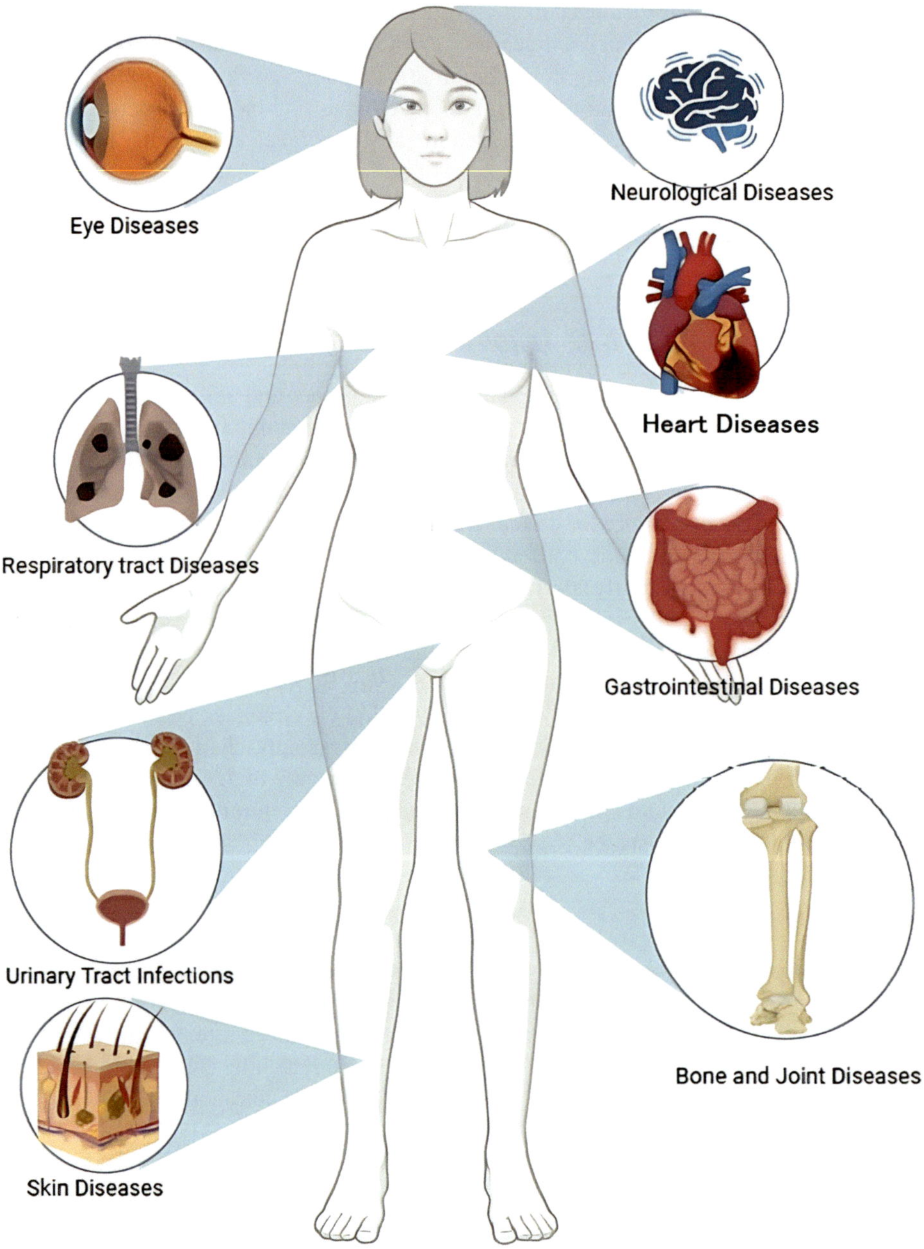

FIGURE 2.3 Diseases caused by *Staphylococcus aureus* in humans. Pathogenic *S. aureus* manifests as several disease conditions, which impact multiple human organ systems.

symptoms like swelling and fever but quickly escalates to a lethal infection defined by multi-organ failure, hypotension, rapid necrosis of soft tissues, and sepsis (Zhou et al., 2025). Due to the wide range of severity of SSTIs, their treatments are also highly variable. Mild skin infections can be treated with hygiene changes, drainage, or topical antibiotics, while deeper tissue infections might require intravenous or oral antibiotics (Tong et al., 2015).

Virulence Factors and Pathogenesis: As described in detail previously, *S. aureus* employs multiple virulence factors to establish infection, including (1) surface-associated factors (i.e., adhesins),

(2) extracellular toxins and enzymes (i.e., hemolysins/cytolysins, superantigens, and exfoliative toxins), (3) biofilm-related genes (i.e., polysaccharide intercellular adhesin), (4) regulatory systems (i.e., *agr* locus), and (5) mobile genetic elements. Cytolysins, such as α-toxin, γ-hemolysin, PVL, and LukED, destroy immune cells by forming pores in cell membranes, which contribute to tissue damage. PVL is linked epidemiologically to abscesses and furuncles but does not invariably correlate with infection severity (Colin et al., 1994). Superantigens and coagulase, as well as surface-associated factors (Protein A) and biofilm-related genes, allow *S. aureus* to evade the host immune system. Additional surface-associated factors, including adhesion factors (namely FnBPs) and PSMs, enable bacterial attachment and aggregation to host tissues. PSMs can also lyse neutrophils and erythrocytes, which enhances evasion of *S. aureus* and promotes its colonization and dissemination. *sasX*, which is commonly found in Asian *S. aureus* clones and most linked with MRSA, encodes a surface protein that promotes aggregation, abscess formation, and nasal colonization (Li et al., 2012). Mobile genetic elements, such as the arginine catabolic mobile element (ACME), provide resistance to acidic conditions and toxic polyamines, thereby enhancing the colonization and biofilm formation of *S. aureus* (Planet et al., 2013). These properties give specific strains, including methicillin-resistant USA300, survival advantages on the skin and promote persistent infection (Planet et al., 2013). Furthermore, regulatory systems and transcription factors are crucial for the proper regulation of virulence factors and the progression of *S. aureus* infection in many different organs, including the lungs and respiratory tract, bones and joints, urinary tract, heart, and brain.

Respiratory Tract Infections. Respiratory infections encompass any acute or long-term infection in the nose, oral cavity, bronchi, or lungs. *S. aureus* colonizes in the respiratory tract, which can be asymptomatic or lead to severe conditions like necrotizing pneumonia, which results in life-threatening necrosis and liquefaction of lung tissue (Labandeira-Rey et al., 2007). Pneumonia is a very common lung infection, culminating in over 2.5 million deaths annually, and is characterized by severe cold-like symptoms such as coughing with mucus, fever, and chest pain. *S. aureus* is the most common bacterial causative agent, particularly MRSA (Pickens & Wunderink, 2022). α-toxin and PVL are highly linked with lung injury (even including pulmonary necrosis) and inflammation in pneumonia (Labandeira-Rey et al., 2007), while Protein A impacts the host immune response to clear the infection. Primary *S. aureus* pneumonia and secondary infections (i.e., following respiratory viral infections) are linked with high morbidity and mortality, especially in children. Empyema is one severe complication stemming from bacterial pneumonia and is characterized by infected fluid in the pleural space between the lungs and chest wall (Nakamura et al., 2024). It necessitates prompt intervention, including antibiotics and pus drainage; if left unmanaged, empyema increases the risk of severe morbidity and/or mortality caused by pneumonia.

Bone and Joint Infections. *S. aureus* also infects the bones and joints, resulting in conditions like osteomyelitis and septic arthritis, which are multifactorial in etiology. Osteomyelitis is an infection of the bone, while septic arthritis impacts the joints – both conditions are associated with pain, swelling, and fever and occur when *S. aureus* enters the blood (Beenken et al., 2025; Kinamon et al., 2023). When chronic, osteomyelitis is associated with necrosis of bone tissues and fractures and can also impact the vertebrae (Kinamon et al., 2023). Biomedical devices like prosthetics and implants are a common reservoir for *S. aureus* infection in bones and joints due to biofilm formation and adherence. *S. aureus* has developed multiple mechanisms, including adhesion to the bone matrix (or implants), promotion of biofilm formation, and invasion of osteoblasts and small-colony variants to overcome natural bone resistance (Collet et al., 2025; Proctor et al., 1995). After uptake by FnBPs, *S. aureus* can live in a low metabolic state in bone, further damaging osteoblasts and allowing persistent infection, making osteomyelitis difficult to treat (Collet et al., 2025). Their slow growth, reduced pigmentation by staphyloxanthin and protection from ROS, and defective electron transport chain mechanisms, especially in small-colony variants, prolong infection or increase its severity (Proctor et al., 1995). These factors also contribute to resistance against cell wall-active antibiotics and aminoglycosides.

Urinary Tract Infections. Although urinary tract infections (UTIs) are mostly caused by uropathogenic *Escherichia coli*, other microbes, including *S. aureus*, are also causative agents. *S. aureus* is identified as the causative agent in less than 10% of diagnosed UTIs, but the exact prevalence varies depending on strain and geography (Gajdács et al., 2020). *S. aureus* more commonly causes UTIs in individuals with urinary catheters, anatomical abnormalities, or weakened immune systems, but it can also be community associated (Alshomrani et al., 2023). It carries a high risk for systemic infection and the development of bacteremia or pyelonephritis (Muder et al., 2006). Symptoms of *S. aureus*-associated UTIs include discomfort when urinating, high frequency of urination, pain, and fever. Treatment for UTIs caused by *S. aureus* is multidisciplinary and includes the careful use of antibiotics, such as trimethoprim-sulfamethoxazole or other combinations (Alshomrani et al., 2023). Furthermore, in complicated and device-associated cases, more aggressive intravenous antibiotics and device/catheter removal are required to reduce the risk of bacteremia or kidney infection.

Cardiovascular Infections. *S. aureus* is associated with many cardiovascular infections, including infective endocarditis (IE), myocarditis, pericarditis, and vascular infections. After entry into the host, *S. aureus* first damages the cardiac endothelium, which can result from direct trauma, including the insertion of intravascular catheters, injection of drugs, or turbulent blood flow stemming from valvular abnormalities or prosthetic valves (Kiefer & Bashore, 2012). Inflammation caused by conditions like rheumatic heart disease or degenerative valvular disease can also induce endothelial damage. This injury exposes subendothelial cells, leading to an influx of ECM proteins, tissue factor expression, and fibrin and protein deposition (Kiefer & Bashore, 2012). Consequently, sterile vegetations form in a prothrombotic milieu, allowing *S. aureus* colonization and rapid progression to IE. *S. aureus* adheres to the cardiac endothelium, ECM protein, fibrin, and platelets, resulting in the formation of atherosclerotic colonies, evasion of the host immune system, and damaged tissues (Panizzi et al., 2011). Research suggests that certain strains of *S. aureus*, particularly those belonging to clonal complexes 8 and 30, have a higher propensity for causing IE (Bowers et al., 2018; Sharma-Kuinkel et al., 2015). Many of these complexes include MRSA strains and other healthcare-associated or hospital-acquired strains and, as such, constitute a major public health concern. ClfB, Cna, and EAP, as well as superantigens (i.e., TSST-1) and leukocidins (i.e., PVL), are associated with aggressive infections and enhance the ability of *S. aureus* to colonize the cardiac endothelium. An increased propensity of clonal complex 30 strains to induce IE has also been demonstrated in an animal study (Bastien et al., 2023). Complications of IE include fever, chills, and bacteremia, due to the presence of *S. aureus* in the bloodstream. Bacteremia is a life-threatening condition, potentially leading to sepsis and other serious complications, including heart failure, perivalvular abscesses, pulmonary embolism, stroke, or aneurysm, which have high mortality rates. Thus, early detection and treatment of bacteremia and *S. aureus*-associated cardiovascular conditions are critical for mitigating these adverse outcomes. Clinical trials suggest that early-stage valve surgery significantly lowers the risk of mortality compared to other treatments (Kang et al., 2012; Kiefer et al., 2011).

Neurological Infections. Although serious, neurological infections caused by *S. aureus* are infrequent and commonly occur secondary to other infective diseases (like IE, bacteremia, or osteomyelitis). *S. aureus* has been linked with brain abscesses, meningitis, subdural empyema, stroke, and encephalopathy. Meningitis is defined by inflammation of the protective meningeal membranes surrounding the brain and spinal cord; it is diagnosed by assessing cerebral spinal fluid, blood, or via imaging, and it has a high mortality rate (Akaishi et al., 2024). Although it is relatively uncommon compared to other bacterial causes (like *Streptococcus pneumoniae* and *Neisseria meningitidis* [meningococcus]), *S. aureus* meningitis can emerge as a complication of neurosurgery or trauma, as well as IE or bacteremia. Brain abscesses also commonly occur due to the spread of *S. aureus* from IE or pneumonia (hematogenous seeding) and are characterized by a localized collection of pus within the brain (Akaishi et al., 2024). They can also form from trauma or direct inoculation from spinal surgery or catheter placement. Epidural abscesses and subdural empyema can both arise from hematogenous seeding or from other infections like sinusitis or osteomyelitis. Spinal epidural abscesses frequently affect the lumbar and lower thoracic spine, while subdural empyema is mostly intracranial (Calfee & Wispelwey, 1999). These infections can disseminate across multiple vertebral levels or, in rare instances, affect the entire spine, resulting in severe back pain, neurological impairments (weakness and sensory changes), or even paralysis. Risk factors include diabetes mellitus, injection drug use, recent spinal surgery or trauma, and epidural device implantation. Diagnosis is made using magnetic resonance imaging in patients with *S. aureus* bacteremia and new or worsening back pain (Calfee & Wispelwey, 1999). A clinical staging system categorizes epidural abscess progression into four sub-stages, as follows: (1) localized back pain and/or fever, (2) radiating nerve root pain, (3) motor and sensory weakness or loss and/or bowel or bladder dysfunction, and (4) paralysis (Velissaris et al., 2009). Other neurological conditions like stroke, cerebral hemorrhaging, and encephalopathy manifest from other infectious conditions like IE or systemic inflammation mediated by *S. aureus* superantigens and can be life-threatening.

Eye Infections. *S. aureus* is associated with several ocular pathologies, including both mild surface infections and severe diseases that affect the tear ducts, eyelid, conjunctiva, cornea, vitreous chamber, and anterior/posterior chambers, and ultimately could impair vision. Conjunctivitis ("pink eye"), blepharitis, and styes are three mild eye pathologies that can be caused by *S. aureus* (including MRSA) but only impact outer eye structures (O'Callaghan, 2018). *S. aureus* is one of the most prominent causative agents of bacterial conjunctivitis, which is characterized by inflammation of the conjunctiva (the delicate membrane covering the eyelid and eye) as well as redness, sticky yellow-green discharge, burning, irritation, and crusting near the eyelashes (Azari & Barney, 2013). Blepharitis is similar and impacts the areas surrounding the eyelids. If *S. aureus* colonization becomes chronic, blepharitis can also develop into secondary conjunctivitis or lead to the development of styes, which are tender nodules surrounding the eyelids (O'Callaghan, 2018). Keratitis is defined as inflammation of the corneal stroma, the clear front part of the eye, which is rich in antimicrobial factors and protects the cornea from infection (Davidson & Kuonen, 2004). Although rather infrequent, individuals with keratitis resulting from *S. aureus* infection experience redness, tearing, and pain in their eyes as well as compromised vision and swelling with discharge (O'Callaghan, 2018). Corneal infections can result in corneal perforation, scarring, endophthalmitis (interior eye infection), or even glaucoma if not treated, increasing the risk of vision impairment. Endophthalmitis is a severe and potentially life-threatening intraocular infection that commonly occurs following surgical interventions, traumatic injury, or through hematogenous seeding of *S. aureus* strains (most notably MRSA) and can rapidly impair eyesight (Ho et al., 2022; O'Callaghan, 2018). There are multiple lines of treatment available – including ointments, antibiotics (i.e., gentamicin), and improved hygiene for superficial infections or systemic/intravitreal antibiotics or even surgery for interior eye infections.

Gastrointestinal Infections. There are two major types of gastrointestinal diseases stemming from *S. aureus* infection. The first is staphylococcal food poisoning and occurs due to ingestion of pre-formed enterotoxins (even at very low concentrations), which can be present in contaminated food or animal products (Argudín et al., 2010). These enterotoxins function like superantigens, non-specifically activating a large T cell repertoire and releasing a massive amount of cytokines, resulting in a rapid onset of nausea, vomiting, diarrhea, and abdominal cramps, which usually subside in 1–2 days (Argudín et al., 2010; Krakauer, 2019). Staphylococcal enterotoxins are a growing concern in the food industry, particularly in dairy products (Campos et al., 2022). TSST-1 has also been reported in cases of food poisoning (Goudsmit et al., 2021). The second manifestation of *S. aureus* in the intestine is direct gastrointestinal infection ("staphylococcal enterocolitis"), where the bacteria proliferate and overgrow in the gastrointestinal tract; MRSA is a highly suspected causative agent (Azimirad et al., 2017). Although much less common, it is often reported in individuals with a dampened immune system or comorbidities and is mostly observed in hospitalized patients. Enterocolitis, abscess formation, and bacteremia are examples of direct infections, which can result from *S. aureus* in the gastrointestinal tract; they develop because of antibiotics modulating the gut microbiota (less common), hematogenous seeding, or from major gastrointestinal surgery and are associated with prolonged gut symptoms like diarrhea and abdominal cramping (Boyce & Havill,

2005; Okada et al., 2018). The most common treatments for these ailments include supportive care or intravenous nutrition/hydration (to replace fluids and electrolytes) until discomforting symptoms subside, but antibiotics (i.e., rifampicin) are also given for direct infections (Okada et al., 2018).

2.4.2 Diseases in Animals

S. aureus also exhibits significant pathogenic potential in other animal species, including dairy cattle, sheep, goats, horses, pigs, poultry (chickens and turkeys), dogs, cats, and rabbits. Mastitis (inflammation of the mammary glands) is diagnosed in multiple animals, including cattle, sheep, and goats, and *S. aureus* is one causative agent. Bovine mastitis (inflammation of the mammary glands) specifically presents in cattle and places a significant economic burden on the dairy industry due to potential contamination and decreased yield of milk from infected cattle (Campos et al., 2022; Girma & Tamir, 2022). This constitutes a public health concern for people who consume dairy because of the risk of enterotoxigenic *S. aureus*. Subclinical manifestations include decreased milk production, while clinical cases are associated with pain, swelling, and heat in the udder, milk clots, and a decline in milk production (Campos et al., 2022). For example, in humans, abscesses and skin infections also manifest in sheep and goats, pigs, rabbits, and in cats and dogs (called pyoderma) (Costa et al., 2022). Chicken and turkeys can develop foot infections (called bumblefoot) (Anjum et al., 2019), while dogs and cats can be diagnosed with outer ear infections linked to *S. aureus* (Costa et al., 2022). Bone and joint infections, like those that manifest in humans, also develop in horses and poultry and sometimes become septic.

2.4.3 Obstacles to Effective Treatment and Future Directions

2.4.3.1 Current Challenges with Treatments

The increasing prevalence of MRSA and other antibiotic-resistant *S. aureus* strains (i.e., VRSA and VISA) poses substantial challenges to effective treatment, highlighting the urgent need for alternative therapies and improved diagnostic methods. As of 2023, the global prevalence of MRSA was approximately 15% – further indicating antibiotic resistance as a public health crisis (Hasanpour et al., 2023). Antibiotic resistance varies depending on demographic features and geography; different genotypes of MRSA have been identified in Asian countries compared to the United States (Wong et al., 2018). MRSA is just one example of antibiotic-resistant *S. aureus*. Emerging studies suggest that almost every strain of *S. aureus* demonstrates some degree of resistance, with almost all strains showing resistance to penicillins (90–95%) due to bacterial production of β-lactamase (Lade & Kim, 2023). Resistance to other antibiotics, including methicillin, oxacillin, vancomycin, erythromycin, clindamycin, fluoroquinolones, and tetracyclines, is also highly prevalent, ranging from 20% to 70%, depending on the antibiotic and strain lineage (Carrel et al., 2024; Wu et al., 2023).

2.4.3.2 Emergence of Effective Antibiotic Strategies

Despite its resistance against methicillin and oxacillin, several conventional antibiotics, such as vancomycin, lipoglycopeptides (dalbavancin, ortavancin), and oxazolidinones (tedizolid, linezolid) are effective alternatives for MRSA and other *S. aureus* infections. Emerging antibiotics target PBP2a to prevent this enzyme from cross-linking peptidoglycan, thus lowering resistance mechanisms. Clinical trials have demonstrated that ceftobiprole and ceftaroline alone or in combination can reduce *S. aureus* infectivity in bacteremia and IE, attenuating disease progression (Caniff et al., 2024; Holland et al., 2023). Emerging literature suggests that antibiotics in combination have stronger effects, especially in MRSA. One classic example is combination aminoglycosides, which have shown efficacy in bacteremia and endocarditis (Lemonovich et al., 2011). In bacteremia, ceftaroline combined with daptomycin has demonstrated enhanced bacterial clearance compared to single therapies alone, and combinatory effects are specific based on which antibiotics are used (Patel et al., 2023; Tsai et al., 2022). However, there are a few caveats to implementing combination antibiotics. Overuse may enhance the risk of developing antibiotic resistance to *S. aureus*, making it even more challenging to treat. Furthermore, antibiotic use is linked with perturbations in the gut microbiome composition (gut dysbiosis), which alters its functions and can dampen the immune system via the microbiota-immune axis (Prajapati et al., 2025). One trial reported that combination antibiotics (vancomycin/daptomycin + β-lactam) did not significantly change outcomes like mortality risk and bacteremia and actually demonstrated an increased risk of kidney injury (Tong et al., 2020). Thus, future preclinical studies are necessary to delineate the precise mechanisms of combinatory antibiotic regimens as well as identify potential pitfalls before implementation in human trials.

2.4.3.3 Ongoing Research and Future Directions

Because of these challenges with conventional antibiotics, current research focuses on designing novel antibiotics with new targets, such as fatty acid synthesis. Furthermore, developing effective receptor analogs (i.e., for cell wall components), employing bacteriophage therapy (Liu et al., 2024) or CRISPR/Cas9 (gene editing) technologies (Tao et al., 2022), utilizing enzymes which disrupt biofilm formation, and implementing nanotechnology (Zhu et al., 2024) are novel approaches which are being explored preclinically and in clinical trials. Recent decades of research have aimed at developing the first *S. aureus* vaccine. This is a challenging feat due to the wide arsenal of virulence factors used by *S. aureus*, its role as a commensal bacterium in many people and asymptomatic nasal colonization, and variability in antigenic components. A few clinical trials have been conducted to determine the safety, feasibility, and overall efficacy of staphylococcal vaccines (targeting both passive and active immunity), but have been unsuccessful in late-phase trials (Fattom et al., 2004; Hassanzadeh et al., 2023). A recent study implemented a novel vaccine that targets SpA and LukAB and showed that this vaccine significantly lowered bacterial load and reduced its dissemination; skin infection was also attenuated (Poolman et al., 2025). As a result, one feasible approach being explored is multi-component vaccines, which could target several virulence factors. mRNA vaccines have been designed to target *S. aureus* toxin and adhesion proteins and are currently experimental (Zhu et al., 2023). Peptide-based antimicrobials like Hp(2–20), which is derived from a ribosomal protein of *Helicobacter pylori*, have demonstrated antibacterial properties in *S. aureus*, through pore formation, lipid interactions, and changes in membrane structure (Liu et al., 2011). However, these molecules are highly sensitive and require additional research to enhance their stability and delivery during active infection. Metal-based therapies are an intriguing area of research, especially in the context of MRSA and antibiotic-resistant *S. aureus*. Zinc and copper and their nanoparticles target bacterial membrane integrity, disrupt biofilms, generate ROS, impair bacterial enzyme functions, and degrade proteins, lipids, and DNA (Cioffi et al., 2005; Raghupathi et al., 2011; Zhu et al., 2024). However, metals must be used with caution due to their high toxicity to host cells at high concentrations.

Another important step is improving diagnostic techniques to facilitate rapid and accurate detection of *S. aureus* and its antibiotic resistance profiles. PCR-based assays (including multiplexing) can be performed quickly and provide high-capacity detection of multiple genes at the same time (Liu et al., 2016). Whole genome sequencing is also a useful approach for determining resistance genotypes and can generate a complete lineage for a specific strain, including clinical isolates. Emergence of these technologies, along with others like nanotechnology, artificial intelligence, and antibiotic susceptibility testing, allows for a more comprehensive and multifaceted approach to target antibiotic resistance in clinical *S. aureus*.

An additional concern with *S. aureus* is hospital-acquired infections. As a hospital-associated pathogen, *S. aureus* colonization (including healthy colonization) constitutes a significant reservoir for transmission through either contact with infected individuals, fomites, or medical equipment. Furthermore, it is a major concern in hospitals because it is highly resistant and persistent and can become lethal in opportunistic conditions. Thus, it is essential to establish practices to control and/or prevent its spread in healthcare settings and communities. Active screening approaches using nasal swabs or diagnostic PCR-based assays are a good approach for early detection of virulence genes. PCR-based diagnostic assays are available in the clinic and are used, but their widespread use is integral to the early detection of virulent and antibiotic-resistant *S. aureus* (Liu et al., 2016). However, high cost is one major factor limiting the widespread integration of PCR detection assays. Furthermore, consistent and proper handwashing, hygiene, and decontamination practices are critical for minimizing the spread of *S. aureus*. Aseptic techniques should always be practiced when handling devices and during surgery. The main goal is to quickly detect virulent *S. aureus* and then isolate these individuals to prevent the bacteria from spreading in hospital and community settings. Standard decontamination and hygiene protocols should be followed to minimize the spread of *S. aureus* and other opportunistic pathogens during routine checkups, device use, surgery, or any other contact with patients or potentially infected surfaces.

2.5 CONCLUSIONS

S. aureus is a commensal bacterium but can become pathogenic under favorable conditions, leading to the development of a wide range of infections in the respiratory tract, bones and joints, cardiac endothelium, and brain in both humans and animals. It has become a global public health concern due to the rising prevalence of antibiotic-resistant strains, which are more challenging to treat. Furthermore, by harnessing multiple virulence factors, including surface proteins, hemolysins, superantigens, toxins, and leukocidins, *S. aureus* can colonize and invade host tissues, evade innate and adaptive immune responses, and form biofilms to enhance its pathogenesis. Implementing novel technologies is critical for early identification of virulence or antibiotic resistance genes, effective treatment, and reduced prevalence and spreading of *S. aureus*, especially clinical isolates.

REFERENCES

Adams, C. E., & Dancer, S. J. (2020). Dynamic transmission of *Staphylococcus aureus* in the intensive care unit. *International Journal of Environmental Research and Public Health*, *17*(6), 2109. https://doi.org/10.3390/ijerph17062109

Akaishi, T., Tarasawa, K., Fushimi, K., Yaegashi, N., Aoki, M., & Fujimori, K. (2024). Demographic profiles and risk factors for mortality in acute meningitis: A nationwide population-based observational study. *Acute Medicine & Surgery*, *11*(1), e920. https://doi.org/10.1002/ams2.920

Alseqely, M., Newton-Foot, M., Khalil, A., El-Nakeeb, M., Whitelaw, A., & Abouelfetouh, A. (2021). Association between fluoroquinolone resistance and MRSA genotype in Alexandria, Egypt. *Scientific Reports*, *11*(1), 4253. https://doi.org/10.1038/s41598-021-83578-2

Alshomrani, M. K., Alharbi, A. A., Alshehri, A. A., Arshad, M., & Dolgum, S. (2023). Isolation of *Staphylococcus aureus* urinary tract infections at a community-based healthcare center in Riyadh. *Cureus*, *15*(2), e35140. https://doi.org/10.7759/cureus.35140

Anjum, M. F., Marco-Jimenez, F., Duncan, D., Marín, C., Smith, R. P., & Evans, S. J. (2019). Livestock-associated methicillin-resistant *Staphylococcus aureus* from animals and animal products in the UK. *Frontiers in Microbiology*, *10*, 2136. https://doi.org/10.3389/fmicb.2019.02136

Argudín, M. Á., Mendoza, M. C., & Rodicio, M. R. (2010). Food poisoning and *Staphylococcus aureus* enterotoxins. *Toxins*, *2*(7), 1751–1773. https://doi.org/10.3390/toxins2071751

Atkins, K. L., Burman, J. D., Chamberlain, E. S., Cooper, J. E., Poutrel, B., Bagby, S., Jenkins, A. T. A., Feil, E. J., & Van Den Elsen, J. M. (2008). *S. aureus* IgG-binding proteins SpA and Sbi: Host specificity and mechanisms of immune complex formation. *Molecular Immunology*, *45*(6), 1600–1611. https://doi.org/10.1016/j.molimm.2007.10.021

Azari, A. A., & Barney, N. P. (2013). Conjunctivitis: A systematic review of diagnosis and treatment. *JAMA*, *310*(16), 1721–1729. https://doi.org/10.1001/jama.2013.280318.

Azimirad, M., Dezfulian, A., Alebouyeh, M., Esfehani, R. B., Shahrokh, S., & Zali, M. R. (2017). Infection with enterotoxigenic *Staphylococcus aureus* as a concern in patients with gastroenteritis. *Journal of Global Antimicrobial Resistance*, *9*, 111–114. https://doi.org/10.1016/j.jgar.2017.02.009

Bastien, S., Meyers, S., Salgado-Pabón, W., Giulieri, S. G., Rasigade, J. P., Liesenborghs, L., Kinney, K. J., Couzon, F., Martins-Simoes, P., Moing, V. L., Duval, X., Holmes, N. E., Bruun, N. E., Skov, R., Howden, B. P., Fowler, V. G., Jr, Verhamme, P., Andersen, P. S., Bouchiat, C., Moreau, K., . . . Vandenesch, F. (2023). All *Staphylococcus aureus* bacteraemia-inducing strains can cause infective endocarditis: Results of GWAS and experimental animal studies. *The Journal of Infection*, *86*(2), 123–133. https://doi.org/10.1016/j.jinf.2022.12.028

Becker, K. A., Fahsel, B., Kemper, H., Mayeres, J., Li, C., Wilker, B., Keitsch, S., Soddemann, M., Sehl, C., Kohnen, M., Edwards, M. J., Grassmé, H., Caldwell, C. C., Seitz, A., Fraunholz, M., & Gulbins, E. (2017). *Staphylococcus aureus* alpha-toxin disrupts endothelial-cell tight junctions via acid sphingomyelinase and ceramide. *Infection and Immunity*, *86*(1), e00606.17. https://doi.org/10.1128/IAI.00606-17

Beenken, K. E., Campbell, M. J., & Smeltzer, M. S. (2025). The ability of sarA to limit protease production plays a key role in the pathogenesis of *Staphylococcus aureus* osteomyelitis irrespective of the functional status of agr. *Infection and Immunity*, *93*(1), e0047324. https://doi.org/10.1128/iai.00473-24

Beenken, K. E., Mrak, L. N., Griffin, L. M., Zielinska, A. K., Shaw, L. N., Rice, K. C., Horswill, A. R., Bayles, K. W., & Smeltzer, M. S. (2010). Epistatic relationships between sarA and agr in *Staphylococcus aureus* biofilm formation. *PLoS One*, *5*(5), e10790. https://doi.org/10.1371/journal.pone.0010790

Bestebroer, J., Poppelier, M. J., Ulfman, L. H., Lenting, P. J., Denis, C. V., van Kessel, K. P., van Strijp, J. A., & de Haas, C. J. (2007). Staphylococcal superantigen-like 5 binds PSGL-1 and inhibits P-selectin–mediated neutrophil rolling. *Blood*, *109*(7), 2936–2943. https://doi.org/10.1182/blood-2006-06-015461

Bleul, L., Francois, P., & Wolz, C. (2022). Two-component systems of *S. aureus*: Signaling and sensing mechanisms. *Genes*, *13*(1), 34. https://doi.org/10.3390/genes13010034

Boles, B. R., & Horswill, A. R. (2008). Agr-mediated dispersal of *Staphylococcus aureus* biofilms. *PLoS Pathogens*, *4*(4), e1000052. https://doi.org/10.1371/journal.ppat.1000052

Bowers, J. R., Driebe, E. M., Albrecht, V., McDougal, L. K., Granade, M., Roe, C. C., Lemmer, D., Rasheed, J. K., Engelthaler, D. M., Keim, P., & Limbago, B. M. (2018). Improved subtyping of *Staphylococcus aureus* clonal complex 8 strains based on whole-genome phylogenetic analysis. *mSphere*, *3*(3), e00464.17. https://doi.org/10.1128/mSphere.00464-17

Boyce, J. M., & Havill, N. L. (2005). Nosocomial antibiotic-associated diarrhea associated with enterotoxin-producing strains of methicillin-resistant *Staphylococcus aureus*. *The American Journal of Gastroenterology*, *100*(8), 1828–1834. https://doi.org/10.1111/j.1572-0241.2005.41510.x

Bubeck Wardenburg, J., Palazzolo-Ballance, A. M., Otto, M., Schneewind, O., & DeLeo, F. R. (2008). Panton-Valentine leukocidin is not a virulence determinant in murine models of community-associated methicillin-resistant *Staphylococcus aureus* disease. *The Journal of Infectious Diseases*, *198*(8), 1166–1170. https://doi.org/10.1086/592053

Calfee, D. P., & Wispelwey, B. (1999). Brain abscess, subdural empyema, and intracranial epidural abscess. *Current Infectious Disease Reports*, *1*(2), 166–171. https://doi.org/10.1007/s11908-996-0025-5

Campos, B., Pickering, A. C., Rocha, L. S., Aguilar, A. P., Fabres-Klein, M. H., de Oliveira Mendes, T. A., Fitzgerald, J. R., & de Oliveira Barros Ribon, A. (2022). Diversity and pathogenesis of *Staphylococcus aureus* from bovine mastitis: Current understanding and future perspectives. *BMC Veterinary Research*, *18*(1), 115. https://doi.org/10.1186/s12917-022-03197-5

Caniff, K. E., Judd, C., Lucas, K., Goro, S., Orzol, C., Eshaya, M., Al Musawa, M., Veve, M. P., & Rybak, M. J. (2024). Heartfelt impact: A descriptive analysis of ceftaroline-containing regimens in endocarditis due to methicillin-resistant *Staphylococcus aureus*. *Infectious Diseases and Therapy*, *13*(12), 2649–2662. https://doi.org/10.1007/s40121-024-01068-0

Carrel, M., Smith, M., Shi, Q., Hasegawa, S., Clore, G. S., Perencevich, E. N., & Goto, M. (2024). Antimicrobial resistance patterns of outpatient *Staphylococcus aureus* isolates. *JAMA Network Open*, *7*(6), e2417199. https://doi.org/10.1001/jamanetworkopen.2024.17199

Cheng, A. G., McAdow, M., Kim, H. K., Bae, T., Missiakas, D. M., & Schneewind, O. (2010). Contribution of coagulases towards *Staphylococcus aureus* disease and protective immunity. *PLoS Pathogens*, *6*(8), e1001036. https://doi.org/10.1371/journal.ppat.1001036

Cheung, G. Y. C., Bae, J. S., & Otto, M. (2021). Pathogenicity and virulence of *Staphylococcus aureus*. *Virulence*, *12*(1), 547–569. https://doi.org/10.1080/21505594.2021.1878688

Cheung, G. Y., Yeh, A. J., Kretschmer, D., Duong, A. C., Tuffuor, K., Fu, C. L., Joo, H. S., Diep, B. A., Li, M., Nakamura, Y., Nunez, G., Peschel, A., & Otto, M. (2015). Functional characteristics of the *Staphylococcus aureus* δ-toxin allelic variant G10S. *Scientific Reports*, *5*(1), 18023. https://doi.org/10.1038/srep18023

Chmielowiec-Korzeniowska, A., Tymczyna, L., Wlazło, Ł., Nowakowicz-Dębek, B., & Trawińska, B. (2020). *Staphylococcus aureus* carriage state in healthy adult population and phenotypic and genotypic properties of isolated strains. *Postepy Dermatologii i Alergologii*, *37*(2), 184–189. https://doi.org/10.5114/ada.2020.94837

Cioffi, N., Torsi, L., Ditaranto, N., Tantillo, G., Ghibelli, L., Sabbatini, L., Bleve-Zacheo, T., D'Alessio, M., Zambonin, P. G., & Traversa, E. (2005). Copper nanoparticle/polymer composites with antifungal and bacteriostatic properties. *Chemistry of Materials*, *17*(21), 5255–5262. https://doi.org/10.1021/cm0505244

Colin, D.A., Mazurier, I., Sire, S., & Finck-Barbancon, V. (1994). Interaction of the two components of leukocidin from *Staphylococcus aureus* with human polymorphonuclear leukocyte membranes: Sequential binding and subsequent activation. *Infection and Immunity*, *62*(8), 3184–3188. https://doi.org/10.1128/iai.62.8.3184-3188.1994

Collet, Q., Velard, F., Laurent, F., & Josse, J. (2025). Intracellular *Staphylococcus aureus* in osteoblasts and osteocytes and its impact on bone homeostasis during osteomyelitis. *Bone*, *198*, 117536. https://doi.org/10.1016/j.bone.2025.117536

Congdon, S. T., Guaglione, J. A., Ricketts, O. M. A., Murphy, K. V., Anderson, M. G., Trowbridge, D. A., Al-Abduladheem, Y., Phillips, A. M., Beausoleil, A. M., Stanley, A. J., Becker, T. J., & Silver, A. C. (2023). Prevalence and antibiotic resistance of *Staphylococcus aureus* associated with a college-aged cohort: Life-style factors that contribute to nasal carriage. *Frontiers in Cellular and Infection Microbiology*, *13*, 1195758. https://doi.org/10.3389/fcimb.2023.1195758

Corbin, B. D., Seeley, E. H., Raab, A., Feldmann, J., Miller, M. R., Torres, V. J., Anderson, K. L., Dattilo, B. M., Dunman, P. M., Gerads, R., Caprioli, R. M., Nacken, W., Chazin, W. J., & Skaar, E. P. (2008). Metal chelation and inhibition of bacterial growth in tissue abscesses. *Science (New York, N.Y.)*, *319*(5865), 962–965. https://doi.org/10.1126/science.1152449

Costa, S. S., Ribeiro, R., Serrano, M., Oliveira, K., Ferreira, C., Leal, M., Pomba, C., & Couto, I. (2022). *Staphylococcus aureus* causing skin and soft tissue infections in companion animals: Antimicrobial resistance profiles and clonal lineages. *Antibiotics (Basel, Switzerland)*, *11*(5), 599. https://doi.org/10.3390/antibiotics11050599

Crosby, H. A., Tiwari, N., Kwiecinski, J. M., Xu, Z., Dykstra, A., Jenul, C., Fuentes, E. J., & Horswill, A. R. (2020). The *Staphylococcus aureus* ArlRS two-component system regulates virulence factor expression through MgrA. *Molecular Microbiology*, *113*(1), 103–122. https://doi.org/10.1111/mmi.14404

Davidson, H. J., & Kuonen, V. J. (2004). The tear film and ocular mucins. *Veterinary Ophthalmology*, *7*(2), 71–77. https://doi.org/10.1111/j.1463-5224.2004.00325.x

Deisenhofer, J. (1981). Crystallographic refinement and atomic models of a human Fc fragment and its complex with fragment B of protein A from *Staphylococcus aureus* at 2.9- and 2.8-A resolution. *Biochemistry*, *20*(9), 2361–2370

Del Giudice, P. (2020). Skin infections caused by *Staphylococcus aureus*. *Acta Dermato-Venereologica*, *100*(9), 5725. https://doi.org/10.2340/00015555-3466

Dinges, M. M., Orwin, P. M., & Schlievert, P. M. (2000). Exotoxins of *Staphylococcus aureus*. *Clinical Microbiology Reviews*, *13*(1), 16–34. https://doi.org/10.1128/cmr.13.1.16-34.2000

Downer, R., Roche, F., Park, P. W., Mecham, R. P., & Foster, T. J. (2002). The elastin-binding protein of *Staphylococcus aureus* (EbpS) is expressed at the cell surface as an integral membrane protein and not as a cell wall-associated protein. *Journal of Biological Chemistry*, *277*(1), 243–250. https://doi.org/10.1074/jbc.M107621200

Edet, U. O., Meseko, C., Okoroiwu, H. U., Elizabeth, M., Mkpuma, N., Cecilia, O.-K. I., & Fagbamila, I. (2025). Profile and evolution of antimicrobial resistance genes in *Staphylococcus aureus* whole genomes in Africa revealed by phylogeography and bioinformatics. *Scientific Reports*, *15*(1), 31476. https://doi.org/10.1038/s41598-025-01398-0

Elasri, M. O., Thomas, J., Skinner, R., Blevins, J., Beenken, K., Nelson, C., & Smelter, M. (2002). *Staphylococcus aureus* collagen adhesin contributes to the pathogenesis of osteomyelitis. *Bone*, *30*(1), 275–280. https://doi.org/10.1016/s8756-3282(01)00632-9

Endres, J. L., Chaudhari, S. S., Zhang, X., Prahlad, J., Wang, S.-Q., Foley, L. A., Luca, S., Bose, J. L., Thomas, V. C., & Bayles, K. W. (2022). The *Staphylococcus aureus* CidA and LrgA proteins are functional holins involved in the transport of by-products of carbohydrate metabolism. *mBio*, *13*(1), e0282721. https://doi.org/10.1128/mbio.02827-21

Fattom, A. I., Horwith, G., Fuller, S., Propst, M., & Naso, R. (2004). Development of StaphVAX™, a polysaccharide conjugate vaccine against *S. aureus* infection: From the lab bench to phase III clinical trials. *Vaccine*, *22*(7), 880–887. https://doi.org/10.1016/j.vaccine.2003.11.034

Fernández, L., Gonzalez, S., Campelo, A. B., Martinez, B., Rodriguez, A., & Garcia, P. (2017). Downregulation of autolysin-encoding genes by phage-derived lytic proteins inhibits biofilm formation in *Staphylococcus aureus*. *Antimicrobial Agents and Chemotherapy*, *61*(5), e02724.16. https://doi.org/10.1128/AAC.02724-16

Feuillie, C., Formosa-Dague, C., Hays, L. M., Vervaeck, O., Derclaye, S., Brennan, M. P., Foster, T. J., Geoghegan, J. A., & Dufrêne, Y. F. (2017). Molecular interactions and inhibition of the staphylococcal biofilm-forming protein SdrC. *Proceedings of the National Academy of Sciences of the United States of America*, *114*(14), 3738–3743. https://doi.org/10.1073/pnas.1616805114

Foster, T. J., Geoghegan, J. A., Ganesh, V. K., & Höök, M. (2014). Adhesion, invasion and evasion: The many functions of the surface proteins of *Staphylococcus aureus*. *Nature Reviews Microbiology*, *12*(1), 49–62. https://doi.org/10.1038/nrmicro3161

Frank, M. W., Yao, J., Batte, J. L., Gullett, J. M., Subramanian, C., Rosch, J. W., & Rock, C. O. (2020). Host fatty acid utilization by *Staphylococcus aureus* at the infection site. *mBio*, *11*(3), e00920.20. https://doi.org/10.1128/mBio.00920-20

Gajdács, M., Ábrók, M., Lázár, A., & Burián, K. (2020). Increasing relevance of Gram-positive cocci in urinary tract infections: A 10-year analysis of their prevalence and resistance trends. *Scientific Reports*, *10*(1), 17658. https://doi.org/10.1038/s41598-020-74834-y

Gaupp, R., Ledala, N., & Somerville, G. A. (2012). Staphylococcal response to oxidative stress. *Frontiers in Cellular and Infection Microbiology*, *2*, 33. https://doi.org/10.3389/fcimb.2012.00033

Ghasemian, A., Peerayeh, S. N., Bakhshi, B., & Mirzaee, M. (2015). The microbial surface components recognizing adhesive matrix molecules (MSCRAMMs) genes among clinical isolates of *Staphylococcus aureus* from hospitalized children. *Iranian Journal of Pathology*, *10*(4), 258

Gillman, A. N., Breshears, L. M., Kistler, C. K., Finnegan, P. M., Torres, V. J., Schlievert, P. M., & Peterson, M. L. (2017). Epidermal growth factor receptor signaling enhances the proinflammatory effects of *Staphylococcus aureus* gamma-toxin on the mucosa. *Toxins*, *9*(7), 202. https://doi.org/10.3390/toxins9070202

Girma, A., & Tamir, D. (2022). Prevalence of bovine mastitis and its associated risk factors among dairy cows in Ethiopia during 2005–2022: A systematic review and meta-analysis. *Veterinary Medicine International*, *2022*(1), 7775197. https://doi.org/10.1155/2022/7775197

Gootz, T. D. (1990). Discovery and development of new antimicrobial agents. *Clinical Microbiology Reviews*, *3*(1), 13–31. https://doi.org/10.1128/cmr.3.1.13

Gordon, R. J., & Lowy, F. D. (2008). Pathogenesis of methicillin-resistant *Staphylococcus aureus* infection. *Clinical Infectious Diseases: An Official Publication of the Infectious Diseases Society of America*, *46 Suppl 5*(Suppl 5), S350–S359. https://doi.org/10.1086/533591

Goudsmit, A., Markowicz, S., Lali, S. E., & Cherifi, S. (2021). Food poisoning due to a TSST1-producing *Staphylococcus aureus*. *IDCases*, *26*, e01272. https://doi.org/10.1016/j.idcr.2021.e01272

Grigg, J. C., Ukpabi, G., Gaudin, C. F., & Murphy, M. E. (2010). Structural biology of heme binding in the *Staphylococcus aureus* ISD system. *Journal of Inorganic Biochemistry*, *104*(3), 341–348. https://doi.org/10.1016/j.jinorgbio.2009.09.012

Harkins, C. P., Pichon, B., Doumith, M., Parkhill, J., Westh, H., Tomasz, A., de Lencastre, H., Bentley, S. D., Kearns, A. M., & Holden, M. T. (2017). Methicillin-resistant *Staphylococcus aureus* emerged long before the introduction of methicillin into clinical practice. *Genome Biology*, *18*(1), 130. https://doi.org/10.1186/s13059-017-1252-9

Hart, M. E., Tsang, L. H., Deck, J., Daily, S. T., Jones, R. C., Liu, H., Hu, H., Hart, M. J., & Smeltzer, M. S. (2013). Hyaluronidase expression and biofilm involvement in *Staphylococcus aureus* UAMS-1 and its *sarA*, *agr* and *sarA agr* regulatory mutants. *Microbiology (Reading, England)*, *159*(Pt 4), 782–791. https://doi.org/10.1099/mic.0.065367-0

Hasanpour, A. H., Sepidarkish, M., Mollalo, A., Ardekani, A., Almukhtar, M., Mechaal, A., Hosseini, S. R., Bayani, M., Javanian, M., & Rostami, A. (2023). The global prevalence of methicillin-resistant *Staphylococcus aureus* colonization in residents of elderly care centers: A systematic review and meta-analysis. *Antimicrobial Resistance & Infection Control*, *12*(1), 4. https://doi.org/10.1186/s13756-023-01210-6

Hassanzadeh, H., Baber, J., Begier, E., Noriega, D. C., Konishi, H., Yato, Y., Wang, M. Y., Le Huec, J. C., Patel, V., Varga, P., Liljenqvist, U., Conly, J., Sabharwal, C., Munjal, I., Cooper, D., Radley, D., Jaques, A., Patton, M., Gruber, W. C., Jansen, K. U., . . . *Staphylococcus aureus* suRgical Inpatient Vaccine Efficacy Investigator Team (2023). Efficacy of a 4-antigen *Staphylococcus aureus* vaccine in spinal surgery: The *Staphylococcus aureus* surgical inpatient vaccine efficacy (STRIVE) randomized clinical trial. *Clinical Infectious Diseases: An Official Publication of the Infectious Diseases Society of America*, *77*(2), 312–320. https://doi.org/10.1093/cid/ciad218

Herman-Bausier, P., Labate, C., Towell, A. M., Derclaye, S., Geoghegan, J. A., & Dufrêne, Y. F. (2018). *Staphylococcus aureus* clumping factor A is a force-sensitive molecular switch that activates bacterial adhesion. *Proceedings of the National Academy of Sciences of the United States of America*, *115*(21), 5564–5569. https://doi.org/10.1073/pnas.1718104115

Herman-Bausier, P., Valotteau, C., Pietrocola, G., Rindi, S., Alsteens, D., Foster, T. J., Speziale, P., & Dufrêne, Y. F. (2016). Mechanical strength and inhibition of the *Staphylococcus aureus* collagen-binding protein Cna. *mBio*, *7*(5), e01529.16. https://doi.org/10.1128/mBio.01529-16

Ho, D., Clayton, N. A., Silverstein, B., & Koff, A. (2022). Endogenous endophthalmitis from methicillin-resistant *Staphylococcus aureus* bacteremia treated with ceftaroline. *Cureus*, *14*(2), e22216. https://doi.org/10.7759/cureus.22216

Holland, T. L., Cosgrove, S. E., Doernberg, S. B., Jenkins, T. C., Turner, N. A., Boucher, H. W., Pavlov, O., Titov, I., Kosulnykov, S., Atanasov, B., Poromanski, I., Makhviladze, M., Anderzhanova, A., Stryjewski, M. E., Assadi Gehr, M., Engelhardt, M., Hamed, K., Ionescu, D., Jones, M., Saulay, M., . . . ERADICATE Study Group (2023). Ceftobiprole for treatment of complicated *Staphylococcus aureus* bacteremia. *The New England Journal of Medicine*, *389*(15), 1390–1401. https://doi.org/10.1056/NEJMoa2300220

Holzinger, D., Gieldon, L., Mysore, V., Nippe, N., Taxman, D. J., Duncan, J. A., Broglie, P. M., Marketon, K., Austermann, J., Vogl, T., Foell, D., Niemann, S., Peters, G., Roth, J., & Löffler, B. (2012). *Staphylococcus aureus* Panton-Valentine leukocidin induces an inflammatory response in human phagocytes via the NLRP3 inflammasome. *Journal of Leukocyte Biology*, *92*(5), 1069–1081. https://doi.org/10.1189/jlb.0112014

Hotchkiss, R. S., Monneret, G., & Payen, D. (2013). Sepsis-induced immunosuppression: From cellular dysfunctions to immunotherapy. *Nature Reviews Immunology*, *13*(12), 862–874. https://doi.org/10.1038/nri3552

Howden, B. P., Giulieri, S. G., Wong Fok Lung, T., Baines, S. L., Sharkey, L. K., Lee, J. Y., Hachani, A., Monk, I. R., & Stinear, T. P. (2023). *Staphylococcus aureus* host interactions and adaptation. *Nature Reviews Microbiology*, *21*(6), 380–395. https://doi.org/10.1038/s41579-023-00852-y

Huseby, M., Shi, K., Brown, C. K., Digre, J., Mengistu, F., Seo, K. S., Bohach, G. A., Schlievert, P. M., Ohlendorf, D. H., & Earhart, C. A. (2007). Structure and biological activities of beta toxin from *Staphylococcus aureus*. *Journal of Bacteriology*, *189*(23), 8719–8726. https://doi.org/10.1128/JB.00741-07

Ibberson, C. B., Jones, C. L., Singh, S., Wise, M. C., Hart, M. E., Zurawski, D. V., & Horswill, A. R. (2014). *Staphylococcus aureus* hyaluronidase is a CodY-regulated virulence factor. *Infection and Immunity*, *82*(10), 4253–4264. https://doi.org/10.1128/IAI.01710-14

Josse, J., Laurent, F., & Diot, A. (2017). Staphylococcal adhesion and host cell invasion: Fibronectin-binding and other mechanisms. *Frontiers in Microbiology*, *8*, 2433. https://doi.org/10.3389/fmicb.2017.02433

Kaempfer, R., Popugailo, A., Levy, R., Arad, G., Hillman, D., & Rotfogel, Z. (2017). Bacterial superantigen toxins induce a lethal cytokine storm by enhancing B7–2/CD28 costimulatory receptor engagement, a critical immune checkpoint. *Receptors & Clinical Investigation*, *4*(1), e1500

Kang, D. H., Kim, Y. J., Kim, S. H., Sun, B. J., Kim, D. H., Yun, S. C., Song, J. M., Choo, S. J., Chung, C. H., Song, J. K., Lee, J. W., & Sohn, D. W. (2012). Early surgery versus conventional treatment for infective endocarditis. *The New England Journal of Medicine*, *366*(26), 2466–2473. https://doi.org/10.1056/NEJMoa1112843

Karlsson, A., & Arvidson, S. (2002). Variation in extracellular protease production among clinical isolates of *Staphylococcus aureus* due to different levels of expression of the protease repressor sarA. *Infection and Immunity*, *70*(8), 4239–4246. https://doi.org/10.1128/IAI.70.8.4239-4246.2002

Kasimir, S., Schönfeld, W., Alouf, J., & König, W. (1990). Effect of *Staphylococcus aureus* delta-toxin on human granulocyte functions and platelet-activating-factor metabolism. *Infection and Immunity*, *58*(6), 1653–1659. https://doi.org/10.1128/iai.58.6.1653-1659.1990

Kengmo Tchoupa, A., Elsherbini, A. M. A., Camus, J., Fu, X., Hu, X., Ghaneme, O., Seibert, L., Lebtig, M., Böcker, M. A., Horlbeck, A., Lambidis, S. P., Schittek, B., Kretschmer, D., Lämmerhofer, M., & Peschel, A. (2024). Lipase-mediated detoxification of host-derived antimicrobial fatty acids by *Staphylococcus aureus*. *Communications Biology*, *7*(1), 572. https://doi.org/10.1038/s42003-024-06278-3

Kiefer, T., Park, L., Tribouilloy, C., Cortes, C., Casillo, R., Chu, V., Delahaye, F., Durante-Mangoni, E., Edathodu, J., Falces, C., Logar, M., Miró, J. M., Naber, C., Tripodi, M. F., Murdoch, D. R., Moreillon, P., Utili, R., & Wang, A. (2011). Association between valvular surgery and mortality among patients with infective endocarditis complicated by heart failure. *JAMA*, *306*(20), 2239–2247. https://doi.org/10.1001/jama.2011.1701

Kiefer, T. L., & Bashore, T. M. (2012). Infective endocarditis: A comprehensive overview. *Reviews in Cardiovascular Medicine*, *13*(2–3), 105–120. https://doi.org/10.3909/ricm0633

Kinamon, T., Dagher, M., Park, L., Ruffin, F., Fowler Jr, V. G., & Maskarinec, S. A. (2023). Risk factors and outcomes of hematogenous vertebral osteomyelitis in patients with *Staphylococcus aureus* bacteremia. *Clinical Infectious Diseases: An Official Publication of the Infectious Diseases Society of America*, *77*(9), 1226–1233. https://doi.org/10.1093/cid/ciad377

Koch, T. K., Reuter, M., Barthel, D., Böhm, S., Van Den Elsen, J., Kraiczy, P., . . . Skerka, C. (2012). *Staphylococcus aureus* proteins Sbi and Efb recruit human plasmin to degrade complement C3 and C3b. *PloS One*, *7*(10), e47638. https://doi.org/10.1371/journal.pone.0047638

Krakauer, T. (2019). Staphylococcal superantigens: Pyrogenic toxins induce toxic shock. *Toxins*, *11*(3), 178. https://doi.org/10.3390/toxins11030178

Krones, D., Rühling, M., Becker, K. A., Kunz, T. C., Sehl, C., Paprotka, K., Gulbins, E., & Fraunholz, M. (2021). *Staphylococcus aureus* α-toxin induces acid sphingomyelinase release from a human endothelial cell line. *Frontiers in Microbiology*, *12*, 694489. https://doi.org/10.3389/fmicb.2021.694489

Kuipers, A., Stapels, D. A., Weerwind, L. T., Ko, Y.-P., Ruyken, M., Lee, J. C., van Kessel, K. P., & Rooijakkers, S. H. (2016). The *Staphylococcus aureus* polysaccharide capsule and Efb-dependent fibrinogen shield act in concert to protect against phagocytosis. *Microbiology (Reading, England)*, *162*(7), 1185–1194. https://doi.org/10.1099/mic.0.000293

Laarman, A. J., Ruyken, M., Malone, C. L., van Strijp, J. A., Horswill, A. R., & Rooijakkers, S. H. (2011). *Staphylococcus aureus* metalloprotease aureolysin cleaves complement C3 to mediate immune evasion. *The Journal of Immunology (Baltimore, Md.: 1950)*, *186*(11), 6445–6453. https://doi.org/10.4049/jimmunol.1002948

Labandeira-Rey, M., Couzon, F., Boisset, S., Brown, E. L., Bes, M., Benito, Y., Barbu, E. M., Vazquez, V., Höök, M., Etienne, J., Vandenesch, F., & Bowden, M. G. (2007). *Staphylococcus aureus* Panton-Valentine leukocidin causes necrotizing pneumonia. *Science (New York, N.Y.)*, *315*(5815), 1130–1133. https://doi.org/10.1126/science.1137165

Lade, H., & Kim, J.-S. (2023). Molecular determinants of β-lactam resistance in methicillin-resistant *Staphylococcus aureus* (MRSA): An updated review. *Antibiotics (Basel, Switzerland)*, *12*(9), 1362. https://doi.org/10.3390/antibiotics12091362

Ladhani, S., Joannou, C. L., Lochrie, D. P., Evans, R. W., & Poston, S. M. (1999). Clinical, microbial, and biochemical aspects of the exfoliative toxins causing staphylococcal scalded-skin syndrome. *Clinical Microbiology Reviews*, *12*(2), 224–242. https://doi.org/10.1128/CMR.12.2.224

Lee, E. Y., Choi, D. Y., Kim, D. K., Kim, J. W., Park, J. O., Kim, S., Kim, S. H., Desiderio, D. M., Kim, Y. K., Kim, K. P., & Gho, Y. S. (2009). Gram-positive bacteria produce membrane vesicles: Proteomics-based characterization of *Staphylococcus aureus*-derived membrane vesicles. *Proteomics*, *9*(24), 5425–5436. https://doi.org/10.1002/pmic.200900338

Lee, H.-J., Woo, Y., Hahn, T.-W., Jung, Y. M., & Jung, Y.-J. (2020). Formation and maturation of the phagosome: A key mechanism in innate immunity against intracellular bacterial infection. *Microorganisms*, *8*(9), 1298. https://doi.org/10.3390/microorganisms8091298

Lemonovich, T., Haynes, K., Lautenbach, E., & Amorosa, V. (2011). Combination therapy with an aminoglycoside for *Staphylococcus aureus* endocarditis and/or persistent bacteremia is associated with a decreased rate of recurrent bacteremia: A cohort study. *Infection*, *39*(6), 549–554. https://doi.org/10.1007/s15010-011-0189-2

Levine, D. P. (2006). Vancomycin: A history. *Clinical Infectious Diseases: An Official Publication of the Infectious Diseases Society of America, 42 Suppl 1*, S5–S12. https://doi.org/10.1086/491709

Li, M., Du, X., Villaruz, A. E., Diep, B. A., Wang, D., Song, Y., Tian, Y., Hu, J., Yu, F., Lu, Y., & Otto, M. (2012). MRSA epidemic linked to a quickly spreading colonization and virulence determinant. *Nature Medicine*, *18*(5), 816–819. https://doi.org/10.1038/nm.2692

Li, Q., Chen, N., Liu, C., Zhao, Z., Huang, M., Li, J., & Yang, G. (2025). *Staphylococcus aureus* β-hemolysin impairs oxygen transport without causing hemolysis. *Virulence*, *16*(1), 2490208. https://doi.org/10.1080/21505594.2025.2490208

Linz, M. S., Mattappallil, A., Finkel, D., & Parker, D. (2023). Clinical impact of *Staphylococcus aureus* skin and soft tissue infections. *Antibiotics (Basel, Switzerland)*, *12*(3), 557. https://doi.org/10.3390/antibiotics12030557

Liu, K., Wang, C., Zhou, X., Guo, X., Yang, Y., Liu, W., Zhao, R., & Song, H. (2024). Bacteriophage therapy for drug-resistant *Staphylococcus aureus* infections. *Frontiers in Cellular and Infection Microbiology*, *14*, 1336821. https://doi.org/10.3389/fcimb.2024.1336821

Liu, L., Fang, Y., Huang, Q., & Wu, J. (2011). A rigidity-enhanced antimicrobial activity: A case for linear cationic α-helical peptide HP (2–20) and its four analogues. *PLoS One*, *6*(1), e16441. https://doi.org/10.1371/journal.pone.0016441

Liu, Y., Zhang, J., & Ji, Y. (2016). PCR-based approaches for the detection of clinical methicillin-resistant *Staphylococcus aureus*. *The Open Microbiology Journal*, *10*, 45–56. https://doi.org/10.2174/1874285801610010045

Lobanovska, M., & Pilla, G. (2017). Penicillin's discovery and antibiotic resistance: Lessons for the future? *The Yale Journal of Biology and Medicine*, *90*(1), 135–145.

Lowy, F. D. (1998). *Staphylococcus aureus* infections. *New England Journal of Medicine*, *339*(8), 520–532. https://doi.org/10.1056/NEJM199808203390806

Matsuo, K., & Iwasa, Y. (2025). Modeling innate immunity causing chronic inflammation and tissue damage. *Bulletin of Mathematical Biology*, *87*(2), 34. https://doi.org/10.1007/s11538-024-01410-0

Mazmanian, S. K., Liu, G., Jensen, E. R., Lenoy, E., & Schneewind, O. (2000). *Staphylococcus aureus* sortase mutants defective in the display of surface proteins and in the pathogenesis of animal infections. *Proceedings of the National Academy of Sciences of the United States of America*, *97*(10), 5510–5515. https://doi.org/10.1073/pnas.080520697

Melish, M. E., & Glasgow, L. A. (1970). The staphylococcal scalded-skin syndrome: Development of an experimental model. *New England Journal of Medicine*, *282*(20), 1114–1119. https://doi.org/10.1056/NEJM197005142822002

Montgomery, C. P., Boyle-Vavra, S., Adem, P. V., Lee, J. C., Husain, A. N., Clasen, J., & Daum, R. S. (2008). Comparison of virulence in community-associated methicillin-resistant *Staphylococcus aureus* pulsotypes USA300 and USA400 in a rat model of pneumonia. *The Journal of Infectious Diseases*, *198*(4), 561–570. https://doi.org/10.1086/590157

Moza, B., Varma, A. K., Buonpane, R. A., Zhu, P., Herfst, C. A., Nicholson, M. J., Wilbuer, A. K., Seth, N. P., Wucherpfennig, K. W., McCormick, J. K., Kranz, D.M., & Sundberg, E. J. (2007). Structural basis of T-cell specificity and activation by the bacterial superantigen TSST-1. *The EMBO Journal*, *26*(4), 1187–1197. https://doi.org/10.1038/sj.emboj.7601531

Muder, R. R., Brennen, C., Rihs, J. D., Wagener, M. M., Obman, A., Obman, A., Stout, J. E., & Yu, V. L. (2006). Isolation of *Staphylococcus aureus* from the urinary tract: Association of isolation with symptomatic urinary tract infection and subsequent staphylococcal bacteremia. *Clinical Infectious Diseases: An Official Publication of the Infectious Diseases Society of America*, *42*(1), 46–50. https://doi.org/10.1086/498518

Müller, I. B., Bergmann, B., Groves, M. R., Couto, I., Amaral, L., Begley, T. P., Walter, R. D., & Wrenger, C. (2009). The vitamin B1 metabolism of *Staphylococcus aureus* is controlled at enzymatic and transcriptional levels. *PLoS One*, *4*(11), e7656. https://doi.org/10.1371/journal.pone.0007656

Mustor, E. M., Frey, A. M., Jobson, M.-E., Chaput, D., & Shaw, L. N. (2025). Development of a new N-terminomic method to study the pathodegradome of the *Staphylococcus aureus* V8 protease in human neutrophils. *mSystems*, *10*(7), e0069725. https://doi.org/10.1128/msystems.00697-25

Nakamura, T., Ishikawa, K., Murata, N., Sato, K., Kitamura, A., Mori, N., & Jinta, T. (2024). Empyema necessitans caused by methicillin-resistant *Staphylococcus aureus*: A case report and literature review. *BMC Infectious Diseases*, *24*(1), 157. https://doi.org/10.1186/s12879-024-09062-0

O'Brien, L., Kerrigan, S. W., Kaw, G., Hogan, M., Penadés, J., Litt, D., Fitzgerald, D. J., Foster, T. J., & Cox, D. (2002). Multiple mechanisms for the activation of human platelet aggregation by *Staphylococcus aureus*: Roles for the clumping factors ClfA and ClfB, the serine–aspartate repeat protein SdrE and protein A. *Molecular Microbiology*, *44*(4), 1033–1044. https://doi.org/10.1046/j.1365-2958.2002.02935.x

O'Callaghan, R. J. (2018). The pathogenesis of *Staphylococcus aureus* eye infections. *Pathogens (Basel, Switzerland)*, *7*(1), 9. https://doi.org/10.3390/pathogens7010009

O'Riordan, K., & Lee, J. C. (2004). *Staphylococcus aureus* capsular polysaccharides. *Clinical Microbiology Reviews*, *17*(1), 218–234. https://doi.org/10.1128/CMR.17.1.218-234.2004

Ohbayashi, T., Irie, A., Murakami, Y., Nowak, M., Potempa, J., Nishimura, Y., Shinohara, M., & Imamura, T. (2011). Degradation of fibrinogen and collagen by staphopains, cysteine proteases released from *Staphylococcus aureus*. *Microbiology (Reading, England)*, *157*(Pt 3), 786–792. https://doi.org/10.1099/mic.0.044503-0

Ojo, K. K., Striplin, M. J., Ulep, C. C., Close, N. S., Zittle, J., Luis, H., Bernardo, M., Leitao, J., & Roberts, M. C. (2006). *Staphylococcus* efflux msr (A) gene characterized in *Streptococcus, Enterococcus, Corynebacterium,* and *Pseudomonas* isolates. *Antimicrobial Agents and Chemotherapy*, *50*(3), 1089–1091. https://doi.org/10.1128/AAC.50.3.1089-1091.2006

Okada, N., Fujita, T., Kanamori, J., Sato, A., Horikiri, Y., Sato, T., Fujiwara, H., & Daiko, H. (2018). A case report of postoperative VRSA enteritis: Effective management of rifampicin for vancomycin resistant *Staphylococcus aureus* enteritis after esophagectomy and colon reconstruction. *International Journal of Surgery Case Reports*, *52*, 75–78. https://doi.org/10.1016/j.ijscr.2018.08.038

Otto, M. (2014). *Staphylococcus aureus* toxins. *Current Opinion in Microbiology*, *17*, 32–37. https://doi.org/10.1016/j.mib.2013.11.004

Panizzi, P., Nahrendorf, M., Figueiredo, J. L., Panizzi, J., Marinelli, B., Iwamoto, Y., Keliher, E., Maddur, A. A., Waterman, P., Kroh, H. K., Leuschner, F., Aikawa, E., Swirski, F. K., Pittet, M. J., Hackeng,

T. M., Fuentes-Prior, P., Schneewind, O., Bock, P. E., & Weissleder, R. (2011). In vivo detection of *Staphylococcus aureus* endocarditis by targeting pathogen-specific prothrombin activation. *Nature Medicine*, *17*(9), 1142–1146. https://doi.org/10.1038/nm.2423

Patel, D., Brown, M. L., Edwards, S., Oster, R. A., & Stripling, J. (2023). Outcomes of daptomycin plus ceftaroline versus alternative therapy for persistent methicillin-resistant *Staphylococcus aureus* (MRSA) bacteraemia. *International Journal of Antimicrobial Agents*, *61*(3), 106735. https://doi.org/10.1016/j.ijantimicag.2023.106735

Peetermans, M., Vanassche, T., Liesenborghs, L., Claes, J., Vande Velde, G., Kwiecinksi, J., Jin, T., De Geest, B., Hoylaerts, M. F., Lijnen, R. H., & Verhamme, P. (2014). Plasminogen activation by staphylokinase enhances local spreading of *S. aureus* in skin infections. *BMC Microbiology*, *14*, 310. https://doi.org/10.1186/s12866-014-0310-7

Periasamy, S., Joo, H.-S., Duong, A. C., Bach, T.-H. L., Tan, V. Y., Chatterjee, S. S., Cheung, G. Y., & Otto, M. (2012). How *Staphylococcus aureus* biofilms develop their characteristic structure. *Proceedings of the National Academy of Sciences of the United States of America*, *109*(4), 1281–1286. https://doi.org/10.1073/pnas.1115006109

Périchon, B., & Courvalin, P. (2009). VanA-type vancomycin-resistant *Staphylococcus aureus*. *Antimicrobial Agents and Chemotherapy*, *53*(11), 4580–4587. https://doi.org/10.1128/AAC.00346-09

Pickens, C. I., & Wunderink, R. G. (2022). Methicillin-resistant *Staphylococcus aureus* hospital-acquired pneumonia/ventilator-associated pneumonia. *Seminars in Respiratory and Critical Care Medicine*, *43*(2), 304–309. https://doi.org/10.1055/s-0041-1740583

Pietrocola, G., Nobile, G., Rindi, S., & Speziale, P. (2017). *Staphylococcus aureus* manipulates innate immunity through own and host-expressed proteases. *Frontiers in Cellular and Infection Microbiology*, *7*, 166. https://doi.org/10.3389/fcimb.2017.00166

Planet, P. J., LaRussa, S. J., Dana, A., Smith, H., Xu, A., Ryan, C., Uhlemann, A. C., Boundy, S., Goldberg, J., Narechania, A., Kulkarni, R., Ratner, A. J., Geoghegan, J. A., Kolokotronis, S. O., & Prince, A. (2013). Emergence of the epidemic methicillin-resistant *Staphylococcus aureus* strain USA300 coincides with horizontal transfer of the arginine catabolic mobile element and speG-mediated adaptations for survival on skin. *mBio*, *4*(6), e00889.13. https://doi.org/10.1128/mBio.00889-13

Poolman, J. T., Torres, V. J., Missiakas, D., Welten, S. P. M., Fernandez, J., DuMont, A. L., O'Keeffe, A., Konstantinov, S. R., Morrow, B., Burghout, P., Grijpstra, J., van Beers, M. M. C., Anish, C., Beurret, M., Geurtsen, J., Rood, P. M. L., Koeberling, O., Shi, M., & van den Dobbelsteen, G. P. J. M. (2025). A SpA+ LukAB vaccine targeting *Staphylococcus aureus* evasion factors restricts infection in two minipig infection models. *NPJ Vaccines*, *10*(1), 78. https://doi.org/10.1038/s41541-025-01119-8

Porayath, C., Suresh, M. K., Biswas, R., Nair, B. G., Mishra, N., & Pal, S. (2018). Autolysin mediated adherence of *Staphylococcus aureus* with fibronectin, gelatin and heparin. *International Journal of Biological Macromolecules*, *110*, 179–184. https://doi.org/10.1016/j.ijbiomac.2018.01.047

Prajapati, S. K., Jain, S., & Yadav, H. (2025). Age-related cognitive decline and dementia: Interface of microbiome-immune-neuronal interactions. *The Journals of Gerontology, Series A: Biological Sciences and Medical Sciences*, *80*(7), glaf038. https://doi.org/10.1093/gerona/glaf038

Prat, C., Haas, P.-J., Bestebroer, J., de Haas, C. J., van Strijp, J. A., & van Kessel, K. P. (2009). A homolog of formyl peptide receptor-like 1 (FPRL1) inhibitor from *Staphylococcus aureus* (FPRL1 inhibitory protein) that inhibits FPRL1 and FPR. *Journal of Immunology (Baltimore, Md: 1950)*, *183*(10), 6569–6578. https://doi.org/10.4049/jimmunol.0801523

Proctor, R. A., van Langevelde, P., Kristjansson, M., Maslow, J. N., & Arbeit, R. D. (1995). Persistent and relapsing infections associated with small-colony variants of *Staphylococcus aureus*. *Clinical Infectious Diseases: An Official Publication of the Infectious Diseases Society of America*, *20*(1), 95–102. https://doi.org/10.1093/clinids/20.1.95

Queck, S. Y., Jameson-Lee, M., Villaruz, A. E., Bach, T.-H. L., Khan, B. A., Sturdevant, D. E., Ricklefs, S. M., Li, M., & Otto, M. (2008). RNAIII-independent target gene control by the agr quorum-sensing system: Insight into the evolution of virulence regulation in *Staphylococcus aureus*. *Molecular Cell*, *32*(1), 150–158. https://doi.org/10.1016/j.molcel.2008.08.005

Raghupathi, K. R., Koodali, R. T., & Manna, A. C. (2011). Size-dependent bacterial growth inhibition and mechanism of antibacterial activity of zinc oxide nanoparticles. *Langmuir: The ACS Journal of Surfaces and Colloids*, *27*(7), 4020–4028. https://doi.org/10.1021/la104825u

Ramos-León, F., Anjuwon-Foster, B. R., Anantharaman, V., Updegrove, T. B., Ferreira, C. N., Ibrahim, A. M., Tai, C. H., Kruhlak, M. J., Missiakas, D. M., Camberg, J. L., Aravind, L., & Ramamurthi, K. S. (2024). PcdA promotes orthogonal division plane selection in *Staphylococcus aureus*. *Nature Microbiology*, *9*(11), 2997–3012. https://doi.org/10.1038/s41564-024-01821-8

Räz, A. K., Andreoni, F., Boumasmoud, M., Bergada-Pijuan, J., Schweizer, T. A., Mairpady Shambat, S., Hasse, B., Zinkernagel, A. S., & Brugger, S. D. (2023). Limited adaptation of *Staphylococcus aureus* during transition from colonization to invasive infection. *Microbiology Spectrum*, *11*(4), e0259021. https://doi.org/10.1128/spectrum.02590-21

Risley, A. L., Loughman, A., Cywes-Bentley, C., Foster, T. J., & Lee, J. C. (2007). Capsular polysaccharide masks clumping factor A-mediated adherence of *Staphylococcus aureus* to fibrinogen and platelets. *The Journal of Infectious Diseases*, *196*(6), 919–927. https://doi.org/10.1086/520932

Roux, A., Todd, D. A., Velázquez, J. V., Cech, N. B., & Sonenshein, A. L. (2014). CodY-mediated regulation of the *Staphylococcus aureus* Agr system integrates nutritional and population density signals. *Journal of Bacteriology*, *196*(6), 1184–1196. https://doi.org/10.1128/JB.00128-13

Saei, H. D. (2012). Distribution of collagen adhesin gene among various types of *Staphylococcus aureus* strains associated with bovine mammary gland. *Comparative Clinical Pathology*, *21*(5), 571–576. https://doi.org/10.1007/s00580-010-1136-9

Schwarz-Linek, U., Werner, J. M., Pickford, A. R., Gurusiddappa, S., Kim, J. H., Pilka, E. S., Briggs, J. A., Gough, T. S., Höök, M., Campbell, I. D., & Potts, J. R. (2003). Pathogenic bacteria attach to human fibronectin through a tandem β-zipper. *Nature*, *423*(6936), 177–181. https://doi.org/10.1038/nature01589

Seilie, E. S., & Wardenburg, J. B. (2017). *Staphylococcus aureus* pore-forming toxins: The interface of pathogen and host complexity. *Seminars in Cell & Developmental Biology*, *72*, 101–116. https://doi.org/10.1016/j.semcdb.2017.04.003

Sharif, S., Singh, M., Kim, S. J., & Schaefer, J. (2009). *Staphylococcus aureus* peptidoglycan tertiary structure from carbon-13 spin diffusion. *Journal of the American Chemical Society*, *131*(20), 7023–7030. https://doi.org/10.1021/ja808971c

Sharma-Kuinkel, B. K., Mongodin, E. F., Myers, J. R., Vore, K. L., Canfield, G. S., Fraser, C. M., Rude, T. H., Fowler Jr, V. G., & Gill, S. R. (2015). Potential influence of *Staphylococcus aureus* clonal complex 30 genotype and transcriptome on hematogenous infections. *Open Forum Infectious Diseases*, *2*(3), ofv093. https://doi.org/10.1093/ofid/ofv093

Shuter, J., Hatcher, V., & Lowy, F. (1996). *Staphylococcus aureus* binding to human nasal mucin. *Infection and Immunity*, *64*(1), 310–318. https://doi.org/10.1128/iai.64.1.310-318.1996

Spaan, A. N., Vrieling, M., Wallet, P., Badiou, C., Reyes-Robles, T., Ohneck, E. A., Benito, Y., de Haas, C. J., Day, C. J., Jennings, M. P., Lina, G., Vandenesch, F., van Kessel, K. P., Torres, V. J., van Strijp, J. A., & Henry, T. (2014). The staphylococcal toxins γ-haemolysin AB and CB differentially target phagocytes by employing specific chemokine receptors. *Nature Communications*, *5*(1), 5438. https://doi.org/10.1038/ncomms6438

Sperber, W. z., & Tatini, S. (1975). Interpretation of the tube coagulase test for identification of *Staphylococcus aureus*. *Applied Microbiology*, *29*(4), 502–505. https://doi.org/10.1128/am.29.4.502-505.1975

Speziale, P., & Pietrocola, G. (2020). The multivalent role of fibronectin-binding proteins A and B (FnBPA and FnBPB) of *Staphylococcus aureus* in host infections. *Frontiers in Microbiology*, *11*, 2054. https://doi.org/10.3389/fmicb.2020.02054

Surewaard, B., De Haas, C., Vervoort, F., Rigby, K., DeLeo, F., Otto, M., Van Strijp, J., & Nijland, R. (2013). Staphylococcal alpha-phenol soluble modulins contribute to neutrophil lysis after phagocytosis. *Cellular Microbiology*, *15*(8), 1427–1437. https://doi.org/10.1111/cmi.12130

Tanaka, M., Kamitani, S., & Kitadokoro, K. (2018). *Staphylococcus aureus* lipase: Purification, kinetic characterization, crystallization and crystallographic study. *Acta Crystallographica. Section F, Structural Biology Communications*, *74*(Pt 9), 567–570. https://doi.org/10.1107/S2053230X18010506

Tao, S., Chen, H., Li, N., & Liang, W. (2022). The application of the CRISPR-Cas system in antibiotic resistance. *Infection and Drug Resistance*, *15*, 4155–4168. https://doi.org/10.2147/IDR.S370869

Tavares, A., Nielsen, J. B., Boye, K., Rohde, S., Paulo, A. C., Westh, H., Schønning, K., de Lencastre, H., & Miragaia, M. (2014). Insights into alpha-hemolysin (Hla) evolution and expression among *Staphylococcus aureus* clones with hospital and community origin. *PLoS One*, *9*(7), e98634. https://doi.org/10.1371/journal.pone.0098634

Ton-That, H., Liu, G., Mazmanian, S. K., Faull, K. F., & Schneewind, O. (1999). Purification and characterization of sortase, the transpeptidase that cleaves surface proteins of *Staphylococcus aureus* at the LPXTG motif. *Proceedings of the National Academy of Sciences of the United States of America*, *96*(22), 12424–12429. https://doi.org/10.1073/pnas.96.22.12424

Tong, S. Y., Davis, J. S., Eichenberger, E., Holland, T. L., & Fowler Jr, V. G. (2015). *Staphylococcus aureus* infections: Epidemiology, pathophysiology, clinical manifestations, and management. *Clinical Microbiology Reviews*, *28*(3), 603–661. https://doi.org/10.1128/CMR.00134-14

Tong, S. Y. C., Lye, D. C., Yahav, D., Sud, A., Robinson, J. O., Nelson, J., Archuleta, S., Roberts, M. A., Cass, A., Paterson, D. L., Foo, H., Paul, M., Guy, S. D., Tramontana, A. R., Walls, G. B., McBride, S., Bak, N., Ghosh, N., Rogers, B. A., Ralph, A. P., . . . Australasian Society for Infectious Diseases Clinical Research Network (2020). Effect of vancomycin or daptomycin with vs without an antistaphylococcal β-lactam on mortality, bacteremia, relapse, or treatment failure in patients with MRSA bacteremia: A randomized clinical trial. *JAMA*, *323*(6), 527–537. https://doi.org/10.1001/jama.2020.0103

Traykovska, M., & Penchovsky, R. (2022). Targeting SAM-I riboswitch using antisense oligonucleotide technology for inhibiting the growth of *Staphylococcus aureus* and *Listeria Monocytogenes*. *Antibiotics (Basel, Switzerland)*, *11*(11), 1662. https://doi.org/10.3390/antibiotics11111662

Tsai, C.-E., Yang, C.-J., Chuang, Y.-C., Wang, J.-T., Sheng, W.-H., Chen, Y.-C., & Chang, S.-C. (2022). Evaluation of the synergistic effect of ceftaroline against methicillin-resistant *Staphylococcus aureus*. *International Journal of Infectious Diseases: IJID: Official Publication of the International Society for Infectious Diseases*, *122*, 230–236. https://doi.org/10.1016/j.ijid.2022.05.057

Tuffs, S. W., Herfst, C. A., Baroja, M. L., Podskalniy, V. A., DeJong, E. N., Coleman, C. E., & McCormick, J. K. (2019). Regulation of toxic shock syndrome toxin-1 by the accessory gene regulator in *Staphylococcus aureus* is mediated by the repressor of toxins. *Molecular Microbiology*, *112*(4), 1163–1177. https://doi.org/10.1111/mmi.14353

Ungureanu, V. (2010). Macrolides, lincosamides, streptogramines (MLS): Mechanisms of action and resistance. *Bacteriologia, Virusologia, Parazitologia, Epidemiologia (Bucharest, Romania: 1990)*, *55*(2), 131–138.

Urso, A., Monk, I. R., Cheng, Y. T., Predella, C., Wong Fok Lung, T., Theiller, E. M., Boylan, J., Perelman, S., Baskota, S. U., Moustafa, A. M., Lohia, G., Lewis, I. A., Howden, B. P., Stinear, T. P., Dorrello, N. V., Torres, V., & Prince, A. S. (2024). *Staphylococcus aureus* adapts to exploit collagen-derived proline during chronic infection. *Nature Microbiology*, *9*(10), 2506–2521. https://doi.org/10.1038/s41564-024-01769-9

Van Boeckel, T. P., Brower, C., Gilbert, M., Grenfell, B. T., Levin, S. A., Robinson, T. P., Teillant, A., & Laxminarayan, R. (2015). Global trends in antimicrobial use in food animals. *Proceedings of the National Academy of Sciences of the United States of America*, *112*(18), 5649–5654. https://doi.org/10.1073/pnas.1503141112

van Wamel, W. J., Rooijakkers, S. H., Ruyken, M., Van Kessel, K. P., & Van Strijp, J. A. (2006). The innate immune modulators staphylococcal complement inhibitor and chemotaxis inhibitory protein of *Staphylococcus aureus* are located on β-hemolysin-converting bacteriophages. *Journal of Bacteriology*, *188*(4), 1310–1315. https://doi.org/10.1128/JB.188.4.1310-1315.2006

Velissaris, D., Aretha, D., Fligou, F., & Filos, K. S. (2009). Spinal subdural *Staphylococcus aureus* abscess: Case report and review of the literature. *World Journal of Emergency Surgery: WJES*, *4*, 31. https://doi.org/10.1186/1749-7922-4-31

Voyich, J. M., Vuong, C., DeWald, M., Nygaard, T. K., Kocianova, S., Griffith, S., Jones, J., Iverson, C., Sturdevant, D. E., Braughton, K. R., Whitney, A. R., Otto, M., & DeLeo, F. R. (2009). The SaeR/S gene regulatory system is essential for innate immune evasion by *Staphylococcus aureus*. *The Journal of Infectious Diseases*, *199*(11), 1698–1706. https://doi.org/10.1086/598967

Wang, H., Mann, P. A., Xiao, L., Gill, C., Galgoci, A. M., Howe, J. A., Villafania, A., Barbieri, C. M., Malinverni, J. C., Sher, X., Mayhood, T., McCurry, M. D., Murgolo, N., Flattery, A., Mack, M., & Roemer, T. (2017). Dual-targeting small-molecule inhibitors of the *Staphylococcus aureus* FMN riboswitch disrupt riboflavin homeostasis in an infectious setting. *Cell Chemical Biology*, *24*(5), 576–588.e6. https://doi.org/10.1016/j.chembiol.2017.03.014

Wang, X., Uppu, D. S., Dickey, S. W., Burgin, D. J., Otto, M., & Lee, J. C. (2023). *Staphylococcus aureus* delta toxin modulates both extracellular membrane vesicle biogenesis and amyloid formation. *mBio*, *14*(5), e0174823. https://doi.org/10.1128/mbio.01748-23

Wanner, S., Schade, J., Keinhörster, D., Weller, N., George, S. E., Kull, L., Bauer, J., Grau, T., Winstel, V., Stoy, H., Kretschmer, D., Kolata, J., Wolz, C., Bröker, B. M., & Weidenmaier, C. (2017). Wall teichoic acids mediate increased virulence in *Staphylococcus aureus*. *Nature Microbiology*, *2*, 16257. https://doi.org/10.1038/nmicrobiol.2016.257

Wilke, G. A., & Wardenburg, J. B. (2010). Role of a disintegrin and metalloprotease 10 in *Staphylococcus aureus* α-hemolysin–mediated cellular injury. *Proceedings of the National Academy of Sciences of the United States of America*, *107*(30), 13473–13478. https://doi.org/10.1073/pnas.1001815107

Wong, J. W., Ip, M., Tang, A., Wei, V. W., Wong, S. Y., Riley, S., Read, J. M., & Kwok, K. O. (2018). Prevalence and risk factors of community-associated methicillin-resistant *Staphylococcus aureus* carriage in Asia-Pacific region from 2000 to 2016: A systematic review and meta-analysis. *Clinical Epidemiology*, *10*, 1489–1501. https://doi.org/10.2147/CLEP.S160595

Woodford, N. (2005). Biological counterstrike: Antibiotic resistance mechanisms of Gram-positive cocci. *Clinical Microbiology and Infection: The Official Publication of the European Society of Clinical Microbiology and Infectious Diseases, 11 Suppl 3*, 2–21. https://doi.org/10.1111/j.1469-0691.2005.01140.x

Wu, X., Wang, C., He, L., Xu, H., Jing, C., Chen, Y., Lin, A., Deng, J., Cao, Q., Deng, H., Cai, H., Chen, Y., Yang, J., Zhang, T., Huang, Y., Hao, J., & Yu, H. (2023). Antimicrobial resistance profile of methicillin-resistant *Staphylococcus aureus* isolates in children reported from the ISPED surveillance of bacterial resistance, 2016–2021. *Frontiers in Cellular and Infection Microbiology, 13*, 1102779. https://doi.org/10.3389/fcimb.2023.1102779

Xu, Z., Li, Y., Xu, A., Soteyome, T., Yuan, L., Ma, Q., Seneviratne, G., Li, X., & Liu, J. (2024). Cell-wall-anchored proteins affect invasive host colonization and biofilm formation in *Staphylococcus aureus. Microbiological Research, 285*, 127782. https://doi.org/10.1016/j.micres.2024.127782

Yamashita, K., Kawai, Y., Tanaka, Y., Hirano, N., Kaneko, J., Tomita, N., Ohta, M., Kamio, Y., Yao, M., & Tanaka, I. (2011). Crystal structure of the octameric pore of staphylococcal γ-hemolysin reveals the β-barrel pore formation mechanism by two components. *Proceedings of the National Academy of Sciences of the United States of America, 108*(42), 17314–17319. https://doi.org/10.1073/pnas.1110402108

Yao, W., Xu, G., Li, D., Bai, B., Wang, H., Cheng, H., Zheng, J., Sun, X., Lin, Z., Deng, Q., & Yu, Z. (2019). *Staphylococcus aureus* with an erm-mediated constitutive macrolide-lincosamide-streptogramin B resistance phenotype has reduced susceptibility to the new ketolide, solithromycin. *BMC Infectious Diseases, 19*(1), 175. https://doi.org/10.1186/s12879-019-3779-8

Yin, S., Daum, R. S., & Boyle-Vavra, S. (2006). VraSR two-component regulatory system and its role in induction of pbp2 and vraSR expression by cell wall antimicrobials in *Staphylococcus aureus. Antimicrobial Agents and Chemotherapy, 50*(1), 336–343. https://doi.org/10.1128/AAC.50.1.336-343.2006

Zhang, X., Xiong, T., Gao, L., Wang, Y., Liu, L., Tian, T., Shi, Y., Zhang, J., Zhao, Z., Lu, D., Luo, P., Zhang, W., Cheng, P., Jing, H., Gou, Q., Zeng, H., Yan, D., & Zou, Q. (2022). Extracellular fibrinogen-binding protein released by intracellular *Staphylococcus aureus* suppresses host immunity by targeting TRAF3. *Nature Communications, 13*(1), 5493. https://doi.org/10.1038/s41467-022-33205-z

Zhou, L., Li, H., Luo, G., CANF experts group representing the Chinese Burn Association, Editorial Board of the Chinese Journal of Burns and Wounds, & Burn Medicine Branch of China International Exchange and Promotion Association for Medical and Healthcare (2025). Consensus on the diagnosis and treatment of adult necrotizing fasciitis (2025 edition). *Burns & Trauma, 13*, tkaf031. https://doi.org/10.1093/burnst/tkaf031

Zhu, F., Ma, S., Wen, H., Rao, M., Zhang, P., Peng, W., Cui, Y., Yang, H., Tan, C., & Chen, J. (2023). Development of a novel circular mRNA vaccine of six protein combinations against *Staphylococcus aureus. Journal of Biomolecular Structure & Dynamics, 41*(20), 10525–10545. https://doi.org/10.1080/07391102.2022.2154846

Zhu, J., Xie, R., Gao, R., Zhao, Y., Yodsanit, N., Zhu, M., Burger, J. C., Ye, M., Tong, Y., & Gong, S. (2024). Multimodal nanoimmunotherapy engages neutrophils to eliminate *Staphylococcus aureus* infections. *Nature Nanotechnology, 19*(7), 1032–1043. https://doi.org/10.1038/s41565-024-01648-8

3 Understanding Drug Resistance Mechanisms in *Staphylococcus aureus*

P. Gnana Deepu[1], *Kalal Akhila*[2], *Teetla Shivani*[2], *and Sudharshan Reddy Dachani*[3]

[1]Center for Biotechnology, University College of Engineering Science and Technology, Hyderabad, JNTUH, Hyderabad, Telangana, India

[2]Mahavir Hospital & Research Centre, Lakdikapul, Hyderabad, Telangana, India

[3]Department of Pharmacy Practice & Pharmacology, College of Pharmacy, Shaqra University, Al-Dawadmi, Saudi Arabia

ABBREVIATIONS

Agr	accessory gene regulator
CLSI	clinical and laboratory standards institute
HGT	horizontal gene transfer
MDR	multidrug resistance
MGEs	mobile genetic elements
MRSA	methicillin-resistant *Staphylococcus aureus*
PBP	penicillin-binding protein
PVL	panton–valentine leukocidin
SCCmec	Staphylococcal cassette chromosome mec
TSS	toxic shock syndrome
TSST-1	toxic shock syndrome toxin-1
VISA	vancomycin-intermediate methicillin-resistant *Staphylococcus aureus*
VRSA	vancomycin-resistant *S. aureus*

3.1 INTRODUCTION

Staphylococcus aureus is an important cause of infections ranging from mild skin infections to serious systemic infections. There are many reasons for its clinical importance. One important point is its impressive ability to develop resistance to antibiotics, and it's a major member of the ESKAPE pathogens. ESKAPE includes *Enterococcus faecium*, *S. aureus*, *Klebsiella pneumoniae*, *Acinetobacter baumannii*, *Pseudomonas aeruginosa*, and *Enterobacter* species that escape antimicrobial treatment through various mechanisms (Mlynarczyk-Bonikowska et al., 2022). One of the major contributors to the growing concern for *S. aureus* is its development of resistance to methicillin and other β-lactam antibiotics. The development of methicillin-resistant *S. aureus* (MRSA) has been particularly troubling. MRSA was originally associated with healthcare-associated infections (HA-MRSA) typically acquired in hospitals and other healthcare settings. These strains of MRSA are usually resistant to multiple antibiotic classes, and, therefore, it can be challenging to treat with conventional therapies (Chambers & DeLeo, 2009). Another alarming trend in the last three

DOI: 10.1201/9781003647072-4

decades has been the emergence of community-associated MRSA (CA-MRSA), which has moved from healthcare facilities in the community to healthy people in the community.

The emergence of CA-MRSA and its increased global burden of *S. aureus* infections, as well as the potential consequences linked to MRSA's ability to acquire and disseminate resistance determinants, have made MRSA a widespread pathogen in distinct populations and a significant public health threat. The ability of *S. aureus* to resist many classes of antibiotics, such as β-lactams, macrolides, and aminoglycosides, was mediated by many mechanisms, most notably the acquisition of the mecA gene, which encodes a second penicillin-binding protein (PBP2a) that is resistant to β-lactam antibiotics, as well as the development of efflux pumps and enzymatic inactivation systems that confer resistance to other classes of antibiotics. MRSA strains may be geographically dispersed as a consequence of various factors, including both genetic mobility and horizontal gene transfer (HGT), and some clones can survive in hospital and community settings, which has contributed to their global dissemination. Epidemic clones, like ST239 and USA300, have been the most successful global spread of MRSA and are prominent contributors to HA-MRSA and CA-MRSA (Chambers & DeLeo, 2009). The intricate dynamics of genetic adaptation and the mechanism of resistance reflect the seriousness of the challenge presented by *S. aureus* and highlight the need for new treatment modalities and methods of infection control.

In both community and healthcare systems, the rise of antibiotic resistance, specifically MRSA, is becoming a growing concern. *S. aureus* used to primarily be associated with HA-MRSA; however, it has transitioned into the community and is now responsible for significant infections in otherwise healthy individuals. This epidemiological shift helps to understand how *S. aureus* continues to evolve and why it is capable of being resistant to multiple classes of antibiotics. The resistance is due to either chromosomal mutations of *S. aureus* or the acquisition of resistance genes horizontally. The mecA gene is responsible for the resistance to methicillin and other β-lactams; therefore, it is a determinant of MRSA resistance to common antibiotic therapy. Furthermore, *S. aureus* can accumulate resistance genes, and it can spread these traits by HGT; if resistance is capable of spreading from strains and patients, our challenge is heightened (Vestergaard et al., 2018; Chambers & DeLeo, 2009). Therefore, this chapter will investigate the multifaceted mechanisms behind the antimicrobial resistance of *S. aureus*, focusing specifically on the evolution of MRSA. The bacterium has become resistant through many methods, including the means of the Staphylococcal Cassette Chromosome mec (SCCmec) element, which carries the mecA gene and allows the pathogen to survive in the presence of methicillin and other β-lactams. Over the years, MRSA has developed resistance to not only β-lactams but also almost every other class of antibiotics, making it increasingly challenging to treat (Vestergaard et al., 2018).

An important aspect of *S. aureus* pathogenicity is biofilm-forming ability, which greatly contributes to its resistance profile. Biofilms are dense aggregates of bacteria embedded in an extracellular matrix. The biofilm community is protected from antibiotics and host immune responses. Biofilm formation is especially problematic in the medical setting, where *S. aureus* can attach to implanted devices (catheters, prosthetic joints, etc.), resulting in chronic, persistent infections which are not easily treated using conventional therapies (Chambers & DeLeo, 2009). In addition to biofilm formation, another essential aspect of *S. aureus* pathogenicity is a form of bacterial communication termed "quorum sensing" (QS). QS allows bacteria to communicate and coordinate their activities in response to the density of the population, and it is critical in regulating biofilm formation, as well as other virulence factors in *S. aureus*. QS allows *S. aureus* to coordinate expressing antibiotic resistance, virulence, and biofilm genes, thus promoting immune evasion and resistance to treatment (Vestergaard et al., 2018). For example, the mecA gene, which is critical for the resistance profile of MRSA, encodes a modified PBP (PBP2a), which has low affinity for β-lactam antibiotics, making them ineffective against *S. aureus*. The SCCmec element, which harbors the mecA gene, is the principal means by which *S. aureus* obtains methicillin resistance. The horizontal typing of bacteria can allow horizontal transfer of the mecA gene, creating methicillin-resistant strains, in both the hospital and community environments, and further challenging treatment options (Chambers & DeLeo, 2009).

Vancomycin, linezolid, daptomycin, clindamycin, and others are the primary drug treatments for MRSA infections. However, since many reports of vancomycin-intermediate MRSA (VISA) and heterogeneous VISA (hVISA) cases are emerging, there is a growing concern about therapies that include VISA as an instance of the virulence of MRSA. This situation raises the need for new antimicrobial agents to effectively target MRSA. Some of the new strategies considered novel therapies for targeting resistant strains include bacteriophage therapy, antibody-based therapies, and immunomodulatory therapies (Vestergaard et al., 2019). In addition to previously mentioned therapeutics, while *S. aureus* is evolving and new virulent strains will continue to emerge over time, better diagnostics, antimicrobial stewardship, and preventive measures, including vaccines, are needed to help curb the growing *S. aureus* burden. As previously mentioned, MRSA will continue to evolve and move forward; to stay ahead of the resistance curve, further studies need to be done, and surveillance needs to be improved in both hospital and community regarding the virulence of *S. aureus* (Chambers & DeLeo, 2009).

3.2 PATHOGENESIS OF *S. AUREUS*

S. aureus has a complicated pathogenesis that gives it the ability to cause a wide variety of infections from superficial skin lesions to serious, potentially fatal systemic infections such as sepsis, pneumonia, and endocarditis. It is thought that most of the pathogenicity of *S. aureus* is due to the extensive array of virulence factors that allow the organism to penetrate host tissues, escape immune defenses, and persist within the host. *S. aureus* has surface proteins, including fibronectin-binding proteins and clumping factor A, which aid in bacterial adhesion to host tissues and cells, as well as virulence factors such as toxins (hemolysins, exfoliative toxins, toxic shock syndrome toxin-1 (TSST-1)) that damage host tissues and contribute to infection severity. *S. aureus* can also form biofilms, which allow the bacteria to escape immune responses as well as antibiotic treatment, leading to chronic infections, particularly in foreign body infection situations (Foster, 1996). The complicated pathogenesis of *S. aureus* is its ability to acquire multidrug resistance (MDR) to many different antibiotics, including vancomycin and methicillin, which complicates treatment of infections, while contributing to the public health problem associated with non-susceptible MRSA strains. Considering the complexity of how these virulence factors interact and how *S. aureus* can adapt to multiple environmental conditions, it is clear that *S. aureus* has the capacity to cause an array of infections and continues to represent a great risk in a variety of clinical and community settings.

3.2.1 Types of Infections and Virulence Factors

Type of Infections: *S. aureus* is an incredibly adaptable infectious agent that causes a wide range of infections, from mild to life-threatening. It often causes superficial complications such as boils, impetigo, cellulitis, and abscesses on our skin or skin and soft tissue infections or infections, which frequently pose as "the first blush" of the infection. These infections often occur at greater depth into the tissue, resulting in infections termed osteomyelitis, septic arthritis, endocarditis, and pneumonia. *S. aureus* is also a leading agent of food poisoning, which results from the ingestion of pre-formed enterotoxins from food that has been handled by an asymptomatic carrier, and the symptoms associated with *S. aureus* infection may vary from nausea, vomiting, abdominal cramps, and diarrhea. Further, some strains of *S. aureus* also have the ability to produce superantigens, and these are capable of eliciting a response related to toxic shock syndrome (TSS). While TSS is rare, it may also be fatal and occurs suddenly, in conjunction with fever, rash, hypotension, and multi-organ failure. The diverse range of infections reflects the bug's versatility; in addition to possessing numerous virulence factors and the ability to evade the host immune response, it possesses a large number of accessible tissues that are susceptible to colonization (Liu, 2009).

Virulence Factors: *S. aureus* harbors numerous virulence factors to facilitate the process of infection and enable it to infect the host. One important part of its pathogenesis is through the surface proteins that *S. aureus* expresses. These proteins, commonly called microbial surface components recognizing adhesive matrix molecules (MSCRAMMs), have a direct role in adherence. These proteins, such as fibronectin-binding proteins and clumping factors, are able to help the bacteria adhere to extracellular matrix components, like fibrinogen, fibronectin, and collagen, which allow the bacteria to colonize and establish their infection. *S. aureus* also produces a variety of toxins to damage host tissues and evade the immune system. Such toxins include cytotoxins, for example, alpha-hemolysin, which is able to lyse red blood cells and white blood cells, or leukocidins, like the panton–valentine leukocidin, which destroys immune cells. In addition, *S. aureus* produces and secretes superantigens, like toxic shock syndrome toxin-1 (TSST-1) and staphylococcal enterotoxins, which overstimulate the immune system to cause systemic effects, such as toxic shock. Alongside its virulence factors, *S. aureus* produces various enzymes, including coagulase, proteases, and hyaluronidase, which help in spreading through the tissues and breaking down host defenses. Together, these virulence factors contribute significantly to the pathogenic potential of *S. aureus*, making it a formidable and often difficult-to-treat pathogen (Liu, 2009).

3.2.1.1 Host Defense and Bacterial Evasion

a. **Host Defense Mechanisms**: The foremost line of defense against staphylococcal infections is phagocytosis carried out by neutrophils and macrophages. These immune cells will detect, engulf, and destroy invading infectious pathogens. Certain species of *S. aureus* employ various methods of avoiding this immune action. Some of the most important methods are the production of a polysaccharide capsule that helps avoid recognition and engulfment by phagocytic immune cells and the expression of protein A, a surface protein that binds to the Fc region of immunoglobulin G (IgG), resulting in a flipped antibody orientation that leads to impaired opsonization and phagocytosis (Liu, 2009). These methods allow *S. aureus* to survive and reproduce, notwithstanding an active immune response.
b. **Survival Mechanisms**: *S. aureus* has a diverse range of virulence factors that enable it to survive in the dangerous environment of the human body. These virulence factors consist of enzymes, toxins, and regulatory elements that allow it to survive in varying environments, resist immune attacks, and obtain essential nutrients. It has evolved to withstand conditions such as oxidative stress, nutrient deprivation, and antimicrobial peptides (AMPs) produced by the host. In addition to this adaptability, *S. aureus* can overcome the resident microbiota in colonized environments by producing bacteriocins, excreting biochemical signals to circumvent host immune recruitment, and manipulating the immune environment to its advantage. These characteristics are critical to its success as a pathogen and as a commensal organism (Pantosti et al., 2007).
c. **Colonization and Invasion**: Nasal Colonization Hurdles: Nasal colonization of humans is essential to the lifecycle of *S. aureus*, as it is often the reservoir for subsequent infections. It's not easy to colonize the nasal cavity successfully, as *S. aureus* must overcome biological barriers. First, *S. aureus* needs to firmly adhere to nasal mucosa epithelial cells, which is mediated by surface adhesins that recognize specific receptors on the host cell surface. In addition to that, *S. aureus* must evade the mechanical clearance and antimicrobial factors generated by the nasal mucosa. One of the greatest challenges is to outcompete the body's resident nasal flora (*Corynebacterium* spp., *Staphylococcus epidermidis*, *Streptococcus pneumoniae*) that can inhibit *S. aureus* colonization by direct antagonism or immune modulation (Pantosti et al., 2007). These challenges require a finely orchestrated balance of virulence and adaptation strategies to colonize and colonize successfully.

d. **Regulation of Virulence Genes**: During colonization, many virulence genes of *S. aureus* are downregulated. The accessory gene regulator (agr) is the main part of virulence gene regulation, regulating the expression of many virulence and colonization factors. Downregulation of agr is linked to colonization, and activation of agr is linked with invading the host (Pantosti et al., 2007).
e. **Invasion Mechanisms**: *S. aureus* uses microbial surface components recognizing adhesive matrix molecules (MSCRAMM) to attach to and invade host epithelial cells. This is the vital first step in initiating infection, particularly with regard to open wounds or bacteria colonizing the skin or mucosal surfaces. The pathogenesis of *S. aureus* is further complicated by its ability to survive both inside and outside host cells, evade host immune responses, and acquire essential nutrients, particularly iron, from the host. Despite the host's adaptive immune response, *S. aureus* can repeatedly infect the human host throughout life, partly due to its ability to alter T cell functions and evade adaptive immune responses (Munita & Arias, 2016). This complexity underscores the challenges in treating and preventing infections caused by *S. aureus*, including MRSA, which exhibit distinct epidemiological profiles and virulence mechanisms (Foster, 1996; Liu, 2009).

3.2.2 Overview and Mechanism of Drug Resistance in *S. aureus*

S. aureus has a formidable set of defenses against antibiotics and a variety of mechanisms to avoid the action of antibiotics. To counter antibiotic resistance, we need to understand the mechanisms that cause it. The main mechanisms are as follows:

- **Enzymatic Inactivation**: Enzymatic inactivation is one of the most significant means by which *S. aureus* resists antibiotic effectiveness. This bacterium can produce specific enzymes that chemically modify or break down antibiotics, thus inactivating the antibiotic therapeutic activity. One of the best examples is the production of a β-lactamase (also called penicillinase), which hydrolyzes the β-lactam ring that is part of the structure of penicillin and other antibiotics with a β-lactam ring structure. The disruption of this structure prevents this antibiotic from binding to its target, also called PBPs, which are required for bacterial cell wall synthesis. The bacterium can continue to grow and divide in the presence of the drug. This is often a plasmid-encoded resistance mechanism, which can be easily transferred between strains, causing the widespread penicillin-resistant *S. aureus*. Sometimes, these enzymes can be inducible, meaning that their expression increases in response to the antibiotic, further increasing the organism's ability to resist antimicrobial agents (Harkins et al., 2017).
- **Target Site Alteration**: Target site alteration is an important mechanism used by *S. aureus* when it develops resistance to different classes of antibiotics. This involves structural alteration of the components that bind the antibiotic, which effectively reduces the drug's ability to identify and bind to the intended target site. Perhaps the most famous target site alteration is the alteration of PBPs, which are essential for bacterial cell wall synthesis. In MRSA, the introduction of the mecA gene instructs the bacterium to produce a modified PBP called PBP2a, which has a much lower affinity for β-lactam antibiotics (methicillin and other penicillins) than native PBPs. The modification allows the bacterium to continue to synthesize its cell wall in the presence of antibiotics that would normally inhibit the process. Target site alterations can also occur with other antibiotic classes. For example, mutations in the genes encoding DNA gyrase (gyrA) and topoisomerase IV (grlA) confer fluoroquinolone resistance and affect the methylation of the 23S rRNA by erm genes that have reduced binding of macrolides, lincosamides, and streptogramins. Such modifications provide effective protection of important bacterial functions from antibiotic action, which clearly complicates therapeutic treatment (Harkins et al., 2017).

- **Antibiotic Trapping**: While antibiotic trapping is less prevalent as a resistance mechanism, it remains an important resistance strategy utilized by *S. aureus* to resist the actions of particular antibiotics. This strategy refers to the bacterium trapping, sequestering or binding the antibiotic, rendering it unable to access its biological target intracellularly. An example of this is VISA strains, where the bacterial cell wall becomes very much thicker. This type of abnormal cell wall as a sponge, trapping numerous vancomycin molecules in the cell wall before it arrives at the intended target sites involved in peptidoglycan synthesis, whereas trapped vancomycin accumulates in the outer layers of the cell wall and becomes ineffective, allowing the inner layers of the cell wall to be synthesized irrespective to the presence of vancomycin in the outer layers of the cell wall. This resistance mechanism also does not involve modification of a target or enzymatic degradation, and this trapping of antibiotics is by way of a physical barrier to antibiotic penetration. In addition, some strains may express decoy molecules known as binding proteins, which will capture the antibiotics in the extracellular space and limit access to essential cellular structures. In such ways, *S. aureus* can increase its chances of survival when exposed to antibiotics, making treatment of infections more complicated and resulting in persistent infections (Harkins et al., 2017).
- **Efflux Pumps**: Efflux pumps are complex protein-based transport systems that are part of the cell membrane of *S. aureus* and are implicated in antibiotic resistance. These pumps directly export antibiotics and other toxic agents from the cell, allowing their intracellular concentration to achieve levels low enough to be essentially nonlethal. The efflux pumps listed earlier provide some antibiotics with the evasion from high-concentrated levels by decreasing their intracellular levels below the effective concentration needed to kill the bacteria; this minimizes the concentration of drug found at its site of action in a manner that allows for bacterial survival and reproduction despite drug exposure, as well as other factors. Some efflux pumps may be specific only for one class of antibiotics, whereas others contain a broader substrate specificity that contributes to MDR. An efflux pump of staphylococci *S. aureus*, which is well-characterized is NorA. NorA provides resistance against fluoroquinolones by actively transporting the drug from the bacterial cell surface. MepA is another previously documented, but less characterized, efflux system for tetracyclines. Other pumps that may actively export antibiotics, or specifically for other classes or types of disinfectants, are TetK and QacA. Efflux pumps are usually regulated by global stress response systems, and their expression can be increased according to an environmental signal and when an antibiotic is used. The expression and increased expression of efflux pumps complicate successful treatment options or the ability to eradicate an infection and may lead to higher levels of resistance development with the combination of other resistance mechanisms, like modifying a target site or enzymatic inactivation (Harkins et al., 2017).

The genetic nature of antibiotic resistance in *S. aureus* is both complex and multifaceted. The primary way that *S. aureus* acquires resistance genes is through HGT, which is the transfer of mobile genetic elements (MGEs) such as plasmids, transposons, and integrons. HGT can occur via transformation, transduction, and conjugation because of this, *S. aureus* is able to "pick up traits" from other bacteria very quickly (Deresinski, 2005). In addition to HGT, chromosomal mutations contribute to the emergence of resistance. Spontaneous mutations exist in the bacterial genome, conferring changes to antibiotic target sites or regulatory pathways, allowing it to live and function better in the presence of antibiotics (Harkins et al., 2017). The two most distinguished examples of genetic elements related to *S. aureus* resistance are the SCCmec and vanA operon. The SCCmec genetic elements contain the mecA gene, which encodes an altered PBP2a that has a reduced affinity to methicillin and other β-lactam antibiotics. The vanA operon causes vancomycin resistance in some *S. aureus* strains (Pantosti et al., 2007; Harkins et al., 2017). It is important to detect these genetic resistance mechanisms for monitoring antibiotic susceptibility and appropriate treatment of

S. aureus infections. The genetic plasticity of *S. aureus*, made possible by mutations and foreign DNA acquisition, makes it difficult to manage antibiotic resistance and highlights how *S. aureus* has adapted as a persistent pathogen (Munita & Arias, 2016).

3.2.3 The Rise of MRSA

The emergence of MRSA can be divided into distinct waves, with individual resistance mechanisms emerging, as well as strains of MRSA, in both the hospital and community settings. The First Wave occurred in the mid-1940s and involved the emergence of penicillin-resistant *S. aureus* strains that produced a plasmid-encoded penicillinase, an enzyme that inactivated penicillin resulting in a lack of effect when treating the bacteria. The Second Wave found its beginnings with the implementation of methicillin in clinical practice, and new forms of resistance developed, leading to the emergence of MRSA strains. This was a marked change in the treatment of *S. aureus* infections (Chambers & DeLeo, 2009). The Third Wave began, in mid-to-late 1970s, with novel MRSA strains containing SCCmec types II and III which contributed to a global pandemic of MRSA, in hospitals. These strains had improved resistance capabilities, and posed a serious threat to public health. The Fourth Wave is shifted to the emergence of MRSA strains in the community, most notably with skin and soft tissue infections as caused in this wave (Chambers & DeLeo, 2009).

The genetic underpinnings of MRSA resistance are complex and involve several key elements:

> SCCmec Elements form the basis of beta-lactam resistance in *S. aureus*, these mobile genomic islands carry the mec gene, which is vital in encoding methicillin resistance. If SCCmec elements are present, it implies that this bacteria is surely able to resist methicillin and some related antibiotics (Mlynarczyk-Bonikowska et al., 2022). Evolutionary History to genetic studies of early MRSA isolates, specifically ST 250 and belonging to clonal complex (CC) (Harkins et al., 2017) – showed that MRSA resistance likely emerged around 1944, and prior to methicillin use in the clinical environment. This genetic analysis also aided in understanding how the ancestral type I SCCmec element was acquired while reinforcing that beta-lactam use led to many resistant strains (Harkins et al., 2017; Hanssen & Sollid, 2006). Spread and prevalence of initial recognition of *S. aureus* was to have occurred in 1961, and MRSA was limited, or restricted to hospital spread initially, and it was only later where community misuse of antibiotics saw MRSA spread from the hospital population, to the community. In the USA, a concerning percentage of *S. aureus* strains isolated from outpatients that were methicillin-resistant, demonstrated the nature of the widespread dissemination of this pathogen. Of concern, and perhaps even scarier, has been the recent increase in infections caused by new novel CA-MRSA strains infecting healthy individuals without healthcare exposure (Scharn et al., 2022; Idrees et al., 2023). The emergence of methicillin-resistant staphylococcal lineages is primarily due to the acquisition of the SCCmec element and its insertion into the chromosome of a susceptible strain, highlighting the dynamic nature of bacterial resistance mechanisms and the continual battle against MRSA infections.
>
> **(Paterson et al., 2013; Bæk et al., 2014)**

3.2.4 Role of SCCmec Elements in MRSA

The SCCmec is an MGE that is crucial for providing methicillin resistance to *S. aureus* via the mecA gene, ultimately encoding the PBP2a that has a lower binding affinity for β-lactam antibiotics. The resistance mechanism, provided by SCCmec, has been integral to the development and dissemination of MRSA as both HA-MRSA and CA-MRSA strains. The SCCmec elements have been recognized to have structural complexity and genetic diversity, classified as types (e.g., I to XIII), which not only correspond with levels of antibiotic resistance but also epidemiological evidence of MRSA strains and ultimately indicate how adaptable *S. aureus* is and how difficult it is to control spread. A fundamental understanding of the architecture and subtype diversity of SCCmec is needed to elucidate MRSA resistance mechanisms, in addition to informing surveillance and infection control (Hanssen & Sollid, 2006; Ito et al., 2009).

3.2.5 Classification and Characteristics of SCCmec Elements in *S. aureus*

SCCmec elements are large MGEs that play an important role in the acquisition and dissemination of methicillin resistance in *S. aureus*. They carry the mecA gene that encodes the altered PBP2a, which has been shown to confer resistance to β-lactam antibiotics. SCCmec elements are categorized into types based on the combinations of the mec gene complex and the ccr (cassette chromosome recombinase) gene complex, since mec is required for the expression of resistance, and ccr is essential for the mobility of the SCCmec. The different SCCmec types were classified based on the presence of specific combinations of genes and their functions (Hanssen & Sollid, 2006).

- **Type I SCCmec**: Type I has a class B mec gene complex and ccrAB type 1 recombinase. This is one of the earliest SCCmec elements identified and carries only mecA, no other antibiotic resistance genes, making it less multidrug resistant than other types (Hanssen & Sollid, 2006).
- **Type II SCCmec**: Type II has a class A mec gene complex and ccrAB type 2. This type also has Tn554, a transposon that encodes resistance to erythromycin and spectinomycin. The presence of multiple resistance genes makes Type II strains particularly difficult to treat, which is most commonly associated with HA-MRSA (Hanssen & Sollid, 2006).
- **Type III SCCmec**: This type has a class A mec gene complex and ccrAB type 3. Type III SCCmec is typically associated with HA-MRSA strains and is distinguished from other SCCmec types by its abundance of resistance genes, some of which mediate cadmium, mercury, and tetracycline resistance, thus contributing to resistance to both heavy metals and multiple drugs (Hanssen & Sollid, 2006).
- **Type IV SCCmec**: Type IV SCCmec has a class B mec gene complex and a ccrAB type 2. Type IV SCCmec is smaller and is not thought to have additional antibiotic resistance genes so that it can be transferred via HGT more freely. Type IV SCCmec is most common to CA-MRSA strains of *S. aureus* and is associated with increased virulence and transmissibility (Hanssen & Sollid, 2006).
- **Type V SCCmec**: A more recently identified type with a class C2 mec gene complex and the ccrC recombinase. Type V SCCmec, similar to front multinouvations, is associated with CA-MRSA and contributes to the changing genetic potential of MRSA, possibly additionally permitting the emergence of new, more virulent or resistant strains (Rice, 2012).

While the presence of the mecA gene, which encodes methicillin resistance, is the primary basis for the definition of SCCmec elements, there is considerable structural and gene content diversity (Figure 3.1). The overall diversity of SCCmec elements is most clearly highlighted in the various SCCmec types and subtypes, which correspond to specific MRSA clones identified by SCCmec type (Bæk et al., 2014) and my understanding of their chromosomal backgrounds (often determined by multi-locus sequence typing, ST). This horizontal transfer of SCCmec between staphylococcal species, while not yet fully understood in terms of how transfers occur and the vectors used, allows for the rapid spread and genetic evolution of MRSA, presenting serious obstacles for infection control (Hanssen & Sollid, 2006; Rice, 2012; Abebe & Birhanu, 2023). SCCmec elements exhibit some stability, but due to their genomic nature, SCCmec elements may undergo structural rearrangements, leading to moaA loss or gain additional resistance determinants. Various mobile elements,

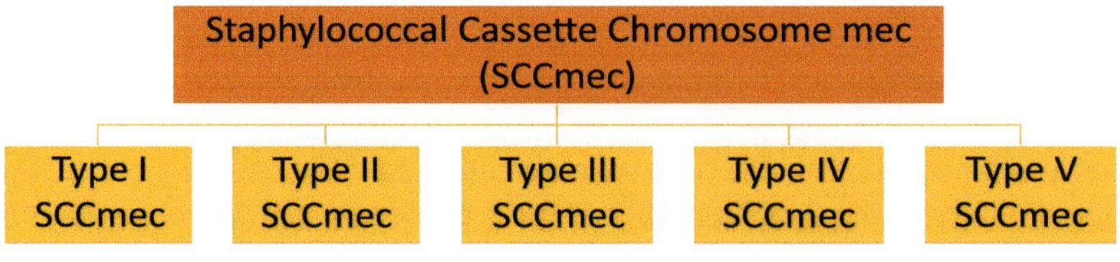

FIGURE 3.1 Classification of SCCmec elements in *Staphylococcus aureus*.

such as IS431 or environmental factors arising from host cells, may also cause such alterations. Based on observations of MRSA isolates in which spa gene deletion was present, loss of SCCmec and greater chromosomal changes raise interesting questions about the genomic and genetic plasticity of MRSA (Scharn et al., 2022; Alexander et al., 2023).

3.2.6 Beta-Lactam Resistance in *S. aureus*

Beta-lactam resistance in *S. aureus* encompasses numerous complex mechanisms, each contributing to the bacterium's resistance to antibiotic action. ClpXP protease and beta-lactam resistance is the inactivation of the components of the ClpXP protease have greatly enhanced the level of beta-lactam resistance in the CA-MRSA USA300 strains 16. Although resistance increased, the cellular levels of PBP2a, which has a lower affinity for beta-lactams, were unchanged in the clp mutant (Blázquez et al., 2014). The clp mutants showed multiple phenotypes that included, but were not limited to, increased thickness of their cell wall, increased peptidoglycan cross-linking, and altered levels of monomeric muropeptide species. In addition, clp mutants showed altered Sle1-mediated peptidoglycan hydrolysis and altered processing of Atl (Bæk et al., 2014). The ClpXP protease is essential for *S. aureus* virulence, and its role in virulence has been observed in both systemic and abscess models of infection. Treatment with daptomycin *in vivo* or vancomycin in vitro selects for mutants with loss-of-function mutations in clpX or clpP. These observations indicate involvement of ClpX and ClpP in the *S. aureus* response to cell wall-active antibiotics (Deresinski, 2005; Bæk et al., 2014; Blázquez et al., 2014).

Cryo-EM structures of full-length *Staphylococcus aureus* BlaR1 reveal that BlaR1 forms a domain-swapped dimer, which stabilizes the signaling loops required to flex during allosteric signal transduction associated with β-lactam resistance. Additionally, BlaR1 undergoes spontaneous autocleavage between Ser283 and Phe284.

The catalytic mechanism and specificity for self-cleavage and BlaI cleavage have been characterized (Mirza, 2017; Alexander et al., 2023). In the case of BlaR1-mediated allosteric signaling, beta-lactam binding stimulates the exclusion of a large extracellular loop contrary to the folded structure of BlaR1, which facilitates the extrusion of the product of the autocalving of BlaR1 from the active site, such that it favors the inactivation of BlaI (Mirza, 2017).

One mechanism of beta-lactam resistance in gram-positive bacteria is typically due to the expression of intrinsic low-affinity PBPs (Tenover et al., 2001). The resistance phenotype is the product of two related systems, the bla and mec operons. Each operon consists of a membrane protein that serves as both a beta-lactam sensor and a signal transducer, a gene repressor, and associated resistance determinants (Soto, 2013). The PC1 beta-lactamase hydrolyzes the beta-lactam ring of antibiotics, while PBP2a is a unique dd-transpeptidase involved in cell wall cross-linking that is not inhibited by beta-lactam antibiotics (Soto, 2013; Blázquez et al. 2014). Understanding these two mechanisms is vital to developing strategies to counter the increasing problem of antibiotic resistance in *S. aureus* and especially MRSA infections.

Glycopeptide resistance in *S. aureus* comprises three different types or levels of resistance, including vancomycin-resistant *S. aureus* (VRSA), VISA, and hVISA that show different levels of tolerance to vancomycin (Mlynarczyk-Bonikowska et al 2022). The emergence of these strains complicates our ability to treat infections that are due to *S. aureus* in a clinical setting. The genetic basis of glycopeptide resistance in *S. aureus* is primarily located in the VanA operon. This operon modifies the antibiotic binding site, reducing the effectiveness of the antibiotic on the bacterium (Mlynarczyk-Bonikowska et al., 2022).

Resistance types are referred to as follows: VRSA: strains of VRSA carry the vanA resistance gene cassette and exhibit high-level resistance to vancomycin. VISA: VISA strains are characterized by having intermediate resistance to vancomycin; they can develop *in vivo* and in vitro related to drug exposure over time and often have nothing to do with the vanA gene (Abebe, 2020). hVISA: these strains demonstrate decreased susceptibility to vancomycin and are not easily detected due to the vague definition and testing (Mirza, 2017).

3.3 DETECTION AND EPIDEMIOLOGY

The laboratory detection of VISA and VRSA strains is based on defined Clinical and Laboratory Standards Institute minimum inhibitory concentration criteria and is, therefore, simple. Conversely, the detection of hVISA is still problematic for most clinical microbiology laboratories without a defined testing approach and a definition (Sharma et al., 2019). The first clinical isolates of VRSA, which have true vancomycin resistance, were identified after nearly 50 years of clinical glycopeptide use. This indicates the impressive adaptive properties of *S. aureus* in the face of antibiotic pressure (Abebe, 2020). These strains of *S. aureus*, as well as VISA and hVISA, have also been isolated in different countries around the world, highlighting the challenge of glycopeptide-resistant *S. aureus* globally (The Role of Bacterial Biofilms in Antimicrobial Resistance I ASM.org, n.d; Tenover et al. 2001).

In response to the emergence of these resistant strains of staphylococci, organizations such as the CDC have issued recommendations for infection control procedures for patients who are infected with glycopeptide-resistant *S. aureus* with the hope of minimizing the dissemination of these strains (Sharma et al, 2019). Our understanding of the molecular mechanisms of glycopeptide resistance, especially in hVISA/VISA strains, is still incomplete and is an area of current investigation (Mirza 2017). Therefore, the complexity of the resistance mechanisms and the difficulty of detection and management of resistant strains suggest that we continue to be vigilant and innovative in our management of *S. aureus* infections.

3.4 EFFLUX PUMP SYSTEMS AND DRUG RESISTANCE

Efflux pumps are important components of bacterial cells responsible for antibiotic resistance, with those that target *S. aureus* receiving considerable focus by the scientific community. Efflux pumps are able to actively pump out a broad range of substances, including antibiotics, from the cell by decreasing the intracellular concentration of a drug to sub-lethal levels, resulting in antibiotic resistance. The two major families of bacterial efflux pumps are the ATP-binding cassette (ABC) superfamily, which uses ATP hydrolysis to transport substrates across cellular membranes. Gram-positive bacteria, such as *S. aureus*, utilize ABC transporters as efflux pumps. The major facilitator superfamily (MFS) uses the proton motive force rather than ATP directly. MFS transporters are also a major efflux mechanism in many gram-positive bacteria, including *S. aureus*. The multidrug and toxic compound extrusion family expels toxins and drugs from the cell, utilizing the electrochemical gradient, and contributes to MDR. The small multidrug resistance family has small proteins conferring resistance to various antiseptics, disinfectants, and antibiotics. The resistance-nodulation-division superfamily is primarily found in gram-negative bacteria and known for broad substrate specificity and high resistance (Pajohesh et al., 2022; Yao et al., 2023). The drug metabolite transporter superfamily is less characterized but is able to transport a wide range of substrate categories, including drugs and metabolites (Soto 2013; Sharma et al 2019; Pajohesh et al. 2022; Yao et al., 2023).

Efflux pumps not only participate in antibiotic resistance but also play a key role in the development of resistance to bacterial biofilms. Their functions include exporting toxic substances, metabolite compounds, and QS signal molecules, and all three of these functions relate to the development and functioning of biofilms. QS is a means of regulation of processes like biofilm development, and efflux pumps were shown to facilitate diffusion of QS molecules across the cell membrane (Soto 2013; Pajohesh et al., 2022). The existence of this relationship between an efflux pump and QS systems reinforces the complexity of bacterial resistance mechanisms, indicating the need to include inhibition of efflux pumps in the development of new therapeutic strategies. Efflux pump inhibitors (EPIs) are a novel mechanism to address antibiotic resistance. Specifically, EPIs work by inhibiting the efflux pump via energy dissipation or direct binding, which will cause increased intracellular concentrations of antibiotics, resulting in the potentiation of antibiotics. EPIs can be classified based on their mechanism of action, either energy dissipating or inhibition by direct binding, and based on their origin, plant products, synthetic compounds, or microorganisms (Yao et al., 2023). Notably, EPIs have a number of hurdles to overcome before there will be development and potential commercialization of EPIs, including, but not limited to, economic feasibility, pharmacokinetics, and specificity. While there are hurdles that need to be addressed, EPIs have performed well in laboratory experiments, suggesting that they may be targets for anti-biofilm drug development (Sharma et al 2019; Yao et al., 2023).

3.4.1 Biofilm Formation and Antibiotic Resistance

Efflux pumps and QS are very important in biofilm establishment and biofilm resistance, and they create a difficult challenge in dealing with many bacterial infections, with *S. aureus* exemplifying a well-defined problem in treating infections. Understanding the biofilm resistance mechanism and biofilm establishment mechanism is key to improving treatment. Efflux pumps are important players in biofilm antibiotic resistance because they transport toxic agents, metabolites, and QS signal molecules essential for the establishment and maintenance of biofilms (Pajohesh et al., 2022). For example, *Pseudomonas aeruginosa* biofilm antibiotic resistance is influenced greatly by the MexAB-OprM efflux pump, which can contribute to antibiotic resistance to a variety of antibiotics, for example, aztreonam and gentamicin. *Candida albicans* deploys ABC transporters and major facilitators encoded by the CDR and MDR genes in biofilm that are upregulated in biofilm, leading to increased fluconazole resistance (Pajohesh et al., 2022).

3.4.1.1 Biofilm Formation

Biofilms are the most dominant life form on Earth and consist of microorganisms forming a cement-like matrix conducive to attachment on surfaces. They have been identified as a problem in medical and food processing environments due to their association with chronic infection and compromised food safety (Abebe, 2020). Biofilms represent the microbial front of battles, as their three-dimensional structures are suited for bacteria as a protective purview for several threats of injury or removal; in fact, this allows colonization and maturation of the bacteria still embedded in the biofilm (The Role of Bacterial Biofilms in Antimicrobial Resistance | ASM.org, n.d., n.d.). A study found that 65.71% of isolated *S. aureus* strains from milk samples were strong biofilm-formers (all biofilm-formers were also penicillin resistant, along with many of the biofilm-formers resistant to ampicillin, oxacillin, and methicillin); thus, there is a recognized relationship between strong biofilm-formers and antibiotic resistance (Pajohesh et al., 2022).

3.4.1.2 Antibiotic Resistance Mechanisms in Biofilms

The antibiotic resistance of biofilms arises from physical, physiological, and gene-related factors, which are different from the typical mechanisms of antibiotic resistance, such as point mutations or enzymes. In combination, these factors reduce – or completely block – the effectiveness of antibiotics, which leads to resistance (Plackett, 2022; Cassir et al, 2014). Mechanisms that allow organisms to survive within biofilms at high concentrations of antibiotics include resistance at the surface of the biofilm, resistance in microenvironments of the biofilms, and resistance of the bacterial "persister" cells in the biofilms (Cassir et al, 2014). The presence of the blaZ gene in a substantial amount of *S. aureus* isolates and the significant association between strong biofilm formation and resistance to multiple antibiotics show the complexity of the relationship between biofilm formation and antibiotic resistance (Kumar & Kaushal, 2021). This thorough description of the mechanisms of biofilm formation and antibiotic resistance points to the complexity of treating infections from biofilm-forming bacteria like *S. aureus*. There are significant challenges in developing successful therapeutic strategies being the interplay between efflux pumps, QS, and structure at the biofilm (Abebe 2020; Pajohesh et al 2022).

3.5 FUTURE DIRECTIONS IN COMBATING *S. AUREUS* DRUG RESISTANCE

Researchers are investigating a range of new strategies and treatments to fight *S. aureus* drug resistance. These efforts are intended to tackle the growing issue of antibiotic resistance, using new and reused interventions (Figure 3.2).

3.5.1 Novel Therapeutic Approaches

Anti-Virulence Strategies and Bacteriophage Therapeutics: The goal is to use approaches that disarm the pathogen (e.g., weaken their virulence) rather than to kill them, to reduce the selective pressure on developing resistance. Bacteriophages, viruses which target bacteria, can be used to target specific bacterial populations to remove the pathogens while leaving other healthy microbiota (Vestergaard et al., 2019). AMPs are another example of an antimicrobial strategy. AMPs such as Chex1-Arg20 have demonstrated broad-spectrum antimicrobial activity with limited resistance development. Proline-rich AMPs appear to be particularly effective against gram-negative bacteria. AMPs may be used to increase the effectiveness of antibiotics through synergism, whereby they work together as an adjuvant (Yao et al. 2023). Field Therapies (e.g., phage therapy and host-directed therapies): Bacteriophage therapy is the utilization of bacteriophages to kill bacteria, whereas host-directed therapies target the host's immune response rather than the pathogen (Sulakvelidze et al., 2001). Both approaches are promising as alternatives to conventional antibiotics and may circumvent resistance.

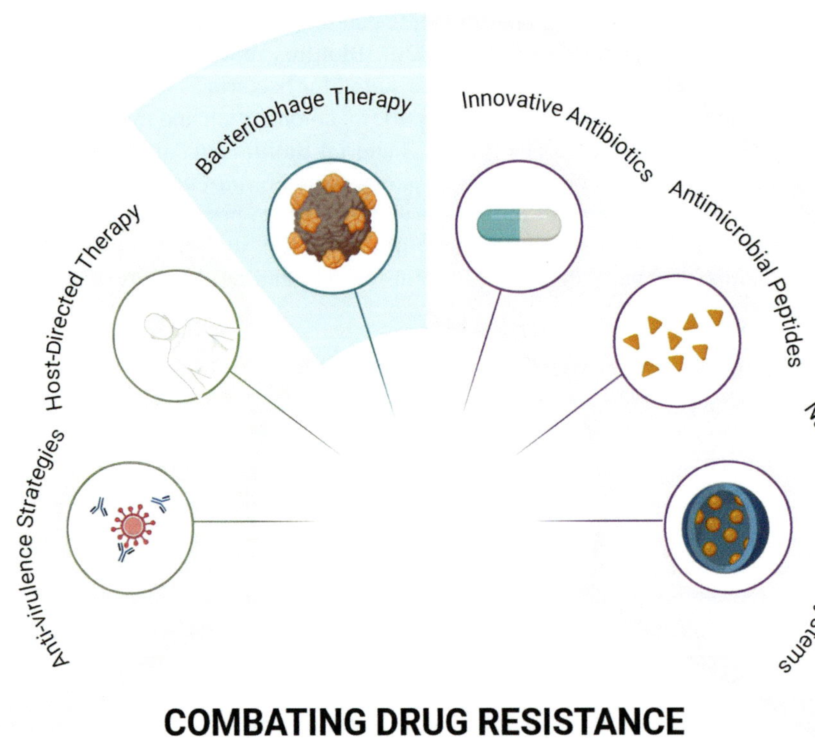

FIGURE 3.2 Future directions in combating *Staphylococcus aureus* drug resistance.

3.5.2 Drug Delivery and Antibiotic Enhancement

EPIs: These agents inhibit bacterial efflux pumps which result in greater intracellular concentrations of antibiotics and greater activity of the antibiotics. EPIs, additionally, inhibit biofilm formation and should be viewed as a target for anti-biofilm drug development. EPI applications still require addressing pharmacokinetics and economic feasibility for clinical use (Pajohesh et al., 2022; Yao et al., 2023). Nano-Drug Delivery Systems (DDSs): DDSs can deliver antibiotics in a more precise manner. This precise antibiotic delivery can enhance efficacy and mitigate resistance. The "Trojan horse" tactic integrates antimicrobial agents with a carrier, whereby the former is delivered in a less effective but more stimulating manner. With any luck, this will prolong the life of a novel antibiotic (Yao et al., 2023).

3.5.3 Innovative Antibiotics and Disinfectants

A newly developed synthetic antibiotic binds tightly to the bacterial ribosome, inhibiting protein synthesis and thus preventing growth. With efficacy in animal models against drug-resistant *S. aureus* and bacteria cresomycin is a good new facet in the fight against drug-resistant pathogens (Designing a New Antibiotic to Combat Drug Resistance, 2024). Gallium Compounds and Plasma-Activated Water: Gallium compounds inhibit MRSA biofilms and bacterial growth, while plasma-activated water offers a new disinfectant possibility. In addition, these simply alternate routes of bacterial population control and, therefore, will begin to reduce resistant populations (Plackett, 2022).

3.5.4 Addressing the Challenge of Antibiotic Discovery

The decrease in the active participation of large drug companies in antibiotic discovery highlights the need to optimize the use of current antibiotics and develop new therapy strategies. The rational application of older antibiotics and the stewardship of antibiotics can maintain current antibiotic treatment efficacy and limit the proliferation of resistance (Kumar & Kaushal, 2021; Cassir et al., 2014). This multifaceted method of combating the drug resistance of *S. aureus* illustrates the need for innovation and cooperation across scientific areas. By exploring a plethora of different strategies, whether new therapeutics or different types of delivery methods, researchers use diverse means to maintain action against antibiotic-resistant pathogens.

3.6 CONCLUSION

In this chapter, we explored the complex ways that *S. aureus* can become resistant to antibiotics, including its medical importance as MRSA and the importance of SCCmec elements that confer methicillin resistance. We also examined beta-lactam and glycopeptide resistance and the role of efflux pumps and biofilm formation in hindering antibiotic efficacy, highlighting the ability of this pathogen to adapt and persist. These discussions provided insight into not only the complexities of drug resistance but also the urgency to develop novel therapeutic innovations and strategies to combat these powerful adversaries. Moving forward, the future study highlighted the necessity of an integrated strategy to fight *S. aureus* resistance, emphasizing a combination of new therapeutic options, including anti-virulence, bacteriophage therapy, AMPs, with new delivery systems, new antibiotics, or disinfectants. The battle against *S. aureus* resistance to antibiotics will require scientific creativity and a united effort from the global health community to prioritize investment in research and development, antiviral stewardship, communication around greater protection of current antibiotics from other uses, and development of easy-to-use delivery systems. The ongoing pursuit of new strategies to combat antibiotic resistance will remain a gesture of hope in the fight against resistant pathogens and to maintain the effectiveness of antibiotics for generations to come.

S. aureus has developed many strategies for resisting antibiotics. Some of these mechanisms involve the creation of a new PBP known as PBP2A that does not act with beta-lactam antibiotics and, therefore, prevents beta-lactam antibiotics from acting on the bacteria. *S. aureus* can also create beta-lactamase enzymes that will degrade the beta-lactam antibiotic or genomic mutations in PBP CA are able to bind the beta-lactam antibiotic, allowing some action from a beta-lactam antibiotic. These same mechanisms for the resistance of *S. aureus* to macrolide antibiotics are also very complex and varied. These are often related to alterations to the target sites, instances where the antibiotic would normally bind (agonism for target sites). Action from macrolide antibiotics can also be inhibited through the use of efflux pumps, which actively remove the antibiotics from the bacterial cell. *S. aureus* works actively and passively to resist antibiotics. The use of acquired resistance genes is one example of the active aspect of resistance that also includes mecA; if this gene is present is a source of methicillin resistance. Other strategies include developing efflux pumps to pump antibiotics out of the cell, producing beta-lactamase enzymes to degrade antibiotics, modifying drug targets to prevent antibiotic binding, and altering the bacterial cell wall to reduce antibiotic permeability. Cells can resist antibiotics by using four broad categories of mechanisms: (1) reducing the entry of the drug into the cell, (2) altering the drug's target site to prevent binding, (3) deactivating the drug directly, and (4) using efflux pumps to actively pump the drug out of the cell. All of these mechanisms protect the cell from the adverse effects of antibiotics.

REFERENCES

Abebe, A., & Birhanu, A. (2023). Methicillin resistant *Staphylococcus aureus*: Molecular mechanisms underlying drug resistance development and novel strategies to combat. *Infection and Drug Resistance*, *16*, 7641–7662. https://doi.org/10.2147/idr.s428103

Abebe, G. M. (2020). The role of bacterial biofilm in antibiotic resistance and food Contamination. *International Journal of Microbiology*, *2020*, 1–10. https://doi.org/10.1155/2020/1705814

Alexander, J. A. N., Worrall, L. J., Hu, J., Vuckovic, M., Satishkumar, N., Poon, R., . . . Strynadka, N. C. J. (2023). Structural basis of broad-spectrum β-lactam resistance in *Staphylococcus aureus*. *Nature*, *613*(7943), 375–382. https://doi.org/10.1038/s41586-022-05583-3

Bæk, K. T., Gründling, A., Mogensen, R. G., Thøgersen, L., Petersen, A., Paulander, W., & Frees, D. (2014). β-lactam resistance in methicillin-resistant *Staphylococcus aureus* USA300 is increased by inactivation of the ClpXP Protease. *Antimicrobial Agents and Chemotherapy*, *58*(8), 4593–4603. https://doi.org/10.1128/aac.02802-14

Blázquez, B., Llarrull, L. I., Luque-Ortega, J. R., Alfonso, C., Boggess, B., & Mobashery, S. (2014). Regulation of the expression of the β-lactam antibiotic-resistance determinants in methicillin-resistant *Staphylococcus aureus* (MRSA). *Biochemistry*, *53*(10), 1548–1550. https://doi.org/10.1021/bi500074w

Cassir, N., Rolain, J., & Brouqui, P. (2014). A new strategy to fight antimicrobial resistance: The revival of old antibiotics. *Frontiers in Microbiology*, *5*. https://doi.org/10.3389/fmicb.2014.00551

Chambers, H. F., & DeLeo, F. R. (2009). Waves of resistance: *Staphylococcus aureus* in the antibiotic era. *Nature Reviews Microbiology*, *7*(9), 629–641. https://doi.org/10.1038/nrmicro2200

Deresinski, S. (2005). Methicillin-resistant *Staphylococcus aureus*: An evolutionary, epidemiologic, and therapeutic odyssey. *Clinical Infectious Diseases*, *40*(4), 562–573. https://doi.org/10.1086/427701

Designing a new antibiotic to combat drug resistance. (2024, March 26). National Institutes of Health (NIH). https://www.nih.gov/news-events/nih-research-matters/designing-new-antibiotic-combat-drug-resistance

Foster, T. (1996). *Staphylococcus. Medical microbiology – NCBI bookshelf*. https://www.ncbi.nlm.nih.gov/books/NBK8448/

Hanssen, A., & Sollid, J. U. E. (2006). SCCmecin staphylococci: Genes on the move. *FEMS Immunology & Medical Microbiology*, *46*(1), 8–20. https://doi.org/10.1111/j.1574-695x.2005.00009.x

Harkins, C. P., Pichon, B., Doumith, M., Parkhill, J., Westh, H., Tomasz, A., De Lencastre, H., Bentley, S. D., Kearns, A. M., & Holden, M. T. G. (2017). Methicillin-resistant *Staphylococcus aureus* emerged long before the introduction of methicillin into clinical practice. *Genome Biology*, *18*(1). https://doi.org/10.1186/s13059-017-1252-9

Idrees, M. M., Saeed, K., Shahid, M. A., Akhtar, M., Qammar, K., Hassan, J.,. . . Saeed, A. (2023). Prevalence of mecA- and mecC-associated methicillin-resistant *Staphylococcus aureus* in clinical specimens, Punjab, Pakistan. *Biomedicines*, *11*(3), 878. https://doi.org/10.3390/biomedicines11030878

Ito, T., Hiramatsu, K., Oliviera, D., De Lencastre, H., Zhang, K., Westh, H., . . . Soderquist, B. (2009). Classification of staphylococcal cassette chromosome mec (SCC mec): Guidelines for reporting novel SCC mec elements. *Antimicrobial Agents and Chemotherapy*, *53*(12), 4961–4967. https://doi.org/10.1128/aac.00579-09

Kumar, A., & Kaushal, M. (2021). Progression of β-lactam resistance in *Staphylococcus aureus*. In *Infectious diseases*. https://doi.org/10.5772/intechopen.100622

Liu, G. Y. (2009). Molecular pathogenesis of *Staphylococcus aureus* infection. *Pediatric Research*, *65*(5 Part 2), 71R–77R. https://doi.org/10.1203/pdr.0b013e31819dc44d

Mirza, H. C. (2017). Glycopeptide resistance in *S. aureus*. In *InTech eBooks*. https://doi.org/10.5772/65471

Mlynarczyk-Bonikowska, B., Kowalewski, C., Krolak-Ulinska, A., & Marusza, W. (2022). Molecular mechanisms of drug resistance in *Staphylococcus aureus*. *International Journal of Molecular Sciences*, *23*(15), 8088. https://doi.org/10.3390/ijms23158088

Munita, J. M., & Arias, C. A. (2016). Mechanisms of antibiotic resistance. *Microbiology Spectrum*, *4*(2). https://doi.org/10.1128/microbiolspec.vmbf-0016-2015

Pajohesh, R., Tajbakhsh, E., Momtaz, H., & Rahimi, E. (2022). Relationship between biofilm formation and antibiotic resistance and adherence genes in *Staphylococcus aureus* strains isolated from raw cow milk in Shahrekord, Iran. *International Journal of Microbiology*, *2022*, 1–10. https://doi.org/10.1155/2022/6435774

Pantosti, A., Sanchini, A., & Monaco, M. (2007). Mechanisms of antibiotic resistance in *Staphylococcus aureus*. *Future Microbiology*, *2*(3), 323–334. https://doi.org/10.2217/17460913.2.3.323

Paterson, G. K., Harrison, E. M., & Holmes, M. A. (2013). The emergence of mecC methicillin-resistant *Staphylococcus aureus*. *Trends in Microbiology*, *22*(1), 42–47. https://doi.org/10.1016/j.tim.2013.11.003

Plackett, B. (2022). Three ways to combat antimicrobial resistance. *Nature*, *612*(7940), S33. https://doi.org/10.1038/d41586-022-04213-2

Rice, L. B. (2012). Mechanisms of resistance and clinical relevance of resistance to β-lactams, glycopeptides, and fluoroquinolones. *Mayo Clinic Proceedings*, *87*(2), 198–208. https://doi.org/10.1016/j.mayocp.2011.12.003

Scharn, C. R., Tickler, I. A., Tenover, F. C., & Goering, R. V. (2022). Characterization of SCC mec instability in methicillin-resistant *Staphylococcus aureus* affecting adjacent chromosomal regions, including the gene for staphylococcal protein A (spa). *Antimicrobial Agents and Chemotherapy*, *66*(4). https://doi.org/10.1128/aac.02374-21

Sharma, A., Gupta, V. K., & Pathania, R. (2019). Efflux pump inhibitors for bacterial pathogens: From bench to bedside. *The Indian Journal of Medical Research*, *149*(2), 129. https://doi.org/10.4103/ijmr.ijmr_2079_17

Soto, S. M. (2013). Role of efflux pumps in the antibiotic resistance of bacteria embedded in a biofilm. *Virulence*, *4*(3), 223–229. https://doi.org/10.4161/viru.23724

Sulakvelidze, A., Alavidze, Z., & Morris, J. G. (2001). Bacteriophage therapy. *Antimicrobial Agents and Chemotherapy*, *45*(3), 649–659. https://doi.org/10.1128/aac.45.3.649-659.2001

Tenover, F. C., Biddle, J. W., & Lancaster, M. V. (2001). Increasing resistance to vancomycin and other glycopeptides in *Staphylococcus aureus*. *DOAJ (DOAJ: Directory of Open Access Journals)*. https://doi.org/10.3201/eid0702.070327

The Role of Bacterial Biofilms in Antimicrobial Resistance. *ASM.org*. (n.d.). ASM.org. https://asm.org/articles/2023/march/the-role-of-bacterial-biofilms-in-antimicrobial-re

Vestergaard, M., Frees, D., & Ingmer, H. (2019). Antibiotic resistance and the MRSA problem. *Microbiology Spectrum*, *7*(2). https://doi.org/10.1128/microbiolspec.gpp3-0057-2018

Yao, J., Zou, P., Cui, Y., Quan, L., Gao, C., Li, Z., . . . Yang, M. (2023). Recent advances in strategies to combat bacterial drug resistance: Antimicrobial materials and drug delivery systems. *Pharmaceutics*, *15*(4), 1188. https://doi.org/10.3390/pharmaceutics15041188

4 Antimicrobial Resistance Trends in *Staphylococcus aureus*

Chittepu Obula Reddy[1], *Srikanth Muthyala*[1], *Perugu Bhavana*[1], *Penchala Disha*[1], *Sreya Palakodeti*[1], *Nikitha Teku*[1], *and Chittepu Sai Tharuni*[2]

[1]Department of Biotechnology, Chaitanya Bharathi Institute of Technology, Gandipet, Hyderabad, Telangana, India

[2]Mamata Academy of Medical Sciences, Bachupally, Hyderabad, Telangana, India

ABBREVIATIONS

ABSSSI	acute bacterial skin and skin structure infections
AI	artificial intelligence
AI	driven artificial intelligence driven
AMR	antimicrobial resistance
CA	MRSA community-associated MRSA
CAMs	cell adhesion molecules
CDC	Centers for Disease Control and Prevention
DNA	deoxyribonucleic acid
EPI	efflux pump inhibitor
EPS	extracellular polymeric substance
ESBL	extended-spectrum beta-lactamase
HA	MRSA healthcare-associated MRSA
HGT	horizontal gene transfer
LA	MRSA livestock-associated MRSA
MATE	multidrug and toxic compound extrusion
MFS	major facilitator superfamily
MIC	minimum inhibitory concentration
ML	machine learning
MRSA	methicillin-resistant *Staphylococcus aureus*
MSSA	methicillin-sensitive *Staphylococcus aureus*
PBP2a	penicillin-binding protein 2a
PCR	polymerase chain reaction
PIA	polysaccharide intercellular adhesin
PVL	panton–valentine leukocidin
RNA	ribonucleic acid
SCCmec	staphylococcal cassette chromosome mec
SSTIs	skin and soft tissue infections
TSST-1	toxic shock syndrome toxin-1
UTIs	urinary tract infections

DOI: 10.1201/9781003647072-5

VISA	vancomycin-intermediate *Staphylococcus aureus*
VRSA	vancomycin-resistant *Staphylococcus aureus*
WHO	World Health Organization

4.1 INTRODUCTION

4.1.1 Overview of *Staphylococcus aureus*

S. aureus is a versatile gram-positive coccus that colonizes about 20–30% of the human population asymptomatically, especially in the nasal passages and on the skin. Regardless of its commensal nature, it is capable of evolving into a potent disease-causing agent, causing diseases that range from superficial skin infections to invasive and lethal conditions like sepsis, pneumonia, osteomyelitis, endocarditis, and toxic shock syndrome. Its pathogenicity is primarily attributed to a diverse array of pathogenic factors, including surface proteins that facilitate adhesion, such as the cell adhesion molecules (CAMs), enzymes that promote tissue invasion, and toxins, such as alpha-toxin, leukocidins, and toxic shock syndrome toxin-1 (TSST-1) (Lowy, 2003).

The development of antimicrobial resistance (AMR) and rapid acquisition is a key feature that sets *S. aureus* apart from many other pathogens (Figure 4.1). The bacterium was originally vulnerable to penicillin, but shortly after its introduction in the 1940s, resistance emerged due to the production of penicillinase, β-lactam ring – an antibiotic which is hydrolyzed by the β-lactamase enzyme (Chambers & DeLeo, 2009). To overcome this challenge, methicillin was developed, but methicillin resistance appeared soon after its clinical use began in 1961, resulting in methicillin-resistant *S. aureus* (MRSA). MRSA strains possess the mecA gene, located on the staphylococcal cassette chromosome mec (SCCmec), which encodes an altered penicillin-binding protein (PBP2a) with a lower affinity for β-lactams (Katayama et al., 2000).

Over time, MRSA has developed into several lineages, including healthcare-associated MRSA (HA-MRSA), community-associated MRSA (CA-MRSA), and livestock-associated MRSA, where each adapted to specific ecological niches. CA-MRSA strains often harbor virulence because of their capacity to produce a wide range of toxins and other factors that contribute to their pathogenicity. The factors can include toxins like panton–valentine leukocidin (PVL), which contribute to their heightened pathogenicity even in otherwise healthy individuals (Otto, 2013).

S. aureus has demonstrated the ability to develop resistance to a variety of other antibiotics, including macrolides, aminoglycosides, tetracyclines, fluoroquinolones and, in rare cases, even glycopeptides such as vancomycin (VISA and VRSA) in addition to β-lactam resistance. This extensive resistance is because of the multiple mechanisms including horizontal gene transfer, biofilm formation, targeted site mutations, and efflux pumps (*Howden* et al., *2010*). The adaptability and virulence of *S. aureus* make it one of the most important bacterial pathogens in both community and hospital settings.

Recent surveillance systems and molecular epidemiology studies have identified an increasing prevalence of MRSA globally, emphasizing the need for continued research and novel therapeutic strategies. The emergence of resistance toward the new agents, such as linezolid and daptomycin, is particularly worrisome and underscores the importance of prudent antibiotic use and the development of alternative therapies.

In conclusion, *S. aureus* illustrates the dynamic interplay between bacterial evolution, antimicrobial use, and public health impact. Understanding its mechanisms of resistance and virulence is vital for guiding clinical treatment and developing effective interventions.

4.1.2 Public Health Impact of AMR

The AMR shown by *S. aureus*, especially MRSA, is a pressing public health challenge with prominent clinical and economic consequences. The World Health Organization (WHO) has mentioned

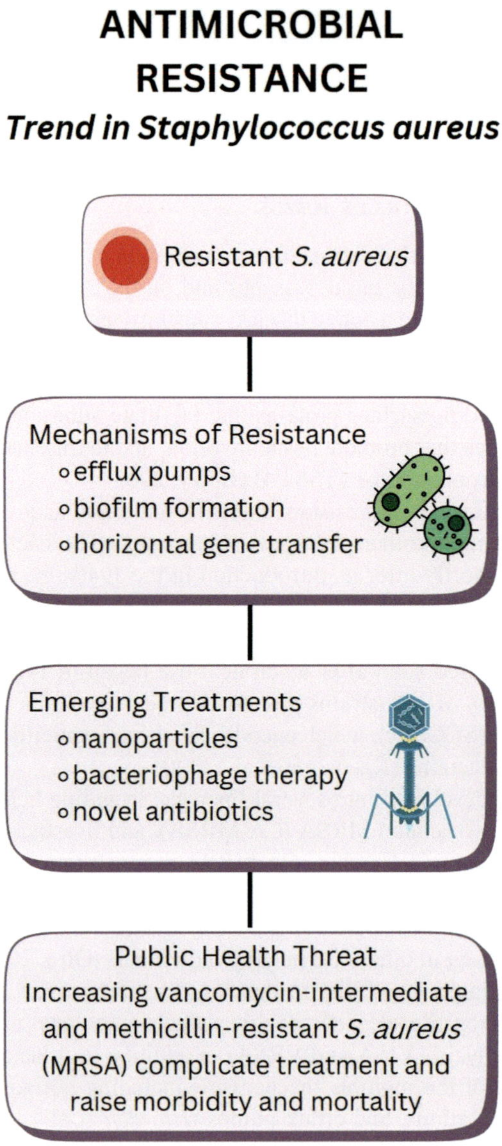

FIGURE 4.1 Visual overview to introduce: antimicrobial resistance trends in *Staphylococcus aureus.*

AMR as one of the top ten global public health threats facing humanity. A significant percentage of infections linked to healthcare are caused by MRSA in particular, and its increasing prevalence in community settings is concerning. The clinical burden of MRSA includes a spectrum of diseases such as bacteremia, pneumonia, endocarditis, surgical-site infections, and skin and soft tissue infections (SSTIs) (Centers for Disease Control and Prevention [CDC], 2019).

MRSA infections are associated with increased morbidity, mortality, and healthcare costs. A study conducted by (Klein et al., 2013) reported that the mortality rate associated with MRSA bacteremia ranges from 20% to 40%, particularly in older people and immunocompromised patients. This is prominently higher than infections caused by methicillin-sensitive *S. aureus* (MSSA). The strong resistance limits the effectiveness of first-line antibiotics, necessitating the use of more toxic, less effective, or more expensive alternatives such as vancomycin, linezolid, or daptomycin. These drugs often require intravenous administration and are associated with long-term hospitalization and intensive monitoring, further straining healthcare systems (Cosgrove et al., 2005).

The economic impact of MRSA is significant. In the United States alone, the CDC estimates that antibiotic-resistant *S. aureus* infections account for more than $1.7 billion in additional healthcare costs annually (*CDC, 2019*). In low- and middle-income countries, the impact is amplified by weaker healthcare infrastructures, limited access to second-line drugs, and inadequate infection prevention measures. Infections in these settings often go unreported or mishandled, contributing to extended outbreaks and high mortality rates.

CA-MRSA has emerged as a major public health issue in recent years. Unlike HA-MRSA, CA-MRSA affects healthy individuals with no prior exposure to healthcare environments. It has been implicated in outbreaks among athletes, military personnel, and schoolchildren, which suggests the widespread transmission in public domains. These strains often harbor virulence factors such as the PVL, which contribute to severe necrotizing infections and abscesses (David & Daum, 2010).

Surveillance studies globally continue to document increasing resistance rates. According to Styers et al. (2006), MRSA prevalence among outpatient isolates in the United States reached 47.9%, while inpatient ICU settings showed resistance rates that are beyond 59%. This trend is reflected in Europe and Asia, further highlighting the global nature of the problem.

4.1.3 Scope and Objective

Focusing on the mechanisms, epidemiological trends, and new interventions to address this growing public health concern, this chapter aims to provide a thorough and analytical overview of AMR trends in *S. aureus*. Clarifying the biological and molecular bases of resistance in *S. aureus*, assessing current resistance profiles in clinical and agricultural settings, and critically analyzing the most recent therapeutic advancements targeted at battling resistant strains are the main objectives.

This chapter covers the historical background of *S. aureus* resistance development, from the early appearance of penicillin resistance to the current ubiquity of strains resistant to methicillin and vancomycin. The mecA gene, which codes for PBP2a, the blaZ gene, which produces β-lactamase, and other genetic components involved in efflux, biofilm formation, and toxin production are among the main resistance determinants covered in this chapter (*Lowy, 2003*). The reader will comprehend how *S. aureus* avoids antimicrobial agents and adjusts to environmental stresses better if these mechanisms are thoroughly explained.

The function of epidemiological and clinical surveillance in tracking resistance trends is also covered in this chapter. To identify high-risk populations, monitor the spread of resistant clones, and direct empirical treatment protocols, for example, regional and global surveillance data from organizations such as the CDC and WHO are useful (Vicetti Miguel et al., 2019). Particularly in environments with limited resources, data-driven insights are essential for guiding public health policies and resource allocation.

This chapter also examines new therapeutic approaches like bacteriophage therapy, which targets particular bacterial strains, the use of nanoparticles to break up biofilms, and the creation of innovative antibiotics like nilofabicin and afabicin that block vital bacterial enzymes. The potential of artificial intelligence (AI) to speed up the identification of new antimicrobial compounds is highlighted in the discussion of AI-driven drug discovery (Theuretzbacher et al., 2020).

The chapter concludes by comparing *S. aureus*'s resistance profile to that of other important pathogens, including *Klebsiella pneumoniae*, *Pseudomonas aeruginosa*, and *Escherichia coli*. These comparisons will shed light on the reasons why *S. aureus* is still a high-priority pathogen, owing to its high virulence and ability to spread quickly, in addition to its resistance capabilities.

This chapter attempts to present a multidisciplinary perspective of *S. aureus* resistance trends by combining clinical data, molecular biology, and creative research, giving researchers, clinicians, and public health professionals knowledge they can use.

TABLE 4.1
Comparative Analysis of Antimicrobial Resistance among Key Pathogens

Bacterial Species	Common Infections	% Methicillin Resistance	% Multidrug Resistance	Primary Concerns	Source
S. aureus	SSTIs, pneumonia, sepsis	40–63%	22–67%	MRSA, VRSA, PVL toxin, biofilm formation	Styers et al. (2006); Vicetti Miguel et al. (2019); Mourabit et al. (2021); Lowy (2003)
Escherichia coli	UTIs, sepsis	NA	~25–40%	ESBL production, plasmid-mediated resistance	Mourabit et al. (2021)
Klebsiella pneumoniae	Pneumonia, bloodstream infections	NA	~50–70%	Carbapenem resistance, hospital-acquired spread	Mourabit et al. (2021)
Pseudomonas aeruginosa	Respiratory and burn infections	NA	~30–60%	Intrinsic resistance, biofilms	Mourabit et al. (2021)
Enterococcus faecalis/faecium	Endocarditis, UTIs	NA	~20–40%	Vancomycin-resistant enterococci (VRE)	Lowy (2003)

4.1.4 Comparative Analyses of *S. aureus* with Other Microorganisms

It is crucial to compare *S. aureus*'s resistance profile, virulence, and prevalence with those of other important bacterial pathogens to put the public health threat it poses into perspective. Although multidrug resistance (MDR) is a well-known trait of gram-negative bacteria such as *E. coli*, *K. pneumoniae*, and *P. aeruginosa*, *S. aureus* stands out for combining strong resistance with strong virulence factors and adaptability in both clinical and community settings.

Based on data from the Surveillance Network-USA (TSN), *S. aureus* was the most frequently isolated pathogen from inpatient specimens (18.7% of isolates), with *E. coli* coming in second (17.3%). *E. coli* was more common in outpatients (38.6%) than *S. aureus* (14.7%) (Styers et al., 2006). With MRSA rates above 50% in many areas, *S. aureus* stands out despite this because of its widespread resistance (Table 4.1).

Although sepsis and urinary tract infections are the main causes of *E. coli* and *K. pneumoniae*, their resistance is mainly limited to hospital settings. *S. aureus*, on the other hand, can cause infections in both hospitals and the community, and CA-MRSA strains frequently cause serious illness in healthy people (David & Daum, 2010).

Additionally, *S. aureus* can grow biofilms on medical equipment, making removal more difficult and frequently necessitating surgery. Gram-negative organisms are a serious threat to resistance, but *S. aureus* is perhaps the most dangerous gram-positive pathogen due to its virulence, adaptability, and widespread distribution.

4.2 MECHANISMS OF AMR IN *S. AUREUS*

4.2.1 Methicillin Resistance and mecA Gene

The acquisition of the mecA gene, which codes for penicillin-binding protein 2a (PBP2a), is the main cause of methicillin resistance in *S. aureus*. PBP2a catalyzes the bacterial cell wall's cross-linking even when beta-lactam antibiotics are present, in contrast to normal PBPs that are inhibited by these medications. All beta-lactam antibiotics, including penicillin, methicillin, and more recent ones like cephalosporins, have little affinity for this protein. The SCCmec, a mobile genetic element (MGE) that carries the mecA gene, varies in size and structure but always has the mec gene complex and ccr (cassette chromosome recombinase) genes that allow it to move (Mlynarczyk-Bonikowska et al., 2022).

Globally, the rise of MRSA strains poses a serious threat to public health. Both HA-MRSA and CA-MRSA strains are now common. HA-MRSA is usually home to SCCmec types I–III, which frequently contain genes that confer resistance to several antibiotics. Types IV and V, on the other hand, are more prevalent in CA-MRSA and typically confer resistance to beta-lactam antibiotics (Abebe & Birhanu, 2023).

Apart from mecA, other genes that contribute to methicillin resistance have been found, such as mecC, which is a homolog of mecA. These genes, which have been isolated from both human and animal sources and are present on variant SCCmec elements, demonstrate the zoonotic potential of resistant strains. For diagnosis and proper treatment planning, these resistance determinants must be found. To determine whether clinical isolates contain the mecA or mecC genes, molecular techniques like PCR are frequently used.

Comprehending the molecular foundation of methicillin resistance is crucial for both infection prevention and the creation of novel treatment approaches. Non-beta-lactam antibiotics like vancomycin, linezolid, and daptomycin, as well as more recent drugs like ceftaroline that exhibit MRSA activity, are currently used to treat MRSA infections. However, constant monitoring and the creation of new antimicrobial agents are required due to the ongoing evolution of resistance mechanisms (Reygaert, 2013).

4.2.2 Vancomycin-Intermediate and Vancomycin-Resistant Strains (VISA and VRSA)

For many years, vancomycin has been regarded as the last resort when dealing with resistant gram-positive bacteria, such as MRSA. Vancomycin-resistant *S. aureus* (VRSA) and vancomycin-intermediate *S. aureus* (VISA) have emerged, though, and this has caused serious clinical concerns. VRSA strains are extremely resistant to vancomycin therapy, with minimum inhibitory concentrations (MICs) of 16 mg/L or greater, whereas VISA strains have MICs ranging from 4 to 8 mg/L (Mlynarczyk-Bonikowska et al., 2022).

The thickness of the bacterial cell wall, which traps vancomycin molecules and keeps them from getting to their target sites, is typically linked to VISA resistance. Unlike mecA-mediated methicillin resistance, this mechanism is not gene-based; rather, it is caused by mutations in genes that control cell wall metabolism, such as the walKR, vraSR, and graSR systems (Reygaert, 2013).

On the other hand, VRSA strains obtain the vanA operon through horizontal gene transfer, usually from *Enterococcus faecalis*. Vancomycin's binding affinity is significantly decreased by the enzymes encoded by the vanA gene cluster, which change the binding site from D-Ala-D-Ala to D-Ala-D-Lac. Plasmids or transposons like Tn1546 are frequently used to mediate this gene transfer (Abebe & Birhanu, 2023).

Because there are few treatment options and effective therapy is delayed, both VISA and VRSA strains are linked to poor clinical outcomes. Because its resistance might not be detected by routine susceptibility testing, VISA is especially pernicious and calls for more accurate diagnostic techniques like population analysis profiling.

These strains' emergence emphasizes *S. aureus*'s capacity for adaptation and the need for different approaches to treatment. These days, linezolid, daptomycin, tigecycline, and the more recent lipoglycopeptides (like dalbavancin and oritavancin) are substitutes. But there is also an increase in resistance to these agents, which calls for close monitoring and stewardship (Mlynarczyk-Bonikowska et al., 2022).

4.2.3 Efflux Pump Mechanisms

Membrane proteins called efflux pumps actively extrude harmful compounds, such as antibiotics, from bacterial cells. Several efflux pump systems in *S. aureus* reduce the intracellular concentration of medications, thereby reducing their effectiveness, which leads to AMR. The NorA, NorB,

NorC, MdeA, MepA, and QacA/B efflux pumps are the most notable of these (Truong-Bolduc et al., 2005). Known for its part in resistance to fluoroquinolones like ciprofloxacin, the NorA efflux pump belongs to the major facilitator superfamily. Global regulatory systems like MgrA control the activity of NorA, and exposure to antibiotics and environmental stress can increase its expression. Likewise, resistance to tigecycline is facilitated by MepA, a member of the multidrug and toxic compound extrusion family.

Although some efflux pump genetic determinants can be plasmid-encoded, which increases their potential for horizontal transmission, most of them are found on the bacterial chromosome. Depending on the regulatory context, efflux pump gene expression levels can be either constitutive or inducible. In addition to causing antibiotic resistance (AR), overexpression of these pumps may affect the fitness and virulence of bacteria.

When paired with additional mechanisms such as target modification or enzymatic inactivation, efflux pumps can serve as a precursor to higher-level resistance. Moreover, they complicate efforts to control infections by fostering resistance to disinfectants and biocides (Mlynarczyk-Bonikowska et al., 2022).

4.2.4 Biofilm Formation and Resistance

One of *S. aureus*'s main mechanisms of virulence and resistance is biofilm formation. Biofilms are organized bacterial communities that stick to biotic or abiotic surfaces and are encased in an extracellular matrix that the bacteria produce on their own. According to Silva et al. (2020), they are frequently linked to persistent and device-related infections like endocarditis, bloodstream infections from catheter use, and infections of the prosthetic joints.

Significant defense against host immune responses and antimicrobial agents is provided by biofilms. The cells in the biofilm are in a variety of metabolic states, such as slow-growing or dormant persister cells, which are naturally resistant to antibiotics. According to Suresh et al. (2019), the extracellular polymeric substance (EPS) matrix can bind antimicrobials and limit their diffusion, lowering their effective concentration.

Numerous genes and environmental cues control the development of biofilms in *S. aureus*. The production of polysaccharide intercellular adhesin, a key constituent of the biofilm matrix, depends on the icaADBC operon. In reaction to environmental factors, regulatory systems like SarA, Agr, and SigB modify the formation of biofilms.

Additionally, biofilms serve as horizontal gene transfer reservoirs, which promote the spread of resistance genes among microbial communities. Treatment attempts are made more difficult by this genetic exchange, which also encourages the development of strains that are resistant to drugs (Figure 4.2).

4.2.5 Horizontal Gene Transfer

One of the main processes causing *S. aureus* to evolve AR is horizontal gene transfer or HGT. HGT allows bacteria to obtain genetic material from other organisms, including those of different species, in contrast to vertical gene transfer, which entails the inheritance of genetic traits from parent to offspring. Multidrug-resistant strains are a result of this process, which also speeds up the spread of resistance genes (Sun et al., 2019). Three main processes underlie HGT in *S. aureus*: conjugation, transduction, and transformation. In *S. aureus*, the transformation of the absorption of free DNA from the surroundings occurs comparatively infrequently. Bacteriophage-mediated transduction is more important. Phages have the ability to encapsulate and transport resistance gene-containing DNA fragments between bacterial cells (Table 4.2). According to Norman et al. (2019), conjugative transposons and integrative conjugative elements frequently aid in conjugation, which is the direct transfer of plasmids through cell-to-cell contact. Important carriers of resistance genes are MGEs, including integrons, transposons, plasmids, and staphylococcal cassette chromosomes (e.g., SCCmec).

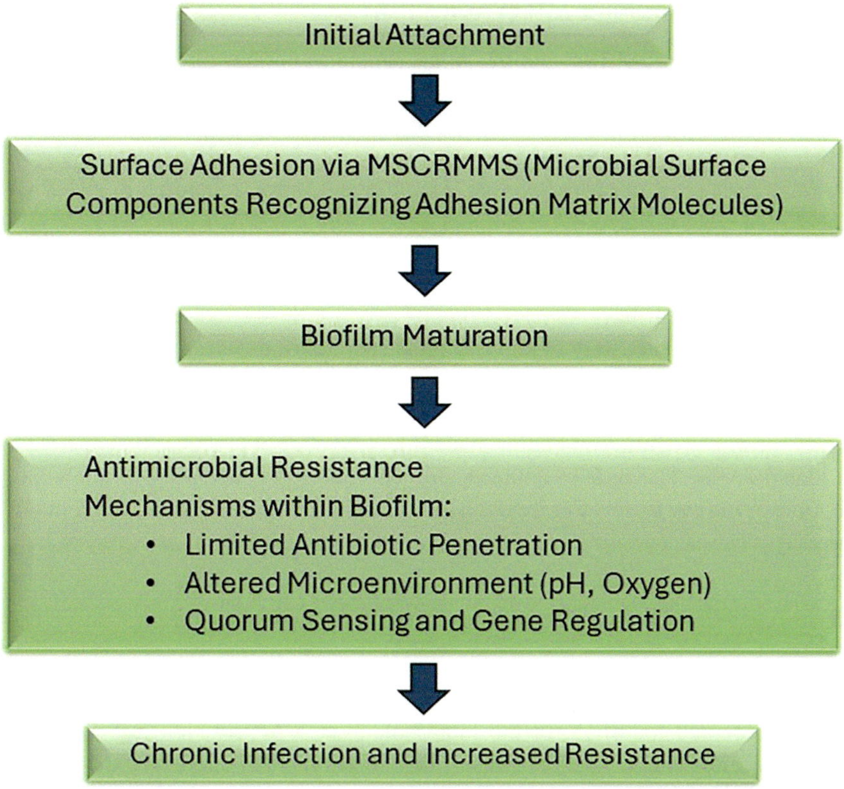

FIGURE 4.2 Flowchart: biofilm formation and its role in antimicrobial resistance in *Staphylococcus aureus*.

TABLE 4.2
Key Resistance Mechanisms in *Staphylococcus aureus*

Resistance Mechanism	Genetic Element(s)	Target Drug Classes	References
Methicillin resistance	mecA, SCCmec	β-lactams	Mlynarczyk-Bonikowska et al. (2022)
Vancomycin resistance	vanA, Tn1546	Glycopeptides	Abebe and Birhanu (2023)
Efflux pump-mediated resistance	tet(K), NorA, NorB	Tetracyclines, fluoroquinolones	Guay et al. (1993); Truong-Bolduc et al. (2005)
Biofilm-mediated resistance	icaADBC, SarA	Multidrug tolerance	Silva et al. (2020); Suresh et al. (2019)
Horizontal gene transfer	Plasmids, transposons	Multidrug resistance	Norman et al. (2019); Sun et al. (2019)

4.3 EPIDEMIOLOGY AND TRENDS OF RESISTANCE

S. aureus has dynamic epidemiology, with varying rates of colonization by geographic attributes, population, and exposure to healthcare. MRSA remains found in both hospital and community environments. Resistance patterns display increasing proportions of multidrug-resistant forms, such as VISA and VRSA isolates. Horizontal gene exchange, biofilm, and antibiotic stress drive resistance evolution. Surveillance evidence supports the importance of region-specific and worldwide surveillance to stem the escalation of resistant *S. aureus* strains (Touaitia et al., 2025).

4.3.1 Global Distribution of MRSA and VISA

The global dissemination of MRSA and VISA has generated serious distress in clinical microbiology. MRSA has emerged globally as a significant pathogen in hospital settings, whereas VISA and heterogeneous VISA (hVISA) strains are increasing dangers since they are less sensitive to vancomycin, which is the last drug to be used for MRSA infections. A study on meta-analysis indicated that VISA cases have seen an overwhelming rise in recent years, according to global data, which indicates that Asia takes the lead in the number at 2.1%, followed by Europe at 1.8% and America at 1.0% (Shariati et al, 2020). VRSA has shown an increasing presence in hospitals across the United States and various regions across Asia despite being less common compared to other forms of resistance. These strains have appeared due to two significant factors that involve the alteration of cell wall production by way of chromosomal mutations, as well as the dissemination of the vanA gene via horizontal gene transfer from enterococci. The patterns identified in these strains indicate necessary requirements for continuous surveillance activities in conjunction with effective antibiotic stewardship programs and tailor-made infection control practices that operate at both the local and global levels. The lack of coordinated action across countries will allow vancomycin-non-susceptible *S. aureus* strains to spread faster, which will cause treatment failure and increased disease severity.

4.3.2 Resistance Patterns in Healthcare versus Community Settings

The line between HA-MRSA and CA-MRSA *S. aureus* has become increasingly blurred, with similar traits existing between resistant organisms in the two settings. In one from Mangalore, India, MRSA accounted for 61.4% of the total isolates, CA-MRSA, not just among outpatients but also among almost half of inpatient cases, demonstrating the inroads of the strains from the community into hospital settings. Molecular typing confirmed that CA-MRSA typically carries SCCmec types IV and V and HA-MRSA types I and III. Both strains were resistant to multiple drugs, although CA-MRSA demonstrated comparatively higher resistance to ciprofloxacin, erythromycin, and clindamycin.

In a Barbados investigation, 90% of hospital MRSA segregated CA-MRSA with predominant PVL gene prevalence and vancomycin, rifampin, and linezolid susceptibility. Isolates from the community shared the same resistance patterns, suggesting ongoing transmission between facilities (Gittens-St Hilaire et al, 2020). A Tanzanian hospital-based study found an 8.5% MRSA carriage rate on admission. MRSA showed increased gentamycin, ciprofloxacin, and kanamycin resistance compared with methicillin-sensitive strains. MRSA colonization was more frequent in those with an acute illness, and most of the MRSA isolates were multidrug resistant (Joachim et al, 2017).

4.3.3 Recent Surveillance Data and Case Studies

MRSA continues to pose a serious public health threat, with increasing prevalence observed in both clinical and community settings across multiple regions.

4.3.3.1 United Arab Emirates

A comprehensive 12-year retrospective analysis (2010–2021) utilizing national AMR surveillance data in the UAE revealed an escalating burden of MRSA infections. The study, which included over 29,000 non-duplicate MRSA clinical isolates, indicated a rise in MRSA prevalence among *S. aureus* isolates from 21.9% in 2010 to 33.5% in 2021. Most infections were associated with skin and soft tissue, with the majority occurring in adult male outpatients. Notably, approximately 40% of inpatient MRSA cases were community-acquired. Antimicrobial susceptibility patterns revealed a growing resistance to fluoroquinolones, macrolides, gentamicin, and trimethoprim-sulfamethoxazole. Conversely, linezolid retained low resistance rates (<1% in most years), and no vancomycin-resistant isolates were confirmed (Thomsen et al., 2023).

4.3.3.2 India (Human Infections)

A systematic review and meta-analysis of 98 studies conducted across India from 2015 to 2020 estimated an overall MRSA prevalence in human populations at approximately 37% (95% CI: 32–41%). Significant geographical variability was evident, with the highest state-level prevalence reported in Jammu and Kashmir (55%) and the lowest in Maharashtra (21%). The findings underscore the need for localized surveillance and strict infection control practices to curtail MRSA transmission in healthcare environments (Patil et al, 2022).

4.3.3.3 India (Veterinary Perspective – Bovine Mastitis)

In the context of animal health, particularly bovine mastitis, MRSA is emerging as a critical concern in India. An investigation involving 80 clinical isolates of *S. aureus* from infected dairy cattle across several Indian states found that 25% of these were methicillin resistant. Among MRSA isolates, 95% exhibited multidrug resistance, highlighting the complexity of treatment. The isolates demonstrated significant genetic diversity, with ST-97 being the predominant sequence type. Furthermore, biofilm formation was common and potentially associated with virulence gene presence, complicating both detection and treatment (Kumar et al, 2021)

4.4 EMERGING THERAPEUTIC STRATEGIES

The unyielding growth of antimicrobial-resistant strains of *S. aureus*, such as MRSA and VRSA, has created an alarming and immediate demand for new and effective treatment. The world is moving toward innovative therapeutic strategies that complement traditional pharmacology due to last-resort antibiotics becoming increasingly scarce. These strategies seek to go beyond simply destroying resistant pathogens and aim to halt the progression of resistance by focusing on bacterial mechanisms. Techniques such as nanotechnology-based drug delivery, bacteriophage therapy, next-generation antibiotics, and synergistic drug combinations are on the rise. Collectively, there is great hope that these approaches will mark a paradigm shift in combating multidrug-resistant *S. aureus*.

4.4.1 Nanoparticles for Disruption of Biofilms and Drug Delivery

S. aureus is infamous for its ability to form biofilms, structured bacterial communities encased in a self-produced extracellular matrix that provides a shield against antimicrobial agents and immune response monitoring. Biofilms are particularly problematic for the treatment of chronic infections, especially in the case of hospital-acquired infections, where they develop on implanted medical appliances. Resistance posed by biofilms has been increasingly effectively addressed by nanoparticles. Due to their small size and large surface area, engineered nanoparticles provide better diffusion into biofilms and selective delivery of antimicrobials directly to bacterial cells (Xie et al., 2025).

Many nanoparticles are also demonstrated to exhibit significant anti-biofilm activity, such as:

- **Silver nanoparticles (AgNPs)**: Destroy the microbial membrane, release Ag^+ ions, and create ROS.
- **Zinc oxide (ZnO) and titanium dioxide (TiO_2) nanoparticles**: Inhibit quorum sensing and increase matrix synthesis.
- **Chitosan nanoparticles**: Surface-modified to increase mucoadhesive properties or damage the bacterial membrane.
- **Mesoporous silica nanoparticles (MSNs)**: As carriers of antibiotics (and drug-loaded) with high doses of antibiotics.

These nanoparticles are operational by destabilizing the EPSs, penetrating the biofilm matrix, and delivering their payload in a controlled manner. Several systems are now purposefully designed to respond to pH or enzymes and allow for site-specific drug release. For example, ciprofloxacin-loaded PLGA nanoparticles were proven to be more efficacious in penetrating *S. aureus* biofilms and reducing bacterial load overall (Yathavan et al., 2023).

Second, there are developing combinatorial approaches utilizing bacteriophages and nanoparticles. In this synergetic approach, the NP's mechanical effect and the bacteria's lytic enzyme can have dual effects. Furthermore, you have the area of stimuli-responsive nanoparticles, ultrasound-activated or guided by magnetic nanoparticles, which offer external control of droplets for drug release and are proposed as a way to improve the local efficacy of the drug without systemic toxicity (Xie et al., 2025). Certainly, this area is promising; however, challenges remain in terms of biocompatibility of nanoparticles, potential cytotoxicity and sensitivity, clinical scalability of nanoparticles, and regulatory standardization. However, the use of nanotechnology in antimicrobial therapy is a paradigm shift in leveraging alternatives in combating life-threatening biofilm-associated *S. aureus* infections.

4.4.2 Bacteriophage Therapy

As multidrug-resistant *S. aureus* – and MRSA in particular – continues to pose an increasing threat, bacteriophage therapy is coming to the forefront as a powerful, highly targeted therapeutic option in the new field of antimicrobial therapies. Bacteriophages (aka phages) are viruses which get their name because they exclusively infect and lyse specific species of bacteria, all while retaining the microflora around the biofilm area and providing great options for targeted therapy in polymicrobial infections (Altamirano et al. 2021; Monteiro et al. 2019).

Whereas antibiotic treatments often lose effectiveness as resistance continues to be widespread, phages co-evolve with bacteria, making them a promising option for the persistent treatment of infections. Regarding biofilm infections, phages penetrate and degrade biofilms, which makes them well-suited against chronic and device-related *S. aureus* infections. Continuing to build on viral biology, developing cocktails (multiple lytic phages) increases activity with a broader range and lessens the risk of developing resistance (Monteiro et al. 2019).

Imagine a therapeutic method that apparently multiplies only when it is located in the target, smartly attacking the infection site while leaving the healthy tissue alone. Phages can be delivered in hydrogels, sprayed as aerosols, or injected intravenously, depending on a particular clinical need. Researchers are already looking at using phages in wound dressing or coating of implants, with many of the studies demonstrating considerable reductions in infection rates.

Now there are still obstacles that lie ahead. Regulatory pathways, patient-specific formulations, and the possibility of selected resistance to phage infection in certain bacteria (rather than broad-spectrum resistance), all these aspects require further research. Fortunately, advances in synthetic biology and CRISPR-based phage editing tools are allowing researchers to engineer phages that are both safe and very likely to work (Altamirano et al., 2021; Monteiro et al., 2019).

Therapeutic workflow of bacteriophage therapy against *S. aureus*:

1. Isolation of target bacterial strain
2. Phage screening and selection (specific lytic phages identified from phage libraries)
3. Formulation of phage cocktail (combination of multiple lytic phages to broaden host range)
4. Phage delivery method chosen (topical gel, IV infusion, inhalable spray, etc.)
5. Application to the infection site (phages localize to infected tissues)
6. Bacterial lysis and biofilm disruption (phages replicate, burst bacterial cells, and degrade biofilm matrix)
7. Reduction of bacterial load and inflammation (tissue recovery and healing)
8. Monitoring and adaptive response (if resistance emerges, new phage variants can be introduced)

4.4.3 Novel Antibiotics: Afabicin, Nilofabicin, and Others

The increasing AMR of *S. aureus*, particularly MRSA and VISA/VRSA had led to investigations into the creation of new antibiotics with mechanisms that act on a different bacterial pathway or avoid the current AMR mechanisms. Of the reported next-generation antimicrobials on the market, Afabicin and Nilofabicin, along with newer compounds such as Corbomycin and STK848198, can be classified into a new class and mechanisms of action with clinical significance.

- **Afabicin**: Afabicin (*Debio 1450*) is a selective staphylococcal antibiotic that targets the enoyl–acyl carrier protein reductase (FabI) enzyme involved in bacterial fatty acid biosynthesis. Upon administration, Afabicin is transformed to its active form desphosphono-Afabicin, which inhibits membrane lipid synthesis – a mechanism not found in current classes of antibiotics, therefore, likely to limit resistance (Wittke et al., 2020).

Key Features: Narrow-Spectrum Selectivity: Afabicin specifically targets *Staphylococcus* species. It does not target the commensal bacteria in the gut, allowing it to not only treat the pathogen but also spare the normal flora that provides vital function. Available orally and intravenously: Afabicin has an oral formulation designed to enable prescriber preference and patient convenience. It is orally administered, but dosing and dosage forms are flexible for the patient's needs and tolerability. Low resistance potential: The role of FabI in *S. aureus* is essential and unique, so spontaneous problems with resistance are rare. Clinical performance: Afabicin is shown to be non-inferior to vancomycin/linezolid for the treatment of acute bacterial skin and skin structure infections in phase II clinical trials with less gastrointestinal toxicity (Wittke et al., 2020).

This precision approach contributes to antibiotic stewardship efforts by preventing unnecessary disruption to the microbial flora and facilitating the reduction of selection pressure for resistance.

- **Nilofabicin**: Nilofabicin (CG400549) is another FabI inhibitor, less structurally similar to Afabicin than the other FabI inhibitors, which target, likewise, the biosynthetic pathway. Created in South Korea, nilofabicin has been beta-tested preclinically and in early-phase clinical studies after being developed to treat MRSA (*CrystalGenomics, n.d.*).

Key features: Oral bioavailability: It has a great pharmacokinetic profile for outpatient or step-down therapy. Very potent against MRSA: Suggested to have in vitro and *in vivo* activity toward MRSA cells that were resistant to standard monotherapies (linezolid and daptomycin). Pipeline status: Still in clinical trials, but early data suggest safety and tolerance (CrystalGenomics, n.d.).

Nilofabicin is developing directly the same Fabi drug receptor as Afabicin but has a different chemical structure and pharmacodynamics that may allow for rotation therapy or resistance mitigation.

- **Other Potential New Antibiotics**: There are many more antibiotics at different stages of discovery and development that exhibit novel mechanisms of action against *S. aureus* and its resistant phenotypes.
- **Corbomycin and Complestatin**: These glycopeptide antibiotics inhibit autolysins (bacterial enzymes which remodel the bacterial cell wall when the bacteria grow and divide) by a different mode of action than standard drugs such as vancomycin (Culp et al., 2020). Non-traditional mechanism: These agents block autolysins, which mediate the turnover of the cell wall, not the synthesis pathways, which do not depend on transpeptidase inhibition and, therefore, inhibit a different essential physiological function. Examples of clinically relevant efficacy against resistant strains: MRSA and biofilm-forming *S. aureus* isolates. Ability to disrupt biofilm: forming substances make them very useful to treat chronic and device-related infections (Culp et al., 2020). These findings push the boundaries of functional classes of antibiotics outside of structural families and provide optimism for treating resistant and chronic infections.

STK848198: Recognized as a result of high-throughput screening (HTS), STK848198 has shown strong bactericidal activity against *S. aureus* and MRSA; however, the mechanism is still unknown.

Key points: Synergistic potential: It has shown synergy with β-lactams and glycopeptides, which may prove useful in combination therapy with those antibiotics. Resistance prevention: Its level of activity across serial passages would indicate a small chance of resistance induction. New structural class: It also provides unique and pharmacological diversity as a new structural class.

Although still early in the research phase, they demonstrate the potential influence that genomics and computational technologies can have on the discovery and development of antimicrobials.

4.4.4 Drug Combinations and Adjuvant Therapies

The emergence of multidrug-resistant (MDR) *S. aureus*, including MRSA and VRSA strains, has revealed the limitations of monotherapy and created interest in novel treatment approaches. Increasingly explored, drug combination therapy and adjuvant therapies are two likely strategies to achieve the restoration or enhancement of antibiotic efficacy through the simultaneous targeting of multiple mechanisms of bacterial survival. These two modalities are being heavily researched for complicated infections and their ability to expand the spectrum of activity of existing antibiotics (Worthington et al., 2013; Liu et al., 2019; Dhanda et al., 2023).

4.4.4.1 Drug Combination Therapies

Drug combinations exploit the power of two or more antimicrobials with different mechanisms of action; their total bactericidal activities are superior to their individual actions when combined in specific ways. They are designed to:

- Overcome resistance by simultaneously attacking different bacterial targets
- Reduce the selective pressure on individual antibiotics
- Reduce the chance of resistance developing

Some examples of combinations that are effective include:

- β-lactams + β-lactamase inhibitors → Amoxicillin/clavulanic acid: clavulanic acid inhibits β-lactamases and gives amoxicillin the opportunity to target cell wall synthesis. → Common in treating penicillinase-producing *S. aureus* strains.
- Trimethoprim + Sulfamethoxazole → Work synergistically by inhibiting sequential steps in the pathway of folate synthesis. → Has been around for years in treating community-acquired MRSA infections.
- Ceftazidime + Avibactam → Avibactam inhibits serine β-lactamases and expands the spectrum of ceftazidime against MDR organisms (Liu et al., 2019).
- Fluoroquinolones + Efflux Pump Inhibitors (EPIs) → EPIs such as reserpine will increase the intracellular accumulation of the drug and reverse resistance.
- Polymyxins + Rifampicin or Carbapenems → Disrupts the bacterial membranes such that any secondary antibiotic is better positioned to reach the intracellular target.

These combinations have provided better clinical outcomes, especially with polymicrobial infections and resistant *S. aureus*. However, it is prudent to check the pharmacokinetics and toxicity of each combination to avoid antagonism or cumulative drug effects.

4.4.4.2 Adjuvant Therapies

Adjuvants are compounds that are not antibiotics but improve the effectiveness of antibiotics without acting as antimicrobials. Adjuvants target mechanisms of AMR, bacterial virulence factors, or permeability barriers.

There are different types of adjuvant action:

- **Efflux Pump Inhibitor**: Verapamil and PAβN inhibit the active expulsion of antibiotics from the bacteria and restoring antibiotic susceptibility in the presence of efflux (Liu et al., 2019).
  Adjuvants that function as efflux pump inhibitors are ideal for the treatment of infections with biofilm-associated resistance. Quorum-Sensing Inhibitor: Furanones and ajoene disrupt the bacterial communication between pathogens, which alters toxin production and virulence (Dhanda et al., 2023). This quorum-sensing inhibition reduces pathogenicity and increases susceptibility to antibiotics. Membrane Permeabilization: EDTA is used to chelate divalent cations as a means to destabilize the outer membrane of gram-negatives. Continued work on adjuvants continues with evidence that EDTA increases the penetration of large or hydrophilic antibiotics. Biofilm Disruption: The biofilm extracellular matrix can be degraded with agents like DNase and dispersin B, which will expose the biofilm bacteria to the antibiotics. Demonstrates efficacy in device-associated infections.
- **Enzymatic inhibition**: β-lactamase inhibitors, like clavulanic acid, sulbactam, and tazobactam, inhibit the β-lactamase enzymes, which in turn provide some level of β-lactam efficacy in resistant strains (Worthington et al., 2013).

These adjuvants may be used in topical formulations, inhalation therapies, or intravenous cocktails, particularly in chronic and biofilm-associated infections. There is ongoing pharmaceutical innovation in their development (Table 4.3).

This strategy, with layers of combinations and adjuvants, not only restores the utility of these legacy antibiotics but also brings a multifactorial defense against the highly adaptable bacterial pathogens we face. Over time, as resistance continues to evolve, these combinations will provide adaptive, mechanism-based responses.

4.5 TECHNOLOGICAL INNOVATIONS IN ANTIMICROBIAL DISCOVERY

4.5.1 AI in Drug Screening and Design

The field of antimicrobial drug discovery has seen a revolution thanks to AI. Quickly and accurately analyzing large biological datasets improves the screening, design, and optimization of new compounds. The drug development process can be streamlined by using machine learning (ML)

TABLE 4.3
Comparison of Common Drug Combinations and Adjuvant Therapies

Therapy Type	Agents Involved	Mechanism	Targeted Resistance	Clinical Use
Combination therapy	Amoxicillin + clavulanic Acid	β-lactamase inhibition	Penicillinase resistance	Respiratory, skin infections
Combination therapy	Trimethoprim + sulfamethoxazole	Folate synthesis blockade	MRSA, CA-MRSA	UTIs, SSTIs
Combination therapy	Ceftazidime + avibactam	Inhibition of ESBLs and AmpC β-lactamases	MDR gram-negatives	Hospital-acquired infections
Adjuvant therapy	Reserpine + ciprofloxacin	Efflux pump inhibition	Fluoroquinolone-resistant strains	Experimental
Adjuvant therapy	Furanones	Quorum-sensing inhibition	Biofilm-associated resistance	Device-related infections
Adjuvant therapy	EDTA	Membrane permeabilization	Outer membrane barrier	Wound irrigation, adjunct IV therapy
Adjuvant therapy	Clavulanic acid	β-lactamase inhibition	β-lactam-resistant *S. aureus*	Combined with penicillins

algorithms to forecast the antibacterial activity, toxicity, and resistance potential of drug candidates (Thalia, 2025). Finding new antimicrobial peptides and compounds is a specialty of deep learning models. Millions of molecules can be screened against bacterial targets using AI-driven methods to find the most effective and least harmful ones. Additionally, using generative models like variational autoencoders and generative adversarial networks, AI systems can create new molecules from scratch that are specific to bacterial targets (Olatunji et al., 2024).

Multi-parameter optimization is made possible by the combination of AI with other technological platforms, such as bioinformatics and HTS. Potential treatment candidates for resistant pathogens, such as MRSA and VRSA, can be found with the aid of platforms like IBM Watson and Atomwise, which use neural networks trained on chemical and biological datasets to predict molecular interactions (Köser et al., 2014). Despite its potential, issues with AI-driven drug discovery include clinical validation, algorithm transparency, and data quality. However, the application of AI to antimicrobial research is speeding up the process of finding promising drug candidates, which lowers development time and cost.

4.5.2 Genomic and Proteomic Approaches

Finding new therapeutic targets and understanding the molecular causes of AR depend heavily on genomic and proteomic technologies. While proteomics examines protein expression and interaction patterns that affect bacterial survival and drug susceptibility, genomics uses bacterial DNA sequencing to comprehend genetic determinants of resistance (Matias et al., 2024).

Real-time surveillance and diagnostic development are supported by whole genome sequencing, which enables researchers to pinpoint the mutations and gene acquisitions causing resistance phenotypes. These insights track the evolution of resistant strains and guide customized treatments (Köser et al., 2014). By identifying functional proteins involved in resistance, such as efflux pumps, beta-lactamases, and modified penicillin-binding proteins, proteomics enhances genomics. Proteomics using mass spectrometry provides high-resolution analysis of these proteins, making it possible to find druggable targets and biomarkers.

Our understanding is further enhanced by transcriptomics and metagenomics, which examine gene expression profiles and microbial communities under drug pressure. These methods enable the functional validation of resistance genes and pathways when paired with CRISPR-Cas9 genome editing tools (Olatunji et al., 2024).

Innovations in genomics and proteomics are propelling the creation of precision antibiotics and next-generation diagnostics, greatly improving our capacity to fight bacteria that are resistant to multiple drugs.

4.5.3 HTS and Target Validation

A potent technique for quickly assessing thousands of compounds against particular bacterial targets is HTS. Because it enables researchers to quickly identify promising lead compounds, it has become crucial in the early stages of antimicrobial drug discovery (Baker et al., 2018). Contemporary HTS platforms use data analytics, microfluidics, and robotics to evaluate antimicrobial activity in a simplified, automated manner. These platforms check for a variety of mechanisms of action, including DNA replication, protein synthesis, and cell wall synthesis inhibition. AI integration reduces false positives and improves hit selection accuracy.

Target validation guarantees that biologically relevant targets interact with identified leads. The importance of these targets for bacterial survival is confirmed by functional genomics, RNA interference (RNAi), and CRISPR-based techniques. Target–ligand interactions can be understood at the atomic level through structure-based drug design, which is guided by crystallography and NMR spectroscopy (Baker et al., 2018). Furthermore, HTS and cellular imaging are combined in high-content screening to track phenotypic reactions and help identify compounds with unique mechanisms. These tools greatly improve the efficacy and dependability of antimicrobial drug discovery when paired with omics data (Table 4.4).

TABLE 4.4
Summary of Technological Innovations in AMR Research

Technology	Purpose	Applications	Reference
Artificial intelligence	Predict drug efficacy and design	Drug screening, toxicity prediction, molecule design	Thalia (2025); Olatunji et al. (2024)
Genomics and proteomics	Identify resistance genes/proteins	Target discovery, diagnostics, surveillance	Matias et al. (2024); Köser et al. (2014)
High-throughput screening (HTS)	Rapid compound screening	Lead identification, mechanism exploration	Baker et al. (2018)

4.6 CLINICAL AND ETHICAL CONSIDERATIONS

4.6.1 Challenges in Clinical Implementation

The clinical use of the methods for regulating MRSA is also made difficult by the sophistication of AMR. One of the distinguishing challenges is the premature detection of MRSA cases, especially in cases where empirical treatment must be initiated before the ultimate microbiological reports. A study conducted in the United States illustrated how inappropriate empiric treatment not only extends stay in hospital but potentially also endangers results, making predictive measures like bedside risk scores necessary to inform early treatment (Zilberberg et al., 2012). Yet another significant obstacle is the occurrence of CA-MRSA breaching hospital environments. A 12-year surveillance study in the UAE documented a gradual increase in CA-MRSA occurrence within the healthcare environment. Over one-third of inpatient MRSA infection was characterized as community-acquired, which reflected a lacuna in early detection and infection control within the community (Thomsen et al., 2023). Diagnostic latencies and local resistance pattern heterogeneities remain significant barriers in countries such as India, where MRSA prevalence ranges between 21% and 55% by state. Inefficient diagnostic capacity in rural and inadequately equipped healthcare facilities may complicate effective antimicrobial selection and exacerbate resistance phenomena (Patil et al., 2022). AMR is a multifaceted and dynamic worldwide health emergency that requires strong surveillance and stewardship systems. Proper containment of resistant organisms such as MRSA and carbapenem-resistant bacteria relies not only on clinical management but also on systematic monitoring and rational antibiotic use.

4.6.2 Resistance Monitoring Systems

Monitoring of AR is the cornerstone of resistance trends understanding and empirical treatment decision-making. Yet global surveillance networks tend to be patchy, differing in their scope, organization, and effectiveness. A systematic review found 71 AR surveillance systems worldwide, the majority of which are national in scope and based in Europe and North America. The systems largely track resistance in humans, but with minimal integration between animal and environmental health sectors, contrary to recommendations for a One Health approach. In an extended scoping review of integrated surveillance systems, Delpy et al. (2024) highlighted that surveillance today is largely limited to high-income nations. Although initiatives like GLASS (Global Antimicrobial Resistance and Use Surveillance System) and DANMAP (Danish Integrated Antimicrobial Resistance Monitoring and Research Programme) are exemplary examples, operational integration between human, animal, and environmental health is still limited. Major challenges are variations in laboratory capacities, non-harmonized methodologies, and a lack of adequate intersectoral governance (Delpy et al., 2024)

Antimicrobial Stewardship Programs (ASPs): ASPs are powerful weapons in reducing inappropriate uses of antibiotics and maintaining the potency of last-line agents like polymyxins and carbapenems. ASPs are especially important in low- and middle-income countries, where the misuse of broad-spectrum antibiotics is more prevalent owing to inadequate diagnostics and limited resources. ASPs use a range of interventions, such as prospective audit and feedback, formulary limitation, education programs, and use of cumulative antibiograms to facilitate earlier de-escalation of antibiotics. Utilization of biomarkers such as procalcitonin is also increasingly being used as a strategy to optimize the duration of treatment. Institutional dedication, departmental cooperation, and ongoing training of medical professionals, however, are essential to the success of these initiatives. Additionally, stewardship interventions need to consider the pharmacokinetics and local resistance patterns of alternative drugs. For example, minocycline and sulbactam have been recognized as potential alternatives for the management of multidrug-resistant *Acinetobacter baumannii* in environments where polymyxin supply is limited or toxicity is problematic.

4.6.3 Ethical Considerations

Ethically, MRSA management combines with concepts of fairness, justice, and stewardship. As (Adebisi, 2023) indicates, equitable access to effective antibiotics is a moral imperative, particularly in impoverished environments where economic and infrastructural constraints restrict treatment access. Rural communities, migrants, and people without access to conventional healthcare are most severely impacted by both under-treatment and over-exposure to antimicrobials (Adebisi, 2023). Nonmaleficence and beneficence moral principles also demand judicious use of antibiotics to avoid harm through the induction of resistance while facilitating patient recovery. The overuse of antibiotics, although generally beneficial, can inadvertently cause long-term harm through promoting resistance, calling for ethical risk–benefit analysis in treatment decisions. Autonomy is also in question when it comes to MRSA treatment, as patients must be appropriately informed and involved in decision-making regarding their treatment. Autonomy must be balanced, however, against public health interests, particularly through antimicrobial stewardship programs encouraging appropriate use of antibiotics. These types of programs, while necessary, must be ethically developed so they don't abridge patient choice of treatment or intrude upon informed consent. Finally, the global imbalance in research spending and drug discovery is an ethical imperative. Nations with lean health budgets disproportionately carry the AMR, including MRSA, cost but commonly end up with underinvestment in the production of new antibiotics and detection networks. The imbalance translates to a call for regionally contextualized global action by way of financial support and technical support to underfunded health systems (Adebisi, 2023).

4.7 FUTURE PERSPECTIVES AND RECOMMENDATIONS

AMR continues to pose a significant global health threat that could undo decades of progress. The WHO has outlined AMR as one of ten global public health threats to humanity. The uptick in multidrug-resistant organisms calls for novel and integrative approaches in the management of this crisis. Future efforts must be varied in technology, policies, and intersectoral activities.

4.7.1 Integrative Approaches for Combating Resistance

- **Advancements in Drug Discovery and Development**: Despite the potential for viable new antimicrobial products to emerge, the traditional antibiotic development pipeline has lacked pace, monetary inappetence, and resulting stagnateable production of new antimicrobial agents. To alleviate the stagnation, we are starting to see integrative methods using AI and ML, which can expedite drug discovery methods and products. These methods, by design, allow for the examination of unbelievable amounts of data to identify potential antimicrobial compounds, predict efficacy for previously designed and structurally characterized compounds, and optimize chemical structures in novel antimicrobial compounds for activity (Szymczak et al., 2023). For example, AI-based bioinformatics platforms and models have been very successful at identifying novel antimicrobial peptides as active against resistant pathogens, with potent interaction (Szymczak et al., 2023). There have been ample opportunities through nanotechnology for innovative methods for use of antibiotic therapy. Due to established engineering advances, nanoparticles can transport antibiotics to the site of infection with perfect concentration that is delivered locally, reducing systemic effects of drug therapy. Additionally, nanomaterials used in these advances offer unique antimicrobial properties that disrupt bacterial cell membranes and biofilms, which enhance the activity of existing antibiotic therapy (Chahar et al., 2023).

- **Alternative Therapeutic Modalities**: In addition to traditional antibiotics, other new therapies are of interest as potential methods for addressing AMR. Bacteriophage therapy, which is a therapy with viruses that specifically infect and lyse bacterial cells, has been shown to be effective in treating infections from multidrug-resistant bacteria. Phages can be directed to specific bacterial strains, allowing a reduction in collateral damage to the patient's microbiota. Clinical trials and compassionate use examples on phage therapy have shown effectiveness against refractory infections (Van den Biggelaar et al., 2024).

  Immunotherapy, for example, monoclonal antibodies and vaccines, is also being examined as methods to improve the host immune response to resistant pathogens. Vaccines can block infections from occurring and hence limit the use of antimicrobials and the development of antimicrobial resistance. Vaccines targeting resistant organisms (i.e., *S. aureus* and *P. aeruginosa*) are being evaluated, and some vaccine candidates have demonstrated promising results in preclinical testing (Bergman et al., 2020).
- **Implementation of Antimicrobial Stewardship Programs**: Antimicrobial stewardship programs (ASPs) are essential in promoting the rational use of antibiotics. ASPs include guidelines for appropriate prescribing, practitioner and patient education, and monitoring systems to assess antibiotic usage and resistance trends. Through endorsing optimal and safe antibiotic use, ASPs conserve the effectiveness of current antibiotics and minimize the risk of resistance development (Majumder et al., 2020).
- **Environmental and Agricultural Interventions**: While there is limited awareness of the environmental aspect of AMR, humans release antibiotics and resistant bacteria in the environment via agricultural runoff, wastewater and protein waste, and industrial discharge. Improved control over antibiotic usage in agriculture and better treatment methods for wastewater are necessary to diminish adverse environmental exposure and subsequent resistance development (Berendonk et al., 2022).
- **Global Policy and Collaborative Efforts**: AMR warrants coordinated, global responses from all sectors, including decision-makers, healthcare providers, researchers, and the wider public. Diverse international initiatives, like the WHO's Global Action Plan on AMR, are underway to streamline initiatives across different geographies, incentivizing surveillance, fostering research, and endorsing sustainable practices and policies. Various economic approaches, such as market entry rewards or public–private collaboration, can stimulate research and development of new antibiotics while ensuring that there will be viable, effective options available (Kessel, 2024).

4.7.2 Policy and Regulatory Measures

AMR is a serious and growing threat to global health security as it demands coordinated and multi-sectoral action. Strong policies and regulatory frameworks are key to dealing with the crisis by accepting responsible antimicrobial use, systems for surveillance, research innovation, and collaboration globally. This section looks at a range of relevant policy initiatives and regulatory approaches being proposed and implemented at a global level and at national levels to combat AMR, including successes and challenges that need immediate attention.

- **Global Policy Frameworks**: Global policies are a useful base for national governments' and health systems' further action. A very evident example has support from the WHO's 2015 Global Action Plan on Antimicrobial Resistance, with important resonance.

  The plan included five ways:

  1. Enhance awareness and understanding of AMR
  2. Strengthen surveillance communities and research
  3. Reduce the occurrence of infection by improving sanitation and hygiene
  4. Optimize the use of antimicrobial agents in human health and animal health
  5. Maintain a global investment in new medicines, new diagnostics, and new vaccines (World Health Organization et al., 2023)

  Furthermore, the One Health framework, supported by WHO, FAO, and OIE, acknowledges the links between human, animal, and environmental health, which is significant because one important means of resistance is antibiotic overuse and misuse in agriculture. Through global multi-species surveillance systems (e.g., WHO's Global Antimicrobial Resistance Surveillance System (GLASS)), member countries will contribute and share their respective data on AMR, which will ultimately better inform evidence-based policymaking for AMR (World Health Organization et al., 2023).
- **National Strategies and Regulations**: Countries across the globe have been aligning with global frameworks by creating and implementing their own National Action Plans (NAPs). In India, the NAP on AMR (2017–2021) was established with the aim of improving antimicrobial use in health and agriculture. The following was included from India's action plan:

  - Limiting antibiotic distribution to once daily over-the-counter sales to directly limit self-medication misuse
  - Creating and enforcing rational use directives in hospitals
  - Training healthcare professionals on stewardship practice
  - Engaging in infection prevention and control occupational services in healthcare (*Government of India, 2017*).

  A regional example is Kerala, which implemented a state-based strategy on prescription restrictions, educational, and tracing systems. These localized strategies show how targeted governance can facilitate behavioral and organizational change (*Government of India, 2017*).
- **Incentivizing Research and Development**: The absence of new antibiotics being launched into the market creates a large barrier for AMR work. From a policy perspective, driving much-needed innovation is crucial, specifically

  - **Public–Private Partnerships (PPPs)**: Working with the pharmaceutical industry to help limit risk and reduce costs of R&D
  - **Market Entry Rewards**: Compensating the pharmaceutical companies which have had success in launching new antibiotics
  - **Increased Market Exclusivity**: Longer patent protection for new antimicrobial products
  - **Regulatory Fast-Tracking**: Fast-tracking processes to expedite new product availability (Smith et al., 2022)

  These steps should help improve the attractiveness of the pharmaceutical innovation environment in a low-profit margin space.

- **Monitoring and Surveillance**: An efficient surveillance infrastructure is essential for AMR containment. Governments are working to:

  - Build robust national monitoring systems to track AMR trends in real time
  - Mandate reporting of resistance patterns in hospitals and clinics
  - Integrate AMR databases with digital health records for seamless data flow
  - Share findings with international bodies like GLASS for coordinated response (World Health Organization, 2023)

  Policy interventions also include enhancing laboratory capacity, improving diagnostic accuracy, and enabling access to rapid diagnostic tests, all of which are critical for informed treatment choices and containment of resistant strains.

4.7.3 Public Awareness and Global Collaboration

The world's crisis of AMR has come to a boiling point, where awareness raising and collaboration are no longer optional; they form an essential crux. Although scientific innovation and medical policymaking are paramount, we cannot disregard the role that public behavior has on AMR, and certainly we cannot forget how being aligned as countries has been helpful too. The basis for the containment of AMR will rely on communities understanding the threat and countries finding a way to collaborate to tackle it with shared resources, data, and responsibility.

- **Driving Public Awareness: Education as a Frontline Tool:** The importance of public awareness is typically the first step in containing the spread of resistance. Misuse and overuse of antibiotics – particularly via self-medication and not completing a prescription – is one of the strongest promoters of resistance around the world. Therefore, awareness raising at any level of society, from rural villages to urban populations, is not a nice thing to do but a core element.

  - National educational campaigns on responsible antimicrobial use that affect schools, colleges, and community centers can be the basis for normalizing the societal culture shift toward a change in the way we view responsible antimicrobial use.
  - Social media channels – especially in low- and middle-income countries – have shown to be an effective and scalable approach to engage younger populations in conversation about AR, quite rapidly.
  - Gaining the attention of community health workers and pharmacists to be potential educators and gatekeepers to discourage all forms of over-prescription and over-the-counter misuse could be valuable once trained (Laxminarayan et al., 2020).

  One clear example is World Antibiotic Awareness Week (WAAW), which occurs in November every year. WAAW brings government, NGOs, healthcare administrators and practitioners, and the general public together as a collective under the umbrella of stewardship and education to drive behavior change associated with the proper use of antibiotics (World Health Organization, 2022).

  It may also be useful to enlist influential people, like faith-based leaders, celebrities, or local politicians, to help emphasize the urgency of AMR and connect with the communities they serve. In an environment where social influence is prevalent and skepticism of science and trust in scientific institutions wanes, they can help amplify the messaging.
- **Strategic Global Collaboration: Breaking the National Barriers:** AMR is global in scope, and no nation alone can tackle the problem. Bacteria that are resistant to antibiotics do not observe borders; they travel with people, food, and animals. Therefore, the problem requires a unified global response, which begins with sharing data, harmonizing standards, and aligning investment priorities (O'Neill et al., 2016). For example:

- The WHO has developed the GLASS to help create data sharing between countries and support the identification of hotspots of resistance to develop sustainable mitigating strategies and interventions (World Health Organization, 2022).
- Initiatives to leverage research, such as the Combating Antibiotic-Resistant Bacteria Biopharmaceutical Accelerator (CARB-X), finance and bring together in-kind support from countries such as the United States, the United Kingdom, and Germany to support early antibiotic-related research.
- Offering cross-border training programs on antibiotic data collection for healthcare workers and microbiologists can introduce consistency to diagnostic and stewardship methods, especially in regions where a health system may be under-resourced (O'Neill et al., 2016).

The One Health paradigm facilitates collaboration between human, animal, and environmental health systems in an interrelated way. Countries leveraging the One Health concept have made progress in better harmonizing policy among the agriculture, pharmaceutical, and clinical health sectors (Laxminarayan et al., 2020). For example:

- Use of antibiotics in livestock is being limited through international agreements, through organizations like the FAO/OIE/WHO tripartite partnership, to reach a standard in monitoring antibiotic use in food-producing animals.
- Environmental monitoring for antimicrobial pollutants in waters and near pharmaceutical manufacturing sites has gained traction within joint environmental health regulatory frameworks (Laxminarayan et al., 2020).

Global funding for AMR through new mechanisms such as the AMR Multi-Partner Trust Fund is also assisting low- and middle-income countries with enhancing capabilities for surveillance and response. These prospects are critical in contexts where antimicrobial overuse and misuse often coincide with poor health systems infrastructure and access to diagnostics.

- **Embedding Awareness into Policy and Culture**: Awareness and action cannot be occasional campaigns; they must be a part of the culture and policy of the country. This will take political will, sustained investment, and interdisciplinary coordination.

  - AMR education in healthcare should be integrated into the curricula as basic education for health professionals, including physicians, veterinarians, and pharmacists.
  - Public health measures should stipulate regular awareness campaigns, with a focus on outbreaks when inappropriate antibiotic use can be expected to rise.
  - National drug authorities could go into partnership with the media houses to promote joint message campaigns and information sharing on AMR.

The importance of translating AMR knowledge into non-scientific areas – economics, education, and employment matter! The consequences of AMR – increased healthcare costs, lower productivity, and the end of everyday surgery – need to be explained to every level of society. Such interdisciplinary understanding can help underpin sustainable, inclusive, and globally aligned, whole-of-society responses to AMR.

4.8 CONCLUSION

The ongoing emergence of AMR in *S. aureus*, especially in MRSA, VISA, and VRSA types, is a global public health issue. In this chapter, complex mechanisms of *S. aureus* resistance, including the gene mutation of the mecA and vanA genes, efflux pump function, biofilm formation, and horizontal gene transfer, have all been elaborated to confer high resistance and treatment complexity.

The epidemiological data indicate an explosive rise of multidrug-resistant strains within the health and community settings. CA-MRSA, being more virulent and versatile, contributes to the control challenge. Surveillance data from around the world repeatedly confirm the global spread of resistant *S. aureus* strains, emphasizing the need for enhanced diagnostic techniques, local surveillance, and international collaboration.

In consideration of these issues, new therapeutic strategies are in the pipeline. Nanotechnology, phage therapy, and new antibiotics like afabicin and nilofabicin promise to circumvent traditional resistance pathways. AI, HTS, and genomics are revolutionizing drug discovery, enabling more, faster, and more targeted creation of antimicrobials. But technology is not sufficient on its own. Effective ASPs, ethical policy agendas, and intersectoral action are needed to ensure equal access, responsible use of antibiotics, and sustainable management of resistance. Educational campaigns and global cooperation must be supplemented to enable behavior change as well as to enhance global action.

In essence, the war against AMR in *S. aureus* has to be a synergistic effort that encompasses scientific expertise, clinical watchfulness, policy compliance, and public participation. By collaborative action over an extended duration, the rising wave of resistance can be stopped, with the effectiveness of life-saving antibiotics guaranteed for future generations.

REFERENCES

Abebe, E., & Birhanu, A. (2023). Methicillin-resistant *Staphylococcus aureus*: Molecular mechanisms underlying drug resistance development and novel strategies to combat. *Infection and Drug Resistance, 16*, 7642–7657. https://doi.org/10.2147/IDR.S428103

Adebisi, Y. A. (2023). Balancing the risks and benefits of antibiotic use in a globalized world: The ethics of antimicrobial resistance. *Globalization and Health, 19*(1), 27.

Altamirano, F. L. G., Forsyth, J. H., Patwa, R., & Sabharwal, D. (2021). Bacteriophage-based therapies for *Staphylococcus aureus* infections: Challenges and opportunities. *Current Opinion in Pharmacology, 60*, 244–251. https://doi.org/10.1016/j.coph.2021.09.004

Baker, S. J., Payne, D. J., Rappuoli, R., & De Gregorio, E. (2018). Technologies to address antimicrobial resistance. *Proceedings of the National Academy of Sciences, 115*(51), 12887–12895. https://doi.org/10.1073/pnas.1717160115

Berendonk, T. U., Manaia, C. M., Merlin, C., Fatta-Kassinos, D., Cytryn, E., Walsh, F., . . . Martinez, J. L. (2022). Tackling environmental antimicrobial resistance: The need for coordinated action. *The Lancet Infectious Diseases, 22*(2), e20–e28.

Bergman, P., Andersson, J., & Björkhem-Bergman, L. (2020). Is vitamin D an antimicrobial agent? A review of the literature. *Antimicrobial Resistance & Infection Control, 9*(1), 1–7.

Centers for Disease Control and Prevention. (2019). *Antibiotic resistance threats in the United States*, Retrieved from https://www.cdc.gov/drugresistance/pdf/threats-report/2019-ar-threats-report–508.pdf

Chahar, K., Sharma, Y., Patel, P., Asati, V., & Das Kurmi, B. (2023). A mini-review on recent strategies and applications of nanomedicines to combat antimicrobial resistance. *CurrentDrug Metabolism, 24*(6). https://doi.org/10.2174/1389200224666230731093319

Chambers, H. F., & DeLeo, F. R. (2009). Waves of resistance: *Staphylococcus aureus* in the antibiotic era. *Nature Reviews Microbiology, 7*(9), 629–641. https://doi.org/10.1038/nrmicro2200

Cosgrove, S. E., Sakoulas, G., Perencevich, E. N., Schwaber, M. J., Karchmer, A. W., & Carmeli, Y. (2005). Comparison of mortality associated with methicillin-resistant and methicillin-susceptible *Staphylococcus aureus* bacteremia: A meta-analysis. *Clinical Infectious Diseases, 36*(1), 53–59. https://doi.org/10.1086/345476

CrystalGenomics. (n.d.). *Nilofabicin – CrystalGenomics*. AdisInsight. Retrieved from https://adisinsight.springer.com/drugs/800021715

Culp, E., Waglechner, N., Wang, W., Fiebig-Comyn, A., Hsu, Y.-P., Koteva, K., . . . Wright, G. D. (2020). Evolution-guided discovery of antibiotics that inhibit peptidoglycan remodelling. *Nature, 578*(7796), 582–587. https://doi.org/10.1038/s41586-020-1990-9

David, M. Z., & Daum, R. S. (2010). Community-associated methicillin-resistant *Staphylococcus aureus*: Epidemiology and clinical consequences. *Clinical Microbiology Reviews, 23*(3), 616–687. https://doi.org/10.1128/CMR.00081-09

Delpy, L., Astbury, C. C., Aenishaenslin, C., Ruckert, A., Penney, T. L., Wiktorowicz, M., . . . Bordier, M. (2024). Integrated surveillance systems for antibiotic resistance in a One Health context: A scoping review. *BMC Public Health*, *24*(1), 1717.

Dhanda, G., Sandhu, R., Kumar, R., & Kaur, J. (2023). Combating multidrug-resistant *Staphylococcus aureus*: Innovative strategies and emerging therapeutics. *Frontiers in Microbiology*, *14*, 1179289. https://doi.org/10.3389/fmicb.2023.1179289

Gittens-St Hilaire, M. V., Chase, E., & Alleyne, D. (2020). Prevalence, molecular characteristics and antimicrobial susceptibility patterns of MRSA in hospitalized and nonhospitalized patients in Barbados. *New Microbes and New Infections*, *35*, 100659.

Government of India. (2017). *National action plan on antimicrobial resistance (NAP-AMR) 2017–2021*. Ministry of Health and Family Welfare. https://ncdc.gov.in/WriteReadData/l892s/File622.pdf

Guay, G. G., Khan, S. A., & Rothstein, D. M. (1993). The tet(K) gene of plasmid pT181 of *Staphylococcus aureus* encodes an efflux protein. *Plasmid*, *30*(2), 163–166. https://doi.org/10.1006/plas.1993.1053

Howden, B. P., Davies, J. K., Johnson, P. D., Stinear, T. P., & Grayson, M. L. (2010). Reduced vancomycin susceptibility in *Staphylococcus aureus*, including vancomycin-intermediate and heterogeneous vancomycin-intermediate strains: Resistance mechanisms, laboratory detection, and clinical implications. *Clinical Microbiology Reviews*, *23*(1), 99–139. https://doi.org/10.1128/CMR.00042-09

Joachim, A., Moyo, S. J., Nkinda, L., Majigo, M., Mmbaga, E., Mbembati, N., . . . Lyamuya, E. F. (2017). Prevalence of methicillin-resistant *Staphylococcus aureus* carriage on admission among patients attending regional hospitals in Dar es Salaam, Tanzania. *BMC Research Notes*, *10*, 1–7.

Katayama, Y., Ito, T., & Hiramatsu, K. (2000). A new class of genetic element, staphylococcal cassette chromosome MEC, encodes methicillin resistance in *Staphylococcus aureus*. *Antimicrobial Agents and Chemotherapy*, *44*(6), 1549–1555. https://doi.org/10.1128/AAC.44.6.1549-1555.2000

Kessel, M. (2024). Actions to end the crisis of antimicrobial resistance. *Financial Times*. Retrieved from https://www.ft.com/content/92670b8d-c660-4719-9127-5e68c6357247

Köser, C. U., Ellington, M. J., & Peacock, S. J. (2014). Whole-genome sequencing to control antimicrobial resistance. *Trends in Genetics*, *30*(9), 401–407. https://doi.org/10.1016/j.tig.2014.07.003

Kumar, R., Yadav, B. R., & Singh, R. S. (2021). Molecular fingerprinting of bovine mastitis-associated *Staphylococcus aureus* isolates from India. *Veterinary World*, *14*(8), 2108–2115

Laxminarayan, R., Matsoso, P., Pant, S., Brower, C., Røttingen, J. A., Klugman, K., & Davies, S. (2020). Access to effective antimicrobials: A worldwide challenge. *The Lancet*, *387*(10014), 168–175. https://doi.org/10.1016/S0140-6736(15)00474–2

Liu, C., Bayer, A., Cosgrove, S. E., Daum, R. S., Fridkin, S. K., Gorwitz, R. J., . . . Chambers, H. F. (2019). Clinical practice guidelines by the Infectious Diseases Society of America for the treatment of methicillin-resistant *Staphylococcus aureus* infections in adults and children. *Clinical Infectious Diseases*, *52*(3), e18–e55. https://doi.org/10.1093/cid/ciq146

Lowy, F. D. (2003). Antimicrobial resistance: The example of *Staphylococcus aureus*. *The Journal of Clinical Investigation*, *111*(9), 1265–1273. https://doi.org/10.1172/JCI18535

Majumder, M. A. A., Rahman, S., Cohall, D., Bharatha, A., Singh, K., Haque, M., & Gittens-St Hilaire, M. (2020). Antimicrobial stewardship: Fighting antimicrobial resistance and protecting global public health. *Infectious Diseases: Research and Treatment*, *13*, 1178633720909390.

Matias, L. L. R., Damasceno, K. S. F. D. S. C., Pereira, A. S., Passos, T. S., & Morais, A. H. D. A. (2024). Innovative biomedical and technological strategies for the control of bacterial growth and infections. *Biomedicines*, *12*(1), 176. https://doi.org/10.3390/biomedicines12010176

Mlynarczyk-Bonikowska, B., Kowalewski, C., Krolak-Ulinska, A., & Marusza, W. (2022). Molecular mechanisms of drug resistance in *Staphylococcus aureus*. *International Journal of Molecular Sciences*, *23*(15), 8088. https://doi.org/10.3390/ijms23158088

Monteiro, R., Pires, D. P., Costa, A. R., & Azeredo, J. (2019). Phage therapy: Going temperate? *Trends in Microbiology*, *27*(4), 368–378. https://doi.org/10.1016/j.tim.2018.10.008

Mourabit, N., Arakrak, A., Bakkali, M., Zian, Z., Bakkach, J., & Laglaoui, A. (2021). Antimicrobial resistance trends in *Staphylococcus aureus* strains carried by poultry in North of Morocco: A preliminary analysis. *Journal of Food Quality*, 2021, 8856004. https://doi.org/10.1155/2021/8856004

Norman, A., Hansen, L. H., & Sørensen, S. J. (2019). Conjugative plasmids: Vessels of the communal gene pool. *Philosophical Transactions of the Royal Society B: Biological Sciences*, *364*(1527), 2275–2289. https://doi.org/10.1098/rstb.2009.0037

Olatunji, A. O., Olaboye, J. A., Maha, C. C., Kolawole, T. O., & Abdul, S. (2024). Next-generation strategies to combat antimicrobial resistance: Integrating genomics, CRISPR, and novel therapeutics for effective

treatment. *Engineering Science & Technology Journal*, *5*(7), 2284–2303. https://doi.org/10.51594/estj.v5i7.1344
O'Neill, J., Davies, S. C., Rex, J. H., White, L. J., & Murray, R. (2016). *Tackling drug-resistant infections globally: Final Report and Recommendations*. Review on Antimicrobial Resistance. https://amr-review.org/
Otto, M. (2013). Community-associated MRSA: What makes them special? *International Journal of Medical Microbiology*, *303*(6–7), 324–330. https://doi.org/10.1016/j.ijmm.2013.02.007
Patil, S. S., Suresh, K. P., Shinduja, R., Amachawadi, R. G., Chandrashekar, S., Pradeep, S., . . . Shivamallu, C. (2022). Prevalence of methicillin-resistant *Staphylococcus aureus* in India: A systematic review and meta-analysis. *Oman Medical Journal*, *37*(4), e440.
Reygaert, W. C. (2013). Antimicrobial resistance mechanisms of *Staphylococcus aureus*. *Journal of Molecular Biology Research*, *3*(2), 1–8. https://doi.org/10.5539/jmbr.v3n2p1
Shariati, A., Dadashi, M., Moghadam, M. T., van Belkum, A., Yaslianifard, S., & Darban-Sarokhalil, D. (2020). Global prevalence and distribution of vancomycin resistant, vancomycin intermediate and heterogeneously vancomycin intermediate *Staphylococcus aureus* clinical isolates: A systematic review and meta-analysis. *Scientific Reports*, *10*(1), 12689.
Silva, V., Capelo, J. L., Igrejas, G., & Poeta, P. (2020). Molecular epidemiology of *Staphylococcus aureus* lineages in wild animals in Europe: A review. *Antibiotics*, *9*(3), 122. https://doi.org/10.3390/antibiotics9030122
Smith, R. D., Coast, J., Millar, M., Wilton, P., & Karcher, A. (2022). Incentivising innovation in antibiotic drug discovery and development: Progress, challenges and next steps. *Journal of Antimicrobial Chemotherapy*, *77*(3), 676–684. https://doi.org/10.1093/jac/dkab456
Styers, D., Sheehan, D. J., Hogan, P., & Sahm, D. F. (2006). Laboratory-based surveillance of current antimicrobial resistance patterns and trends among *Staphylococcus aureus*: 2005 status in the United States. *Annals of Clinical Microbiology and Antimicrobials*, *5*, 2. https://doi.org/10.1186/1476-0711-5-2
Sun, D., Jeannot, K., Xiao, Y., & Knapp, C. W. (2019). Editorial: Horizontal gene transfer-mediated bacterial antibiotic resistance. *Frontiers in Microbiology*, *10*, 1933. https://doi.org/10.3389/fmicb.2019.01933
Suresh, M. K., Biswas, R., & Biswas, L. (2019). An update on recent developments in the prevention and treatment of *Staphylococcus aureus* biofilms. *International Journal of Medical Microbiology*, *309*, 1–12. https://doi.org/10.1016/j.ijmm.2018.11.002
Szymczak, P., & Szczurek, E. (2023). Artificial intelligence-driven antimicrobial peptide discovery. *arXiv preprint arXiv:2308.10921*.
Thalia, O. P. (2025). Emerging technologies in antimicrobial resistance research. *Research Invention Journal of Engineering and Physical Sciences*, *4*(1), 30–35. https://doi.org/10.59298/RIJEP/2025/413035
Theuretzbacher, U., Outterson, K., Engel, A., & Karlén, A. (2020). The global preclinical antibacterial pipeline. *Nature Reviews Microbiology*, *18*(5), 275–285. https://doi.org/10.1038/s41579-019-0288-0
Thomsen, J., Abdulrazzaq, N. M., UAE AMR Surveillance Consortium, Menezes, G. A., Ayoub Moubareck, C., Everett, D. B., & Senok, A. (2023). Methicillin resistant *Staphylococcus aureus* in the United Arab Emirates: A 12-year retrospective analysis of evolving trends. *Frontiers in Public Health*, *11*, 1244351.
Touaitia, R., Mairi, A., Ibrahim, N. A., Basher, N. S., Idres, T., & Touati, A. (2025). *Staphylococcus aureus*: A review of the pathogenesis and virulence mechanisms. *Antibiotics*, *14*(5), 470.
Truong-Bolduc, Q. C., Dunman, P. M., Strahilevitz, J., Projan, S. J., & Hooper, D. C. (2005). MgrA is a multiple regulator of two new efflux pumps in *Staphylococcus aureus*. *Journal of Bacteriology*, *187*(7), 2395–2405. https://doi.org/10.1128/JB.187.7.2395-2405.2005
van den Biggelaar, R. H., Geurts, M. H., & Beekman, J. M. (2024). Bacteriophage therapy for respiratory infections: A promising alternative to antibiotics? *Frontiers in Microbiology*, *15*, 1455759.
Vicetti Miguel, C. P., Mejias, A., Leber, A., & Sanchez, P. J. (2019). A decade of antimicrobial resistance in *Staphylococcus aureus*: A single center experience. *PLOS ONE*, *14*(2), e0212029. https://doi.org/10.1371/journal.pone.0212029
Wittke, F., Wagner, F., Essack, S., Sprenger, M., & Roch, M. (2020). Afabicin, a first-in-class antistaphylococcal antibiotic, in the treatment of acute bacterial skin and skin structure infections: Clinical noninferiority to vancomycin/linezolid. *Antimicrobial Agents and Chemotherapy*, *64*(10), e00250.20. https://doi.org/10.1128/AAC.00250-20
World Health Organization. (2022). *World antibiotic awareness week campaign toolkit*. https://www.who.int/campaigns/world-antibiotic-awareness-week
World Health Organization. (2023). *Global action plan on antimicrobial resistance*. World Health Organization. https://www.who.int/publications/i/item/9789241509763
Worthington, R. J., Richards, J. J., & Melander, C. (2013). Small molecule control of bacterial biofilms. *Organic & Biomolecular Chemistry*, *10*(37), 7457–7474. https://doi.org/10.1039/C2OB25094A

Xie, Y., Liu, H., Teng, Z., Ma, J., & Liu, G. (2025). Nanomaterial-enabled anti-biofilm strategies: New opportunities for treatment of bacterial infections. *Nanoscale*, *17*, 5605–5628. https://doi.org/10.1039/D4NR04774E

Yathavan, B., Chhibber, T., Steinhauff, D., Pulsipher, A., Alt, J. A., Ghandehari, H., & Jafari, P. (2023). Matrix-mediated delivery of silver nanoparticles for prevention of *Staphylococcus aureus* and *Pseudomonas aeruginosa* biofilm formation. *Pharmaceutics*, *15*(10), 2426. https://doi.org/10.3390/pharmaceutics15102426

Zilberberg, M. D., Chaudhari, P., Nathanson, B. H., Campbell, R. S., Emons, M. F., Fiske, S., . . . Shorr, A. F. (2012). Development and validation of a bedside risk score for MRSA among patients hospitalized with complicated skin and skin structure infections. *BMC Infectious Diseases*, *12*, 1–9.

5 Genetic and Molecular Insights into *Staphylococcus aureus* Pathogenesis and Therapeutic Potential of Bioactive Compounds

S.S. Sravanthi Pammi[1] *and Archana Giri*[2]

[1]Department of Botany, SVR Government Degree College, Nidadavole, E.G.DT., Nidadavole, Andhra Pradesh, India

[2]Centre for Biotechnology, University College of Engineering and Science and Technology Hyderabad, Jawaharlal Nehru Technological University, Hyderabad, India

ABBREVIATIONS

agr	accessory gene regulator
CA	community-associated MRSA
ClfA, ClfB	clumping factors A and B
EGCG	epigallocatechin gallate
EPS	extracellular polymeric substances
FICI	fractional inhibitory concentration index
FnBPs	fibronectin-binding proteins
icaA	icaADBC intercellular adhesin genes
MDR	multidrug-resistant
MGEs	mobile genetic elements
MRSA	methicillin-resistant *Staphylococcus aureus*
norA	norA efflux pump gene
PBP2a	penicillin-binding protein 2a
PIA	polysaccharide intercellular adhesion
PVL	panton–valentine leukocidin
QS	quorum sensing
ROS	reactive oxygen species
SaeRS, VraSR, WalKR	specific TCS systems
SaPIs	staphylococcal pathogenicity islands
SarA/SarA	staphylococcal accessory regulator A
SCCmec	staphylococcal cassette chromosome mec
SCIN	staphylococcal complement inhibitor
SNPs	single-nucleotide polymorphisms
SpA	staphylococcal protein A
sRNAs	small regulatory RNAs

DOI: 10.1201/9781003647072-6

SSTIs	skin and soft tissue infections
T7SS	type VII Secretion System
TCS	two-component systems
vanA	vancomycin resistance gene A
VRSA	vancomycin-resistant *Staphylococcus aureus*
WHO	World Health Organization

5.1 INTRODUCTION

Antibiotics have been considered as the basis for the elimination of transmissible diseases since few years and are found to have an immense role in the prevention of bacterial infections. However, these infections seem to be the main cause of deaths over the past few decades in spite of great advancements in diagnostic and therapeutic aspects. Usage of extra dosage of antibiotics in recent times has contributed to the resistance of bacterial strains to antibiotics and their prevalence in the environment. This causes a threat to the treatment of bacterial infections, which in turn necessitates the novel alternative approach. New drugs have to be developed for high-risk groups of bacteria such as methicillin- and vancomycin-resistant strains of *S. aureus*, per the report submitted by WHO (Pacyga et al., 2024). *S. aureus* is the foremost cause of infections related to bacteremia, osteoarticular, and ventilator-associated pneumonia (Gopikrishnan et al., 2024). Naturally occurring plant-derived bioactive molecules are the compounds which play a key role in the biological activities of plants. The attribution of health benefits and healing properties of various medicinal plants is due to the presence of these bioactive molecules in plants. Phytochemicals exhibit good biological activity in humans and other organisms, as these organic molecules produced by plants influence several physiological processes. Hence, these phytochemicals are a huge and varied group of natural products, which had a wide range of biological activities and applications. Wide-ranging evolutionary adaptations are reflected by diverse compounds present in plants, which play an important role in defense, communication, and survival. *S. aureus* is a notorious gram-positive bacterium responsible for infections ranging from common skin infections to life-threatening diseases.

The severity of infections caused by this pathogen lies in its ability to develop resistance to antibiotics, enabling it to cause a wide range of diseases, including bone and joint infections, respiratory infections, skin and soft tissue infections (SSTIs), sepsis, endocarditis, and infections associated with medical devices such as catheters and prosthetic implants.

Infections caused by methicillin-resistant *Staphylococcus aureus* (MRSA) are difficult to treat because this strain of *S. aureus* is resistant to most of the common antibiotics that are being used and is one of the unique concerns at present. A concise summary of a wide range of infections caused by *S. aureus* is highlighted in Table 5.1, which provides information about their common site of infection, symptoms of that infection, and ongoing treatment challenges that are presented by antibiotic resistance (Figure 5.1).

5.2 VIRULENCE FACTORS OF *S. AUREUS*

Modern research on virulence factors of *S. aureus* has extended our introspection over various mechanisms, which were employed by refined properties such as the following:

a. **Adhesins**: The ability of *S. aureus* to colonize is reinforced by the expression of various surface proteins such as fibronectin-binding proteins (FnBPs) and clumping factors (ClfA, ClfB), which was revealed by Costa et al. (2022). Adhesion of bacteria to the host surface is the first crucial step of infection that is facilitated by FnBPs, which help in the strong adhesion of corneocyte and loricrin, increasing bacterial capacity for colonization.

TABLE 5.1
Common Infections Caused by *Staphylococcus aureus* and Their Resistance Patterns

Infection Type	Common Sites	Symptoms	Treatment Challenges	References
Skin and soft tissue	Skin, subcutaneous tissue	Redness, swelling, pus, pain	High prevalence of MRSA strains complicates empirical treatment and increases recurrence.	Peng et al. (2023)
Respiratory	Lungs	Cough, fever, difficulty breathing	Resistance to β-lactams and emergence of multidrug-resistant clones (e.g., ST239).	Peng et al. (2023)
Bone and joint	Bones (osteomyelitis), joints	Pain, swelling, fever	Biofilm formation on bone surfaces reduces antibiotic penetration and promotes persistence.	Lamret et al. (2021)
Endocarditis	Heart valves	Fever, heart murmur, fatigue	Biofilm-mediated colonization of heart valves complicates clearance, leading to chronic infection.	Peng et al. (2023)
Sepsis	Bloodstream	Fever, chills, tachycardia, hypotension	High virulence and resistance to frontline antibiotics increase treatment failure.	Peng et al. (2023)
Device-related	Catheters, prosthetics	Localized pain, inflammation, and signs of systemic infection	Biofilm formation on abiotic surfaces necessitates device removal and long-term antibiotics.	Von Eiff et al. (2006); Lamret et al. (2021)

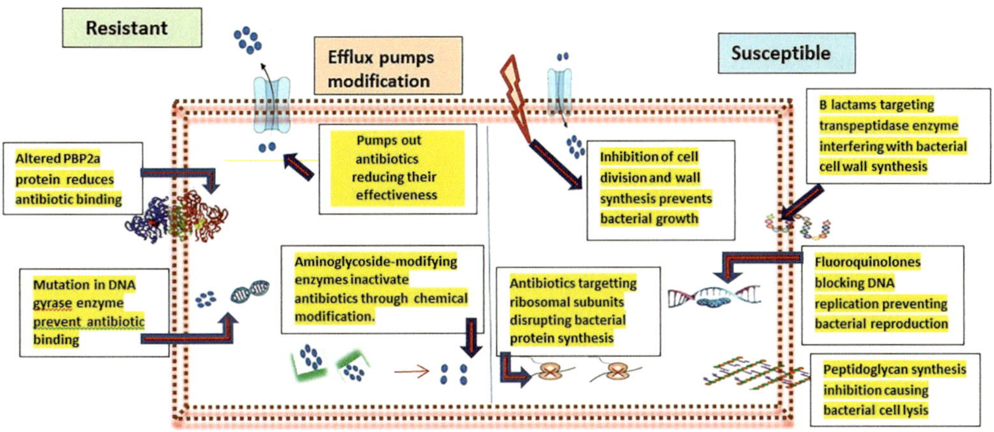

FIGURE 5.1 Split-cell schematic diagram explaining antibiotic resistance and susceptibility mechanisms in *Staphylococcus aureus*.

b. **Immune Evasion Factors**: Immune-modulatory proteins were being secreted by *S. aureus* to avoid host immunity. Opsonophagocytosis is hindered by the binding of protein A (SpA) to the Fc region of IgG. Complement activation is blocked by the staphylococcal complement inhibitor (SCIN). The role of these factors and bacterial mechanisms of survival of bacteria within the host were elaborated by Cheng et al. (2023).

c. **Toxins**: Khoramrooz et al. (2023) elaborated that the occurrences of virulence genes emphasize their role in public health. Hemolysins (α, β, γ, δ) were found to play an important role in the cytolysis of red and white blood cells. Panton–valentine leukocidin (PVL) was known to be the cause of severe SSTIs and necrotizing pneumonia. Enterotoxins are heat-stable and are associated with staphylococcal food poisoning.
d. **Biofilm Formation**: Development of bacterial biofilm increases its persistence on medical devices and tissues. Polysaccharide intercellular adhesin (PIA) helps in building biofilm matrix, which is encoded by the *ica* operon. Arciola et al. (2022) demonstrated that biofilm-forming strains intensified antibiotic resistance and immune tolerance, which poses clinical treatment challenges.
e. **Small Regulatory RNAs (sRNAs)**: Caldelari et al. (2021) discussed that sRNAs, such as RNAIII, modulate post-transcriptional expression of virulence genes, and these molecules play an important role in pathogenesis by regulating quorum sensing (QS) and biofilm formation.
f. **Type VII Secretion System (T7SS)**: Williams et al. (2020) reported that T7SS plays an evolving role in host colonization and competition among bacteria, which contributes to survival under stress by maintaining membrane system integrity and antimicrobial fatty acid resistance.
g. **Staphyloxanthin**: Oxidative killing of *S. aureus is* protected by Staphyloxanthin, which is a golden carotenoid pigment. Its importance is studied by Campbell et al. (2023) in causing persistent chronic infections and delayed wound healing, which is enhanced by its antioxidant capacity.

5.3 *S. AUREUS* PATHOGENESIS AND ITS GENETIC AND MOLECULAR INSIGHTS

Virulence in *S. aureus* is regulated by a QS mechanism known as the accessory gene regulator (agr) system, which perceives cyclic peptide signaling molecules. Bacteria can able to sense the density of their population through this system and can accordingly modulate gene expression, thereby influencing the production of various virulent factors such as adhesins, toxins, and extracellular enzymes, as studied by Girma (2024). *S. aureus* also employs various two-component systems along with the agr system to respond to environmental stimuli such as alternative sigma factors, including the SarA protein family. These regulatory networks ensure adaptive responses to changing environmental conditions, thus integrating the host with these environmental signals (Girma, 2024). The genetic diversity and virulence of *S. aureus* are significantly contributed to by mobile genetic elements (MGEs) such as transposons, plasmids, insertion sequences, prophages, and pathogenicity islands, which help in carrying genes for virulence and antibiotic resistance. Toxins such as PVL can be encoded by prophages, which destroy the white blood cells for targeting the host immune system and enhancing bacteria's evading ability (Girma, 2024). When antimicrobial agents were exposed to *S. aureus*, adaptive strategies of bacteria came into light through recent transcriptomic analysis. Upregulation of the mntABC operon was found to be responsible for manganese transport. Hence, enhanced expression of the mntABC operon helps in combating antibiotic-induced oxidative stress, proving manganese to be crucial in stress response, activating enzymes and metabolism (Zhang et al., 2022). Sbi, a gene which encodes an IgG-binding protein, inhibits innate and adaptive immune responses, playing an important role in immune evasion. Enhanced expression of the sbi gene under antimicrobial pressure is found by Zhang et al. (2022), indicating the importance of this gene as a strategy for the survival of bacteria. Particulars of these mechanisms have been revealed by recent studies as follows:

a. **Accessory Gene Regulator (agr) System**: Expression of various virulence factors in *S. aureus* is modulated by a QS mechanism called "the arg system." This system regulates a very crucial step of infection during adhesion and invasion (Girma, 2024). Patel and Rawat

(2023) found that any disruption in the agr system leads to remarkable changes, such as increased biofilm formation and alteration in the production of toxins, and has also shown an impact on the disease-causing ability of the pathogen.

b. **Staphylococcal Accessory Regulator (sarA)**: This is a locus which influences the expression of multiple virulence genes and is a global regulator playing a substantial role in biofilm formation. This is possible by repressing the formation of extracellular proteases, which can destroy the biofilm matrix. When there are mutations in the sarA gene, changes such as impaired biofilm formation and reduction of virulence have been observed (Patel and Rawat, 2023).
c. **Small Regulatory RNAs (sRNAs)**: In *S. aureus*, sRNAs play a pivotal role in regulating post-transcriptional gene expression. Various physiological processes, such as stress responses, metabolism, and virulence, have been modulated by sRNAs. Modern research has emphasized the role of sRNAs in acclimatizing to changing environmental conditions and in regulating virulence determinants, although targets of specific sRNAs seem to be elucidated (Girma, 2024).
d. **Two-Component Systems (TCS)**: There are multiple TCSs in *S. aureus* which can recognize the environmental stimuli and respond. First, the SaeRS System is found to regulate the genes and factors encoding toxins and immune evasion, which is crucial for the pathogen in causing invasive diseases and for evading immune responses of the host. Second, the VraSR System gets activated when antibiotics target the cell wall, which in turn contributes to antibiotic resistance by upregulating genes involved in cell wall synthesis and cell wall modification. The next prominent TCS is the WalKR System, which is essential in maintaining cell wall homeostasis by regulating enzymes that damage cell wall components, namely, autolysins. Any mutation in the WalKR system was found to affect cell wall turnover, and there is increased susceptibility to antibiotics (Patel and Rawat, 2023).
e. **MGEs**: These contribute to the genetic diversity and adaptability of *S. aureus* by facilitating horizontal transfer of genetic material. Staphylococcal cassette chromosome mec (SCCmec) is an element which carries the mecA gene, which confers resistance to antibiotics such as methicillin and other β-lactams. Different types of SCCmec have been recognized, which influence variations in the epidemiology of methicillin-resistant *S. aureus* (MRSA) strains. Panton–valentine leukocidin (PVL) genes are encoded by prophages, which get integrated into the genome of *S. aureus*. Severe skin infections and necrotizing pneumonia are associated with strains harboring PVL. Pathogenicity islands (SaPIs) are genomic islands which transmit genes for superantigens and other virulence factors. These are mobilized by helper phages, which facilitate the spread of virulence traits among inhabitants of *S. aureus*.
f. **Evolutionary Insights from Comparative Genomics**: Genetic diversity in strains of S. *aureus* has been significantly exposed by a comparative genomic analysis, which reveals their adaptations to various host niches. The core genome comprises vital housekeeping genes, while MGEs that provide adaptive benefits were present in the accessory genome. Phylogenetic studies reveal that the evolution of MRSA strains took place many times in an independent manner, and resistance genes were acquired through horizontal gene transfer (Girma, 2024). Rapid adaptation to host environments as a part of *S. aureus* pathogenesis is possible by sophisticated interaction between regulatory systems and genetic elements. Research is in a continuous process to reveal novel regulatory pathways through potential therapeutic intervention. Effective strategies for combating *S. aureus* infections are possible only with a comprehensive understanding of these mechanisms.

5.4 CLINICAL MANAGEMENT AND CHALLENGES

Obstacles in clinical management occur due to the presence of diverse virulence factors in *S. aureus*. Antibiotics such as beta-lactams (e.g., methicillin), glycopeptides (e.g., vancomycin), and novel agents such as linezolid play a major role in its common treatment. But widespread resistance of *S. aureus*, particularly MRSA strains, complicates its therapeutic treatment, which necessitates the need for effective vaccines by the discovery of novel antimicrobials (Turner et al., 2020).

a. **Antibiotic Resistance Mechanisms**: Among many resistance mechanisms, the strains which carry the *mecA* gene encode a low-affinity penicillin-binding protein 2a (PBP2a) for beta-lactams, which render them ineffective (Turner et al. 2020). Treatment resistance in MRSA necessitates the costly use of glycopeptides or oxazolidinones, which are associated with toxicity (Lee et al. 2021). VRSA strains of *S. aureus* has *vanA* gene, which alters the target site of vancomycin and reduces binding efficacy (Osei Sekyere 2020), stressing the need for agents like daptomycin for clinical management (Osei Sekyere 2020). MDR strains of S. *aureus* exhibit resistance by various mechanisms such as target site alterations, antibiotic-modifying enzymes, and efflux pumps (Lakhundi and Zhang 2020). This resistance potential of bacteria increases the cost of healthcare and creates complicated treatment regimens (Aslam et al. 2021).
b. **Genetic and Molecular Factors**: Shariati et al. (2020) studied that in the horizontal gene transfer mechanism of *S. aureus*, resistance genes are transferred via plasmids and transposons, which helps the bacterium to transform into a resistant strain. It was found that MGEs such as SCCmec element, which carries the *mecA* gene and several resistance genes, facilitate the spread of resistance by integrating those genes into the genome (Lakhundi and Zhang 2020).
c. **Diagnostic and Therapeutic Challenges**: Conventional culture-based methods may not notice resistance quickly and are also time-consuming. Hence, modern diagnostic methods such as PCR and sequencing techniques, having increased speed and accuracy, help in resistance detection at the molecular level (Liu et al., 2021). Treating resistant *S. aureus* infections is a therapeutic challenge in recent times, as antibiotics which are less costly are toxic and less effective and show adverse effects. Hence, there is a need for the development of new antibiotics and adjunctive therapies in healthcare (Aslam et al., 2021).
d. **Infection Control and Prevention**: Spread of infection can be controlled by healthy practices such as personal hygiene, regular cleaning of hands, and isolation from crowds, which limits the spread of resistant strains of *S. aureus*. Community healthcare programs should be initiated in public, stressing the need for health education, which can diminish the prevalence of community-associated MRSA (Reta et al., 2020).
e. **Future Directions**: Novel research strategies to overcome resistance are under examination for developing next-generation antimicrobials such as vaccines and phage therapy (Aslam et al., 2021). Resistance trends are under tracking using advanced surveillance systems for notifying treatment and prevention strategies (World Health Organization 2020).

5.5 EVOLUTIONARY INSIGHTS FROM COMPARATIVE GENOMICS

Insights into comparative genomics advancements have significantly improved our understanding of the evolutionary dynamics of *S. aureus*, which reveals its adapting ability to many environmental and host niches. Large-scale genome analyses highlighted the information about the conserved and variable genome elements of *S. aureus*. The genome of *S. aureus* consists of a conserved core genome and an accessory genome. Core genome controls vital metabolic and replication processes, whereas accessory genome harbors MGEs, which play active role in antibiotic resistance and virulence adaptation. Yang et al (2024) performed 558 large-scale genome studies, and ten major sequence clusters were identified in *S. aureus* strains, which highlights differences in accessory genome composition. It was identified that variations in MGEs is responsible strain-specific traits in antibiotic resistance profiles and niche adaptation. Polyphyletic emergence of MRSA is found to be one of the types of evidence for phylogenetic analyses using single-nucleotide polymorphisms. Yang et al (2024) proved that the evolution of MRSA strains is not as a result of emergence of single ancestor but must have evolved parallelly within different clades by the addition or deletion of accessory gene sequences, which supports the concept of multiple independent resistance events. According to Raghuram et al (2022), *S. aureus* was tracked through longitudinal studies for its adaptation in the host, and after analyzing 40,000 genomes, it was found that more than 5% of strains carried frame shift mutations in the *agr* operon. This operon is a QS regulator, which is crucial for the expression of virulence genes. These mutations were scattered across multiple clonal complexes, signifying that inactivation of *agr* is a recurrent adaptive event in chronic infections such as cystic fibrosis or prosthetic joint infections.

5.6 MOLECULAR TARGETS IN *S. AUREUS*

Antibiotics have proven to play a critical role in the treatment of bacterial infections. The rise and spread of drug-resistant bacteria pose a global health risk. Natural products offer an adaptable alternative for drugs, particularly in the field of antibiotics. Plants, as the most abundant source of biodiversity on Earth, have historically played a pivotal role in drug discovery. Despite their rich and diverse metabolic fingerprints, herbal medicines have been basically ignored for a long time. Phytochemicals exhibit a myriad of therapeutic properties, and these natural products are also precious for the development of new drug candidates. Metabolic fingerprint and biological properties of various plants, along with their mechanisms of action, were reviewed by Sravanthi et al. (2023). Abundant studies have confirmed their ability to inhibit antibiotic resistance through diverse mechanisms, namely, bacterial cell wall synthesis inhibition, bacterial cell membrane integrity disruption, efflux pump inhibition, and deterrence of biofilm formation. These highlight the potential of bioactive compounds as effective agents for controlling antibiotic resistance. Furthermore, phytochemicals have demonstrated the ability to reverse antibiotic resistance and enhance the effectiveness of antibiotics, which opens a new path for harnessing the potential of these plant-derived bioactive compounds as promising antimicrobial agents (Liu et al 2025). Plant-derived bioactive compounds represent a promising avenue for combating antibiotic-resistant *S. aureus* (including MRSA strains) due to their ability to target multiple molecular pathways in bacterial cells (Table 5.2). The antimicrobial mechanisms operated by phytochemicals are as follows:

1. **Cell Wall Synthesis Inhibition**: The disruption of the synthesis of the cell wall is one of the crucial antibacterial mechanisms. Peptidoglycan is an essential component of the cell wall, and its synthesis is interfered with by some of the phytochemicals. Li et al. (2020) identified some downregulating genes, which are associated with the biosynthesis of peptidoglycan in MRSA strains, leading to weakening of cell wall structure, thereby causing bacterial lysis. This is possible by specific compounds present in medicinal plants.
2. **Membrane Integrity Disruption**: Isothiocyanates and phenolic acids are phytochemicals which have the capacity to integrate into bacterial cell membrane, thereby causing cell death as a result of leakage of intracellular material by membrane disruption. Al-Huqail et al. (2021) reported that cruciferous vegetables caused membrane damage in *S. aureus* by increasing membrane permeability through potassium ion leakage in the presence of isothiocyanate compounds. Similarly, lipid peroxidation and membrane potential collapse

TABLE 5.2
Phytochemicals and Their Antibacterial Mechanisms

Mechanism	Compound/Plant Source	Key Effect on *S. aureus*	References
Cell wall synthesis inhibition	Flavonoids from Chinese herbs	Suppressed peptidoglycan biosynthesis	Li et al. (2020)
Membrane integrity disruption	Isothiocyanates from crucifers	K^+ leakage, membrane rupture	Al-Huqail et al. (2021)
Membrane integrity disruption	Garlic-derived sulfur compounds	ATP leakage, lipid peroxidation	Xiaoyan Wang et al. (2021)
DNA synthesis inhibition	Ellagic acid from *P. granatum*	Inhibited DNA gyrase activity	Chen et al. (2023)
Quorum-sensing interference	*Paederiafoetida*, *P. guajava*	Downregulated *agr* and *icaA* operons	Bhandari et al. (2023); Mahmud et al. (2023)

in *S. aureus* were reported by Xiaoyan Wang et al. (2021) due to the presence of sulfur-containing compounds.

3. **Inhibition of DNA Synthesis**: DNA gyrase plays an important role in the antibacterial mechanism by inhibiting the DNA replication process. Quercetin and catechins are flavonoids which exhibit such property. Chen et al. (2023) revealed that in MRSA strains, DNA replication is inhibited and the cell cycle is arrested by using ellagic acid extracts from *Punica granatum*.
4. **Interference with QS**: Expression of virulence factors and formation of biofilms in *S. aureus* can be controlled by QS controls. Bacterial pathogenicity can be reduced by phytochemicals, which can interfere with the aforementioned process. Bhandari et al. (2023) reported that *agr* QS system is inhibited in *S. aureus* by decreasing the production of hemolysins and proteases using ethanolic extracts of *Paederiafoetida*. Flavonoids from *Psidium guajava* were also found to interrupt biofilm-associated gene expression as reported by Mahmud et al. (2023).

5.7 ROLE OF PHYTOCHEMICALS IN INHIBITION OF BIOFILMS

S. aureus biofilm formation is found to be inhibited by phytochemicals through several mechanisms such as the following:

a. **Disruption of Biofilm Formation**: Preliminary stages of biofilm formation were found to be prevented by some phytochemicals. Zhou et al. (2023) found that natural isoflavonoid biochanin A decreases the production of PIA, which is indispensable for biofilm integrity and thus disrupts the EPSs, thereby inhibiting biofilm development. Quercetin's anti-biofilm effectiveness and its validation by in silico modeling against drug-resistant *S. aureus* were studied by Anjaneyulu et al. (2024).
b. **Virulence Factors Suppression**: Virulence factors of *S. aureus* were targeted by certain phytochemicals. Production of important virulence factors such as staphyloxanthin, lipase, and α-hemolysin was found to be suppressed by a monounsaturated fatty acid, petroselinic acid, which plays an important role in the pathogenicity of *S. aureus*. Kim et al. (2022) found that this suppression is due to the downregulation of genes associated with these three factors.
c. **QS Inhibition**: Phytochemicals interferes with bacterial communication system known as QS, which synchronizes virulence and biofilm formation. Husain et al. (2021) proved that flavonoid compounds like naringenin present in fruits of grapes

exhibit their QS inhibiting potential in *Pseudomonas aeruginosa* signifying its similar effects in *S. aureus*.

d. **Synergistic Effects with Antibiotics**: A combination of conventional antibiotics and phytochemicals was found to improve antibacterial efficacy. Lee et al. (2023) found that oleuropein, which is a phenolic compound rich in olive leaf extract, increases the activity of antibiotics against biofilm formation in *S. aureus*, overcoming its antibiotic resistance capacity.

5.8 SIGNIFICANCE OF PLANT-DERIVED BIOACTIVE COMPOUNDS

In recent years, bioactive compounds derived from plants have gained substantial consideration as they exhibit extraordinary health potential benefits, predominantly in their antimicrobial properties. Alkaloids, phenolic compounds, terpenoids, and essential oils are some of these compounds exhibiting their efficacy against a wide range of pathogens, such as *S. aureus*, which are resistant to antibiotics.

a. **Flavonoids**: These are polyphenolic secondary metabolites widely prevalent in the plant kingdom. They exhibit diverse antibacterial mechanisms. One such mechanism is the disruption of the bacterial cell membrane, which causes leakage of cytoplasmic contents. Quercetin and apigenin are some flavonoid compounds, and according to Borges et al. (2020), quercetin is found to increase the permeability of the membrane, causing membrane dissipation in MRSA strains. Wu et al. (2020) confirmed that green tea has the epigallocatechin gallate (EGCG) compound, which inhibits NorA efflux pumps in *S. aureus*, enhancing intracellular antibiotic concentration such as ciprofloxacin. Flavones were found to suppress *hla* and *sigB* gene expression, decreasing the production of α-hemolysin and staphyloxanthin, as defined by Alves et al. 2021. This is an important mechanism for decreasing the pathogenicity of bacteria without killing them, thereby lowering the development of their resistance and suppressing virulence factor suppression.
b. **Terpenoids**: These are a huge class of natural compounds with substantial antimicrobial properties. T. L. C. de Oliveira et al. (2020) found that monoterpene compounds such as carvacrol and thymol were present in oils of oregano and thyme, which destabilize their structure by integration into membranes of bacteria. *norA* gene expression was downregulated in MRSA strains by compounds such as carvacrol and thymol, as observed by Raut et al. (2022), which is an efflux pump inhibition mechanism potentiating the activity of fluoroquinolones. Zhou et al. (2020) proved that thymol is a compound which prevents the adhesion of bacteria initially by inhibiting matrix synthesis genes *icaA* and *sarA*, thereby hindering bacterial biofilms.
c. **Alkaloids**: Berberine and matrine are examples of plant alkaloids which have broad-spectrum antibacterial potential. Fang et al. (2021) reported that berberine obtained from *Berberis vulgaris* impairs DNA replication by intercalating into the DNA of bacteria and inhibiting the activity of DNA gyrase. *agrA* and *sarA* are QS genes which are downregulated by Matrine, a compound which weakens the resistance associated with biofilm formation, as studied by Liu et al. (2020).
d. **Phenolic Compounds**: Caffeic acid and rosmarinic acid are phenolic acids that are found to exhibit antibacterial activity either through loss of permeability by their integration into the membrane of bacteria, as studied by Nostro et al. (2021) or through the inhibition of enzymes that are required for the integrity of the cell wall, particularly during the biosynthesis of fatty acids and peptidoglycans.
e. **Cyclic Peptides**: Gurmarin is a cyclic peptide obtained from *Gymnema sylvestre* and exhibits anti-virulence properties without affecting bacterial growth. This compound downregulates the expression of the *icaADBC* operon and interferes with initial adhesion, disrupting the development of biofilms in *S. aureus*, as studied by Swargiary et al. (2022).

5.9 MECHANISMS OF ACTION OF PLANT-DERIVED COMPOUNDS

Plant-derived bioactive compounds act through multiple antibacterial mechanisms to inhibit the growth and virulence of *S. aureus* (Figure 5.2). These mechanisms offer an effective alternative to traditional antibiotics, especially in the era of increasing resistance (Figure 5.3). Continued investigation into the molecular pathways affected by phytochemicals is essential for the development of novel antimicrobial therapies. The growing incidence of MRSA poses a serious public health threat due to the reduced efficacy of conventional antibiotics. The search for alternative therapies has shifted attention to natural products, particularly plant-derived bioactive compounds, known for their antimicrobial and anti-biofilm properties.

Plant-derived compounds present a multifaceted approach to combating *S. aureus*, especially antibiotic-resistant strains. These natural products target bacterial membranes, enzymes, virulence factors, and communication systems. Synergistic effect of these phytochemicals with conventional antibiotics was found to increase their therapeutic potential, which is vital for the expansion of next-generation antimicrobials (Table 5.3), and hence continuous exploration of these compounds is very much necessary.

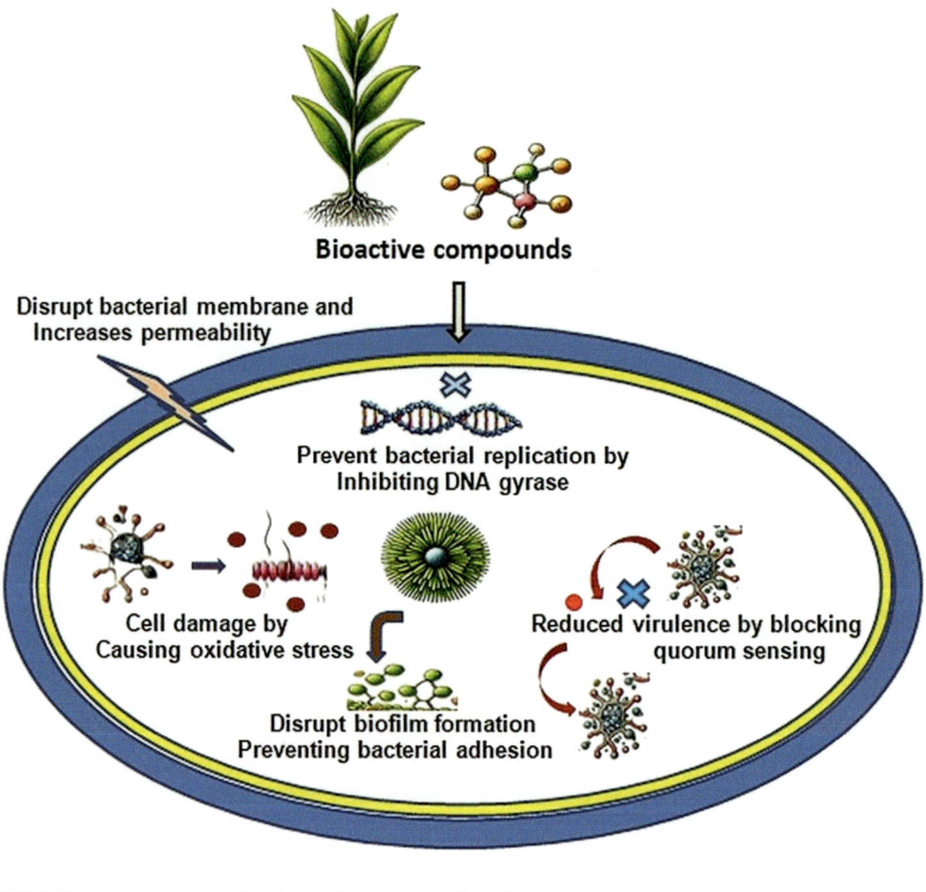

FIGURE 5.2 Secondary metabolite mechanisms against *S. aureus*.

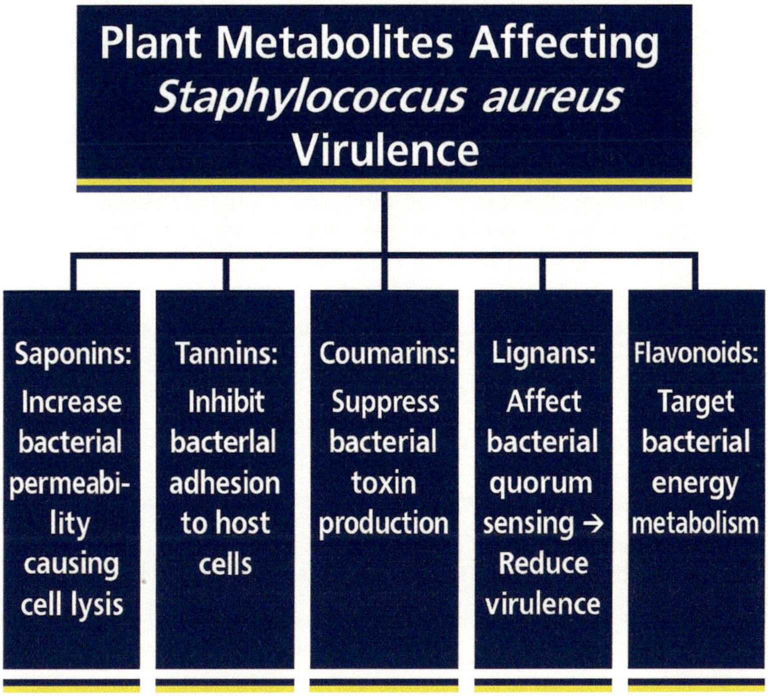

FIGURE 5.3 Potential of plant metabolites in affecting the virulence *of Staphylococcus aureus.*

TABLE 5.3
Mode of Action of Various Bioactive Compounds toward Combating *S.aureus*

Bioactive Compound	Plant Source	Mechanism of Action	References
Quercetin	Various fruits/vegetables	Membrane disruption	Borges et al. (2020)
EGCG	Green tea	NorA efflux pump inhibition	Wu et al. (2020)
Thymol	*Thymus vulgaris*	Biofilm inhibition, efflux pump downregulation	Zhou et al. (2020)
Carvacrol	*Origanum vulgare*	Membrane disruption, efflux pump inhibition	Raut et al. (2022)
Berberine	*Berberis vulgaris*	DNA gyrase inhibition	Fang et al. (2021)
Matrine	*Sophora flavescens*	Biofilm gene suppression	Liu et al. (2020)
Rosmarinic acid	Rosemary	Membrane perturbation, enzyme inhibition	Nostro et al. (2021)
Gurmarin	*Gymnemasylvestre*	Anti-biofilm activity	Swargiary et al. (2022)

5.10 PLANT-DERIVED COMPOUNDS INVOLVED IN SYNERGISTIC EFFECTS AND COMBINATION THERAPIES AGAINST *S. AUREUS*

Antibiotic resistance is an important universal concern at present for the protection of health against various pathogens, predominantly *S. aureus*. Modern research is focusing on the importance of bioactive compounds, which are derived from plants, as they have the capacity to interact synergistically with conventional antibiotics. This type of approach helps in combating infections which are very resistant, and there will be progress in therapeutic outcomes by exploiting the natural medicinal properties of plants. Some plant extracts were found to increase the antibiotic potential against MRSA strains of *S. aureus.* Mayyas et al. (2021) have studied the synergistic effects of ethanolic extracts from 12 plant species combined with clindamycin antibiotic against MRSA strains and evaluated their antimicrobial potential. It was found that a blend of these plant extracts with clindamycin antibiotic initiated a strong synergistic effect, resulting in a fractional inhibitory concentration index of less than 0.5. Antagonistic activity was also not observed by using these combinations, which suggests that synergistic interactions between plant extracts and antibiotics can play an effective role in decreasing the consumption of antibiotics by delaying the commencement of resistance. The basic mechanisms attributed to these synergistic effects include inhibition of bacterial efflux pumps,

TABLE 5.4
Synergistic Effects of Plant Compounds with Antibiotics against *Staphylococcus aureus*

Plant/Compound	Antibiotic Combined With	Target Pathogen	Mechanism of Synergy	References
Camellia sinensis extract	Clindamycin	MRSA	Membrane disruption; increased permeability; enhanced intracellular antibiotic concentration	Mayyas et al. (2021)
Thymus vulgaris extract	Clindamycin	MRSA	Disruption of the cell membrane; inhibition of efflux pumps	Mayyas et al. (2021)
Rosmarinus officinalis	Clindamycin	MRSA	Synergistic ROS production and membrane destabilization	Mayyas et al. (2021)
Salvia officinalis	Clindamycin	MRSA	Inhibition of resistance genes; membrane permeability increase	Mayyas et al. (2021)
Thymol + carvacrol	Conventional antibiotics	MRSA	Lipophilic interactions with the membrane; efflux pump inhibition	Silveira et al. (2022)
Centaurea damascena	Gentamicin, vancomycin	*E. coli*, *Klebsiella pneumonia*	Inhibition of ribosomal binding proteins and protein synthesis	Silveira et al. (2022)
Coriandrum sativum	Mezlocillin, cefoperazone, levofloxacin	MDR *Pseudomonas aeruginosa*	Membrane permeabilization and inhibition of DNA gyrase	Silveira et al. (2022)
Pithecellobium clypearia	Imipenem, tetracycline	MDR *Acinetobacter baumannii*, *P. aeruginosa*	Biofilm disruption and resistance reversal	Silveira et al. (2022)
Curcumin + piperine	Ciprofloxacin	MRSA	Piperine enhances curcumin bioavailability and disrupts bacterial quorum sensing	Wadhwa et al. (2021)
Quercetin + kaempferol	Ampicillin	MRSA	Inhibits penicillin-binding proteins and increases ROS production	D. M. de Oliveira et al. (2020)
Berberine + tetracycline	Tetracycline	MRSA	Inhibition of efflux pumps (NorA); membrane integrity alteration	Yanhong Wang et al. (2021))
EGCG (green tea polyphenol) + oxacillin	Oxacillin	MRSA	Disruption of biofilm matrix and β-lactamase inhibition	Zhao et al. (2020)

which blocks expelling of antibiotics from bacterial cells. This is responsible for increasing the intracellular concentration of antibiotics by plant compounds for enhancing their efficacy. Silveira et al. (2022) found that this mechanism prevents exposure of bacteria to sub-therapeutic antibiotic concentrations, which are a common cause of acquired resistance. Some phytochemicals enhance the membrane permeability and allow more penetration of antibiotics, disrupting the bacterial cell membrane. Silveira et al. (2022) reported that thymol, which is a monoterpene phenol present in thyme, can cause membrane depolarization, which facilitates antibiotic entry into bacterial cells, thereby increasing the antibacterial activity. Phytochemicals also exhibit a mechanism of inhibiting bacterial enzymes, which can degrade antibiotics. This can help to maintain antibiotic potential against resistant bacteria by protecting the antibiotic from inactivation, as studied by Silveira et al. (2022). Silveira et al. (2022) demonstrated the use of some plant-antibiotic synergistic combinations. *Salvadora persica*, when combined with metronidazole, doubled its antibiotic activity against *Aggregatibacter actinomycetemcomitans. Ficus nitida* enhanced antibacterial activity in the presence of tetracycline against *Bacillus cereus*, *S. aureus*, *Escherichia coli*, and *P. aeruginosa. Centaurea damascene e*xhibited synergistic effects when combined with gentamicin, vancomycin, ampicillin, and chloramphenicol against *E. coli* and *Klebsiella pneumoniae. Pithecellobium clypearia* showed synergistic effects with imipenem and tetracycline against multidrug-resistant *Acinetobacter baumannii* and *P. aeruginosa. Coriandrum sativum* exhibited synergism with antibiotics such as mezlocillin, cefoperazone, cefotaxime, and levofloxacin against multidrug-resistant *P. aeruginosa.* Several plant species have been identified to exhibit synergistic effects when combined with conventional antibiotics (Table 5.4).

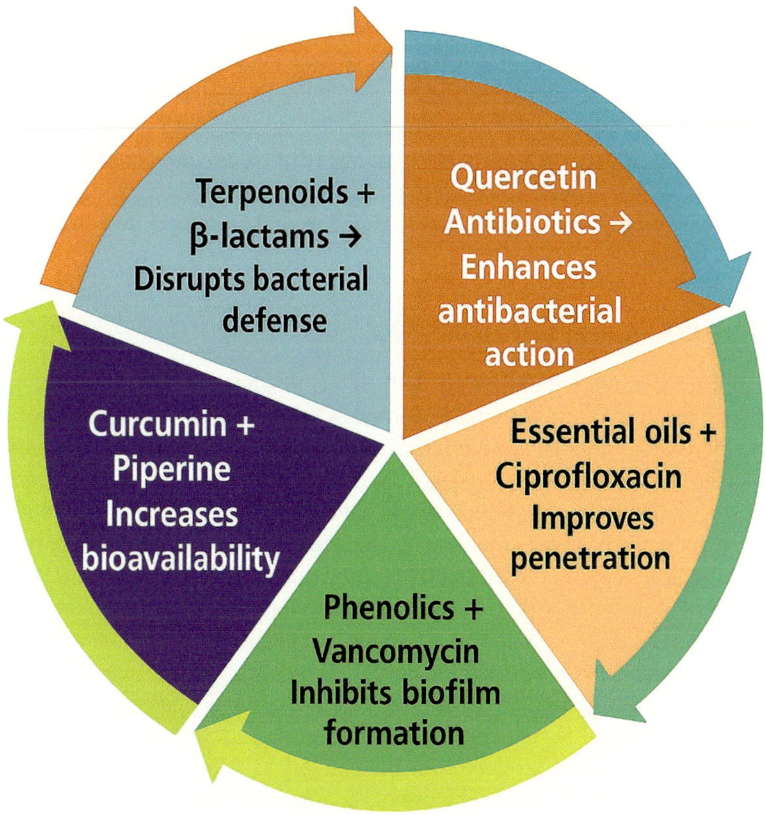

FIGURE 5.4 Synergistic effects of plant compounds and antibiotics.

The integration of plant-derived bioactive compounds with conventional antibiotics offers a promising strategy to combat antibiotic-resistant infections (Figure 5.4). Researchers can focus on more effective combination therapies by explicating the molecular mechanisms underlying these synergistic interactions, such as efflux pump inhibition, membrane permeability enhancement, and enzyme inhibition. It is also essential to completely understand these interactions and has to be extended further to interpret these findings into clinical applications.

5.11 CONCLUSION

Developing novel and effective therapeutic strategies against *S. aureus* is possible only by understanding the molecular mechanisms of various bioactive compounds that are derived from plants. Multifaceted mechanisms of these phytochemicals decrease bacterial resistance safely compared to the use of traditional antibiotics. Synergistic use of phytochemicals with conventional antibiotics minimizes the antibiotic dosage and increases their efficiency, which is very much essential for overcoming the problems caused by antibiotic-resistant strains of *S. aureus*, such as MRSA. Phytochemicals target the virulence factors to boost the efficacy of immune responses and also reduce the pathogenicity caused by *S. aureus*. Modulation of the immune system of the host is also possible by compounds derived from plants, which help in boosting the host's ability to fight off infections caused by *S. aureus*. This dual action of phytochemicals provides a valuable and consolidative approach for infection management, as they have the capacity of antimicrobial activity directly and support for host immune boosting indirectly. But the clinical advancement of phytochemicals expresses significant challenges in spite of their high potential, as they are in the form of complex and variable mixtures without any standardization. Hence, there is a need for the maintenance of quality control standards of in vitro clinical findings, as they are difficult to operate *in vivo* in terms of their efficiency. Latest technology advancements and methodologies should be in coordination with underlying research not only to understand the molecular mechanisms of these compounds but also to overcome current challenges. One of the future directions is the integrative medicine approach, which combines conventional treatments with phytochemicals for treating infections caused by *S. aureus*. Deep understanding of molecular mechanisms of plant-derived bioactive compounds is very much essential to address the challenges caused during their development, and their applications have the potential to transform into an important compound for fighting against *S. aureus* infections.

5.12 FUTURE RESEARCH IMPLICATIONS

The focus of future research should be on delineation of molecular pathways of specific phytochemicals which exert their antimicrobial action on *S. aureus*, which opens up several avenues for addressing the increasing tasks of antibiotic resistance. Identification of specific targets inside the bacterial cell should be traced out by using advanced bioinformatic techniques such as proteomics, transcriptomics, and metabolomics, which can be studied at the molecular level and gene level for understanding the cascade of events that lead to the death of bacteria. Rationally designing phytochemical-based therapies can be more effectively implemented by such deep mechanistic insights. Systematic exploration of synergistic effects of several combinations of phytochemicals and antibiotics helps us understand synergy mechanisms, which results in the identification of the most effective combinations. This paves the way for the development of more effective combination therapies, which minimizes the development of resistance. In vitro clinical findings critically challenge the stability and bioavailability of various phytochemicals during their *in vivo* research process. Novel formulation strategies should be developed for the delivery of these compounds through the use of the nano-encapsulation technique or by the use of bioenhancers. Pharmacokinetics and pharmacodynamics investigations should also be taken up for optimizing their efficacy in human subjects. The interaction of plant-derived compounds with the host immune system should also be studied in detail to understand their influence on host–pathogen dynamics during *S. aureus* infections. New therapeutic strategies could be revealed by understanding these interactions to improve overall treatment outcomes, which targets on pathogen as well as the modulation of host immune responses. It is very crucial to understand how *S. aureus* is continuously evolving resistance to antibiotics by exploring strategies such as rotating phytochemicals or using resistance inhibitors to mitigate this risk and how phytochemicals are overcoming such resistance mechanisms. Many clinical trials are directed at the treatment of *S. aureus* infections to determine the optimal dosage of phytochemicals without disturbing their safety and efficacy. Translational research shall act as a bridge in filling the gap between in vitro lab findings and clinical trials to reveal the benefits of phytochemicals to real-world applications in healthcare. Ethical and environmental aspects of medicinal plants should also be taken into account for sustainable usage of phytochemicals, which have gained interest in recent times. This is a crucial practice for the cultivation of medicinal plants and extracting their phytochemicals to ensure that these resources remain accessible for coming generations without causing ecological imbalance. By addressing these challenges in research, scientists can investigate and reveal the latest advances about plant-derived bioactive compounds to fight against *S. aureus* infections as a new frontier for contributing to more effective and sustainable therapeutic processes.

5.13 FINAL REMARKS

Molecular mechanisms exhibited by phytochemicals represent a promising frontier in the era of rising antibiotic resistance against the ongoing battle of *S. aureus* infections. As phytochemicals have diverse modes of action, they are compelled to be used as an alternative to conventional antibiotics. Phytochemicals target *S. aureus* infections by a multifaceted approach for enhancing host immune responses to potentially overcome bacterial resistance mechanisms using different pathways. However, practically, the journey from laboratory findings to clinical trials development is complex and requires constant research efforts. Hence, future research through rigorous clinical trials should focus on the optimization of compounds for their delivery and bioavailability by elucidating their comprehensive mode of action to ensure their efficacy. Additionally, ethical considerations in sourcing and environmental sustainability must guide the development and use of these natural resources. In conclusion, plant-derived bioactive compounds hold significant potential in transforming the treatment landscape for *S. aureus* infections. With continued research and innovation, these natural agents could become integral components of modern therapeutic strategies, offering hope for more effective, sustainable, and accessible treatments in the future.

REFERENCES

Al-Huqail, Abdullah Abdulrahman, Ehab Azab, Manal T. Wabaidur, Mona H. Alharbi, Reem K. Ibrahim, Hamed A. Ghramh, Mohammad S. Al-Buriahi, and Mohd Sayeed Akhtar. 2021. “Biogenic isothiocyanates disrupt membrane integrity and induce oxidative stress in *Staphylococcus aureus*.” *Plants* 10 (9): 1894. https://doi.org/10.3390/plants10091894

Alves, L. A., N. C. de Souza, T. M. Lima, and M. F. L. de Araújo. 2021. “Flavonoids as antivirulence agents against *Staphylococcus aureus*.” *Frontiers in Microbiology* 12: 663935. https://doi.org/10.3389/fmicb.2021.663935.

Arciola, Carla Renata, Stefano Campoccia, Luca Speziale, and Lucio Montanaro. 2022. “Biofilm formation in *Staphylococcus aureus* device-associated infections: A review of molecular mechanisms and therapeutic perspectives.” *International Journal of Molecular Sciences* 23 (1): 369. https://doi.org/10.3390/ijms23010369.

Aslam, Bushra, Mohsin Rasool, Mamoona Naz, and Muhammad Iqbal. 2021. “Antibiotic resistance: One health one world outlook.” *Frontiers in Cellular and Infection Microbiology* 11: 771510.

Bhandari, Parash, Dharmendra Singh, Harendra Kumar, Umesh Yadav, Bhaskar Sharma, and Ravi Kant. 2023. “Anti-quorum sensing and anti-biofilm potential of *Paederiafoetida* against MRSA.” *Frontiers in Microbiology* 14: 11273848. https://doi.org/10.3389/fmicb.2023.11273848.

Borges, A., S. Ferreira, M. J. Saavedra, and M. Simões. 2020. "Antibacterial activity and mode of action of ferulic and gallic acids against pathogenic bacteria." *Microbial Drug Resistance* 26 (3): 199–206. https://doi.org/10.1089/mdr.2019.0287.

Caldelari, Isabelle, Jana Chicher, Pierre Allard, Anne-Marie Arraou, and Brice Felden. 2021. "Small RNAs in *Staphylococcus aureus*: Key roles in regulation and pathogenesis." *Frontiers in Microbiology* 12: 706690. https://doi.org/10.3389/fmicb.2021.706690

Campbell, Jacquelyn, Jennifer L. Edwards, M. Fata Moradali, Yue Qian, and Ian Marriott. 2023. "Staphyloxanthin production by *Staphylococcus aureus* promotes persistent cutaneous infection and delays wound healing." *mBio* 14 (6): e01747.23. https://doi.org/10.1128/mbio.01747-23.

Chen, Jingwen, Xiangyu Li, Wenqiang Ma, Xiaoyu Deng, Jiahui Zhang, Min Wang, and Yanping Li. 2023. "Ellagic acid inhibits DNA gyrase activity and biofilm formation in methicillin-resistant *Staphylococcus aureus*." *Antibiotics* 12 (8): 746. https://doi.org/10.3390/antibiotics12080746.

Cheng, Guang-Hua, Li-Min Zhao, Wei Zhang, and Li-Li Guo. 2023. "The immune evasion roles of *Staphylococcus aureus* protein A and staphylococcal complement inhibitor." *Frontiers in Immunology* 14: 834711. https://doi.org/10.3389/fimmu.2022.834711.

da Costa, Thaina M., Deborah J. De Oliveira, Mariane C. Borges, et al. 2022. "Fibronectin binding protein B binds to loricrin and promotes corneocyte adhesion by *Staphylococcus aureus*." *Nature Communications* 13 (1): 2517. https://doi.org/10.1038/s41467-022-30271-1.

de Oliveira, Daniel Mendes, Juliana C. D. F. Ferreira, Patrícia S. de Sousa, Laura T. N. da Silva, Felipe T. Carneiro, and Reginaldo G. de Oliveira. 2020. "Synergistic interactions between flavonoids and antibiotics against *Staphylococcus aureus*." *Microbial Pathogenesis* 142: 104056. https://doi.org/10.1016/j.micpath.2020.104056.

de Oliveira, T. L. C., et al. 2020. "Antimicrobial mechanism of carvacrol against *Staphylococcus aureus*." *Food Control* 113: 107163. https://doi.org/10.1016/j.foodcont.2020.107163.

Fang, J., X. Li, X. Zhang, and W. Zhang. 2021. "Berberine as a potential antibacterial agent against *Staphylococcus aureus*." *Biomedicine & Pharmacotherapy* 133: 111025. https://doi.org/10.1016/j.biopha.2020.111025.

Girma, Abayeneh. 2024. "*Staphylococcus aureus*: Current perspectives on molecular pathogenesis and virulence." *The Cell Surface* 13: 100137. https://doi.org/10.1016/j.tcsw.2024.100137.

Gopikrishnan, M., S. Haryini, and G. Priya Doss. 2024. "Emerging strategies and therapeutic innovations for combating drug resistance in *Staphylococcus aureus* strains: A comprehensive review". *Journal of Basic Microbiology* 64 (5): eLoc, 2300579.

Husain, Fohad M., Mohd Saeed, Mohammad S. Alshahrani, and Huda A. Alajmi. 2021. "Naringin inhibits the biofilms of metallo-β-lactamases producing *Pseudomonas* species isolated from camel meat." *Saudi Journal of Biological Sciences* 28 (1): 333–341. https://doi.org/10.1016/j.sjbs.2020.10.004.

Khoramrooz, Seyed Sadegh, Mahdi Sedigh Ebrahim-Saraie, Mohammad Javad Nasiri, and Akram Baghani. 2023. "Molecular characterization of virulence genes and toxin production in clinical isolates of *Staphylococcus aureus* recovered from foodborne illnesses." *Toxins* 15 (11): 701. https://doi.org/10.3390/toxins15110701.

Kim, Jae, Ji-Young Lee, Hyun-Young Kim, and Dong-Gyun Kim. 2022. "Inhibition of *Staphylococcus aureus* biofilm formation and virulence factor expression by petroselinic acid." *Microbiology Spectrum* 10 (5): e01330.22. https://doi.org/10.1128/spectrum.01330-22.

Lakhundi, Saeed, and Kunyan Zhang. 2020. "Methicillin-resistant *Staphylococcus aureus*: Molecular characterization, evolution, and epidemiology." *Clinical Microbiology Reviews* 33 (4): e00020.19.

Lamret, Fabien, Jennifer Varin-Simon, Frédéric Velard, Christine Terryn, Céline Mongaret, Marius Colin, Sophie C. Gangloff, and Fany Reffuveille. 2021. "*Staphylococcus aureus* strain-dependent biofilm formation in bone-like environment." *Frontiers in Microbiology* 12: 714994. https://doi.org/10.3389/fmicb.2021.714994.

Lee, Arlene S., Jacqui Reilly, Benedikt Huttner, Stephan Harbarth, and Matthew H. Holmes. 2021. "Methicillin-resistant *Staphylococcus aureus*." *Nature Reviews Disease Primers* 7 (1): 10.

Lee, Ji Hoon, Jae-Hyung Lee, and Jintae Lee. 2023. "Antibiofilm effects of plant extracts against *Staphylococcus aureus*." *International Journal of Molecular Sciences* 24 (2): 607. https://doi.org/10.3390/ijms24020607.

Li, Changhong, Zhaoying Zhang, Rui Zhao, Qinghua Wang, Jiayi Tang, Jingxuan Zhang, and Liangzhi Shen. 2020. "Traditional Chinese medicine compounds suppress peptidoglycan biosynthesis genes in MRSA." *Frontiers in Pharmacology* 11: 563243. https://doi.org/10.3389/fphar.2020.563243.

Liu, Y., H. Liu, and X. Zhu. 2020. "Inhibition of MRSA biofilm formation by Matrine through *agr* and *sarA* pathways." *Microbial Pathogenesis* 149: 104312. https://doi.org/10.1016/j.micpath.2020.104312.

Liu, Y., J. Zhu, Z. Liu, Y. Zhi, C. Mei, and H. Wang. 2025. "Flavonoids as promising natural compounds for combating bacterial infections." *International Journal of Molecular Sciences*, 26: 2455. https://doi.org/10.3390/ijms26062455

Liu, Yuan, Chuanqi Zhang, and Ying Huang. 2021. "Rapid diagnostic technologies for *Staphylococcus aureus* infections: Current status and future perspectives." *Frontiers in Microbiology* 12: 649222.

Mahmud, Shahinur Rahman, Fahmida Akhter, Shaheda Islam, Tasmia Taznin, Tanjina Jahan Mou, Mohammad Shoeb, and Md. Mahbubur Rahman. 2023. "Quercetin from *Psidium guajava* attenuates biofilm formation by downregulating *icaA* expression in *Staphylococcus aureus*." *Heliyon* 9 (9): e18624. https://doi.org/10.1016/j.heliyon.2023.e18624.

Mayyas, Amal, Hanan Azzam, Iman Tayseer, Nidaa Al-Karablieh, and Talal Aburjai. 2021. "Evaluation of the synergistic antimicrobial effect of folk medicinal plants with clindamycin against methicillin-resistant *Staphylococcus aureus* strains." *Letters in Applied Microbiology* 73 (6): 735–740. https://doi.org/10.1111/lam.13565.

Musini, Anjaneyulu, Himanshu Narayan Singh, Jhansi Vulise, S. S. Sravanthi Pammi, and Archana Giri. 2024. "Quercetin's antibiofilm effectiveness against drug resistant *Staphylococcus aureus* and its validation by *in silico* modeling." *Research in Microbiology* 175 (3): 104091. https://www.sciencedirect.com/science/article/pii/S0923250823000669

Nostro, A., D. Cellini, M. Di Giulio, and L. Alonzo. 2021. "Antimicrobial and antibiofilm activity of caffeic and rosmarinic acids against *Staphylococcus aureus*." *Phytotherapy Research* 35 (6): 3371–3380. https://doi.org/10.1002/ptr.6998.

Osei Sekyere, John. 2020. "Current state of resistance to antibiotics of last-resort in South Africa: A review from a public health perspective." *Frontiers in Public Health* 8: 568.

Pacyga, K., P. Pacyga, E. Topola, S. Viscardi, and A. Duda-Madej. 2024. "Bioactive compounds from plant origin as natural antimicrobial agents for the treatment of wound infections." *International Journal of Molecular Sciences* 25: 2100. https://doi.org/10.3390/ijms25042100.

Patel, Hardi, and Seema Rawat. 2023. "A genetic regulatory see-saw of biofilm and virulence in MRSA pathogenesis." *Frontiers in Microbiology* 14: 1204428. https://doi.org/10.3389/fmicb.2023.1204428.

Peng, Qi, Xiaohua Tang, Wanyang Dong, Ning Sun, and Wenchang Yuan. 2023. "A review of biofilm formation of *Staphylococcus aureus* and its regulation mechanism." *Antibiotics* 12 (1): 12. https://doi.org/10.3390/antibiotics12010012.

Raghuram, Vishnu, Ashley M. Alexander, Hui Qi Loo, Robert A. Petit III, Joanna B. Goldberg, and Timothy D. Read. 2022. "Species-wide phylogenomics of the *Staphylococcus aureus* agr operon revealed convergent evolution of frameshift mutations." *Microbiology Spectrum* 10 (1): e01334.21. https://doi.org/10.1128/spectrum.01334-21.

Raut, J. S., S. R. Karuppayil, and R. B. Amaral. 2022. "Plant-derived terpenes and their synergistic effects with antimicrobials against *Staphylococcus aureus*." *Current Microbiology* 79 (1): 22. https://doi.org/10.1007/s00284-021-02643-7.

Reta, Alemayehu, Melaku Ashagrie Belete, and Feleke Moges. 2020. "Antimicrobial resistance and its association with biofilm formation in clinical isolates of *Staphylococcus aureus* from Ethiopia." *Infection and Drug Resistance* 13: 3991–4000.

Shariati, Alireza, Ebrahim Dadashi, and Ali-Mohammad Banar. 2020. "Horizontal gene transfer and antibiotic resistance in *Staphylococcus aureus*." *Microbial Drug Resistance* 26 (9): 1096–1105.

Silveira, Cíntia S., Ana C. Guterres, and Eduardo L. Foletto. 2022. "Plant spices as a source of antimicrobial synergic molecules to treat bacterial and viral co-infections." *Frontiers in Microbiology* 13: 973747. https://doi.org/10.3389/fmicb.2022.973747.

Sravanthi Pammi, S. S., B. Suresh, and A. Giri. 2023. "Antioxidant potential of medicinal plants." *Journal of Crop Science and Biotechnology* 26: 13–26. https://doi.org/10.1007/s12892-022-00159-z.

Swargiary, A., et al. 2022. "Plant-derived peptide gurmarin targets biofilm formation in *Staphylococcus aureus*." *Frontiers in Microbiology* 13: 957837. https://doi.org/10.3389/fmicb.2022.957837.

Turner, Nicholas A., Amee R. Sharma-Kuinkel, Henry F. Chambers, Arnold S. Holland, Frank R. Fowler, and Vance G. Fowler Jr. 2020. "Methicillin-resistant *Staphylococcus aureus*: An overview of basic and clinical research." *Nature Reviews Microbiology* 18 (4): 203–218.

von Eiff, Christoph, Christoph Heilmann, and Georg Peters. 2006. "Biofilm formation in medical device-related infection." *Clinical Microbiology and Infection* 12 (12): 117–123. https://doi.org/10.1111/j.1469-0691.2006.01592.x.

Wadhwa, Ranjan, Himanshu Sharma, and Vivek Kaushik. 2021. "Synergistic antibacterial activity of curcumin and piperine against multidrug resistant pathogens." *Frontiers in Pharmacology* 12: 723219. https://doi.org/10.3389/fphar.2021.723219.

Wang, Xiaoyan, Mengyang Liu, Xiangdong Yin, Yang Liu, Xiaoqing Song, and Yanwei Li. 2021. "Antibacterial effect of garlic sulfur compounds on *Staphylococcus aureus*: Membrane disruption and oxidative damage." *Microbial Pathogenesis* 150: 104713. https://doi.org/10.1016/j.micpath.2020.104713.

Wang, Yanhong, Hui Li, Xinying Wang, Xiaoyan Yang, Yanyan Wang, and Qianqian Ma. 2021. "Berberine combined with antibiotics inhibits the growth of *Staphylococcus aureus* by regulating the expression of resistance genes." *Frontiers in Pharmacology* 12: 721559. https://doi.org/10.3389/fphar.2021.721559.

Williams, Jessica, Mayumi Nakayama, and Jonathan D. Lippiat. 2020. "The type VII secretion system of *Staphylococcus aureus* contributes to membrane integrity and resistance to antimicrobial fatty acids." *Scientific Reports* 10: 14516. https://doi.org/10.1038/s41598-020-71653-z.

World Health Organization. 2020. *Global antimicrobial resistance and use surveillance system (GLASS) report: 2020. fourth annual report.* Geneva: World Health Organization. https://www.who.int/publications/i/item/9789240005587.

Wu, Z., J. Xu, X. Liu, and H. Zhong. 2020. "Epigallocatechin gallate inhibits the efflux pumps of *Staphylococcus aureus* and potentiates antibiotic activity." *Journal of Medical Microbiology* 69 (6): 864–871. https://doi.org/10.1099/jmm.0.001222.

Yang, Jiayu, Lu Cao, and Ming Yan. "Evolutionary dynamics of the accessory genomes of *Staphylococcus aureus*." *Frontiers in Microbiology* 15 (2024): 1293845. https://doi.org/10.3389/fmicb.2024.1293845.

Zhang, Xinyu, Yujie Wang, and Xiaofang Li. 2022. "A general map of transcriptional expression of virulence, metabolism, and biofilm formation adaptive changes of *Staphylococcus aureus* when exposed to different antimicrobials." *Frontiers in Microbiology* 13: 825041. https://doi.org/10.3389/fmicb.2022.825041.

Zhao, Wen, Xiaofeng Tao, Hui Chen, Ying Liu, and Bo Liu. 2020. "Epigallocatechin gallate enhances the antibacterial activity of β-lactam antibiotics against methicillin-resistant *Staphylococcus aureus*." *Frontiers in Microbiology* 11: 1721. https://doi.org/10.3389/fmicb.2020.01721.

Zhou, M., Y. Ren, and X. Wang. 2020. "Antibiofilm activity of thymol against clinical MRSA isolates through gene regulation." *Microbial Pathogenesis* 149: 104547. https://doi.org/10.1016/j.micpath.2020.104547.

Zhou, Yan, Yanyan Li, Yaping Zhang, and Zhihong Sun. 2023. "*Staphylococcus aureus* biofilm: Formulation, regulatory, and therapeutic strategies." *Journal of Microbial Pathogenesis* 165: 105472. https://doi.org/10.1016/j.micpath.2022.105472.

Section II

Host-Pathogen Interactions and Immune Evasion Mechanisms

6 Investigating Host Immune Response to Chronic *Staphylococcus aureus* Infection

Chittepu Obula Reddy[1], *Pothkanoori Tanishca*[2], *Manasa Dasari*[2], *Gunavardhini Dandu*[2], *Chittepu Sai Tharuni*[3], *and Yaso Deepika Mamidisetti*[4]

[1]Department of Biotechnology, Chaitanya Bharathi Institute of Technology, Gandipet, Hyderabad, Telangana, India

[2]Centre for Biotechnology, University College of Engineering Science and Technology (UCESTH), Jawaharlal Nehru Technological University, Hyderabad, Telangana, India

[3]Mamata Academy of Medical Sciences, Bachupally, Hyderabad, Telangana, India

[4]Department of Pharmacology, School of Allied and Healthcare Sciences, Malla Reddy University, Maisammaguda, Hyderabad, Telangana, India

ABBREVIATIONS

AMPs	antimicrobial peptides
APCs	antigen-presenting cells
AP1	activator protein 1
CWA	cell wall-anchored proteins
CP	capsular polysaccharides
CREB	cAMP response element-binding protein
CHIPs	chemotaxis inhibitory proteins
CTLA-4	cytotoxic T lymphocyte antigen 4
Can	collagen adhesin
ECM	extracellular matrices
Efb	extracellular fibrinogen-binding protein
Eap	extracellular adherence protein
ETs	exfoliative toxins
EGFR	epidermal growth factor receptor
FLIPr	FPRL1 inhibitory protein
IEC2	immune evasion cluster 2
MSSA	methicillin-sensitive *S. aureus*
MRSA	methicillin-resistant *S. aureus*
MSCRAMMs	microbial surface components recognizing adhesive matrix molecules
MyD88	myeloid differentiation primary response 88
NOD-1	nucleotide-binding oligomerization domain-containing protein 1

DOI: 10.1201/9781003647072-8

NF-kB	nuclear factor kappa-light-chain-enhancer of activated B cells
NET	neutrophil extracellular trap
PFTs	pore-forming toxins
PSMs	phenol-soluble modulins
PAMPs	pathogen-associated molecular patterns
PRRs	pattern-recognition receptors
PGRPs	peptidoglycan recognition proteins
PVL	panton–valentine leukocidin
SE	*Staphylococcus* enterotoxins SAK – staphylokinase
SCIN	staphylococcal complement inhibitor
Sags	superantigens
SSSS	staphylococcal scalded skin syndrome
SCIN	*Staphylococcus* complement inhibitor
TSST-1	toxic shock syndrome toxin-1
TSS	toxic shock syndrome
TLRs	toll-like receptors
VRSA	vancomycin-resistant strains

6.1 INTRODUCTION

Staphylococcus aureus is an opportunistic pathogen that normally colonizes the human anterior nares and also causes a wide range of clinical infections. An asymptomatic colonizer of the human nares, it is permanently colonized in about 30% of individuals. It is also the principal agent of skin and soft tissue infections, particularly in previously colonized individuals. Often self-resolving, local skin infections sometimes serve as a portal for this pathogen to deeper tissues and the bloodstream; in fact, it is the most commonly recognized source for *S. aureus* bacteremia (Kwiecinski and Horswill, 2020). It is a leading cause of bacteremia and infective endocarditis, as well as infections related to osteoarticular, skin and soft tissue, pleuropulmonary, and device-related infections (Tong et al., 2015). In addition, they cause the most common bacterial infections in humans and act as the causative agents of various human infections, including infective endocarditis, skin and soft tissue infections (e.g., impetigo, folliculitis, furuncles, carbuncles, cellulitis, scalded skin syndrome and others), osteomyelitis, septic arthritis, prosthetic device infections, pulmonary infections (e.g., pneumonia and empyema), gastroenteritis, meningitis, toxic shock syndrome (TSS), and urinary tract infections (Tong et al., 2015).

S. aureus is a non-motile, gram-positive, catalase-positive, non-spore-forming, facultative aerobic–anaerobic bacterium. It is a member of normal human flora, colonizing the skin and upper respiratory tract. It mainly colonizes in the anterior nares and exists with a high incidence as a commensal; about 20–25% of the population would be found always colonized by *S. aureus*, while about 60% have it as an intermittent carrier. A review from 2012 states that *S. aureus* bacteremia accounts for 20–50 cases per year per 100,000 inhabitants and that 10–30% of these patients risk death from the infection. A new study from 2017 shows that the number of people who die each year due to *S. aureus* bacteremia in the U.S. is 20,000 (Cheung et al., 2021).

S. aureus can form biofilm with a hydrated matrix of polysaccharides and proteins that allow cell adhesion. Adherent biofilms, especially in nasal polyps, may promote *S. aureus* since the defective epithelial barrier cannot remove bacterial colonization. *S. aureus* can replicate within the epithelial cells of nasal polyps, invading the mucosa and releasing chemical mediators that further add to the inflammation (Cheung et al., 2021).

Approximately 70% of isolates of *S. aureus* secrete a family of structurally related exotoxins, including TSS toxin-1 (TSST-1) and staphylococcal enterotoxin (SEs) serotypes SE-A, SE-B, SE-C, SE-D, SE-E, and SE-G, which are powerful T cell superantigens.

S. aureus infections are notably more difficult when resistance to antibiotics occurs, often along with isolated *S. aureus* strains. Most importantly, among these, methicillin-resistant *S. aureus* (MRSA) is an antibiotic-resistant *S. aureus*, which causes several hospital-acquired infections with high mortality and morbidity compared with infections caused by methicillin-sensitive *S. aureus* (Cheung et al., 2021).

S. aureus is resistant not only to methicillin but also to various other antibiotics. For instance, the resistance to conventional beta-lactam antibiotics – penicillin and its derivatives, which can be inactivated by beta-lactamases – is almost universal among *S. aureus*. The *S. aureus* then presents resistance to nearly all antibiotics available, nearly always in a mixed form. Vancomycin continues to be the last-in-line antibiotic available against MRSA infections, although cases of highly vancomycin-resistant strains have been reported, such strains not having spread, probably due to the acute fitness cost associated with vancomycin-resistance genes. In addition to specific antibiotic resistance, nonspecific antibiotic resistance by biofilm formation plays a role in many *S. aureus* infections that are biofilm-associated. These include prosthetic joint and all other indwelling medical device-associated infections, endocarditis, osteomyelitis, conjunctivitis, and other infections in which biofilm-associated *S. aureus* is involved (Cheung et al., 2021).

6.2 PATHOGENESIS OF *S. AUREUS*

S. aureus has an amazing ability to produce virulence factors, so it survives in the most extreme conditions in the human host. A common colonizer of the human skin, *S. aureus* easily becomes an opportunistic pathogen to cause pyogenic and systemic infections, both acute and chronic conditions, toxin-mediated states, and even cause syndromes in households and health institutions where it penetrates beyond the skin barrier. In contrast to the classical view of the microbe acting extracellularly, *S. aureus* has also been considered an intracellular pathogen. This allows persistence, protection against antibiotics, and immune avoidance, which are thought to be relevant to this feature. It has genetic plasticity which makes it potential for adaptation and versatile which enables this species to express strain-dependently. The virulence factors found in Staphylococci often replicate each other's roles in different pathogenic processes, leading to adherence to and aggression against host structures, followed by internalization, intracellular persistence, and immune evasion (Becker, 2018).

Most infections of *S. aureus* come from their asymptomatic colonization or perhaps even infrequently and in particular from infected fomites or other individuals in the hospital setting. Conventionally, the primary site for *S. aureus* colonization has been deemed the nares; however, *S. aureus* has the ability to colonize many skin sites apart from the intestine. Another common mode of infection in the hospital setting is by the contamination of indwelling medical devices. The primary mechanism of this route of infection is that *S. aureus* can quickly attach to the plastic of the devices and matrix molecules that soon cover the devices after insertion; thereafter, it forms biofilms in the device. This may also be the reason for menstrual staphylococcal TSS, wherein specific *S. aureus* strains producing TSS toxin-1 (TSST-1) form biofilms on high-absorbency tampons. A special type of acute *S. aureus* infection is food poisoning; this occurs when contaminated food containing SEs is eaten. The mechanism by which SEs induce emesis is poorly understood; it is thought to involve histamine release from intestinal mast cells. Like TSST-1, these toxins are superantigenic in that they activate T cells primarily in a nonspecific manner, resulting in an exaggerated immune response characterized by polyclonal T cell activation and rampant cytokine release. Acute staphylococcal food poisoning rarely leads to systemic infections. Lastly, *S. aureus* opportunistically exploits primary injury by other pathogens or predisposing conditions. *S. aureus* has also been shown to play a role in the causation of atopic dermatitis by specific toxins.

6.2.1 Virulence Factors of *S. aureus*

Based on their mechanisms, the virulence factors of *S. aureus* are classified into two groups. The first group consists of the invasion and inflammation-related virulence factors involved in establishment or colonization, the rendering of extracellular molecules supporting adhesion, and evasion of host defensive mechanisms. The second one includes them as toxins, which are secreted extracellularly to damage host tissue and increase the spreading of infectious bacteria and biofilm formation, which play a major role in certain infections (Faza et al., 2021).

6.2.1.1 Surface Proteins

S. aureus has various surface-bound adhesin proteins, which are covalently attached to the peptidoglycan layer and are involved in the adhesion and colonization step. Based on the presence of motifs identified through structure–function analysis, cell wall-anchored proteins are divided into four groups (Foster et al., 2014). The first and most abundant group is microbial surface components recognizing adhesive matrix molecules (MSCRAMMs), which includes collagen-binding protein, fibronectin-binding proteins A and B, clumping factor A and B proteins, and Sdr proteins (from SD Repeat) (Foster et al., 2014). The MSCRAMMs help in the adherence of bacteria to the plasma and to different host extracellular matrices, which include collagen, fibrinogen, and fibronectin. In addition to adhesion, the MSCRAMMs promote invasion of host cells and tissue, escaping immunological responses, and biofilm establishment by bacteria.

6.2.1.2 Polysaccharide Capsule

Capsular polysaccharides (CP) of *S. aureus* are typically classified depending on their immunological specificity. CP is suggested to provide increased virulence to *S. aureus*, permitting the bacteria to resist phagocytosis and destruction by polymorphonuclear phagocytes (Faza et al., 2021).

6.2.1.3 Biofilm Formation

A biofilm is a microbial population in which the cells exist in a surface-attached or cell-junction relationship and is embedded in an extracellular polymeric matrix protecting them. Such biofilms are not uniform but are heterogeneous and consist mainly of extracellular polysaccharides, proteins, DNA, and lipids. The biofilm matrix can also block or prevent certain classes of immune defense, including the macrophages, from getting into the biofilm matrix. *S. aureus* can attach and create biofilm on both biotic and abiotic surfaces. These staphylococci generally infect medical devices that are indwelling artifacts, such as implanted artificial heart valves, catheters, and joint prostheses, because of their presence on the human skin and mucosal surfaces. The biofilm structure can range from a single layer of clustered isolated cells to a thick, multilayered mucus structure with channels that permit the entry of both liquid and gas, as well as the transport of nutrients and waste components (Faza et al., 2021).

6.2.1.4 Immunomodulatory Proteins

Excreted proteins by *S. aureus* were observed to play a critical role in the innate and adaptive immune systems of the host. The major examples of these proteins include staphylokinase (Sak), extracellular fibrinogen-binding (Efb) protein, chemotaxis inhibitory protein of *S. aureus* (CHIPS), staphylococcal complement inhibitor (SCIN), formyl peptide receptor-like-1 inhibitory protein, and extracellular adherence protein (Rooijakkers et al., 2005).

6.2.1.5 Toxins

Main toxins from *S. aureus* can be divided into three major classes: pore-forming toxins, exfoliative toxins, and superantigens. Pore-forming toxins can further be classified into four types: hemolysins, which comprise hemolysin-α (Hla or α-toxin), hemolysin-β, leukotoxins, and phenol-soluble modulins (PSMs) (Grumann et al., 2014). This system of exoprotein secretion is in response to numerous and different conditions, and understanding this mechanism would be very important in gaining better control over staphylococcal infections.

The toxins of *S. aureus* can be involved in diseases such as TSS, staphylococcal scalded skin syndrome, necrotizing pneumonia, or deep-seated skin infections. They can damage the host cells' membranes by either reducing intercellular connections or modulating the immune response (Grumann et al., 2014).

6.3 HOST IMMUNE SYSTEM RESPONSE AGAINST *S. AUREUS*

S. aureus is one opportunistic, versatile pathogen that may equally exist harmlessly as part of normal flora or cause serious infections. During this colonization stage *S. aureus* is part of a polymicrobial community primarily of other staphylococci, corynebacteria, and propionibacteria. *In vivo*, however, *S. aureus* biofilms are more typically monomicrobial communities. These biofilms are established by surface-exposed proteins called MSCRAMMs; they essentially help the bacteria in attaching to the surfaces. This is further complicated by the addition of polysaccharides, proteins, and extracellular DNA into an extracellular matrix around the bacteria.

Biofilm-associated infections tend to become chronic because of the increased resistance to antimicrobial agents and evasion from immune responses. Such infections are related to implanted medical devices, and so they are an important pathogenetic factor in *S. aureus* infections in native tissues. Their acquisition of antimicrobial resistance through the mobile genetic elements or mutation in a chromosome has complicated staphylococcal infections in many ways and also the therapeutic measures (Chambers and Deleo, 2009). Penicillin first revolutionized the treatment of severe *S. aureus* infections, yet the acquisition of the β-lactamase encoding gene blaZ soon led to resistance. The mecA gene, responsible for resistance to anti-staphylococcal penicillins, has accounted for the worldwide spread of MRSA, now a major public health threat (Turner et al., 2019).

After breaking through the epithelial barrier and invading the host system, the successful establishment of an *S. aureus* infection depends on its ability to outsmart the host defense systems. Therefore, *S. aureus* escapes from the bloodstream, with the overwhelming majority of the cellular and humoral immune defenses, and invades organs and tissues to develop encapsulated abscesses. In skin or lung infections, abscesses may develop immediately when the epithelium is ruptured.

While still present in the bloodstream, *S. aureus* uses many methods to avoid being eliminated. These are production of toxins that include, among others, leukocidins, which kill phagocytes; mechanisms inducing phagocyte apoptosis; inhibiting complement factors; and agglutination and thrombosis (Cheung et al., 2021). This, therefore, enables *S. aureus* to be resistant to multiple strategies to combat the host and persist in establishing an effective infection.

To these infections, innate and adaptive immune responses are raised by the host immune system at different ends of the spectrum. The innate immune system offers immediate defense, characterized by physical barriers, phagocytic cells, and the release of antimicrobial peptides (AMPs), whereas the adaptive immune system builds a more specific and long-lasting response in which particular antibodies are produced and T cells activated for the eradication of infection and for immunologic memory so that the next exposure can be dealt with quickly, as shown in Figure 6.1.

The identification of this intricate host–*S. aureus* interplay may be the key to devising effective treatments and preventive measures for these infections.

6.3.1 Innate Immune Response

Nonspecific or innate immunity, also called natural immunity, acts as the first line of defense in the host's immune responses. It contains cells that identify and routinely destroy pathogens which enter the body. Innate immune responses are the primary defense mechanisms of the host under these conditions. These responses are then rapidly induced through pattern recognition routes, which detect nonspecific markers that are characteristic of bacterial infections (Kawai et al., 2010). A key

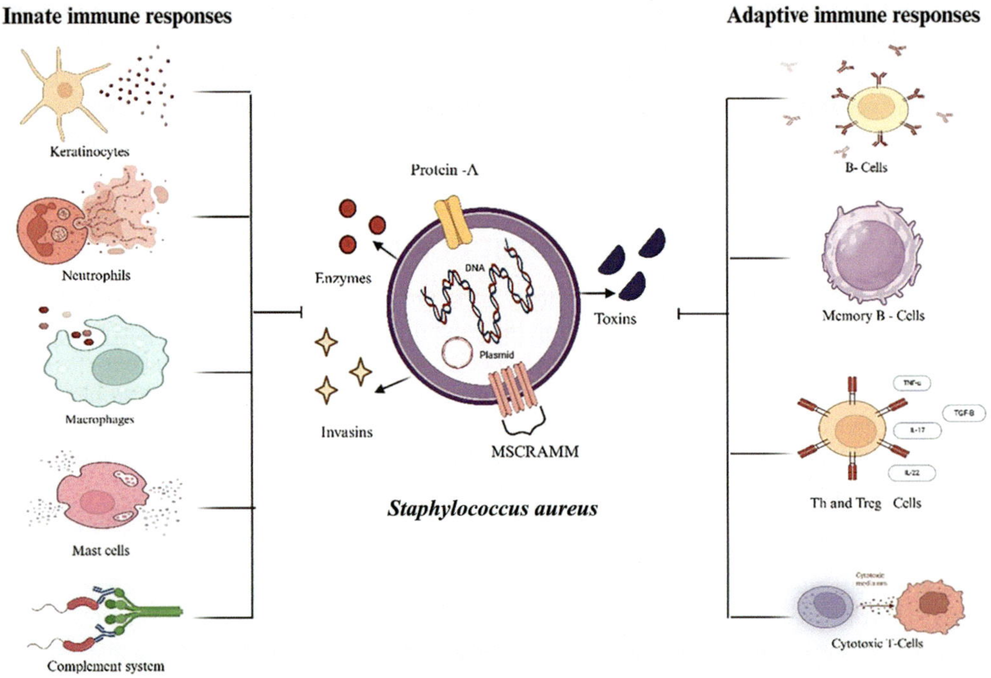

FIGURE 6.1 Various innate and adaptive immune responses against *Staphylococcus aureus* and its various antigens.

outcome of these responses is the activation of phagocytic immune cells such as neutrophils and macrophages. Innate immune cells are activated through toll-like receptors (TLRs) to enhance antibacterial responses, induce an inflammatory reaction, and activate effector cells, leading to the production of interferons and pro-inflammatory cytokines that induce fever.

The different receptor systems present in the innate immune system recruit a variety of cells that are on constant alert to detect invading pathogens and eliminate them. *S. aureus* breaches the body's barrier, such as skin and mucosa, through exposed epithelial surfaces. Their initial interaction with epithelial cells in this context is TLR2 mediated, inducing local secretion of soluble mediators, including cytokines, chemokines, and AMPs, to favor a proper balance of immune mediators and colonizing bacteria. Clearance of bacteria after cell and tissue invasion requires intracellular microbial sensors and the induction of inflammasomes. Effector local mast cells and macrophages recruit neutrophils, macrophages, and NK cells that, in turn, lead to the generation of IL-17-producing NK T cells, γδ T cells, and T helper cells. Local dendritic cells move to the draining lymph nodes to finalize the fine-tuning of the adaptive immune responses (Ding et al., 2015).

The skin is considered the largest organ, forming the primary physical barrier while playing an active role in local immune and inflammatory responses. Melanocytes, keratinocytes, Langerhans cells, and T lymphocytes are some of the important cells found in the organ. The immature dendritic cells, which are Langerhans cells, trap antigens that invade the host skin. An epidermal layer consists of keratinocytes at various levels of maturity; T lymphocytes also accompany them, as well as Langerhans cells. The dermis contains the extracellular matrix components, namely connective tissues, elastin fibers, and collagen, which support the blood vessels, fibroblasts, adipocytes, macrophages, dendritic cells, mast cells, T cells, B cells, plasma cells, and NK cells. The resident immune cells, which are abundant in the skin, assist it greatly in the control of infection caused by *S. aureus*. The ones which are important in the epidermis are keratinocyte cells, which help in protecting against outside exposures, including bacteria. They would, therefore, detect pathogen-associated molecular patterns (PAMPs) using various pattern recognition receptors (PRRs) like NOD-1 (nucleotide-binding oligomerization domain-containing protein 1), NOD-2, TLRs, and scavenger receptors such as CD36. The signaling of these receptors would activate transcription factors like AP1(activator protein 1), NF-kB (nuclear factor kappa-light-chain-enhancer of activated B cells), and CREB (cAMP response element-binding protein) to generate chemokines and cytokines, as mentioned in Table 6.1, and antimicrobial effectors such as AMPs and inducible nitric oxide synthase (iNOS). The TLR-1, TLR-2, and TLR-6 are receptors that can perceive cell wall components of *S. aureus* and especially peptidoglycan and lipopeptides (Fournier et al., 2013). These TLRs use the signaling adapter MyD88 (myeloid differentiation primary response 88) whereby, in this manner, they start strong transcriptional programs that initiate various inflammatory responses (Verstak et al., 2014).

6.3.1.1 Pattern Recognition Receptors

Keratinocytes play an important role in pathogen sensing. They detect PAMPs with the use of PRRs, including nucleotide-binding oligomerization domains NOD-1 and NOD-2, TLRs, and scavenger receptors like CD36 (Ibrahim et al., 2015). This signaling activates transcription factors AP1(activating protein-1), NF-kB (nuclear factor kappa B), and CREB (cAMP response element-binding protein) for the production of chemokines, cytokines, and antimicrobial effectors (Pasparakis et al., 2014). TLR-1, TLR-2, and TLR-6 are involved in recognizing some elements of the *S. aureus* cell wall, like peptidoglycan and lipopeptides. TLR-2 is particularly important in mediating the local production of soluble mediators (cytokines, chemokines, and AMPs) but also provokes antibacterial reactions to maintain the balance between immune mediators and the colonizing bacteria. Mice engineered to be deficient in TLR-1, TLR-2, TLR-6, and MyD88 were found to be more susceptible to infections from *S. aureus*, evidenced by higher bacterial loads and likely poorer inflammatory responses (Feuerstein et al., 2015).

6.3.1.2 Neutrophils

Neutrophils, one of the many different types of phagocytic white blood cells, are the primary elements participating in an acute response in humans against *S. aureus* infections. They are known to induce chemotaxis and perform phagocytosis and internalization, followed by the activation of degradative enzymes and antigen presentation. Neutrophils are potent to recognize the microbial component peptidoglycan through NOD-1 and NOD-2, as well as other PRRs, which stimulates the synthesis of AMPs, inflammation, and phagocytic activity. Here, NOD-2 molecules serve to detect

TABLE 6.1
Cytokine Responses in *Staphylococcus aureus* Infections

Cytokine	Source	Role in Immune Response	Effect on *S. aureus*
IL-1β	Macrophages, neutrophils	Pro-inflammatory, fever induction	Enhances neutrophil recruitment
IL-6	T cells, macrophages	Acute-phase response, fever, inflammation	Supports neutrophil proliferation and activation
IL-17	Th17 cells	Neutrophil recruitment, enhances antimicrobial peptides	Promotes abscess formation for bacterial clearance
TNF-α	Macrophages, T cells	Pro-inflammatory, enhances endothelial permeability	Increases immune cell infiltration at the infection site
IFN-γ	Th1 cells, NK cells	Activates macrophages, promotes phagocytosis	Inhibits the intracellular survival of *S. aureus*
IL-10	Regulatory T cells	Anti-inflammatory, suppresses overactivation	Limits excessive tissue damage during infection

muramyl-dipeptide from *S. aureus* peptidoglycan. At the very initial stages of infection, neutrophils are recruited by chemokines toward the site of infection, exerting bactericidal action. CHIPs and FPRL1 inhibitory protein (FLIPr) are some of the *S. aureus* virulence factors which are secreted in the body and have the ability to inhibit the neutrophil chemotaxis (de Haas et al., 2004).

Neutrophil extracellular traps (NET), where DNA is enveloped by histones, elastase, myeloperoxidase, and cathepsin G, are released from neutrophils. NETs capture and clear *S. aureus* at sites of infection; however, they also injure endothelial cells and other organs by virtue of their trapping function. Neutrophils kill pathogens by releasing toxic components from their granules or synthesizing and secreting AMPs such as LL37, defensins, and cathelicidin-related peptides. Degranulation also includes cathepsin, azurocidin, lysozymes, lactoferrin, elastase, and proteinase-3. The antimicrobial agents restrict the spread of bacteria by NETs, but *S. aureus* can degrade NETs, and, as a byproduct, it creates 2′-deoxyadenosine, which can trigger apoptosis in macrophages while also promoting the survival of bacteria in abscesses (Yipp et al., 2012).

6.3.1.3 Macrophages

Macrophages are found distributed in different parts of the body and play a very important role in the innate immune response against *S. aureus* (Sultan et al., 2019). The skin-resident macrophages thereby participate in the initial clearance of the pathogen, and they arguably act together with perivascular macrophages to modulate monocyte and neutrophil recruitment to infection sites. Dermal macrophages phagocytose and kill *S. aureus* via the generation of AMPs, reactive nitrogen and oxygen species, and chelating proteins to starve microbes of essential micronutrients (Prinz et al., 2014). They also release chemoattractants for recruiting neutrophils in a MyD88 and IL-1-R dependent fashion. Macrophages also help in the clearance of dead cells at the site of infection, which is quite important in the resolution process of the infection (Kobayashi et al., 2015).

6.3.1.4 Antimicrobial Proteins

Alpha-defensins, dermcidins, beta-defensins, cathelicidins, and RNase7 are various human skin AMPs showing anti-*S. aureus* activity (Prasad et al., 2019). Neutrophils release high amounts of HNPs, in which HNP2 peptides show the utmost antibacterial activity. A human skin constituent, keratinocytes produce RNase7, which inhibits *S. aureus* skin colonization. Stimulation of the epidermal growth factor receptor by skin wounding enhances hBD3 production, increasing antibacterial activity against *S. aureus*. The four beta-defensins in humans include hBD1, hBD2, hBD3, and hBD4, made by keratinocytes, macrophages, and dendritic cells. Among these, hBD3 has very potent activity against *S. aureus* (Prasad et al., 2019; van der et al., 2019).

6.3.1.5 Peptidoglycan Recognition Proteins

Humans possess four secreted proteins that are peptidoglycan recognition proteins (PGRPs) (PGRP1–4), involved in the detection and response to infections by bacteria. PGRP1 is synthesized in neutrophils and can bind within tertiary granules to *S. aureus* peptidoglycan, exerting bactericidal effect. PGRP2 is produced by keratinocytes and has active amidases, which cleave peptidoglycan of *S. aureus* (Dziarski et al., 2016). PGRP3 and PGRP4 are located in hair follicles, epidermis, sebaceous glands, and sweat glands, but their relevance to *S. aureus* infection is unknown (Dziarski and Gupta, 2010).

6.3.1.6 Complement System

Activation of the complement system is an important part of the innate immune defense that promotes cytolytic reactions and acute inflammatory responses and regulates the adaptive immune defenses. They are the three types of pathways that activate the complement formation; they are classical, alternative, and lectin pathways, and they form protease complexes, C3 convertase, at the surface of bacteria. As a result, microbial antigens are opsonized with C3b and iC3b, augmenting their ingestion by phagocytic cells through the complement receptors on phagocytic cells. A special factor, termed SCIN, secreted by *S. aureus*, binds to and inhibits the function of C3 convertases on the staphylococcal surface, thereby hindering effective immune responses (Wanke et al., 2011).

6.3.1.7 Mechanisms of Host Resistance at the Molecular Level

At a molecular level, various components and mechanisms contribute to the host's resistance against *S. aureus* infection:

1. **Toll-Like Receptors (TLRs)**: TLRs initiate signaling cascades through the recognition of microbes and the resultant production of pro-inflammatory cytokines and type I interferons. TLR2 is particularly involved in processing the wall components of *S. aureus* cells to activate NF-kB, leading to inflammatory mediator expression (Verstak et al., 2014).
2. **NOD-Like Receptors (NLRs)**: Intracellular peptidoglycan fragments are recognized by NLRs, specifically NOD1 and NOD2. Their activation stimulates AMP production and the biosynthesis of pro-inflammatory cytokines via pathways involving Rip2 and NF-kB.
3. **NETs**: Neutrophils discharge NETs composed of DNA, histones, and AMPs. NETs kill and imprison *S. aureus*. However, if these traps are not adequately regulated, they may also cause tissue injury. NET formation requires reactive oxygen species (ROS) and an enzyme called myeloperoxidase (Yipp et al., 2012).
4. **AMPs**: Produced by many cell types, including keratinocytes and neutrophils, AMPs such as LL-37-cathelicidin, defensins, and RNase7 have been found to have antimicrobial properties against microbes by rupturing microbial membranes and compromising their viability (Prasad et al., 2019).
5. **Reactive Oxygen and Nitrogen Species**: Reactive nitrogen species and ROS are produced within phagocytic cells during a respiratory burst and are then used for deactivating engulfed bacteria. Such an event mainly depends on the activities of the enzymes NADPH oxidase and iNOS (Rigby and DeLeo, 2012).
6. **PGRPs**: PGRP molecules make attachments to bacterial peptidoglycan and show bactericidal activity. PGRP1 is produced by neutrophils and binds peptidoglycan of *S. aureus*. On the other hand, PGRP2 is derived from keratinocytes and has amidase activity, cleaving peptidoglycan (Wolf and Underhill, 2018).
7. **Complement System**: The complement system acts as a bridge between the cellular and humoral systems, facilitating opsonization through which phagocytosis is augmented, as well as bacterial lysis by formation of the membrane attack complex. However, to *S. aureus*, SCIN is produced, which blocks the action of C3 convertases and thus inhibits the efficiency of the complement response (Wanke et al., 2011).
8. **Cytokine and Chemokine Production**: These are the cytokines and chemokines that the infected cells and immune cells emit to recruit and activate other immune cells. IL-1, TNF-α, and IL-17 are the primary cytokines involved in the recruitment of neutrophils and macrophages to sites of infection and amplification of the inflammatory response (Ding et al., 2015).

This way, the system of immunity in a host can efficiently combat *S. aureus* infection by suppressing the growth of bacteria and preventing the establishment of an extensive infection. But *S. aureus* has developed varied strategies to elude and defend itself from these mechanisms of immunity, thus making persistent infection a serious concern for diagnosis and treatment.

6.3.2 Adaptive Immune Response

The innate immune system provides the very first barricade in defending against microbial pathogens. Adaptive immune responses, however, supervene in the late stages of infection and mediate memory responses as well. The adaptive immune system consists of antibody- and cell-mediated immune responses performed by B cells and T cells, respectively, as shown in Figure 6.2.

6.3.2.1 B Cell Responses

The production of specific antibodies against *S. aureus* antigens is considered to be in the B-cell-mediated immune response against *S. aureus*. These antibodies take an active part in opsonizing the bacterium and facilitating antibody-mediated (and complement-mediated) ingestion of the bacteria by phagocytes such as neutrophils and macrophages (Holtfreter et al., 2010).

Inside the body, an *S. aureus* or MRSA infection leads to the production of multiple antibodies, some of which are directed against toxic agents (α-toxin, β and γ hemolysins, as well as panton–valentine leukocidin (PVL) and enterotoxins), virulence factors (e.g., aureolysin, IsdA, and superantigens), cell wall proteins (clumping factors ClfA/B and fibronectin-binding protein), and miscellaneous non-protein antigens, including CPs, lipoteichoic acid (LTA), and peptidoglycan (Holtfreter et al., 2010).

That is, the presence of *S. aureus* protein A, which binds immunoglobulins in the wrong orientation and allows *S. aureus* to escape antibody recognition and antibody-mediated phagocytosis (Krishna and Miller, 2012), exemplifies the importance of B cell responses and antibodies in host defense against *S. aureus* infection.

B lymphocytes secrete two types of antibodies (ASN-1 and ASN-2) that can neutralize the actions of major virulence factors secreted by *S. aureus*. ASN-1 neutralizes alpha-hemolysin (Hla), panton–valentine leukocidin, leukocidin ED (LukED), and γ-hemolysin, and ASN-2 neutralizes LukAB/LukGH (Yingqi et al., 2023). However, *S. aureus* further harbors mechanisms of immune evasion under the section of humoral immunity. The virulence factor SpA inhibits antibody-mediated phagocytosis by binding to antibodies and inducing apoptosis of B cells. IgG cleavage and degradation are caused by the enzyme Sak, thereby drastically nullifying IgG-mediated phagocytosis by neutrophils (Kurosaki et al., 2015).

The association of B cells and *S. aureus* leads to the interaction of CR2, which B lymphocytes express, with C3d that the antigen exhibits at the surface, which has multiple functions that serve to enhance antibody action. In this case, for cognitive B cells, the simultaneous binding of CR2 with C3b and the B lymphocyte antigen receptor complex decreases the activation threshold of B cells by several orders of magnitude, which is required for proper B cell activation (Yingqi et al., 2023). Non-cognitive B lymphocytes in lymph nodes and spleens of mice have been shown to have an important role in transferring immune complexes containing immunoglobulin M to follicular dendritic cells, necessary for affecting proper formation of germinal centers and antibody affinity maturation (Yingqi et al., 2023). This is receptor- and complement-dependent, but "unseen" CR2 B cells attached to antigen-containing immune complexes may also transfer these complexes to antigen-specific T cells to enable proper T cell activation. Molecules secreted by *S. aureus*, inhibiting the association of C3d-containing immune complexes with CR2, could hence impact the host's adaptive immune response against this bacterium (Yingqi et al., 2023).

Through humoral immunity, B lymphocytes develop and create antibodies to pathogens and thus play a significant role in recognizing and eliminating certain foreign agents (Cooper and Alder, 2006). Antigen-deficient lymphocytes act against the antigen as antigen-presenting cells and develop into memory cells when activated through antigen–antibody interactions. These memory cells again evoke the rapid production of antibodies on subsequent encounters with the same antigen. The main activity of B lymphocytes is immunoglobulin production that neutralizes target proteins or opsonizes microbial pathogens for optimized clearance and phagocytosis (Kurosaki et al., 2015).

Increased susceptibility to infection when B lymphocyte maturation is absent shows the importance of antibody-mediated protection; however, the absence of increased susceptibility to *S. aureus* argues for the unimportance of antibodies in protection against this pathogen. But early cutaneous *S. aureus* infection develops secondary antibody-mediated protection against subsequent infections in certain strains of mice, and preclinical studies show at least partial support for the elicited antibody stimulation induced by vaccination (Montgomery et al., 2014).

6.3.2.2 T Cell Responses

T cells perform a significant role within the cutaneous host immune defense against the *S. aureus* bacterium, particularly the T helper (Th) cell subsets (CD4+ T cells), which include Th1, Th2, and Th17 cells and have been implicated in the condition of *S. aureus*-caused skin infections (Cooper and Alder, 2006). The Th1 produces IFN-γ and thereby aids the cell-mediated immune response; Th2 produces IL-4 and IL-13 and helps in antibody-mediated immune response; and the Th17 cell produces IL-17 (i.e., IL-17A and IL-17F), IL-21, IL-22, and IL-26, inducing neutrophil recruitment and abscess formation.

Dendritic cells are important for T cell activation through antigen presentation, but in the case of *S. aureus*, its virulence factor, leukocidin AB (LukAB), kills the dendritic cells and prevents T cell activation (Montgomery et al., 2014). Cytotoxic T lymphocyte antigen 4 (CTLA-4) is an inhibitory molecule produced primarily by T cells; it attaches to CD80/CD86 on APCs and prevents their binding to CD28 on T cells, also interfering with the signaling processes of T cell activation and preventing pathogen clearance, thereby promoting infection persistence by inhibiting normal T cell development (Williams et al., 2011).

Transition from the acute to chronic phase of infection results in a loss of function in CD8+ T cells. Thus, they contribute to the development of a chronic infection. An example of a chronic infection could be osteomyelitis, where, in such cases, the cytotoxic T cells kill poorly, with a sustained antigen and inflammation stimuli. The reduced functioning compared to an effector CD8+ T cell is due to the manifestation of an exhausted CD8+ T phenotype. As a result, there is a decreased production of effector mechanisms (e.g., granzyme perforin) within an effector profile, as well as fewer cytokines such as IFN-γ, TNF-α, IL-12, and IL-18, causing improper pathogen elimination (Hirahara et al., 2011). Exhausted CD8+ T cells express many suppressor receptors, including PD-1, LAG-3, 2B4, CD160, TIM3, and TIGHT, which are immunosuppressive in persistent infections (Hirahara et al., 2011).

With regard to CD4+ T cell responses, where the Th1 cytokine IFN-γ would be associated with recruiting neutrophils, the effect of Th2-type cytokines like IL-4, IL-13, or even IL-10 is the enhancement of skin colonization or superinfection with *S. aureus* with skin lesions (Cooper and Alder, 2006). Thus, Th1 responses are protective, and Th2 responses vacillate negatively toward *S. aureus* skin disease. Again, Th17 cells are mediators promoting the recruitment of neutrophils and host defense activity against the skin disease caused by *S. aureus*. Cytokines like IL-17A and IL-17F made by Th17 cells generalize neutrophil recruitment and the formation of abscesses by stimulating chemokines (e.g., CXCL1, CXCL2, CXCL5, CXCL8) and granulopoiesis factors (G-CSF, GM-CSF). IL-17A, IL-17F, and IL-22 also induce keratinocytes to produce AMPs such as LL-37 and hBD2 (Spaulding et al., 2013).

Recruitment of neutrophils to the *S. aureus* site of infection in the skin, which is IL-17 mediated, is initiated by the stimulation of TLR2 and IL-1α/β and the production of IL-23, which activates different subtypes of T and NK cells (Th17 and γδ T cells, NK cells, and NKT cells) to produce IL-17A and IL-17F, and may even include neutrophils and mast cells. Thus, through the joint action of IL-17A and IL-17F, IL-17R on the keratinocytes is made to stimulate their activity via the production of pro-inflammatory cytokines, chemokines, and adhesion molecules subject to neutrophil recruitment (Cooper and Alder, 2006).

Adaptive immune responses to *S. aureus* infection involve B and T cells, possibly affecting the susceptibility to infections and determining the chronic infection outcome. B lymphocytes produce antibodies that neutralize toxins and opsonize bacteria for phagocytosis, whereas T cells, under the guidance of Th subsets, coordinate different modalities of immune response, including cytokine production, inflammation regulation, and memory formation. The interface for the cooperating activities between innate and adaptive immunity is thereby crucial for effective immune responses against *S. aureus*.

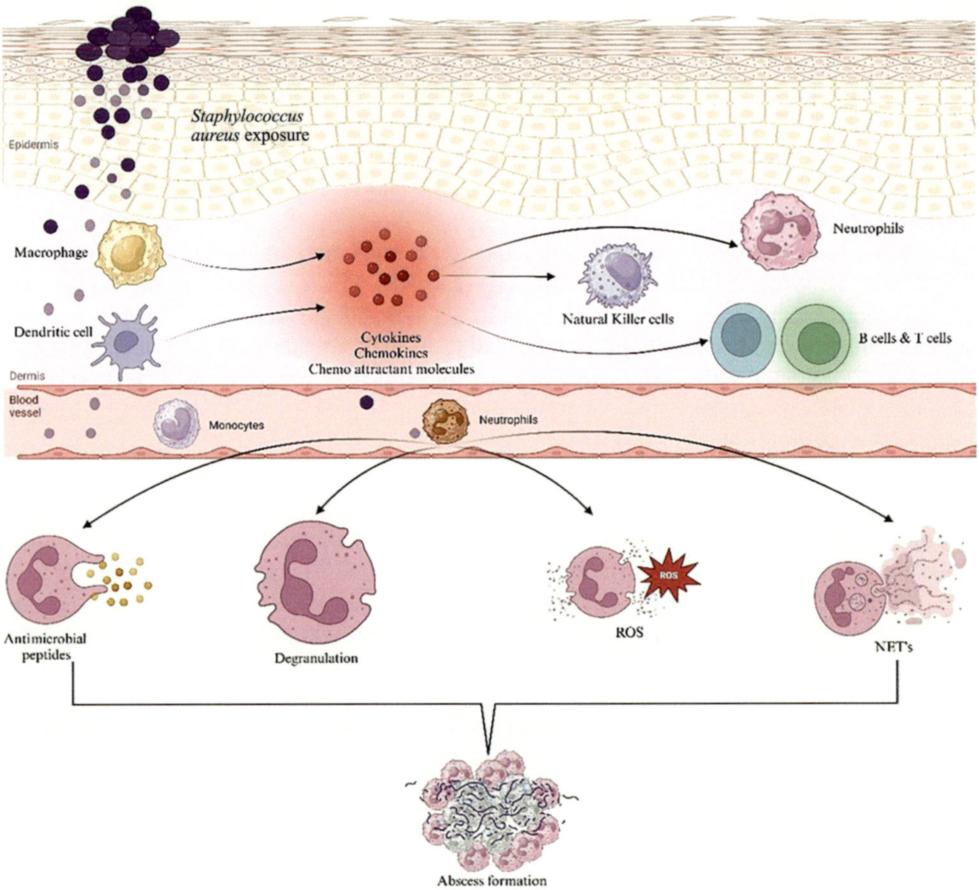

FIGURE 6.2 The immune responses shown by various immune cells against *Staphylococcus aureus* inside the host body.

6.3.2.3 T and B Cell Memory

Long-term effective agents of the adaptive immune responses involve memory T and B cells since they would provide long-standing protection against re-infection. After re-exposure to the antigen, memory B cells rapidly produce the antibodies, while activated memory T cells quickly mount an immune response. During *S. aureus* infections, memory T cells, especially the Th17, are instrumental in developing early individualized immune responses via their production of cytokines that enhance neutrophil recruitment and function. Memory B cells keep an active state to provide specific antibody production to *S. aureus* antigens.

6.3.2.4 Molecular Mechanisms in Adaptive Immunity Against *S. aureus*

1. **Antibody-Mediated Neutralization**:
   - **Binding and Opsonization**: Antibodies produced by B cells bind to *S. aureus* antigens, marking the bacteria for opsonization and subsequent phagocytosis by neutrophils and macrophages. The interaction of CR2 on B cells with C3d on the antigen surface enhances the efficiency of antibody-mediated responses (Yingqi et al., 2023).

- **Neutralization of Toxins**: Antibodies such as ASN-1 and ASN-2 are known to neutralize these virulence factors, such as alpha-hemolysin and leukocidins, which damage the host cells by these toxins (Yingqi et al., 2023).

2. **T Cell-Mediated Immunity**:

- **Production of Cytokines**: All T helper cells, that is, Th1, Th2, and Th17, secrete various cytokines that orchestrate processes in the immune response. For example, Th1 cells produce IFN-γ, which induces macrophages and enhances phagocytosis; on the other hand, Th17 cells secrete IL-17 and IL-22, which mainly function to attract neutrophils and stimulate the production of AMPs (O'Shea et al., 2010; Krishna and Miller, 2012).
- **Regulation by Treg Cells**: Regulatory T cells (Tregs) are important modulator components in immune homeostasis and inflammation. They are responsible for the secretion of some cytokine agents, such as IL-10 and TGF-β, which reduce the effector T cell activity and damage to tissues and organs.

3. **Memory Cell Formation and Function**:

- **Memory B Cells**: Once activated, B cells can differentiate into memory B cells, which show a quick and strong antibody response upon secondary exposure to *S. aureus* antigens. The immunity thus generated can remain for long periods and is extremely valuable for protection against re-infection (Rooijakkers, van Wamel et al., 2005).
- **Memory T Cells**: Memory T cells, including Th17 cells, persist after the primary infection and respond quickly to re-encounters by producing cytokines that stimulate recruitment and activity of neutrophils and other immune cells (Krishna and Miller, 2012).

4. **Immune Evasion by *S. aureus***:

- **Inhibition of Antibody Function**: *S. aureus* uses various immune evasion mechanisms, including protein A (SpA) production, which binds the antibodies non-functionally and

TABLE 6.2
Molecular Mechanisms of Host Resistance to *Staphylococcus aureus*

Molecule/Protein	Source	Function	Impact on *S. aureus*
TLR2	Keratinocytes, immune cells	Recognizes peptidoglycan, initiates NF-κB signaling	Promotes inflammatory cytokine production
MyD88	Immune cells	Adapter for TLR signaling	Activates downstream inflammatory response pathways
AMPs (Defensins, LL-37)	Keratinocytes, neutrophils	Disrupt bacterial membranes	Directly kills *S. aureus*
NOD1, NOD2	Keratinocytes, macrophages	Detect intracellular peptidoglycan	Activates antimicrobial responses
NETs	Neutrophils	Traps and kills bacteria	Degraded by *S. aureus* nucleases, leading to immune evasion
IL-17A, IL-22	Th17 cells	Enhances neutrophil recruitment and AMP production	Key to controlling skin infections and abscess formation

Sak for degrading the IgG antibodies, which reduces their efficiency (Rooijakkers, van Kessel et al., 2005).

- **T Cell Inhibition**: *S. aureus* secretes toxins such as leukocidin AB (LukAB), which kill dendritic cells and prevent T cell activation, thereby impairing adaptive immune responses.

Overall, such an adaptive immune response against *S. aureus* involves coordination between B and T cells, where both of them play crucial roles in making the pathogen ineffective and preventing infection, as mentioned in Table 6.2. Understanding these responses will make effective vaccines and therapeutics available to prevent *S. aureus* infection.

6.4 IMMUNE EVASION STRATEGIES BY *S. AUREUS*

S. aureus employs several sophisticated mechanisms to evade the host immune system, ensuring its persistence and virulence.

6.4.1 Mechanisms to Avoid Host Immune Detection

S. aureus has developed certain modes of action to escape detection or destruction by the host immune system, especially by phagocytes. Organisms can build protective encasements such as the CPs or biofilms and, inside these, secrete enzymes that help them evade immune recognition. Along with it, the organisms use molecules such as CHIPS and SCIN to block phagocyte receptors from formylated peptide recognition and complement function (Joubert et al., 2007). Getting into the phagocytes, *S. aureus* neutralizes ROS through the action of enzymes such as catalase and superoxide dismutase, with the production of staphyloxanthin against oxidative stress. To resist AMPs, it uses processes that include D-alanylation of teichoic acids, modification of surface charge by increasing the incorporation of cationic phospholipids to reduce negative surface charge, and increased expression of VraFG transporters. It generally produces a variety of toxins; that is, among all, leukocidins, such as α-toxin and PSMs, directly attack human leukocytes, improving its virulence (Shi et al., 2021).

6.4.1.1 Surface Protein Interactions

Staphylococcal Protein A (SpA) is a superantigen that is a cell wall-conserved protein and is able to induce anaphylactic responses upon cross-binding with anti-VH3 idiotype antibodies. Structurally, SpA consists of five Ig-binding domains binding to both Fc and Fab regions of antibodies with a preference for the VH3 region. It has a variable-number tandem-repeat region contributing to the fitness of the pathogen. Mutations in SpA may include loss of its anchorage to the cell wall or reduced superantigenic effect. An example of regulation of the process includes insertion into the membrane and spatial deposition via LTA (Zhang et al., 2021).

Clumping Factor A (ClfA) is an adhesin which binds to fibrinogen and also controls the complement system. It binds to the C-terminal γ-chain of fibrinogen/fibrin and, together with coagulase and vWbp, protects *S. aureus* from opsonophagocytic killing. Mutants missing ClfA are compromised in the pathogenesis of lethal bloodstream infection.

Staphylococcal Protein G (Sbi) is a secreted protein with two Ig-binding domains that interfere with the binding of C1q and prevent complement activation by the classical pathway. Sbi can also bind other proteins, such as C3 and factor H, to inhibit the alternative pathway. Additionally, Sak forms active complexes with plasminogen, degrading immune components and preventing the activation of the complement. Collagen adhesin (Cna) binds C1q to inhibit the activation of the classical pathway by blocking C1q–C1r association (Kang et al., 2013).

6.4.1.2 Inhibition of Immune Cell Functions

The various toxins and molecules secreted by *S. aureus* directly impinge on immune cell functions. Mimicking the action of superantigens, the Staphylococcal superantigen-like proteins (SSLs) impede immunological functions. SSL genes can be found in genomic island α (GIα) and immune evasion cluster 2 (IEC2), some of which are universally present in all *S. aureus* isolates (McCarty and Lindsay, 2013). SSL5 had such roles as the inhibition of neutrophil activation and enhancement of platelet adhesion, whereas SSL3 has the function of binding to TLR2 and blocking its immune recognition of staphylococcal components. SSL10 interferes with CXCR4-mediated responses on lymphocytes (McCarty and Lindsay, 2013).

Leukocidins, like PVL, can be defined as β-pore-forming toxins that target myeloid cells and erythrocytes by creating pore structures. These include HlgAB, HlgCB, and LukAB, which are very important for immune evasion. LukAB binds to CD11b on myeloid cells to induce the formation of NETs, thus assisting the escape of bacteria from phagosomes. Commonly found in community-acquired MRSA, PVL binds to C5aR on neutrophils and monocytes so that recruitment of the aforementioned immune cells into sites of infection occurs in the event of inflammation. In GIβ, LukED binds CCR5, CXCR1, and CXCR2 to lyse macrophages, dendritic cells, and T cells.

Alpha-toxin (Hla) induces pore formation in the membranes of neutrophils, macrophages, and epithelial cells, leading to the malfunctioning of these cells and contributing to tissue injury. Thus, it is one of the major pathological actions affecting the lungs during *S. aureus* infections in experimental models.

Sak is the other molecule that is known to cause the dissolution of fibrin clots by converting plasminogen into plasmin that would carry out the process of dissolution and thus bails out the bacteria from the traps caused by fibrin clots. Plasmin also degrades immunoglobulins and complement proteins, further hindering the immune responses.

Biofilm development is another mechanism of evasion of the immune system. A wide range of biofilm architectures are composed of proteins, polysaccharides, and extracellular DNA that act as a physical barrier against the immune response and the action of antibiotics. Secretions of components, including hemolysins and nucleases within the biofilms, can promote such immune response toward anti-inflammatory states (Ricciardi et al., 2018). Biofilms recruit immune-suppressive cells like MDSCs and M2-like macrophages, pushing the host to an endurance rather than a deficiency state. Host metabolites such as itaconate inhibit *S. aureus* glycolysis and ensure biofilm production and immune evasion. Long-term exposure to itaconate augments biofilm-forming abilities in *S. aureus* isolates (Tomlinson et al., 2021). Such metabolic adaptation promotes immune evasion, like in antibiotic resistance.

REFERENCES

Becker K. *Chapter 2 – Pathogenesis of Staphylococcus aureus*, Academic Press, 2018; pp. 13–38. ISBN 9780128096710. https://doi.org/10.1016/B978-0-12-809671-0.00002-4

Bekeredjian-Ding I, Stein C, Uebele J. The innate immune response against *Staphylococcus aureus*. In: Bagnoli F, Rappuoli R, Grandi G (Eds) *Staphylococcus aureus. Current topics in microbiology and immunology*, vol 409. Springer 2015. https://doi.org/10.1007/82_2015_5004

Chambers HF, Deleo FR. Waves of resistance: *Staphylococcus aureus* in the antibiotic era. *Nat Rev Microbiol* 2009; **7**: 629–641.

Cheung GYC, Bae JS., Otto M. Pathogenicity and virulence of *Staphylococcus aureus*. *Virulence* 2021; 12(1): 547–569. https://doi.org/10.1080/21505594.2021.1878688

Cooper MD, Alder MN. The evolution of adaptive immune systems. *Cell* 2006; 124(4): 815–822.

de Haas CJ, Veldkamp KE, Peschel A, Weerkamp F, Van Wamel WJ, Heezius EC, et al. Chemotaxis inhibitory protein of *Staphylococcus aureus*, a bacterial antiinflammatory agent. *J Exp Med* 2004; 199(5): 687–695. https://doi.org/10.1084/jem.20031636

Dziarski R, Gupta D. Review: Mammalian peptidoglycan recognition proteins (PGRPs) in innate immunity. *Innate Immun* 2010; 16(3): 168–174.

Dziarski R, Royet J, Gupta D. Peptidoglycan recognition proteins and lysozyme. *Mol Immunol* 2016; 389–403.

Feuerstein R, Seidl M, Prinz M, Henneke P. MyD88 in macrophages is critical for abscess resolution in staphylococcal skin infection. *J Immunol* (Baltimore, Md: 1950) 2015; 194: 2735–2745.

Foster TJ, Geoghegan JA, Ganesh VK, Höök M. Adhesion, invasion and evasion: The many functions of the surface proteins of *Staphylococcus aureus*. *Nat Rev Microbiol* 2014; 12: 49–62.

Fournier B. The function of TLR2 during staphylococcal diseases. *Front Cell Infect Microbiol* 2013; 2: 167

Grumann D, Nübel U, Bröker BM. *Staphylococcus aureus* toxins – Their functions and genetics. *Infect Genet Evol* 2014; 21: 583–592. https://doi.org/10.1016/j.meegid.2013.03.013.

Hirahara K, Vahedi G, Ghoreschi K, et al. Helper T-cell differentiation and plasticity: Insights from epigenetics. *Immunology* 2011; 134(3): 235–245.

Holtfreter S, Kolata J, Broker BM. Towards the immune proteome of *Staphylococcus aureus* – the anti-*S. aureus* antibody response. *Int J Med Microbiol* 2010; 300: 176–192.

Ibrahim F, Khan T, Pujalte GG. Bacterial skin infections. *Prim Care* 2015; 42(4): 485–499.

Joubert O, Voegelin J, Guillet V, Tranier S, Werner S, Colin DA, et al. Distinction between pore assembly by staphylococcal alpha-toxin versus leukotoxins. *J Biomed Biotechnol* 2007. https://doi.org/10.1155/2007/25935.200725935.

Kang M, et al. Collagen-binding microbial surface components recognizing adhesive matrix molecule (MSCRAMM) of Gram-positive bacteria inhibit complement activation via the classical pathway. *J Biol Chem* 2013; 288: 20520–20531. https://doi.org/10.1074/jbc.M113.454462.

Kobayashi SD, Malachowa N, DeLeo FR. Pathogenesis of *Staphylococcus aureus* abscesses. *Am J Pathol* 2015; 185(6): 1518–1527

Krishna S, Miller LS Innate and adaptive immune responses against *Staphylococcus aureus* skin infections. *Semin Immunopathol* 2012; 34: 261–280. https://doi.org/10.1007/s00281-011-0292-6

Kurosaki T, Kometani K, Ise W. Memory B cells. *Nat Rev Immunol* 2015; 15(3): 149–159.

Kwiecinski JM, Horswill AR. *Staphylococcus aureus* bloodstream infections: Pathogenesis and regulatory mechanisms. *Curr Opin Microbiol* 2020; 53: 51–60. https://doi.org/10.1016/j.mib.2020.02.005

McCarty AJ, Lindsay JA. *Staphylococcus aureus* innate immune evasion is lineage-specific: A bioinformatics study. *Infect Genet Evol* 2013; 19: 7–14. https://doi.org/10.1016/j.meegid.2013.06.012.

McHeyzer-Williams M, Okitsu S, Wang N, McHeyzer-Williams L. Molecular programming of B cell memory. *Nat Rev Immunol* 2011; 12(1): 24–34.

Montgomery CP, Daniels M, Zhao F, Alegre M-L, Chong AS, Daum RS. Protective immunity against recurrent *Staphylococcus aureus* skin infection requires antibody and interleukin-17A. *Infect Immun* 2014; 82(5): 2125–2134.

O'Shea JJ, Paul WE. Mechanisms underlying lineage commitment and plasticity of helper CD4+ T cells. *Science* 2010; 327:1098–1102.

Pasparakis M, Haase I, Nestle FO. Mechanisms regulating skin immunity and inflammation. *Nat Rev Immunol* 2014; 14(5): 289–301.

Prasad SV, Fiedoruk K, Daniluk T, Piktel E, Bucki R. Expression and function of host defense peptides at inflammation sites. *Int J Mol Sci* 2019; 21(1): 104.

Prinz M, Priller J. Microglia and brain macrophages in the molecular age: From origin to neuropsychiatric disease. *Nat Rev Neurosci* 2014; 15(5): 300–312.

Ricciardi BF, et al. *Staphylococcus aureus* evasion of host immunity in the setting of prosthetic joint infection: Biofilm and beyond. *Curr Rev Musculoskelet Med* 2018;11: 389–400. https://doi.org/10.1007/s12178-018-9501-4.

Rigby KM, DeLeo FR. Neutrophils in innate host defense against *Staphylococcus aureus* infections. *Semin Immunopathol* 2012; 34(2): 237–259.

Rooijakkers SH, van Kessel KP, van Strijp JA. Staphylococcal innate immune evasion. *Trends Microbiol* 2005; 13: 596–601.

Rooijakkers SH, van Wamel WJ, Ruyken M, van Kessel KP, van Strijp JA. Anti-opsonic properties of staphylokinase. *Microbes Infect* 2005; 7(3): 476–484. https://doi.org/10.1016/j.micinf.2004.12.014.

Shi M, Chen X, Sun Y, Kim HK, Schneewind O, Missiakas D. A protein A based *Staphylococcus aureus* vaccine with improved safety. *Vaccine* 2021a; 39(29): 3907–3915. https://doi.org/10.1016/j.vaccine.2021.05.072.

Spaulding AR, Salgado-Pabón W, Kohler PL, Horswill AR, Leung DY, Schlievert PM. Staphylococcal and streptococcal superantigen exotoxins. *Clin Microbiol Rev* 2013; 26(3): 422–447.

Sultan AR, Hoppenbrouwers T, Lemmens-den Toom NA, et al. During the early stages of *Staphylococcus aureus* biofilm formation, induced neutrophil extracellular traps are degraded by autologous thermonuclease. *Infect Immun* 2019; 87(12): 87.

Tomlinson KL, et al. *Staphylococcus aureus* induces an itaconate-dominated immunometabolic response that drives biofilm formation. *Nat Commun* 2021; 12: 1399. https://doi.org/10.1038/s41467-021-21718-y.

Tong SYCDavis JS, Eichenberger E, Holland TL, Fowler VG. *Staphylococcus aureus* infections: Epidemiology, pathophysiology, clinical manifestations, and management. *Clin Microbiol Rev* 2015; 28. https://doi.org/10.1128/cmr.00134-14

Turner NA, et al. Methicillin-resistant *Staphylococcus aureus*: An overview of basic and clinical research. *Nat Rev Microbiol* 2019; 17: 203–218.

van der Does AM, Hiemstra PS, Mookherjee N. Antimicrobial host defence peptides: Immunomodulatory functions and translational prospects antimicrobial peptides. *Adv Exp Med Biol* 2019; 149–171.

Verstak B, Stack J, Ve T, et al. The TLR signaling adaptor TRAM interacts with TRAF6 to mediate activation of the inflammatory response by TLR4. *J Leukoc Biol* 2014; 96(3): 427–436.

Wanke I, Steffen H, Christ C, et al. Skin commensals amplify the innate immune response to pathogens by activation of distinct signaling pathways. *J Invest Dermatol* 2011; 131(2): 382–390.

Wolf AJ, Underhill DM. Peptidoglycan recognition by the innate immune system. *Nat Rev Immunol* 2018; 18(4): 243–254.

Yehia FAZA, Yousef N, Askoura M. Exploring *Staphylococcus aureus* virulence factors; special emphasis on Staphyloxanthin. *Microbiol Biotechnol Lett* 2021; 49: 467–477. https://doi.org/10.48022/mbl.2107.07008

Yingqi C, Zixian L, Zexin L, Mincheng L, Yong F, Guanqiao L, Bin L. The effect of *Staphylococcus aureus* on innate and adaptive immunity and potential immunotherapy for *S. aureus*-induced osteomyelitis. *Front Immunol* 2023; 14. https://doi.org/10.3389/fimmu.2023.1219895.

Yipp BG, Petri B, Salina D, et al. Infection-induced NETosis is a dynamic process involving neutrophil multitasking *in vivo*. *Nat Med* 2012; 18(9): 1386–1393.

Zhang R, Shebes MA., Kho K., Scaffidi SJ., Meredith TC, Yu W. Spatial regulation of protein A in *Staphylococcus aureus*. *Mol Microbiol* 2021; 116(2): 589–605. https://doi.org/10.1111/mmi.14734.

7 Navigating *Staphylococcus aureus* Mastitis

Detection, Management, and Prevention

Aysha Sherieff[1], *Palasala Siddhartha*[2], *K. Rashmitha*[2], *Vishwas Somadri*[2], *Samshritha Veluri*[2], and *Zahoorullah S.*[3]

[1]Department of Biotechnology, Jawaharlal Nehru Technological University Anantapur (JNTUA), Andhra Pradesh, India

[2]Centre for Biotechnology, University College of Engineering and Science & Technology Hyderabad, Jawaharlal Nehru Technological University Hyderabad, Kukatpally, Hyderabad, Telangana, India

[3]Atal Incubation Centre, ALEAP WEHUB, Supported by AIM NITI Aayog GOI, ALEAP Industrial Area, Gajularamaram, Hyderabad, Telangana, India

ABBREVIATIONS

AMR	antimicrobial resistance
ELISA	enzyme-linked immunosorbent assay
FnBP	fibronectin-binding protein
MAA	milk amyloid A
NSAIDs	non-steroidal anti-inflammatory drugs
PCR	polymerase chain reaction
SCC	somatic cell count

7.1 INTRODUCTION

Mastitis is a disease that affects the mammary glands of cattle and can also affect breastfeeding humans. It is a common and challenging disease that causes inflammation of the udder tissue. Udder tissue supports the mammary gland, which comprises several other tissues, including the skin, superficial fascia, coarse areolar tissue, and parenchyma (Stanek et al., 2024). Mastitis may be caused by physical trauma, chemical irritation, or infection by pathogens such as bacteria, viruses, fungi, or mycoplasma. *Staphylococcus aureus* and *Streptococcal* bacteria are the two most common causative agents of this disease. Mastitis affects a variety of cattle, including dairy cows, beef cattle, goats, and sheep. Although it is typically a disease in females, occasional cases of mastitis also occur in male animals. Mastitis can be clinical, where symptoms are clear; subclinical, with few or no symptoms; and chronic, in which the inflammatory process lasts for several months (Cheng & Han, 2020).

Symptoms include decreased milk production, increased rectal temperature, lethargy, and anorexia. Stripping milk from a cow and examining it before milking is the best method for detecting most early cases of clinical mastitis. This is because changes in the milk are often the first sign

DOI: 10.1201/9781003647072-9

of mastitis. Mastitis can be managed using antibiotics, fluid therapy, and anti-inflammatory drugs. Milking routine hygiene, housing, bedding management, and culling chronically infected cows can all help limit the extent of the disease (Li et al., 2023).

7.2 CAUSATIVE AGENT

S. aureus, a non-motile, facultative anaerobe belonging to the Staphylococcaceae family, is the primary gram-positive pathogen linked to various clinical and subclinical mastitis cases (Tong et al., 2015). Characterized by its spherical or coccal shape, thick peptidoglycan layer, and inner lipid membrane layer, this bacterium is surrounded by a capsule that facilitates resistance to phagocytosis. Its genetic makeup includes circular DNA and plasmids, often carrying resistant genes, and it grows in grape-like clusters that typically exhibit a golden or yellow color, as depicted in Figure 7.1 (Ritchie et al., 2014).

S. aureus benefits significantly from causing mastitis in cattle because the infection creates an environment that supports its survival, growth, and replication (Dego & Vidlund, 2024). The mammary gland offers abundant nutrients, including lactose, proteins, and lipids, which are essential for bacterial growth. Additionally, the toxins and enzymes produced by *S. aureus* degrade host tissues, releasing cellular contents that serve as supplementary nutrients. Once inside the mammary gland, *S. aureus* is shielded from environmental stressors such as UV light and temperature fluctuations. The bacterium's ability to form biofilms provides further protection, making it highly resistant to antibiotics and immune system attacks, thus establishing a stable niche for long-term survival. Moreover, *S. aureus* employs immune evasion strategies, such as producing Protein A and capsules, to avoid destruction by immune cells. This prolongs its survival and enables persistent infections.

These chronic infections allow the bacteria to continuously thrive and exploit the mammary gland's resources over extended periods, ensuring a consistent supply of nutrients and favorable conditions for replication (Demontier et al., 2021). The primary reservoir of *S. aureus* is the chronically

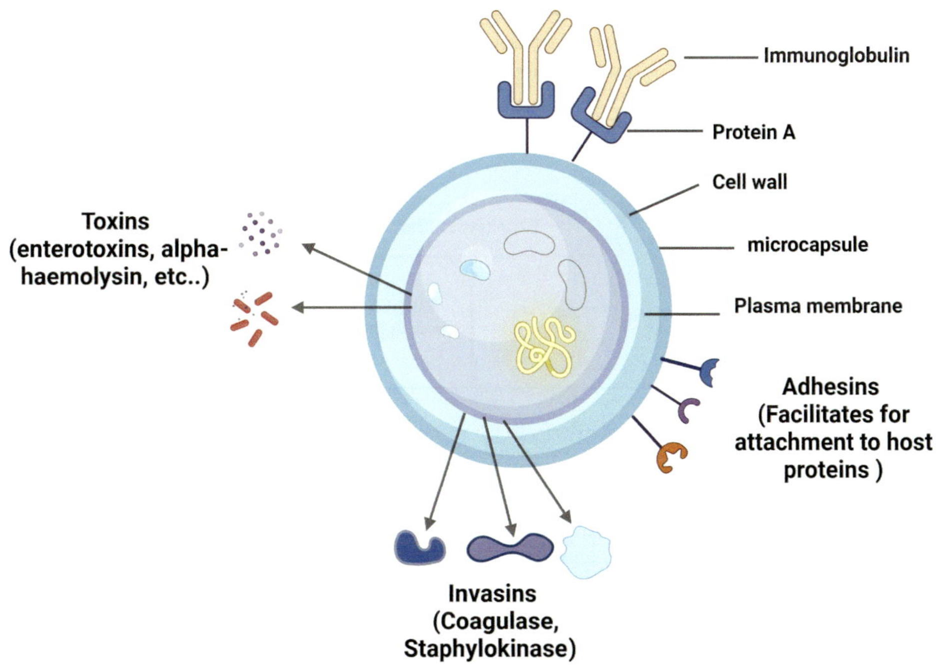

FIGURE 7.1 Structural and virulence factors of *Staphylococcus aureus.*

infected mammary gland. Therefore, maintaining proper udder and milking hygiene is crucial to preventing the transmission of infection from infected cows to healthy ones. Unlike *Escherichia coli* or endotoxins, *S. aureus* does not elicit a strong immune response in cows. As a result, infections caused by *S. aureus* are typically milder but tend to become chronic, lasting for several months. Although *S. aureus* infections rarely cause severe abnormalities or fatalities, the bacteria produce degradative enzymes and toxins that irreversibly damage milking tissue, leading to reduced milk production. Furthermore, *S. aureus* produces an exopolysaccharide called glycocalyx, which facilitates adhesion to mammary epithelial cells and nutrient acquisition (Vargová et al., 2023). This glycocalyx enables the bacteria to survive in high-shear environments and protects against environmental stresses, such as antibiotics, disinfectants, and immune system defenses like phagocytosis.

The production of glycocalyx is a significant factor that makes *S. aureus*, a gram-positive bacterium, difficult to treat with antibiotics. The glycocalyx prevents the antibiotic from reaching the bacteria at the minimum inhibitory concentration (MIC), leaving only sub-MIC levels that are insufficient to kill the pathogen. As a result, alternative therapies that specifically target the biofilm-forming ability of *S. aureus* are necessary for effective treatment (Notcovich et al., 2018).

7.3 MODE OF TRANSMISSION

S. aureus is one of the contagious pathogens that causes intramammary inflammation in cattle (Sharif et al., 2009). The pathogen that causes the disease usually resides in the mammary glands, rectum, rumen, and genital areas of the cattle. The infection is spread during milking, when the infected milk encounters an uninfected mammary gland. The bacteria then penetrate the teat canal (Petersson-Wolf et al., 2010).

The contagious spread of the pathogen occurs during milking through the milkers' hands (in Figure 7.2), even through the milking unit. Mastitis caused by *Mycoplasma* species spreads through the air. This type of transmission is termed aerosol transmission. The other ways of transmission include shedding of bacteria from infected udders, bedding materials, and host-adapted virulence factors. Contaminated teat dips, intramammary infusions, water used for udder preparations, water ponds, skin lesions, teat trauma, and flies also serve as the source of infection (Wieland, 2024). Another route of transmission is via oral to udder, in which the calves which feed on milk may introduce some pathogens in the oral cavity, where they will stay dormant until

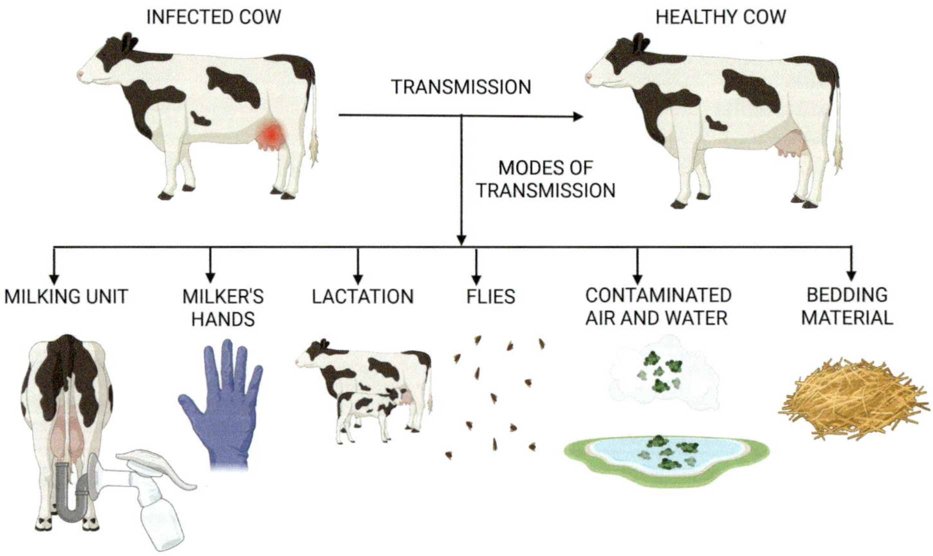

FIGURE 7.2 Modes of transmission of bovine mastitis.

transmitted elsewhere. These calves transmit the pathogen to other calves while stimulating suckling. The bacteria which are transmitted stay dormant in the udder until the calves grow and begin to lactate. During lactation, the bacteria become active and cause mastitis (Hopkin, 2002). Udders infected by pathogens are the common source of infection. The disease is mostly transmitted due to a lack of hygiene during milking, and this may be because the infected cows were not milked last (Ndahetuye et al., 2019).

Infections spread among cows, emphasizing multiple transmission pathways such as milking equipment, contact through humans, environmental factors, and other vectors like flies.

7.4 EPIDEMIOLOGY AND PATHOGENICITY

7.4.1 Epidemiology

Mastitis results in the inflammation of the mammary glands, which affects mostly all kinds of domestic animals and may finally lead to economic loss. It is very common in bovine animals in developing and developed countries. Mastitis can be classified into three types based on the inflammation rate, namely, clinical mastitis, subclinical mastitis, and chronic mastitis. Clinical mastitis can be more evident with symptoms like fever, red and swollen udder, and watery milk with clots and flakes, and it can sometimes be fatal.

Based on the degree of inflammation, it can be further classified into peracute, acute, and subacute. Clinical mastitis can be fatal. On the other side, subclinical mastitis is an asymptomatic mammary gland infection which increases the somatic cell count (SCC)and significantly causes economic losses in the dairy industry. Chronic mastitis is an inflammatory response which leads to the formation of hard lumps in the udder (Kurjogi & Kaliwal, 2014). Based on the origin of mastitis, it can also be classified into environmental and contagious.

Environmental mastitis can be spread through the pathogens present in faeces, soil, and water, whereas contagious mastitis can be spread through various factors, such as from cow to cow during milking and unhygienic dairy practices. Pathogens which use the contagious mode of transmission are mainly *S. aureus*, *Streptococcus agalactiae*, and *Corynebacterium bovis*. Subclinical mastitis is mainly caused by gram-negative rods such as *Klebsiella* spp., *Serratia marcescens*, *Pseudomonas aeruginosa*, and other atypical pathogens such as *Candida* spp. and *Prototheca zopfii* in herds of cattle (Dekker, n.d.). Major pathogens which are responsible for causing mastitis can be transmitted by repeated contact with contaminated milk machine, and environmental factors are also responsible for transmitting bovine mastitis by pathogen *S. aureus* (Deng et al., 2020).

7.4.2 Pathogenicity

Bacterial pathogenesis mainly involves the colonization of the host. Pathogenesis of *S. aureus* bovine mastitis infection mainly involves three specific phases: (1) adhesion to host cells and extracellular matrix, (2) invasion or penetration into tissues and cells, and (3) evasion of the host immune system, as shown in Figure 7.3.

Mainly, these phases are associated with the expression of different virulence factors at varying stages of infection. The expression is regulated by global regulators like Agr (accessory gene regulator) and the alternative transcriptional sigma factor SigB (Côté-Gravel & Malouin, 2018).

S. aureus has undergone several adaptations to adhere to different types of host cells. One such adaptation is that certain species of *S. aureus* consist of microbial surface component recognizing adhesive matrix molecules (MSCRAMMs), which have a similar structure to the N-terminal-folded domain, which is mainly responsible for ligand binding, and a spanning region, followed by a C-terminal sorting signal, secures the protein to the cell wall. Fibronectin-binding protein (FnBP), MSCRAMM, promotes the invasion of bovine mammary epithelial cells. Fibronectin mainly assists

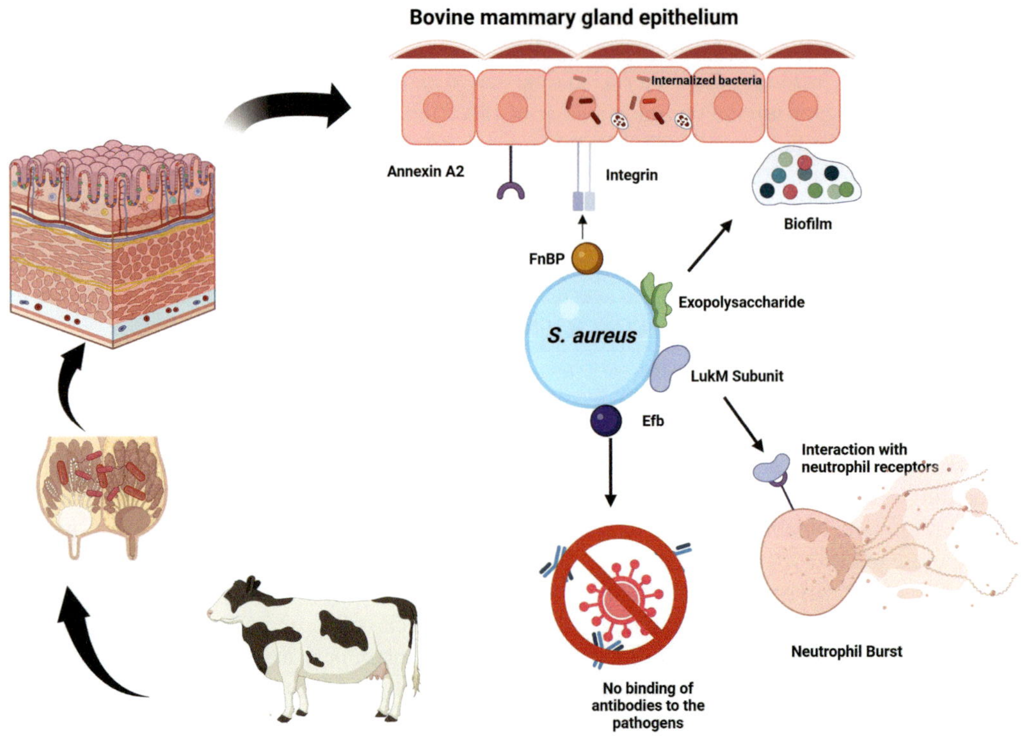

FIGURE 7.3 Modes of transmission of bovine mastitis infections among cows, emphasizing multiple transmission pathways such as milking equipment, contact through humans, environmental factors, and other vectors like flies.

in forming a linkage between FnBPA and $\alpha_5\beta_1$ integrin, which is present on the cell surface. This interaction leads to the induction of the assembly of a cytosolic protein complex, which can mediate bacterial uptake.

Not only to fibronectin, but *S. aureus* can also adhere to fibrinogen, elastin, and plasminogen in an FnBP-dependent manner, which promotes the colonization and spreading to other anatomical sites. FnBP is a key virulence factor of mastitis pathogenicity, which can mainly assist in the colonization of bacteria. Overexpression of *fnb* genes can result in increased invasion of bovine mammary epithelial cells. Especially, overexpression of FnBPB rather than the FnBPA can be related to an increase in invasiveness. On the contrary, these genes are not a prerequisite for the cell adhesion capability of bacteria, as their ability to invade is not lacking even when both genes are absent (Campos et al., 2022). Several exotoxins and enzymes are secreted after adhesion to desired host cells. These will aid in invading, penetrating, and destroying cells and tissues. For example, exotoxins like hemolysin and enzymes like exoenzymes contribute to the breakdown of epithelial tissue in the ducts and alveoli of the mammary glands, which mainly leads to the loss of milk production during mastitis in dairy cattle.

The pathogenesis of *S. aureus* induced bovine mastitis begins with the entry of the pathogen into the bovine mammary gland, where it establishes infection. *S. aureus* interacts with the bovine mammary gland epithelium via annexin A2 and integrin, which further facilitates bacterial internalization. Bacteria produce exopolysaccharides, aiding in biofilm formation and immune evasion. FnBPs help bacterial adhesion, while extracellular FnBP (Efb) prevents antibody binding, halting the host immune response. LukM subunit interacts with neutrophil receptors, inducing a neutrophil burst, which contributes to tissue damage and inflammation.

S. aureus exhibits several evasion strategies to evade the host immune response, that is, engulfing of *S. aureus* by the host's phagocytic cells of the innate immune system. One such evasion is by utilizing evasion proteins like Efb and the leukotoxin subunit (LukM) by many strains of Staphylococci. Efb protein can cover up the surface of *S. aureus* with a capsule-like shield to avoid binding of antibodies and subsequently removes the phagocytic neutrophils. On the contrary LukM subunit can interact with the target receptor on the surface of neutrophils and can eventually neutralize them. Certain strains of *Staphylococcus*, like *Staphylococcus epidermidis*, form biofilms with the help of their exopolysaccharide, which can evade phagocytosis by phagocytic cells of the innate immune system. This will assist in the protection and preservation of bacteria to thrive in the host system under harsh conditions (Vidlund et al., 2024).

7.5 DIAGNOSIS

Several methods have been developed for diagnosing mastitis in cattle (Figure 7.4). Diagnosis of mastitis is mainly based on observing the abnormal milk secretions, symmetry of teats and udder, and palpitation of the udder. Early identification of mastitis helps prevent udder and milk loss.

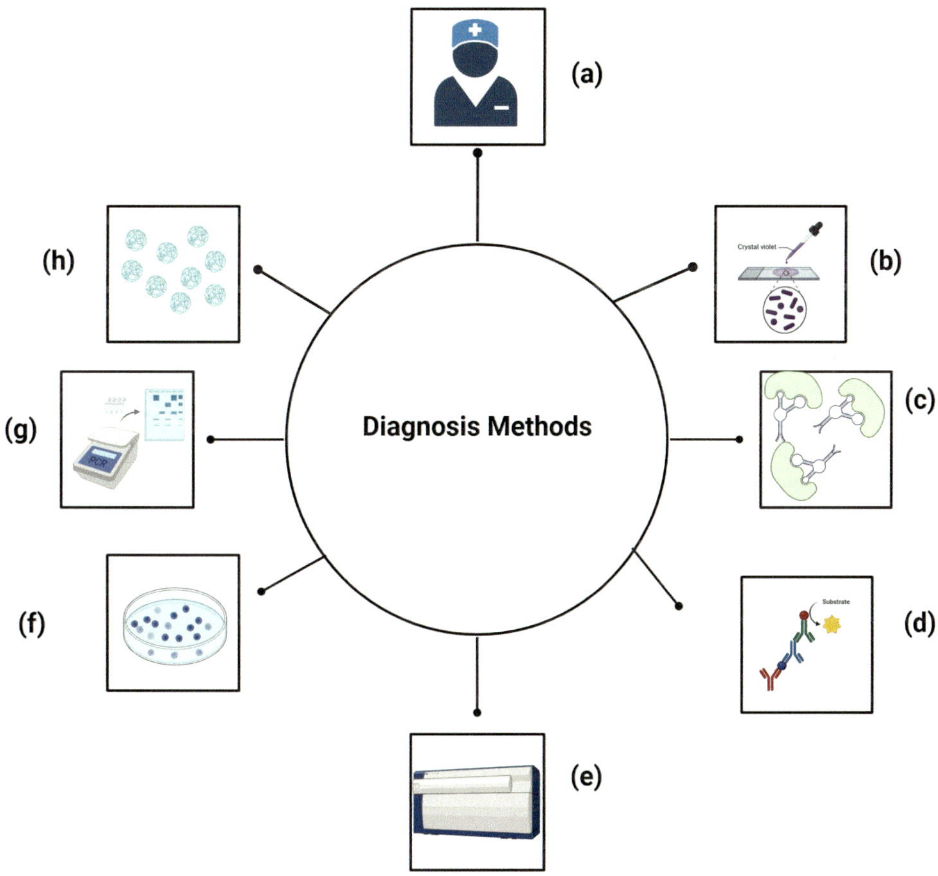

FIGURE 7.4 Diagnosis methods for bovine mastitis.

(a) Clinical signs; (b) bacteriological culture methods; (c) MAA biomarker test; (d) ELISA; (e) MALDI-TOF MS; (f) somatic cell count; (g) PCR; (h) nanotechnology and biosensor-based methods.

Mastitis in cattle can be diagnosed by visual observations. For clinical diagnosis, it is essential to see abnormalities under the udder, where the teat cups are attached. If any kind of changes are being observed in the quarter or in the milk production, it can be considered to have clinical mastitis (Lam et al., 2009). These kinds of changes in the udder tissue are caused by damage to the blood–milk barrier in the gland epithelial lining and breakage of tight junctions, leading to the leakage of blood, cells, and extracellular fluid components. Pus formation is also observed when foremilk is stripped out as the leukocytes are engaged toward the udder glands for fighting against the bacterial infection. The pus usually consists of dead bacterial cells and mammary gland cells. The change in udder tissue is characterized by swelling, redness or hyperemia, and the loss of milk yield (Dego & Vidlund, 2024).

Bacteriological culture method is the most feasible and widely used method for diagnosing the presence of bacteria, causing mastitis in the milk of quarter cattle. Samples from the normal cow and quarter cow are collected and sent to the laboratory for culturing. The collected samples can also be stored for further usage. These samples are then cultured on suitable media (mostly blood agar media is used), incubated at the required conditions, and then the cultured plates are examined. The shape, color, and size of the colonies were observed to identify the bacteria. Antibiotic sensitivity tests can also be done to find out the treatment strategies if the cow is confirmed for infection (Pyörälä, 2003).

Milk amyloid A (MAA) is an important biomarker used for the early diagnosis of clinical and subclinical bovine mastitis. It is an acute-phase protein that is released into the milk due to the inflammation of the mammary glands and, thus, acts as a marker of inflammation. In the milk of cows with mastitis, the levels of MAA are high, and thus they help in the detection of the disease (Hany Ahmed Hussein et al., 2018).

Enzyme-linked immunosorbent assay (ELISA) is a widely used diagnostic tool for detecting mastitis-causing pathogens and immune responses in dairy cattle. It is an immunological assay that identifies specific antigens or antibodies in biological samples based on antigen–antibody interactions. Aptamer-based ELISA (ALISA) is an advanced variation used for detecting bovine mastitis pathogens, including *S. aureus*. Aptamers, like antibodies, specifically recognize target proteins and can form sandwich-like complexes, enabling highly sensitive and specific pathogen detection. Similar to conventional ELISA, ALISA employs aptamer-coated surfaces to capture target antigens (Vivekananda & Kiel, 2006).

Cheng et al. developed an enhanced detection method using aptamer-coated magnetic beads combined with antibiotic-capped gold nanoclusters. Additionally, Baumstummler et al. identified ssDNA-based slow off-rate modified aptamers (SOMAmers), which incorporate pyrimidine residues and exhibit slow dissociation rates, for detecting *S. aureus*. Specifically, SpA-SOMAmers and ClfA-SOMAmers selectively recognize *S. aureus* proteins. Another related technique, the enzyme-linked oligonucleotide assay (ELONA), detects protein A-binding aptamers (PA#2/8) in *S. aureus* (Stoltenburg et al., 2016).

When compared to traditional methods, the MALDI-TOF MS method is faster and more accurate and requires less amount of sample. It is a proteomic-based approach, whose principle is based on ionizing the chemical compounds and generating charged molecules, followed by measuring the mass-to-charge ratio. It is used to identify bacterial isolates from the milk samples and then compare the results with those which are obtained by conventional biochemical methods (Randall et al., 2015; Barreiro et al., 2010). It has the ability to differentiate *S. aureus* from other bacteria (Dego & Vidlund, 2024).

SSC is an important indicator for intramammary infections (Lam et al., 2009). It provides the details of the quality of the milk and measures the number of somatic cells, particularly leucocytes, and epithelial cells of the mammary gland lining in normal cows, and then the details are compared with those of infected cows.

The cells can be counted by using either a normal microscope or an electronic cell counter. The electronic cell counter works on the principle of flow cytometry and provides rapid and accurate results (Sharma et al., 2011; Adkins & Middleton, 2018). A healthy cow has SSC less than 100,000 cells/mL, whereas infected cows have SSC more than 100,000 cells/mL or may be more than 350,000 cells/mL, depending upon whether the pathogens are major or minor (Dego and Vidlund, 2024). The California mastitis test is a semi-quantitative test, which is one of the widely used tests on farms to detect mastitis-causing pathogens. This method is relatively cheap and easy and gives quick results. It involves the indirect mixing of milk samples with reagents like 3% sodium lauryl sulfate, which facilitates the formation of gels (Dego & Vidlund, 2024). The degree of gel formation gives the approximate range of somatic cells present in the milk sample, which helps to diagnose the disease.

The presence of a bacterial DNA sample can be visualized using the PCR technique, various molecular markers are used for identifying the pathogens causing mastitis. Their genetic material (DNA/RNA) is isolated and amplified using primers, which are used for identifying the pathogens (Almeida et al., 2012). The milk samples are collected from the quarters, bacterial samples are extracted from them, and thermocycling is performed after adding the components required to carry out PCR. The amplified DNA samples are then run on an agarose gel and visualized (Lam et al., 2009). Real-time PCR gives faster results with high sensitivity. Apart from that, RT-PCR is used for quantifying the DNA of bacteria (Mahmmod et al., 2013; Koskinen et al., 2009).

Nanotechnology and biosensor-based methods have gained significant attention in the detection of mastitis, offering rapid and precise identification of pathogens responsible for the disease. These advanced techniques have introduced the concept of a laboratory "on a chip" (Driskell & Tripp, 2009). Mujawar et al. developed a nanoparticle-based assay utilizing three-dimensional nitrocellulose combined with microarray diagnostics. This method employs black carbon nanoparticles and protein conjugates to generate secondary signals upon tagging specific antibodies immobilized on the membrane. Using flatbed scanning, this assay enables the detection of mastitis-causing pathogens, including *S. aureus*, *Corynebacterium bovis*, *Mycoplasma bovis*, *Streptococcus agalactiae*, *Streptococcus dysgalactiae*, and *Streptococcus uberis*, within three hours.

A colorimetric assay for protease activity has also been utilized for mastitic milk identification, as demonstrated. This method employs magnetic nanoparticles conjugated with a plasmin substrate, forming a black self-assembled monolayer on a gold sensor surface. When the substrate is cleaved, peptide fragments linked to magnetic beads are released and attracted to a magnet on the sensor strips, leading them to the gold surface in the presence of elevated plasmin levels (Chinnappan et al., 2017).

7.6 TREATMENT

Mastitis caused by *S. aureus* is one of the significant challenges in the dairy farming industry as it shows a great impact on animals' health, milk production, and economic losses. Various treatment strategies have been developed to manage this condition, which include antibiotics, vaccines, and phage therapy. Treatment of mastitis involves varied approaches like antibiotic treatments, alternative antibacterial agent treatments, phage therapy, nanoparticles, immunomodulators, and herbal remedies (represented in Table 7.1). Antibiotics still remain the mainstay in the treatment of mastitis, whereas alternative treatments still show few challenges like antimicrobial resistance and biofilm formation, which need to be rectified. Further research is needed to upgrade these treatments so that we can include them in effective management protocols for mastitis.

7.6.1 Antibiotic Treatments

An extended cefquinome treatment regimen of five days did not significantly enhance bacteriological cure rates compared to the standard 1.5-day treatment. However, the longer treatment duration resulted in improved clinical cure rates. Additionally, pretreatment SCC was identified as a

TABLE 7.1
Treatment Strategies for *Staphylococcus aureus*-Induced Mastitis

Treatment Category	Specific Treatments	Mechanism/Effectiveness	Reference
Antibiotic treatments	Cefquinome Ampicillin and cloxacillin Penicillin and novobiocin	Extended five-day treatment improves clinical cure rates but not bacteriological cure rates Higher cure rates for subclinical mastitis when combined with enrofloxacin Helps in the intermittent isolation of *S. aureus* from milk and tissue samples	Swinkels et al. (2013) Garcia et al. (2022) Owens et al. (1990)
Alternative bacteriological agents	Recombinant endolysin (Trx-SA1) Specific egg yolk immunoglobulin (IgY) Phage therapy Nanoparticles Immunomodulatory treatments	Effective against mild clinical mastitis, including MRSA strains More effective than penicillin in reducing bacterial and somatic cell counts *StaphLyse* phage cocktail reduces *S. aureus* loads in mammary glands Improves drug penetration; reduces biofilm formation and antimicrobial resistance Cytokines like IL-1 and IL-2 enhance the immune response	Fan et al. (2016) Yu-Hong Zhen et al. (2009) Eric Brouilette et al. (2023) Algharib et al. (2020) Michael Daley et al. (1991)
NSAID therapy	COX-2 inhibitors (meloxicam, carprofen) COX-1 inhibitor (flunixin, meglumine) Ketoprofen	Reduces inflammation with fewer side effects than COX-1 inhibitors FDA-approved but may cause retained placenta issues Rapid action, reduces SCC, and improves milk quality	Wagner et al. (2021) Giammarco et al. (2018) Dan D et al. (2018)
Antimicrobial peptide therapy	Nisin	Produced by *Lactococcus lactis*, effective against gram-positive bacteria	Wu et al. (2007)
Herbal remedies	Betel leaf extract Red ginger	Antimicrobial activity varies based on the extraction methods Disrupts bacterial membranes, effective against *S. aureus*	Jannata et al. (2023) Poeloengan (2011)
Alternative therapeutic approaches	Short-chain fatty acids (SCFAs) Bacteriophage therapy	Strengthens mucosal immunity against *S. aureus* Recombinant endolysins (Trx-SA1) show promise against *S. aureus*	Akhtar et al. (2022) Fan et al. (2016)

predictor of bacteriological cure, with lower SCC values linked to higher success rates (Swinkels et al., 2013). Intramammary treatment with ampicillin and cloxacillin shows higher cure rates for subclinical mastitis when combined with another antibiotic called enrofloxacin (Garcia et al., 2022). Penicillin and novobiocin help in intermittent isolation of *S. aureus* from milk and tissue samples. This indicates that there are various challenges in achieving a complete bacteriological cure (Owens et al., 1990).

7.6.2 Alternative Bacteriological Agents

Recombinant endolysin (Trx-SA1) is a novel agent found mainly in IME-SA1 bacteriophage, which effectively controls mild clinical mastitis caused by the bacteria, including the methicillin-resistant strains (MRSA) (Fan et al., 2016). When compared to penicillin, specific egg yolk immunoglobulin (IgY) cures the disease effectively. It also reduces the SCCs and bacterial cell count in treated cows (Zhen et al., 2008). In a murine model, the StaphLyse phage cocktail shows bactericidal activity against *S. aureus*. Phage therapy is considered to be a viable alternative to antibiotics as it has both prophylactic and therapeutic uses, by which it can control bacterial loads in the mammary glands of cows (Brouillette et al., 2023). Various nanoparticles, like liposomes and polymeric nanoparticles, improve drug penetration, reduce antimicrobial resistance, and prevent biofilm formation, thus enhancing the treatment of *S. aureus* (Algharib et al., 2020). Cytokines like interleukin-1 and interleukin-2 enhances immune system, thus moderately aiding in treating this disease (Daley et al., 1991).

NSAIDs, which inhibit cyclooxygenase (COX) enzymes and reduce prostaglandin production, are increasingly used for managing mastitis-associated inflammation and pain (Ricciotti & Fitz Gerald, 2011). COX-2 selective NSAIDs, such as meloxicam and carprofen, have better therapeutic effects with fewer side effects compared to COX-1 inhibitors like flunixin meglumine, which is FDA-approved for fever and endotoxemia in mastitis but may increase the risk of retained placenta and digestive issues (Wagner et al., 2021). Meloxicam has been shown to reduce subclinical mastitis, improve milk production, and alleviate inflammation (Shock et al., 2018). Ketoprofen, with rapid action and no milk withdrawal, reduces SCC and blood–milk barrier damage (Dan et al., 2018). Carprofen, a long-acting COX-2 selective NSAID, reduces rectal temperature, heart rate, and udder swelling in mastitis cases (Lohuis et al., 1991).

Antimicrobial peptides are one of the best alternatives for replacing antibiotics. Many cells produce these antimicrobial peptides, which are the key components of the innate immunity (Zhang & Gallo, 2016). Nisin is one of such antimicrobial peptides produced by the bacteria *Lactococcus lactis*, which shows antibacterial activity against gram-positive bacteria like *S. aureus* (Sears et al., 1992). Nisin reduced SSC and also cured the disease effectively (Wu et al., 2007).

7.6.3 Herbal and Natural Remedies

Betel leaf extract has antimicrobial properties against *S. aureus*, due to which it can be used in the treatment of mastitis. The effectiveness of this extract varies based on the processing techniques, concentration of extract used, and also the adjuvants which are added to increase the activity of the extract. Red ginger shows good antibacterial effects on *S. aureus* by disrupting its extracellular membrane (Poeloengan, 2011).

7.7 PREVENTION AND CONTROL

7.7.1 Vaccine Development and Efficacy

Research efforts to create vaccines for *S. aureus* mastitis continue in response to the limitations of antibiotic treatments and the growing issue of antibiotic resistance. Despite extensive attempts, no vaccine currently on the market has proven to be both highly protective and cost-effective. This challenge stems from factors such as a lack of effective vaccine targets, significant strain variability, and inconsistent immune responses in cattle (Cote-Gravel & Malouin, 2019). While DNA-recombinant protein vaccines and bacterin-toxoid vaccines have shown encouraging results, differences in study methodologies prevent clear conclusions about their overall effectiveness (Pereira et al., 2011).

7.7.2 Management Practices

Effective management practices are crucial for controlling *S. aureus* mastitis. Key practices include post-milking teat disinfection, which is widely adopted and has been shown to reduce the prevalence of *S. aureus* in bulk tank milk (Maréchal et al., 2011; Ruegg, 2017). Blanket dry-cow treatment with antibiotics is associated with a lower probability of isolating *S. aureus* (Maréchal et al., 2011; Greg Keefe, 2012).

Strategic culling of infected cows can help manage and reduce the spread of infection (Goodger et al., 1987; Greg Keefe, 2012). Regular checks of milking equipment by independent technicians are recommended to ensure proper functioning and reduce infection risks (Maréchal et al., 2011). Given the resistance of *S. aureus* to many antibiotics, alternative treatments are being explored: for example, short-chain fatty acids (SCFAs) produced by gut microbiota have demonstrated potential as immunomodulators, aiding in the management of mastitis by strengthening mucosal immunity (Akhtar et al., 2022). Recombinant endolysins derived from bacteriophages, like Trx-SA1, have shown promising results in early trials for treating mild clinical mastitis caused by *S. aureus*, including MRSA (Fan et al., 2016).

7.7.3 Research Priorities and Knowledge Gaps

Several knowledge gaps and research priorities have been identified to improve the control of *S. aureus* mastitis: Research into the molecular mechanisms of *S. aureus* pathogenesis and the identification of protective antigens is of paramount importance. Investigating the cell-mediated immune responses to infection and vaccination can provide insights into developing more effective vaccines. Improved diagnostic methods for early detection and intervention are needed to manage infections more effectively (Rainard et al., 2017). Preventing and controlling *S. aureus* mastitis requires a multifaceted approach, including effective management practices, development of alternative therapeutic agents, and ongoing research to address knowledge gaps. While significant progress has been made, continued efforts are essential to develop more effective and sustainable solutions for this persistent dairy industry challenge.

7.8 CONCLUSION

Mastitis remains a significant concern in both dairy farming and veterinary medicine, affecting not only the health and well-being of animals but also the economic stability of farmers. This inflammatory condition, primarily caused by bacterial infections, presents a complex challenge that requires a multifaceted approach to prevention, diagnosis, and treatment. Over the years, continuous research has advanced our understanding of mastitis, leading to improved diagnostic methods and more effective treatment strategies. However, the battle against mastitis remains challenging to date due to the acquired resistance against antibiotics, which has paved the way for the development of alternative therapeutic approaches, including using probiotics, herbal remedies, and novel biotechnological solutions. Additionally, early detection and preventive measures, such as improved farm hygiene, proper milking techniques, and genetic selection for mastitis-resistant breeds, played a pivotal role in reducing the risk of spread of infections among the livestock. Beyond its impact on milk production and quality, mastitis also underscores the need for sustainable farming practices. The well-being of dairy animals directly influences productivity, making it essential for farmers, veterinarians, and researchers to work together in developing comprehensive solutions. By integrating scientific advancements with traditional knowledge and practices, we can lay the future where mastitis is no longer a major constraint in dairy farming. Ultimately, addressing mastitis is not just about treating an infection; it is about ensuring the health of livestock, the livelihood of farmers, and the quality of dairy products consumed worldwide. As we continue to explore innovative strategies, a collective effort of farmers and researchers is required to minimize the burden of pathogenic infections and build a more sustainable and efficient dairy industry.

REFERENCES

Adkins, P. R. F., & Middleton, J. R. (2018). Methods for diagnosing mastitis. *Veterinary Clinics of North America: Food Animal Practice*, *34*(3), 479–491. https://doi.org/10.1016/j.cvfa.2018.07.003.

Akhtar, M., Naqvi, S. U.-A.-S., Liu, Q., Pan, H., Ma, Z., Kong, N., Chen, Y., Shi, D., Kulyar, M. F.-e.-A., Khan, J. A., & Liu, H. (2022). Short-chain fatty acids (SCFAs) are the potential immunomodulatory metabolites in controlling *Staphylococcus aureus*-mediated mastitis. *Nutrients*, *14*(18), 3687. https://doi.org/10.3390/nu14183687

Algharib, S. A., Dawood, A., & Xie, S. (2020). Nanoparticles for treatment of bovine *Staphylococcus aureus* mastitis. *Drug Delivery*, *27*(1), 292–308. https://doi.org/10.1080/10717544.2020.1724209

Almeida, P., Mezzomo, N., & Ferreira, S. (2012). Extraction of Mentha spicata L. volatile compounds: Evaluation of process parameters and extract composition. *Food and Bioprocess Technology*, *5*(2), 548–559. https://doi.org/10.1007/s11947-010-0356-y.

Barreiro, J. R., Ferreira, C. R., Sanvido, G. B., Kostrzewa, M., Maier, T., Wegemann, B., Böttcher, V., Eberlin, M. N., & dos Santos, M. V. (2010). Short communication: Identification of subclinical cow mastitis pathogens in milk by matrix-assisted laser desorption/ionization time-of-flight mass spectrometry. *Journal of Dairy Science*, *93*(12), 5661–5667. https://doi.org/10.3168/jds.2010-3614.

Brouillette, E., Millette, G., Chamberland, S., Roy, J., Ster, C., Kiros, T., Hickey, S., Hittle, L., Woolston, J., & Malouin, F. (2023). Effective treatment of *Staphylococcus aureus* intramammary infection in a murine model using the bacteriophage cocktail StaphLyseTM. *Viruses*, *15*(4), 887. https://doi.org/10.3390/v15040887

Campos, B., Pickering, A. C., Rocha, L. S., Aguilar, A. P., Fabres-Klein, M. H., De Oliveira Mendes, T. A., Fitzgerald, J. R., & De Oliveira Barros Ribon, A. (2022). Diversity and pathogenesis of *Staphylococcus aureus* from bovine mastitis: Current understanding and future perspectives. *BMC Veterinary Research*, *18*(1). https://doi.org/10.1186/s12917-022-03197-5

Cheng, W. N., & Han, S. G. (2020). Bovine mastitis: Risk factors, therapeutic strategies, and alternative treatments – A review. *Asian-Australasian Journal of Animal Sciences*, *33*(11), 1699–1713. https://doi.org/10.5713/ajas.20.0156

Chinnappan, R., Al Attas, S., Kaman, W. E., Bikker, F. J., & Zourob, M. (2017). Development of a magnetic nanoparticle-based colorimetric assay for the detection of bovine mastitis in cow milk. *Analytical Biochemistry*, *523*, 58–64. https://doi.org/10.1016/j.ab.2017.02.009.

Côté-Gravel, J., & Malouin, F. (2018). Symposium review: Features of *Staphylococcus aureus* mastitis pathogenesis that guide vaccine development strategies. *Journal of Dairy Science*, *102*(5), 4727–4740. https://doi.org/10.3168/jds.2018-15272

Daley, M. J., Coyle, P. A., Williams, T. J., Furda, G., Dougherty, R., & Hayes, P. W. (1991). *Staphylococcus aureus* Mastitis: Pathogenesis and treatment with bovine interleukin-1β and interleukin-2. *Journal of Dairy Science*, *74*(12), 4413–4424. https://doi.org/10.3168/jds.s0022-0302(91)78637-2

Dan, D., Bruckmaier, R. M., & Wellnitz, O. (2018). Ketoprofen affects the mammary immune response in dairy cows in vivo and in vitro. *Journal of Dairy Science*, *101*, 11321–11329. https://doi.org/10.3168/jds.2018-15034

Dego, O. K., & Vidlund, J. (2024). *Staphylococcal* mastitis in dairy cows. *Frontiers in Veterinary Science*, *11*. https://doi.org/10.3389/fvets.2024.1356259

Dekker, A. (n.d.). *Mastitis pathogens in cows: Contagious or environmental?* https://europe.pahc.com/blog/mastitis-pathogens-in-cows-contagious-or-environmental

Demontier, E., Dubé-Duquette, A., Brouillette, E., Larose, A., Ster, C., Lucier, J., Rodrigue, S., Park, S., Jung, D., Ruffini, J., Ronholm, J., Dufour, S., Roy, J., Ramanathan, S., & Malouin, F. (2021). Relative virulence of *Staphylococcus aureus* bovine mastitis strains representing the main Canadian spa types and clonal complexes as determined using in vitro and in vivo mastitis models. *Journal of Dairy Science*, *104*(11), 11904–11921. https://doi.org/10.3168/jds.2020-19904

Deng, Y., Huang, Y., Ning, P., Ma, S., He, P., & Wang, Y. (2020). Maternal risk factors for lactation mastitis: A meta-analysis. *Western Journal of Nursing Research*, *43*(7), 698–708. https://doi.org/10.1177/0193945920967674

Driskell, J. D., & Tripp, R. A. (2009). Emerging technologies in nanotechnology-based pathogen detection. *Clinical Microbiology Newsletter*, *31*(18), 137–144. https://doi.org/10.1016/j.clinmicnews.2009.08.003.

Fan, J., Zeng, Z., Mai, K., Yang, Y., Feng, J., Bai, Y., Sun, B., Xie, Q., Tong, Y., & Ma, J. (2016). Preliminary treatment of bovine mastitis caused by *Staphylococcus aureus* with trx-SA1, recombinant endolysin of *S. aureus* bacteriophage IME-SA1. *Veterinary Microbiology*, *191*, 65–71. https://doi.org/10.1016/j.vetmic.2016.06.001

Garcia, B. L., Pinheiro, E. S., Fidelis, C. E., Freu, G., Leite, R. F., Moreno, A. M., & Santos, M. V. (2022). Efficacy of antimicrobial therapy in association with vaccination on the bacteriological cure of subclinical mastitis caused by *Staphylococcus aureus* in lactating cows. *Pesquisa Veterinária Brasileira*, *42*. https://doi.org/10.1590/1678-5150-pvb-7064

Goodger, W. J., & Ferguson, G. (1987). Benefits and costs of a control program for an epizootic of *Staphylococcus aureus* mastitis. *Journal of the American Veterinary Medical Association*, *190*(10), 1284–1287. https://pubmed.ncbi.nlm.nih.gov/3108206/

Hopkin, E. (2002). International dairy federation. In *Encyclopedia of Dairy Sciences* (pp. 1411–1413). Elsevier. https://doi.org/10.1016/b0-12-227235-8/00226-1.

Hussein, H. A., Abd El-Razik, K. A., Gomaa, A. M., Elbayoumy, M. K., Abdelrahman, K. A., & Hosein, H. I. (2018). Milk amyloid A as a biomarker for diagnosis of subclinical mastitis in cattle. *Veterinary World*, *11*(1), 34–41. https://doi.org/10.14202/vetworld.2018.34-41.

Keefe, G. (2012). Update on control of *Staphylococcus aureus* and *Streptococcus agalactiae* for management of mastitis. *Veterinary Clinics of North America: Food Animal Practice*, *28*(2), 203–216. https://doi.org/10.1016/j.cvfa.2012.03.010

Koskinen, M. T., Holopainen, J., Pyörälä, S., Bredbacka, P., Pitkälä, A., Barkema, H. W., Bexiga, R., Roberson, J., Sølverød, L., Piccinini, R., et al. (2009). Analytical specificity and sensitivity of a real-time polymerase chain reaction assay for identification of bovine mastitis pathogens. *Journal of Dairy Science*, *92*(3), 952–959. https://doi.org/10.3168/jds.2008-1549.

Kurjogi, M. M., & Kaliwal, B. B. (2014). Epidemiology of bovine mastitis in cows of Dharwad District. *International Scholarly Research Notices*, *2014*, 1–9. https://doi.org/10.1155/2014/968076

Lam, T., Olde Riekerink, R., & Sampimon, O. (2009). Mastitis diagnostics and performance monitoring: A practical approach. *Irish Veterinary Journal*, *62*(Suppl 4), S34. https://doi.org/10.1186/2046-0481-62-S4-S34.

Li, X., Xu, C., Liang, B., Kastelic, J. P., Han, B., Tong, X., & Gao, J. (2023). Alternatives to antibiotics for treatment of mastitis in dairy cows. *Frontiers in Veterinary Science*, *10*. https://doi.org/10.3389/fvets.2023.1160350

Lohuis, JA, van Werven, T, Brand, A, van Miert, AS, Rohde, E, Ludwig, B, et al. (1991). Pharmacodynamics and pharmacokinetics of carprofen, a non-steroidal anti-inflammatory drug, in healthy cows and cows with *Escherichia coli* endotoxin-induced mastitis. *Journal of Veterinary Pharmacology and Therapeutics*, *14*, 219–229. https://doi.org/10.1111/j.1365-2885.1991.tb00830.x

Mahmmod, Y. S., Toft, N., Katholm, J., Grønbæk, C., & Klaas, I. C. (2013). Bayesian estimation of test characteristics of real-time PCR, bacteriological culture, and California mastitis test for diagnosis of intramammary infections with *Staphylococcus aureus* in dairy cattle at routine milk recordings. *Preventive Veterinary Medicine*, *112*(3–4), 309–317. https://doi.org/10.1016/j.prevetmed.2013.07.021.

Maréchal, C. L., Thiéry, R., Vautor, E., & Loir, Y. L. (2011). Mastitis impact on technological properties of milk and quality of milk products—A review. *Dairy Science and Technology*, *91*(3), 247–282. https://doi.org/10.1007/s13594-011-0009-6

Ndahetuye, J. B., Persson, Y., Nyman, A., Tukei, M., Ongol, M. P., & Båge, R. (2019). Aetiology and prevalence of subclinical mastitis in dairy herds in pre-urban areas of Kigali in Rwanda. *Tropical Animal Health and Production*, *51*, 2037–2044. https://doi.org/10.1007/s11250-019-01905-2.

Notcovich, S., DeNicolo, G., Flint, S., Williamson, N., Gedye, K., Grinberg, A., & Lopez-Villalobos, N. (2018). Biofilm-forming potential of *Staphylococcus aureus* isolated from bovine mastitis in New Zealand. *Veterinary Sciences*, *5*(1), 8. https://doi.org/10.3390/vetsci5010008

Owens, W., Watts, J., Greene, B., & Ray, C. (1990). Minimum inhibitory concentrations and disk diffusion zone diameter for selected antibiotics against Streptococci isolated from bovine intramammary infections. *Journal of Dairy Science*, *73*(5), 1225–1231. https://doi.org/10.3168/jds.s0022-0302(90)78786-3

Pereira, U. P., Oliveira, D. G. S., Mesquita, L. R., Costa, G. M., & Pereira, L. J. (2011). Efficacy of *Staphylococcus aureus* vaccines for bovine mastitis: A systematic review. *Veterinary Microbiology*, *148*(2–4), 117–124. https://doi.org/10.1016/j.vetmic.2010.10.003

Petersson-Wolfe, C. S., Mullarky, I. K., & Jones, G. M. (2010). *Staphylococcus aureus* mastitis: Cause, detection, and control. *Virginia Cooperative Extension*, *404*, 1–7.

Poeloengan, M. (2011). The effect of red ginger (*zingiber officinale* roscoe) extract on the growth of mastitis causing bacterial isolates. *African Journal of Microbiology Research*, *5*, 382–388. https://doi.org/10.5897/AJMR10.776

Pyörälä, A. (2003). Diagnostic tools and methods for the detection of mastitis in dairy cattle. *Revue Scientifique et Technique*, *22*(3), 565–578. https://doi.org/10.20506/rst.22.3.1416

Pyörälä, S. (2003). Indicators of inflammation in the diagnosis of mastitis. *Veterinary Research*, *34*(5), 565–578.

Rainard, P., Foucras, G., Fitzgerald, J. R., Watts, J. L., Koop, G., & Middleton, J. R. (2017). Knowledge gaps and research priorities in *Staphylococcus aureus* mastitis control. *Transboundary and Emerging Diseases*, *65*(1), 149–165. https://doi.org/10.1111/tbed.12698

Randall, L. P., Lemma, F., Koylass, M., Rogers, J., Ayling, R. D., Worth, D., Klita, M., Steventon, A., Line, K., Wragg, P., et al. (2015). Evaluation of MALDI-ToF as a method for the identification of bacteria in the veterinary diagnostic laboratory. *Research in Veterinary Science*, *101*, 42–49. https://doi.org/10.1016/j.rvsc.2015.05.018.

Ricciotti, E., & Fitz Gerald, G. A. (2011). Prostaglandins and inflammation. *Arteriosclerosis, Thrombosis, and Vascular Biology*. *31*, 986–1000. https://doi.org/10.1161/ATVBAHA.110.207449

Ritchie, S. R., Thomas, M. G., & Rainey, P. B. (2014). The genetic structure of *Staphylococcus aureus* populations from the Southwest Pacific. *PLoS ONE*, *9*(7), e100300. https://doi.org/10.1371/journal.pone.0100300

Ruegg, P. L. (2017). A 100-year review: Mastitis detection, management, and prevention. *Journal of Dairy Science*, *100*(12), 10381–10397. https://doi.org/10.3168/jds.2017-13023 Medscape+2

Sears, P. M., Smith, B. S., Stewart, W. K., Gonzalez, R. N., Rubino, S. D., Gusik, S. A., et al. (1992). Evaluation of a nisin-based germicidal formulation on teat skin of live cows. *The Journal of Dairy Science*, *75*, 3185–3190. https://doi.org/10.3168/jds.S0022-0302(92)78083–7

Sharif, A., Umer, M., & Muhammad, G. (2009). Mastitis control in dairy production. *Journal of Agriculture and Social Sciences*, *5*, 102–105.

Sharma, N., Singh, N. K., & Bhadwal, M. S. (2011). Relationship of somatic cell count and mastitis: An overview. *Asian-Australasian Journal of Animal Sciences*, *24*(4), 429–438. https://doi.org/10.5713/ajas.2011.10233.

Shock, D. A., Renaud, D. L., Roche, S. M., Poliquin, R., Thomson, R., & Olson, M. E. (2018). Evaluating the impact of meloxicam oral suspension administered at parturition on subsequent production, health, and culling in dairy cows: A randomized clinical field trial. *PLoS One*, 13, e0209236. https://doi.org/10.1371/journal.pone.0209236

Stanek, P., Żółkiewski, P., & Januś, E. (2024). A review on mastitis in dairy cows research: Current status and future perspectives. *Agriculture*, *14*(8), 1292. https://doi.org/10.3390/agriculture14081292

Stoltenburg, R., Krafčiková, P., Víglaský, V., & Strehlitz, B. (2016). G-quadruplex aptamer targeting protein A and its capability to detect *Staphylococcus aureus* demonstrated by ELONA. *Scientific Reports*, *6*, 33812. https://doi.org/10.1038/srep33812.

Swinkels, J., Cox, P., Schukken, Y., & Lam, T. (2013). Efficacy of extended cefquinome treatment of clinical *Staphylococcus aureus* mastitis. *Journal of Dairy Science*, *96*(8), 4983–4992. https://doi.org/10.3168/jds.2012-6197

Tong, S. Y. C., Davis, J. S., Eichenberger, E., Holland, T. L., & Fowler, V. G. (2015). *Staphylococcus aureus* infections: Epidemiology, pathophysiology, clinical manifestations, and management. *Clinical Microbiology Reviews*, *28*(3), 603–661. https://doi.org/10.1128/cmr.00134-14

Vargová, M., Zigo, F., Výrostková, J., Farkašová, Z., & Rehan, I. F. (2023). Biofilm-producing ability of *Staphylococcus aureus* obtained from surfaces and milk of Mastitic cows. *Veterinary Sciences*, *10*(6), 386. https://doi.org/10.3390/vetsci10060386

Vidlund, J., Gelalcha, B. D., Gillespie, B. E., Agga, G. E., Schneider, L., Swanson, S. M., Frady, K. D., & Kerro Dego, O. (2024). Efficacy of novel staphylococcal surface-associated protein vaccines against *Staphylococcus aureus* and non-aureus staphylococcal mastitis in dairy cows. *Vaccine*, *42*(6), 1247–1258.

Vidlund, J., Gelalcha, B. D., Swanson, S., Fahrenholz, I. C., Deason, C., Downes, C., & Dego, O. K. (2022). Pathogenesis, diagnosis, control, and prevention of bovine staphylococcal mastitis. In *IntechOpen eBooks*. https://doi.org/10.5772/intechopen.101596

Vivekananda, J., & Kiel, J. L. (2006). Anti-Francisella tularensis DNA aptamers detect tularemia antigen from different subspecies by aptamer-linked immobilized sorbent assay. *Laboratory Investigation*, *86*(6), 610–618. https://doi.org/10.1038/labinvest.3700417.

Wagner, B. K., Nixon, E., Robles, I., Baynes, R. E., Coetzee, J. F., & Pairis-Garcia, M. D. (2021). Non-steroidal anti-inflammatory drugs: Pharmacokinetics and mitigation of procedural-pain in cattle. *Animals (Basel)*, *11*: 282. https://doi.org/10.3390/ani11020282

Wieland, M., Spellman, M. E., Case, K. L., Geary, C. M., & Sipka, A. (2024). Comparison of sensor-based and audible detection of milking liner slips during machine milking of dairy cows. *Sensors*, *24*(5), 1361. https://doi.org/10.3390/s24051361

Wu, J., Hu, S., & Cao, L. (2007). Therapeutic effect of nisin Z on subclinical mastitis in lactating cows. *Antimicrob Agents Chemother*, *51*, 3131–3135. https://doi.org/10.1128/AAC.00629-07

Zhang, L. J., & Gallo, R. L. (2016). Antimicrobial peptides. *Current Biology*, *26*, 14–9. https://doi.org/10.1016/j.cub.2015.11.01

Zhen, Y., Jin, L., Li, X., Guo, J., Li, Z., Zhang, B., Fang, R., & Xu, Y. (2008). Efficacy of specific egg yolk immunoglobulin (IgY) to bovine mastitis caused by *Staphylococcus aureus*. *Veterinary Microbiology*, *133*(4), 317–322. https://doi.org/10.1016/j.vetmic.2008.07.016

8 Chronic *Staphylococcus* Infections and Cancer

Mechanisms and Treatments

Omkar Dhurat[1, 2], *Surisetty Navyatha*[3], *Jogipeta Harihara*[3], *and Sharathchandra Arandkar*[1, 2]

[1]Advanced Centre for Treatment, Research and Education in Cancer (ACTREC), Tata Memorial Centre, Kharghar, Navi Mumbai, India

[2]Homi Bhabha National Institute, Mumbai, Maharashtra, India

[3]Centre for Biotechnology, University College of Engineering Science and Technology, Jawaharlal Nehru Technological University Hyderabad, Kukatpally, Telangana, India

ABBREVIATIONS

cfDNA	cell-free DNA
CRP	C-reactive protein
CT	computed tomography
CTCL	cutaneous T cell lymphoma
EPS	extracellular polymeric substance
ESR	erythrocyte sedimentation rate
FMT	fecal microbiota transplantation
IL-6	interleukin-6
MRI	magnetic resonance imaging
MRSA	methicillin-resistant *Staphylococcus aureus*
NF-κB	nuclear factor kappa-light-chain-enhancer of activated B cells
PCR	polymerase chain reaction
PET	positron emission tomography
S. aureus	*Staphylococcus aureus*
S. epidermidis	*Staphylococcus epidermidis*
SCV	small-colony variant
TGF-β	transforming growth factor-beta
TME	tumor microenvironment
TNF-α	tumor necrosis factor-alpha
VEGF	vascular endothelial growth factor
Wnt/β-catenin	Wnt/β-catenin pathway

8.1 INTRODUCTION

Staphylococcus aureus is one of the life-threatening pathogens responsible for more than one million deaths worldwide in 2019 ("Global burden of bacterial antimicrobial resistance in 2019: a systematic analysis" 2022). *Staphylococcus* is a species of gram-positive cocci that are commensal colonizers on our skin and mucous membranes, which can become pathogenic under certain

DOI: 10.1201/9781003647072-10

circumstances, depending on host–microbe interactions. *S. aureus* and *Staphylococcus epidermidis* are the most prevalent staphylococci responsible for disease in the human host. They are responsible for a spectrum of infections ranging from abscesses to serious systemic human diseases such as bacteremia, endocarditis, and osteomyelitis. They are a common inhabitant of the skin and mucous membranes and are frequent colonizers of the anterior nares. *Staphylococcus* infections are very common at surgical sites, especially in patients undergoing fracture fixation (Lowy 1998; Ahmann et al. 2025). The bacteria dwell on medical devices and predispose the host to severe infections. Chronic *Staphylococcus* infections typically present nonspecific symptoms that include low-grade fever, malaise, localized swelling, and persistent pain (Tong et al. 2015).

Acute infections are well-characterized and effectively treated, while chronic *Staphylococcus* infections constitute a more complicated and persistent clinical issue due to the development of immune evasion and antimicrobial resistance (Tong et al. 2015). Acute *Staphylococcus* infections have the general features of rapid onset, obvious symptoms and relatively brief disease course with prompt response to proper antibiotic therapy. Chronic infections take weeks to months to evolve, with frequent onset and offset of symptoms, thus hindering their diagnosis. One distinguishing feature of chronic infection is the development of biofilms, intricate communities of bacteria adhered and encased in an extracellular matrix secreted by the bacterial colonies to shield themselves against host defenses and antibiotics (Schilcher and Horswill Alexander 2020).

Oxidative stress, subsequent chronic inflammation, and autophagy are some of the common host responses against *Staphylococcus* infections. The bacteria undergo several changes and modulate the host system to overcome such protective responses. During these adaptations, the host undergoes alterations in signaling and autophagy pathways that lead to DNA damage and evasion of host immune responses. Chronic inflammation is one of the hallmarks of cancer, while autophagy acts as a tumor suppressor. Some of the autophagy pathway alterations are pro-tumorigenic in nature (Hanahan and Weinberg 2011). Chronic inflammation, oxidative stress, metabolic reprogramming, DNA damage, and induction of mutations are some of the ways microbes promote tumorigenesis. Most of the above mechanisms are prevalent in a majority of the *Staphylococcus* infections (Becker, Heilmann, and Peters 2014; El Tekle and Garrett 2023; Odunitan et al. 2024).

Many recent studies highlight the role of resident microbes in tumorigenesis, cancer progression, and fatal complications in different cancers. The *Staphylococcus* infections are associated with cancer in several different aspects. These include misdiagnosis of staphylococcal infections as cancer, the potential role of *Staphylococcus* in tumorigenesis and metastasis, and, in contrast, the use of staphylococcal toxins against cancer cells as therapy. In this review, we probe the association of chronic *Staphylococcus* infections with cancer and discuss the mechanism of *Staphylococcus*-mediated tumor progression, metastasis, and the use of staphylococcal toxins in cancer treatment.

8.2 CHRONIC *STAPHYLOCOCCUS* INFECTIONS: VIRULENCE, SUSCEPTIBILITY, AND MECHANISM OF PERSISTENCE

The genus *Staphylococcus* comprises many commensals and opportunistic species of gram-positive cocci. Of these, *S. aureus* and *S. epidermidis* are the clinically most important. *S. aureus* is virulent, as it produces a variety of toxins and enzymes that aid in tissue destruction and immune system evasion. *S. epidermidis*, though less virulent, is one of the most common causes of device-associated infections owing to its capacity to produce biofilms. These microorganisms switch from commensals to pathogens under favorable conditions, such as disruption of the skin barrier or immunosuppression. They are adaptable and genetically plastic, enabling them to cause persistent infections. It has been proven that low-virulence strains can also produce chronic infections with the formation of biofilms (Otto, 2009). Their ubiquity within hospital settings and their correlation with surgery make them an ongoing risk, particularly in immunocompromised hosts.

8.2.1 Virulence and Mechanism of Persistence

The most important mechanisms of persistence employed by *S. aureus* are immune evasion, biofilm formation, toxin production, intracellular survival, and hyperimmune activation (Figure 8.1). Several factors, including suboptimal antibiotic therapy, compromised immune function, and bacterial intracellular survival, are responsible for the conversion of acute *Staphylococcus* infections into chronic infections. Within host tissue, *Staphylococcus* is able to survive in phagocytes, evading immune detection and the effects of antibiotics. Yet another significant consideration is the expression of surface adhesins and exopolysaccharides, which aid in attachment to host tissue and medical devices. Cumulative tissue damage and inflammation are conducive to bacterial persistence and chronicity. The ability of *S. aureus* to form abscesses also increases the difficulty in treatment and elimination. Genetic mutations acquired by the bacteria during the course of infection can also increase bacterial fitness in the host environment (Proctor et al., 2006).

S. aureus produces super antigens (SAg), leading to the indistinct activation of T cells and production of cytokine storm, again adding to chronic inflammation (Zhang et al. 2024). The interaction between chronic inflammation and immune evasion plays an important role in allowing long-term infection persistence (Foster, 2005).

8.2.2 Susceptibility to Infection

Environmental and genetic factors play a vital role in susceptibility to *Staphylococcus* infections. Environmental factors, including substandard hospitalization, in dwelling medical devices, poor post-operative measures, and poor hygiene, predispose the host to *Staphylococcus* infection. Biofilm formation is facilitated by moist and nutrient-rich conditions, such as wounds and implanted medical devices. Additionally, horizontal gene transfer allows *Staphylococcus* species

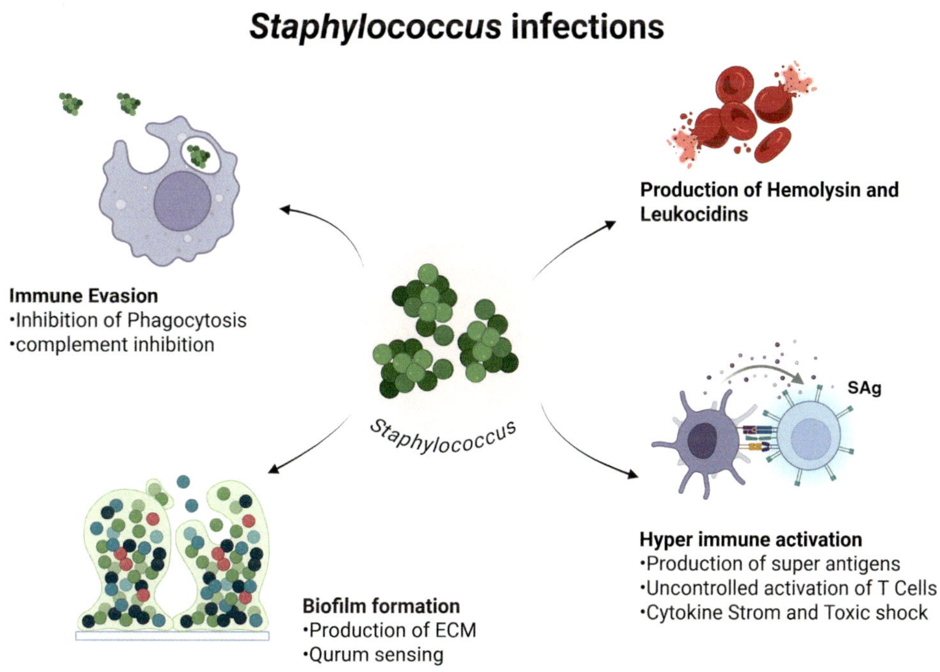

FIGURE 8.1 *Staphylococcus aureus* mechanisms of persistence.

to gain virulence and resistance genes that help them adapt. Substandard disease control strategies within healthcare settings also facilitate the transmission of resistant and virulent strains (Andersen et al. 2012). Infection susceptibility depends on cell adhesion, invasion, and exploitation of innate and adaptive immunity. Genetic polymorphisms in immune response genes, for example, TLR2, IL-1β, and IL-6, potentially contribute to impaired host response and inability to clear infections. Along with environmental risk factors, the susceptibility to infections is reliant on dysfunction in host epidermis integrity and possible defects in defense pathways (Shukla, Rose, and Schrodi 2015).

8.3 MISDIAGNOSIS OF CHRONIC *STAPHYLOCOCCUS* INFECTIONS AS CANCER

In several cases, chronic bacterial infections, including *Staphylococcus* infections, are misdiagnosed as cancer. For instance, a pyoderma harboring coagulase-negative *Staphylococcus* was incorrectly diagnosed with squamous cell carcinoma due to histopathological representation as pseudoepitheliomatous hyperplasia (Yan, Kus, and Sant 2025). Chronic *Staphylococcus* infections may mimic symptoms of malignancy or other chronic diseases, leading to diagnostic confusion. For instance, symptoms such as weight loss, fatigue, chronic inflammation, and localized swelling may overlap with cancer presentations. Inflammatory cytokines released during chronic infection can cause systemic effects similar to paraneoplastic syndromes. Additionally, persistent infection-related inflammation can lead to cellular changes resembling dysplasia or neoplasia. Radiographic findings of bone destruction in chronic osteomyelitis on several occasions resemble bone tumors. Such overlap complicates the differentiation between infection and malignancy, especially in immunocompromised individuals (Tat et al. 2023).

8.4 ROLE OF CHRONIC *STAPHYLOCOCCUS* INFECTION IN TUMORIGENESIS

Chronic *Staphylococcus aureus* infections are often associated with cancer progression through many interrelated mechanisms (Figure 8.2). The oxidative stress, inflammation, DNA damage, and alterations in signaling pathways associated with *Staphylococcus* infections may act as precursors to cancer initiation and progression. *Staphylococcus* releases many superantigens and exotoxins that induce massive cytokine storm, including pro-inflammatory cytokines like IL-6, resulting in a constant inflammatory microenvironment that potentially promotes tumor formation (Zhao et al. 2021; Liao et al. 2021; Goda et al. 2021). In addition to chronic inflammation, *S. aureus* is known to modulate the MAPK signaling pathway in osteoblast cells by the activation of ERK1/2, leading to sustained proliferative signaling in host cells (Ellington et al. 2001). Furthermore, to neutralize the host defense mechanism, the bacteria alter the p38 MAPK signaling, resulting in the suppression of autophagy and evasion of phagophores. Altered proliferative signaling and autophagy pathways play a vital role in cancer development and progression (Neumann et al. 2016).

In breast cancer, *S. aureus* disrupts the DNA damage response mechanisms for double-strand DNA breaks. This compromised DNA damage response does not result in G1 arrest or apoptosis. The bacteria further induce the RIG-1-like receptor activation in breast cancer cells, leading to NF-κB pathway activation. The NF-κB p65 activates CCL5–CCR5 signaling that results in breast cancer metastasis (Haile et al. 2024). Also, in oral cancer, *S. aureus*-induced activation of NF-κB and TNF pathways enhances the cell proliferation and transformation of keratinocytes (Wang et al. 2019). The bacterial secretome modulates different signaling pathways in host cells and tumor cells that boost their tumorigenic and metastatic potential.

Most of these studies are assessing the association of *Staphylococcus* infections with cancer using cell culture-based or DNA-based techniques to identify resident bacteria. The *Staphylococcus* genus is known to be a resident of primary lung adenocarcinoma lesions and their metastatic lymph

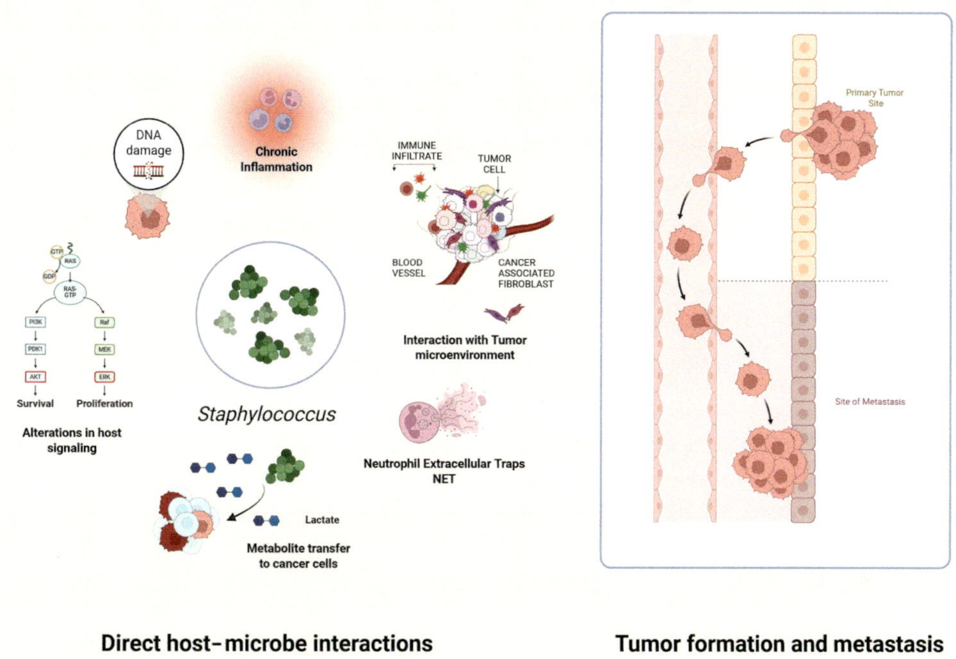

FIGURE 8.2 Pro-tumorigenic *Staphylococcus* – tumor interactions.

nodes. It enhances tumorigenesis and metastatic potential. Yu et al. showed that *S. nepalensis* or *S. capitis* infection significantly increases the metastatic foci in organoid-derived xenografts in mice. The bacterial cells secrete lactate and modulate the lactate uptake in tumor cells through the MCT1-pseudophypoxia-NDRG1 axis (Figure 8.2). The MCT1 inhibition attenuated metastasis in mice that underwent tail vein injections of bacteria-treated cancer cell organoids (Yu et al. 2025). *S. aureus* is also strongly associated with squamous cell carcinoma in skin (Kullander, Forslund, and Dillner 2009). *S. aureus*-secreted factors induce melanoma clustering, thus promoting tumor invasion as demonstrated in the zebrafish melanoma model (Giese et al. 2024).

Along with bacterial residence at the tumor site and its pro-tumorigenic nature, *Staphylococcus* infections are known to complicate the management of several cancers, including lung cancer and leukemia (Li, Zheng, and Leung 2023). *S. aureus* is the most predominant pathogenic population found in acute myeloid leukemia (AML) patients. The enterotoxins produced by the bacteria are known to enhance cell proliferation and migration. The enterotoxins induce the overexpression of immune regulation gene sets in AML cell lines, helping them to evade the fatal effects of the toxins through ER stress pathways. It also promotes resistance to cytarabine, a common chemotherapeutic agent used in the management of AML (Seyhan et al. 2023). Most of the cutaneous T cell lymphoma (CTCL) patients are vulnerable to *Staphylococcus* infections. In Sézary syndrome (SS), a rare leukemic variant of CTCL, the *Staphylococcus* enterotoxins are known to provide drug resistance in malignant T cells through the activation of TCR, JAK-STAT, and NF-κB signaling. These toxins also induce resistance to several HDAC inhibitors in CTCL (Vadivel et al. 2024).

The mechanisms of *Staphylococcus*-associated tumor progression differ from different cancers, and it depends highly on the bacterial secretome and the microenvironment. Most of these mechanisms are part of immune evasion or neutralization of the host defense response. In summary, the inflammatory microenvironment formed during *Staphylococcus* infections and modulation of several signaling pathways in host cells leads to tumor promotion and progression (Figure 8.2).

8.5 ANTI-TUMORIGENIC *STAPHYLOCOCCUS* TOXINS AND THEIR EMERGING ROLE IN CANCER THERAPY

Most of the studies state that *Staphylococcus* infections are pro-tumorigenic and enhance invasion and metastasis, yet there are several reports establishing the anti-tumorigenic potential of *Staphylococcus*-secreted toxins. Although several reports suggest the role of *Staphylococcus* infections in promoting cancer drug resistance, Liu et al. show that immunomodulatory effects of *Staphylococcus* hemolysin α sensitize TNBC to immune checkpoint blockade (ICB) therapy. A subpopulation of TNBC devoid of tumor-infiltrating lymphocytes (TILs) or PD-L1 expression shows poor response to ICB therapy. Interestingly, the intra-tumoral administration of *Staphylococcus*-secreted factors elevates the levels of TIL and reduces tumor size. The study demonstrates that hemolysin α is responsible for an increase in $CD8^+$ T cells and PD-L1 expression, thus making ICB effective in this subset of TNBC (Liu et al. 2024). Also, hemolysin α inhibits the growth of MCF 7 cells and induces apoptosis (Swofford et al. 2014). Another *Staphylococcus*-secreted toxin, enterotoxin type A, inhibits the proliferation and invasion of cervical cancer cells (Afzali et al. 2021). In lung cancer, the cytoplasmic extract of *S. aureus*-induced apoptosis in the A549 lung cancer cell line, implicating the cytotoxic effects of staphylococcal proteins (Mehrshad et al. 2024). Although some of the known bacterial toxins manifest anti-tumor properties in preclinical studies, there is no evidence for the use of these toxins in cancer therapy. Overall toxicity of these toxins and their general immunomodulatory effects should be evaluated thoroughly.

On the other hand, some toxins, including α-toxin, show tumor-promoting properties. α-Toxin selectively kills $CD8^+$ T cells in SS. The malignant T cells survive through the α-toxin cytotoxicity, subsequently evading the $CD8^+$ T cells' immune response (Edda et al. 2020). Table 8.1 lists the role of different *Staphylococcus* toxins in pro- and anti-tumorigenic properties. However, some of these toxins play an anti-tumorigenic role to further understand their implications on the overall recovery and health of the patient. There is a possibility of the use of these bacterial toxins as anti-cancer agents, but more research and preclinical examinations are necessary to endorse them in the clinic.

TABLE 8.1
Role of Different *Staphylococcus* Toxins

Cancer	Toxin/Bacterial Cell Extracts	Nature	Effect	References
Breast cancer	α-Hemolysin	Anti-tumorigenic	Cytotoxic effect in the MCF-7 cell line	Swofford et al. (2014)
			Immunomodulatory effects, sensitizes TNBC to immune checkpoint blockade	Liu et al. (2024)
Lung cancer	Cytoplasmic extract	Anti-tumorigenic	Cytotoxic effect and induction of apoptosis in A549 cells	Mehrshad et al. (2024)
Breast and cervical cancer	Enterotoxin type A	Anti-tumorigenic	Cytotoxic effect and induction of apoptosis in the HeLa cell line	Afzali et al. (2021)
Sézary syndrome, CTCL	α-Toxin	Pro-tumorigenic	Selective cytotoxic activity on CD8+ T cells	Edda et al. (2020)

8.6 *STAPHYLOCOCCUS* INFECTIONS AND TUMOR MICROENVIRONMENT

The tumor microenvironment (TME) is a tumor niche formed by the surrounding stromal cells, including fibroblasts, immune cells, endothelial cells, pericytes, and deposited ECM that helps tumor cells in tumor progression, metastasis, and therapy resistance. The activated stromal cells, such as cancer-associated fibroblasts and tumor-associated macrophages, secrete several factors that are pro-tumorigenic and even responsible for an immunosuppressive microenvironment (Crouigneau et al. 2024). The *Staphylococcus* bacterium is known to interact with such stromal cells in the TME to promote tumor progression and metastasis.

In CTCL, the *Staphylococcus* interacts with the resident keratinocytes and induces IL-2Rγ/Jak3-dependent proliferation in malignant and non-malignant T cells (Ziao et al. 2024). During any bacterial infection, as a response, the host neutrophils secrete DNA, histone, and antimicrobial proteins that form a web-like network to trap and kill the pathogens. Such traps are called neutrophil extracellular traps (NETs) (Meyers et al. 2022). *In vivo* and in vitro studies evidently demonstrated that the regulation of cancer cell metastasis through autophagy-mediated NET formation is provoked by *Staphylococcus* in lung cancer (Qi et al. 2020). There are several reports suggesting that the interaction of bacteria with TME promotes tumor activity, but there are instances in which the bacteria and immune cells interact, leading to tumor suppression. An investigation done in 2019 found that the intra-tumoral injection of *S. aureus* in glioma of C57BL/6 mice leads to the reduction of tumor volume and improvement in overall survival. In glioblastoma, the microglial cells within the TME contribute toward tumor proliferation, migration, and metastasis. Surprisingly, in this study, the bacteria instigate a phenotypic switch in microglial cells, leading to a shift from pro-tumorigenic functions to anti-tumorigenic functions (Zhang et al. 2019). The bacteria indeed may form complex relationships with stromal and immune cells in the TME that may promote or suppress the tumor formation and metastasis. To gain a deeper understanding, more extensive research needs to be conducted to explore the complex relationship between the TME and bacteria and reveal the fate of cancer development and progression in the host.

8.7 TREATMENT STRATEGIES

The primary treatment of *S. aureus* infections has been antibiotic therapy in the past. The development and prevalence of methicillin-resistant *S. aureus* (MRSA) have nonetheless made treatment options very challenging. Chronic infections, particularly those that include biofilms, are notoriously insensitive to treatment because the metabolic activity of bacteria trapped in the biofilm matrix is minimized, as is their susceptibility to antibiotics (Otto, 2018). Conventional antibiotics like beta-lactams do not work very well on MRSA, leading to the usage of alternative antibiotics such as vancomycin, linezolid, or daptomycin. These drugs do have major side effects and might not penetrate biofilm-related infections properly. Additionally, sub-inhibitory levels of antibiotics can induce biofilm growth and resistance. Therefore, it is a priority to develop individualized antimicrobial therapy that takes into account the biofilm mode of growth and resistance type. Antibiotic stewardship and surveillance of resistance are still essential parts of successful long-term control (Kalu, Kao, and Fritz 2022; Tong et al. 2015).

When chronic *Staphylococcus* infections are incidental in cancer development or co-exist with the malignancies, it not only complicates the prognosis of the patient but also complicates the management of the disease. Surgical resection of infected or cancerous tissue is often the first line of intervention, especially in localized lesions. However, surgery may carry additional risks in infected or inflamed tissue and may require concurrent antibiotic therapy. Chemotherapy is commonly employed in cancers but can suppress the immune system and predispose patients to new or recurrent *Staphylococcus* infections, creating a vicious cycle (Naylor et al., 2013). Immunotherapy, particularly immune checkpoint inhibitors, has revolutionized cancer treatment but may exacerbate infection-associated inflammation or autoimmunity. The bacterial infection or colonization at cancer sites worsens the disease and makes the treatment unresponsive. Treatment identification for the bacterial infection may aid in a better prognosis for the patients. Here, we need to comprehend that cancer is a multifaceted developmental disease and is influenced by several factors, of which co-existing or chronic bacterial infections are one of the many stressors (Stanger and Wahl 2024). The management of these bacteria may or may not work based on the condition and severity of the disease in the patient. Using anti-cancer bacteria is one of the emerging solutions in the treatment aspect, but much more penetrating knowledge and research are required.

8.8 CONCLUSION AND IMPLICATIONS

This chapter summarizes the association of chronic *Staphylococcus* infections with cancer. It also engenders many intriguing questions about the implications of these host–microbe interactions in the cancer context. Staphylococci, a genus of bacteria, are common commensal and opportunistic pathogens of the human nares, respiratory system, skin, and breast. The manifestation of Staphylococci infections highly depends on host–microbe interactions, and it is the most frequent etiology found in infection-related deaths worldwide. *Staphylococcus* infections lead to chronic inflammation that cascades into DNA damage and disruption in DNA damage response pathways in the host. It also induces oxidative stress and modulates cell proliferative pathways. The infection leads to immune modulation and production of cytokine storm, achieving an immunosuppressive microenvironment. All these factors and consequences of the infection are closely related to hallmarks of cancer and generate a pro-tumorigenic microenvironment. The major direct interactions of the host and microbe leading to tumor progression or metastasis include the transfer of metabolites and the induction of metabolite uptake in the host. This further influences the immune modulation of the host microenvironment, induction of compromised DNA damage response, and alterations in host cell proliferative pathways. Several direct reports of pro-metastatic interactions are the transfer of metabolites like lactate in lung cancer and the modulation of pattern recognition receptors (PRR) like RIG-1 in breast cancer. Additionally, the modulation of NF-κB signaling and activation of kinases like ERK1 and ERK2 are the major proliferative interactions known. *Staphylococcus* secretome comprises various metabolites and toxins. Staphylococci toxins are known to play a pleiotropic role in cancer. The fate of the toxin response is highly context-dependent manner. Some of the *Staphylococcus*-derived toxins play an anti-tumorigenic role, while toxins like α-toxin help tumor cells to evade T cell-mediated cytotoxicity. These bacteria not only interact with the host cancer cells but also abrogate the function of the stromal cells in the TME. It is known to interact with keratinocytes in CTCL and microglial cells in glioblastoma to promote tumor progression and metastasis. The NET formation by the host system as a defense response plays a supportive role in tumor formation.

The bacteria and tumor–microenvironment interaction can shift toward a pro-tumorigenic phenotype and therapy resistance under favorable conditions. Interestingly, several evidences demonstrate the anti-tumorigenic role in the host TME, which includes the bacterial toxins as anti-tumorigenic. A comprehensive understanding of host–microbe interactions and the overall toxicity of therapeutic interventions is essential before such therapies can be confidently endorsed. The outcomes of these interactions are unlikely to be strictly binary – exhibiting either pro- or anti-tumor effects. Notably, *Staphylococcus* infections not only hinder effective cancer treatment but also exacerbate the clinical complexity of the disease.

ACKNOWLEDGMENTS

The Capacity Building and Development of Novel Cutting Edge Research Activities grant (DPR 1/3(4)/2021/TMC/R&D-II/15063) by the Department of Atomic Energy, Government of India. SA supported by the DBT-Ramalingaswami Fellowship (BT/RLF/Re-entry/09/2018) and SERB grant (SRG/2021/001870-G). OD supported by the ACTREC PhD-JRF fellowship. The authors do not have any competing interests.

REFERENCES

Afzali, S., A. Doosti, M. Heidari, N. Babaei, P. Keshavarz, Z. Nadem, and A. Kahnamoei. 2021. 'Effects of *Staphylococcus aureus* enterotoxin type A on inducing apoptosis in cervical cancer cell line', *Gene Rep*, 25: 101397.

Ahmann, M., J. Compton, J. Pottinger, R. Uhlenhopp, M. Ward, A. Haleem, M. Willey, M. Schweizer, and L. Herwaldt. 2025. '*Staphylococcus aureus* colonization and surgical site infections among patients undergoing surgical fixation for acute fractures', *Infect Control Hosp Epidemiol*, 46: 1–7.

Andersen, Paal Skytt, Jacob Krabbe Pedersen, Peder Fode, Robert L. Skov, Vance G. Fowler, Jr, Marc Stegger, and Kaare Christensen. 2012. 'Influence of host genetics and environment on nasal carriage of *Staphylococcus aureus* in Danish middle-aged and elderly twins', *J Infect Dis*, 206: 1178–1184.

Becker, K., C. Heilmann, and G. Peters. 2014. 'Coagulase-negative staphylococci', *Clin Microbiol Rev*, 27: 870–926.

Crouigneau, Roxane, Yan-Fang Li, Jamie Auxillos, Eliana Goncalves-Alves, Rodolphe Marie, Albin Sandelin, and Stine Falsig Pedersen. 2024. 'Mimicking and analyzing the tumor microenvironment', *Cell Rep Methods*, 4.

Edda, Blümel, Ahmad Shamaila Munir, Nastasi Claudia, Willerslev-Olsen Andreas, Gluud Maria, Fredholm Simon, Hu Tengpeng, G. J. Surewaard Bas, M. Lindahl Lise, Fogh Hanne, B. Koralov Sergei, Gjerdrum Lise Mette Rahbek, A. Clark Rachael, Iversen Lars, Krejsgaard Thorbjørn, Bonefeld Charlotte Menné, Geisler Carsten, C. Becker Jürgen, Woetmann Anders, Andersen Mads Hald, Buus Terkild Brink, and Ødum Niels. 2020. '*Staphylococcus aureus* alpha-toxin inhibits CD8+ T cell-mediated killing of cancer cells in cutaneous T-cell lymphoma', *OncoImmunology*, 9: 1751561.

El Tekle, Geniver, and Wendy S. Garrett. 2023. 'Bacteria in cancer initiation, promotion and progression', *Nat Rev Cancer*, 23: 600–618.

Ellington, J. K., A. Elhofy, K. L. Bost, and M. C. Hudson. 2001. 'Involvement of mitogen-activated protein kinase pathways in *Staphylococcus aureus* invasion of normal osteoblasts', *Infect Immun*, 69: 5235–5242.

Foster, T. J. 2005. 'Immune evasion by staphylococci', *Nat Rev Microbiol*, 3: 948–958.

Giese, Morgan A., Gayathri Ramakrishnan, Laura H. Steenberge, Jerome X. Dovan, John-Demian Sauer, and Anna Huttenlocher. 2024. '*Staphylococcus aureus* lipid factors modulate melanoma cell clustering and invasion', *Dis Models Mech*, 17.

Goda, K., T. Kenzaka, M. Hoshijima, A. Yachie, and H. Akita. 2021. 'Toxic shock syndrome with a cytokine storm caused by *Staphylococcus simulans*: A case report', *BMC Infect Dis*, 21: 19.

Haile, Zhao, Zhang Linzhe, Du Dongdong, Mai Lisu, Liu Yaping, Morigen Morigen, and Fan Lifei. 2024. 'The RIG-I-like receptor signaling pathway triggered by *Staphylococcus aureus* promotes breast cancer metastasis', *Int Immunopharmacol*, 142: 113195.

Hanahan, Douglas, and Robert A Weinberg. 2011. 'Hallmarks of cancer: The next generation', *Cell*, 144: 646–674.

Kalu, I. C., C. M. Kao, and S. A. Fritz. 2022. 'Management and prevention of *Staphylococcus aureus* infections in children', *Infect Dis Clin North Am*, 36: 73–100.

Kullander, J., O. Forslund, and J. Dillner. 2009. '*Staphylococcus aureus* and squamous cell carcinoma of the skin', *Cancer Epidemiol Biomarkers Prev*, 18: 472–478.

Li, Jiaqi, Huangliang Zheng, and Sharon Shui Yee Leung. 2023. 'Potential of bacteriophage therapy in managing *Staphylococcus aureus* infections during chemotherapy for lung cancer patients', *Sci Rep*, 13: 9534.

Liao, F., W. Gu, X. Fu, B. Yuan, and Y. Zhang. 2021. 'Community-acquired methicillin-resistant *Staphylococcus aureus* provoked cytokine storm causing severe infection on BALB/c mice', *Mol Immunol*, 140: 167–174.

Liu, Chin-Chih, Matthew Wolf, Ruth Ortego, Dennis Grencewicz, Tammy Sadler, and Charis Eng. 2024. 'Characterization of immunomodulating agents from *Staphylococcus aureus* for priming immunotherapy in triple-negative breast cancers', *Sci Rep*, 14: 756.

Lowy, F. D. 1998. '*Staphylococcus aureus* infections', *N Engl J Med*, 339: 520–532.

Mehrshad, Ahmadi, Hajikhani Bahareh, Shamosi Atefeh, Yaslianifard Somayeh, Sameni Fatemeh, Qorbani Mostafa, Mohammadzadeh Mohammad, and Dadashi Masoud. 2024. 'Cytotoxic and apoptosis-inducing properties of *Staphylococcus aureus* cytoplasmic extract on lung cancer cells: Insights from MTT assay and bax/bcl-2 gene expression analysis', *Gene Rep*, 36: 101955.

Meyers, S., M. Crescente, P. Verhamme, and K. Martinod. 2022. '*Staphylococcus aureus* and neutrophil extracellular traps: The master manipulator meets its match in immunothrombosis', *Arterioscler Thromb Vasc* Biol, 42: 261–276.

Naylor, M. D., K. H. Bowles, K. M. McCauley, M. C. Maccoy, G. Maislin, M. V. Pauly, and R. Krakauer. 2013. 'High-value transitional care: Translation of research into practice', *J Eval Clin Pract*, 19: 727–733.

Neumann, Y., S. A. Bruns, M. Rohde, T. K. Prajsnar, S. J. Foster, and I. Schmitz. 2016. 'Intracellular *Staphylococcus aureus* eludes selective autophagy by activating a host cell kinase', *Autophagy*, 12: 2069–2084.

Odunitan, T. T., B. T. Apanisile, M. W. Akinboade, W. O. Abdulazeez, A. O. Oyaronbi, T. M. Ajayi, S. A. Oyekola, N. O. Ibrahim, T. Nafiu, H. O. Afolabi, D. M. Olayiwola, O. T. David, S. F. Adeyemo, O. D. Ayodeji, E. M. Akinade, and O. A. Saibu. 2024. 'Microbial mysteries: *Staphylococcus aureus* and the enigma of carcinogenesis', *Microb Pathog*, 194: 106831.

Otto, M. 2009. 'Staphylococcus epidermidis – The "accidental" pathogen', *Nat Rev Microbiol*, 7: 555–567.

Otto, M. 2018. 'Staphylococcal Biofilms', Microbiol Spectrum, 6(4).

Proctor, R. A., C. von Eiff, B. C. Kahl, K. Becker, P. McNamara, M. Herrmann, and G. Peters. 2006. 'Small colony variants: A pathogenic form of bacteria that facilitates persistent and recurrent infection', *Nat Rev Microbiol*, 4: 295–305.

Qi, J. L., J. R. He, C. B. Liu, S. M. Jin, R. Y. Gao, X. Yang, H. M. Bai, and Y. B. Ma. 2020. 'Pulmonary *Staphylococcus aureus* infection regulates breast cancer cell metastasis via neutrophil extracellular traps (NETs) formation', *MedComm*, 1: 188–201.

Schilcher, Katrin, and R. Horswill Alexander. 2020. 'Staphylococcal biofilm development: Structure, regulation, and treatment strategies', *Microbiol Mol Biol Rev*, 84: 10.1128/mmbr.00026-19.

Seyhan, Turk, Yanpar Hatice, Baesmat Ayriana Safari, Canli Secil Demirkol, Cinar Olgu Erkin, Malkan Umit Yavuz, Turk Can, Haznedaroglu Ibrahim Celalettin, and Ucar Gulberk. 2023. 'Enterotoxins A and B produced by *Staphylococcus aureus* increase cell proliferation, invasion and cytarabine resistance in acute myeloid leukemia cell lines', *Heliyon*, 9: e19743.

Shukla, Sanjay K., Warren Rose, and Steven J. Schrodi. 2015. 'Complex host genetic susceptibility to *Staphylococcus aureus* infections', *Trends Microbiol*, 23: 529–536.

Stanger, B. Z., and G. M. Wahl. 2024. 'Cancer as a disease of development gone awry', *Annu Rev Pathol*, 19: 397–421.

Swofford, C. A., A. T. St Jean, J. T. Panteli, Z. J. Brentzel, and N. S. Forbes. 2014. 'Identification of *Staphylococcus aureus* α-hemolysin as a protein drug that is secreted by anticancer bacteria and rapidly kills cancer cells', *Biotechnol Bioeng*, 111: 1233–1245.

Tat, W. Yau, D. Franz Bennet, A. Osman Hanadi, C. Rivera Jessica, and Azhar Ashaur. 2023. 'Bad to the bone. Not all bone tumors are cancer: Case of long bone osteomyelitis', *IDCases*, 34: e01897.

Tong, S. Y., J. S. Davis, E. Eichenberger, T. L. Holland, and V. G. Fowler, Jr. 2015. '*Staphylococcus aureus* infections: Epidemiology, pathophysiology, clinical manifestations, and management', *Clin Microbiol Rev*, 28: 603–661.

Vadivel, C. K., A. Willerslev-Olsen, M. R. J. Namini, Z. Zeng, L. Yan, M. Danielsen, M. Gluud, E. M. H. Pallesen, K. Wojewoda, A. Osmancevic, S. Hedebo, Y. T. Chang, L. M. Lindahl, S. B. Koralov, L. J. Geskin, S. E. Bates, L. Iversen, T. Litman, R. Bech, M. Wobser, E. Guenova, M. R. Kamstrup, N. Ødum, and T. B. Buus. 2024. '*Staphylococcus aureus* induces drug resistance in cancer T cells in Sézary syndrome', *Blood*, 143: 1496–1512.

Wang, Yuxia, Shiyu Liu, Bolei Li, Yaling Jiang, Xinxuan Zhou, Jing Chen, Mingyun Li, Biao Ren, Xian Peng, Xuedong Zhou, and Lei Cheng. 2019. '*Staphylococcus aureus* induces COX-2-dependent proliferation and malignant transformation in oral keratinocytes', *J Oral Microbiol*, 11: 1643205.

Yan, A., J. V. Kus, and N. Sant. 2025. 'The *Staphylococcus aureus* complex: Implications for the clinical microbiology laboratory', *J Clin Microbiol*, 63: e0127624.

Yu, Huan, Yang Du, Yuling He, Yifan Sun, Junfeng Li, Bo Jia, Jiale Chen, Xinya Peng, Tongtong An, Jianjie Li, Yujia Chi, Man Wang, Lihua Cao, Yidi Tai, Xiaoyu Zhai, Reyizha Nuersulitan, Sheng Li, Nan Wu, Jia Wang, Hongchao Xiong, Shujie Wan, Jiaming Liu, Xuelian Yang, Xingsheng Hu, Dongmei Lin, Wei Sun, Yue Wang, Guo An, Xumeng Ji, Lingdong Kong, Lu Ding, Yunan Ma, Zhihua Tian, Bin Dong, Yujing Bi, Jianmin Wu, and Ziping Wang. 2025. 'Lactate production by tumor-resident *Staphylococcus* promotes metastatic colonization in lung adenocarcinoma', *Cell Host & Microbe*, 33: 1089–105.e7.

Zhang, B., J. Zhang, S. Fang, M. Zhang, S. Liu, Y. Tian, M. Ma, F. Liu, and G. Jin. 2019. 'Inflammatory activation of microglia by *Staphylococcus aureus* caused phenotypic alterations and affected glioblastoma growth', *Cell Biochem Funct*, 37: 331–339.

Zhang, Heran, Ian R. Monk, Jessica Braverman, Claerwen M. Jones, Andrew G. Brooks, Timothy P. Stinear, and Linda M. Wakim. 2024. 'Staphylococcal superantigens evoke temporary and reversible T cell anergy, but fail to block the development of a bacterium specific cellular immune response', *Nat Commun*, 15: 9872.

Zhao, Huakan, Lei Wu, Guifang Yan, Yu Chen, Mingyue Zhou, Yongzhong Wu, and Yongsheng Li. 2021. 'Inflammation and tumor progression: Signaling pathways and targeted intervention', *Signal Transduct Target Ther*, 6: 263.

Ziao, Zeng, Vadivel Chella Krishna, Gluud Maria, R. J. Namini Martin, Yan Lang, Ahmad Sana, Hansen Morten Bagge, Coquet Jonathan, Mustelin Tomas, B. Koralov Sergei, Bonefeld Charlotte Menne, Woetmann Anders, Geisler Carsten, Guenova Emmanuella, R. Kamstrup Maria, Litman Thomas, R. Gjerdrum Lise-Mette, B. Buus Terkild, and Ødum Niels. 2024. 'Keratinocytes present *Staphylococcus aureus* enterotoxins and promote malignant and nonmalignant T cell proliferation in cutaneous T-cell lymphoma', *J Investig Dermatol*, 144: 2789–2804.e10.

Section III

Biofilm Formation and Quorum Sensing

9 Exploring Quorum Sensing and Biofilm Formation in *Staphylococcus aureus*

P. Gnana Deepu[1], *Ruchitha K*[2], *Anugum Manish*[1], *and Bikshu Jaga*[3]

[1]Centre for Biotechnology, University College of Engineering Science and Technology, Jawaharlal Nehru Technological University, Hyderabad, India

[2]Department of Pharmacy Practice, MNR College of Pharmacy, MNR Nagar, Fasalwadi Narsapur Road, Sangareddy, Telangana, India

[3]School of Biomedical Sciences, Charles Sturt University, Wagga Wagga, NSW, Australia

ABBREVIATIONS

Agr	accessory gene regulator
AIP	autoinducing peptide
MRSA	methicillin-resistant *Staphylococcus aureus*
EPS	extracellular polymeric substance
PSMs	phenol-soluble modulins
QS	quorum sensing
RIP	RNAIII-inhibiting peptide

9.1 INTRODUCTION

Staphylococcus aureus is a major opportunistic pathogen of humans, capable of causing a wide spectrum of diseases. It commonly colonizes the skin and nasal passages without harm but can invade sterile tissues to cause skin and soft tissue infections, pneumonia, endocarditis, osteomyelitis, and sepsis (Cheung et al., 2021). *S. aureus* is notorious in hospital settings: for example, methicillin-resistant *S. aureus* (MRSA) strains cause hundreds of thousands of infections and thousands of deaths annually in the United States (Ghasemian et al., 2015). It ranks among the world's most frequent causes of infection-related morbidity and mortality, driven largely by the emergence of antibiotic-resistant strains such as MRSA and vancomycin-resistant *S. aureus*. These strains complicate treatment, and no effective human vaccine currently exists (Jordan et al., 2022).

S. aureus modulates its pathogenic behavior through sophisticated regulatory systems. Among these, quorum sensing (QS) is a critical mechanism of cell–cell communication. In general, bacterial QS relies on the production and sensing of small signaling molecules (autoinducers) that accumulate as population density increases. In *S. aureus*, the accessory gene regulator (agr) system is the central QS circuit. At low cell density, *S. aureus* emphasizes attachment and colonization behaviors; at high density, agr-mediated QS triggers a switch to toxin and enzyme production (Yarwood & Schlievert, 2003; Paulander et al., 2018). For example, when cell density is low, RNAIII (the agr effector RNA) is not induced, and *S. aureus* preferentially produces surface adhesins and immune

DOI: 10.1201/9781003647072-12

evasion factors. As the cell density increases, agr senses its autoinducing peptide (AIP) signal and upregulates a regulatory RNA (RNAIII) that represses adhesins and activates exotoxins (such as α-hemolysin and proteases). Thus, QS coordinates *S. aureus* gene expression for cohesive group behaviors (Williams et al., 2023).

Biofilms represent another key pathogenic strategy. A biofilm is a multicellular community attached to a surface and embedded in a self-produced extracellular matrix (ECM). Staphylococcal biofilms (on tissues or medical devices) are central to chronic and device-related infections (Yarwood & Schlievert, 2003). The biofilm matrix – comprising polysaccharides, extracellular DNA (eDNA), and proteins (including amyloid fibers and adhesins) – shields bacteria from antibiotics and the immune system (Arciola et al., 2015). Consequently, biofilm-associated *S. aureus* infections (e.g., endocarditis, osteomyelitis, catheter infections) are notoriously difficult to eradicate. In short, QS and biofilm formation are two intertwined programs that allow *S. aureus* to sense its environment, regulate virulence, and persist in the host.

9.2 QS IN *S. AUREUS*

9.2.1 The Agr System

The primary QS circuit in *S. aureus* is encoded by the agr (accessory gene regulator) locus, which harbors two divergent operons (P2 and P3). The P2 operon (RNAII transcript) encodes four key proteins: AgrB, AgrD, AgrC, and AgrA (Wang & Muir, 2016). AgrD is a precursor peptide that is processed by AgrB (a membrane-bound protease) into a cyclic AIP. The mature AIP is secreted into the environment. When AIP accumulates to a threshold, it binds to the sensor kinase AgrC on the bacterial surface. AgrC autophosphorylates and transfers the phosphate to the response regulator AgrA. Phosphorylated AgrA then binds to the P2 promoter, positively autoregulating the agr operon (creating a feed-forward loop) (Zhao et al., 2022). This positive feedback ensures a rapid QS response when cell density is high. Figure 9.1 illustrates the schematic of the agr quorum-sensing circuit in *S. aureus*. At low cell density (left), the agr circuit is minimally active: cells express colonization factors (adhesins) and attach to surfaces. As cell density increases (right), secreted AIP accumulates. AIP binds the sensor AgrC, activating the AgrA response regulator. Phosphorylated AgrA drives expression of the agr operon and the P3 transcript (RNAIII), which coordinate the shift to toxin production (Lee et al., 2020). The schematic illustrates the molecular steps of the agr quorum-sensing circuit in *S. aureus*, shown in Figure 9.1.

Transcription from the P3 operon produces RNAIII, a regulatory RNA that is the major effector of agr QS. RNAIII encodes δ-hemolysin (a small toxin) and, more importantly, acts as an mRNA regulator. RNAIII base pairs with target mRNAs to downregulate surface proteins (e.g., protein A, fibronectin-binding proteins [FnBPs]) and indirectly upregulate secreted factors. In parallel, AgrA directly upregulates phenol-soluble modulins (PSMs), a family of amphipathic peptides. In sum, AgrA/RNAIII signaling switches *S. aureus* gene expression from an "adhesion mode" to an "invasion mode" (Wang & Muir, 2016; Paulander et al., 2018). When the arg system is activated, production of toxins (α-, β-, γ-hemolysins, proteases, lipases, leukocidins, etc.) and secreted virulence enzymes surges, while surface adhesins and immune evasion factors diminish. This population-dependent regulation allows coordinated tissue invasion: low-density cells focus on attachment, while high-density cells promote dissemination. Table 9.1 provides an overview of the molecular components within the agr QS system, highlighting each protein's specific role in the regulation cascade.

9.2.2 Regulation of Virulence Factors by QS

The agr QS system profoundly impacts *S. aureus*'s pathogenic arsenal. During the transition to high cell density, in vitro studies demonstrate that agr activity suppresses surface-associated colonization factors while upregulating exo-virulence proteins. In effect, agr signaling promotes adhesion and biofilm-promoting proteins, while strongly enhancing toxins and degradative enzymes (Yarwood &

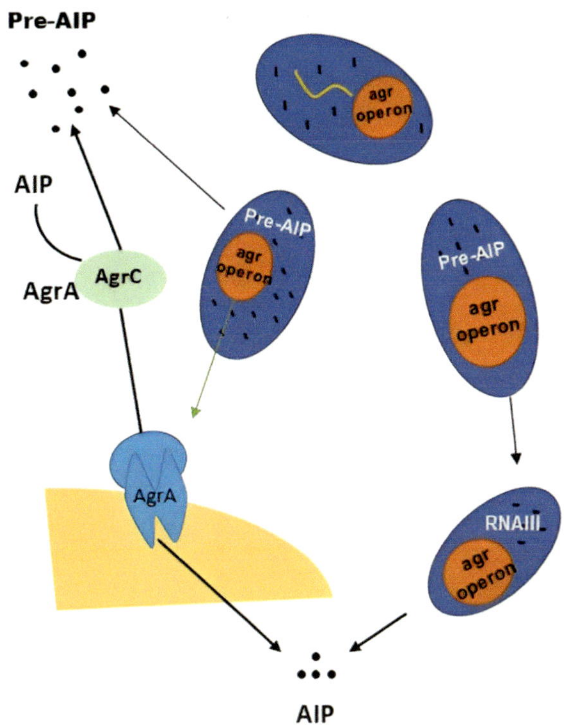

FIGURE 9.1 Molecular schematic of the agr quorum-sensing circuit in *Staphylococcus aureus*. Precursor peptide AgrD (Pre-AIP) is processed and exported by AgrB into active autoinducing peptide (AIP). Once extracellular AIP reaches threshold levels, it binds the sensor kinase AgrC, which phosphorylates the response regulator AgrA. Activated AgrA promotes transcription from the P2 and P3 operons, leading to autoregulation of the agr system and RNAIII synthesis. RNAIII further modulates virulence gene expression, orchestrating the switch between adhesion and toxin production.

TABLE 9.1

Key Components of the agr Quorum-Sensing System in *Staphylococcus aureus*

Component	Function	Molecular Type
AgrD	Precursor peptide that is processed into AIP	Peptide
AgrB	Processes AgrD into AIP and exports it	Membrane protease
AgrC	Sensor kinase that detects AIP	Transmembrane histidine kinase
AgrA	Response regulator that activates target gene expression	DNA-binding response regulator
RNAIII	Regulatory RNA that modulates virulence gene expression	Effector RNA

Schlievert, 2003; Wang & Muir, 2016). For example, agr activation represses protein A (Spa) and FnBPs while inducing α-toxin (Hla), β-toxin (Hlb), δ-hemolysin (Hld, encoded in RNAIII), leukocidins, aureolysin, and coagulase. The PSMs are also agr-dependent effectors; these peptides serve both as cytolysins and as surfactants facilitating tissue invasion and, as discussed in the following text, biofilm structuring (Thammavongsa et al., 2015). In short, agr QS serves as a switch between "adhesion mode" (low virulence) and "invasion mode" (high virulence).

This regulatory pattern has temporal and functional relevance. Early in infection or growth, when bacteria are few, surface adhesins and immune evasion proteins dominate. As bacterial numbers rise, agr triggers a coordinated burst of toxin and enzyme production that helps breach host barriers and promotes dissemination. For instance, α-toxin (hla) and δ-hemolysin (hla/Hld) are maximally expressed at high density, mediating host cell lysis and immune subversion, while adhesins like Protein A and capsule biosynthesis are repressed through RNAIII-mediated regulation (Yarwood & Schlievert, 2003; Paulander et al., 2018). This dynamic regulation underlies *S. aureus*'s ability to cause rapid tissue necrosis (notably in high-density abscesses) following an initial adherent stage. Table 9.2 summarizes key virulence genes modulated by QS, highlighting how agr activation influences pathogenic traits. Figure 9.2 depicts how QS regulates key virulence factors, balancing adhesion proteins and secreted toxins. Figure 9.3 shows a simplified overview of RNAIII-mediated gene regulation that shifts the pathogen from colonization to invasion.

TABLE 9.2
Agr-Regulated Virulence Factors in *Staphylococcus aureus*

Gene/Protein	Role in Pathogenesis	Regulation by QS
Protein A (spa)	Immune evasion via Fc binding	Downregulated
Fibronectin-binding proteins (FnBPs)	Adhesion to host tissues	Downregulated
Alpha-toxin (hla)	Lysis of host cells	Upregulated
Phenol-soluble modulins (PSMs)	Biofilm structuring and cytotoxicity	Upregulated
Leukocidins	Destruction of immune cells	Upregulated
Proteases (e.g., aureolysin)	Tissue invasion and immune evasion	Upregulated

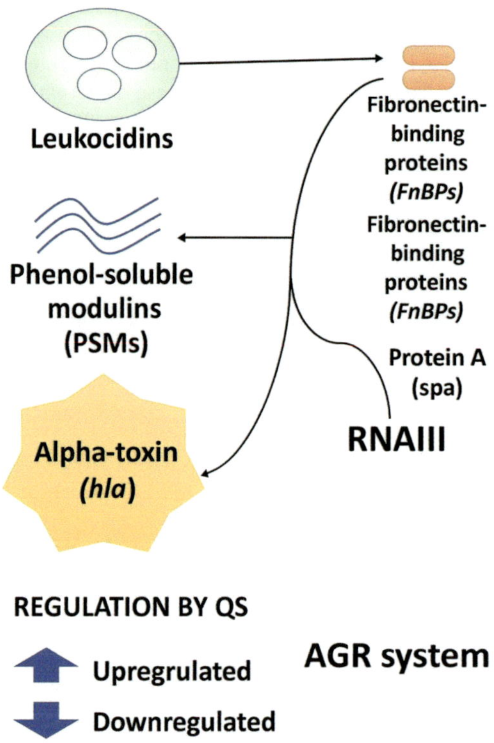

FIGURE 9.2 Agr quorum sensing regulates expression of major *Staphylococcus aureus* virulence determinants. RNAIII acts as a central regulatory effector, downregulating surface adhesins such as Protein A (*spa*) and fibronectin-binding proteins (FnBPs), while upregulating secreted toxins including α-toxin (*hla*), PSMs, and leukocidins. This population-dependent regulation enables *S. aureus* to transition from colonization to tissue invasion.

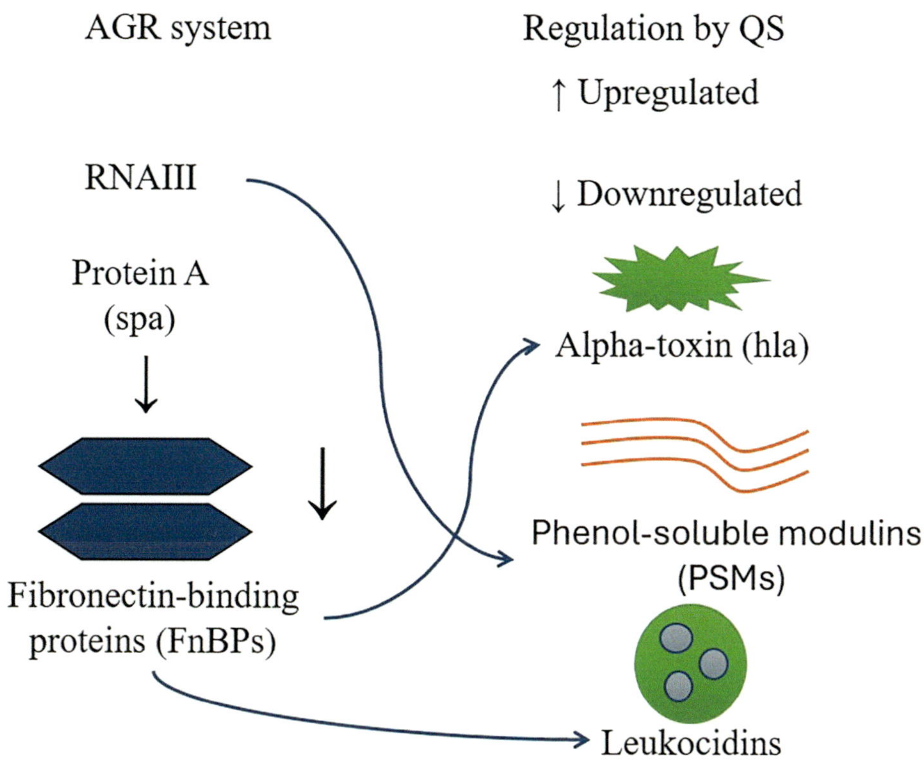

FIGURE 9.3 Simplified representation of agr-dependent virulence regulation in *Staphylococcus aureus*. RNAIII suppresses surface proteins (protein A, FnBPs) while promoting toxin and enzyme expression, including α-toxin (*hla*), PSMs, and leukocidins. This schematic highlights the functional "switch" from adhesion and immune evasion at low cell density to aggressive virulence at high density (Lee et al., 2020; Williams et al., 2023).

9.2.3 Examples of QS Influencing Pathogenicity

QS in *S. aureus* has been linked to differences in infection outcomes. In experimental models, agr positive strains cause severe acute disease, whereas agr-null mutants often show attenuated virulence (Podkowik et al., 2023). For example, agr-deficient mutants are less lethal in mouse models of sepsis, skin abscess, and pneumonia (consistent with lower toxin output). Conversely, agr activation is required for full virulence in many acute infection models. Intriguingly, clinical observations reveal a paradox: many persistent or chronic staphylococcal infections harbor agr-impaired isolates. In bloodstream infections and endocarditis, agr-defective *S. aureus* is frequently isolated and correlates with persistent bacteremia and poor outcomes (Yarwood & Schlievert, 2003; Paulander et al., 2018). Notably, agr-inactive MRSA strains are linked to longer hospital stays and higher mortality relative to agr-functional strains. A similar pattern holds in cystic fibrosis lung infections and device-related biofilm infections: agr-negative variants tend to dominate chronic colonization, presumably because they form more resilient biofilms (Turner et al., 2019). Thus, QS activity appears to tilt the balance between acute aggressive infection and chronic persistence. In essence, active agr signaling promotes acute toxin-driven disease, while agr dysfunction (or pharmacological inhibition) shifts the bacterium toward biofilm development and chronic persistence. These clinical examples illustrate the dual role of agr QS in *S. aureus* pathogenicity: necessary for fulminant virulence, but often downmodulated in persistent infections to promote biofilm-associated persistence.

9.3 BIOFILM FORMATION BY *S. AUREUS*

9.3.1 The Biofilm Formation Process

Biofilm development by *S. aureus* is a multi-step, dynamic process that progresses through four main stages: initial attachment, accumulation (microcolony formation), maturation, and eventual dispersal. During initial attachment, free-swimming (planktonic) bacteria adhere to a surface (biological or inert). Adhesion can be passive (nonspecific charge or hydrophobic interactions) or active. In active attachment, *S. aureus* uses a suite of surface-anchored adhesins (MSCRAMMs) to bind host matrix molecules (Peng et al., 2022). Key adhesins include FnBPA/B, clumping factors (ClfA/B), and collagen-binding protein (Cna) (Ghasemian et al., 2015). These MSCRAMMs (covalently attached via sortase) allow *S. aureus* to anchor to host tissue or device surfaces.

Following attachment, the cells multiply and aggregate into microcolonies. Proliferation and close cell–cell adhesion begin a multilayered community. At this point, bacteria start producing the extracellular polymeric substance (EPS) matrix. The EPS (biofilm matrix) is a complex mix of polymers that cements the cells together and to the surface. Over time, the biofilm matures into a three-dimensional structure with channels and heterogeneous regions (Arciola et al., 2015). During maturation, cells produce large amounts of polysaccharides (mainly poly-*N*-acetylglucosamine, PNAG, also known as polysaccharide intercellular adhesin or PIA), as well as secreted and cell wall proteins, eDNA from autolysis, and amyloid fibers, and provides protection while serving as a scaffold for the biofilm architecture.

Eventually, *S. aureus* biofilms can undergo dispersal. Environmental or regulatory signals (such as nutrient depletion or accumulation of waste) trigger subpopulations to detach and spread. Agr-mediated QS plays a role here: at high density, agr activation leads to the production of dispersal factors (like PSM surfactants and matrix-degrading enzymes), prompting biofilm cells to escape and colonize new sites. Initial Attachment: Single cells adhere to a surface via nonspecific interactions and MSCRAMMs (e.g., ClfA/B, FnBPs, Cna) that bind host proteins (Ghasemian et al., 2015). Accumulation (Multiplication): Attached cells divide, form microcolonies, and begin secreting EPS components (polysaccharides, proteins, DNA). Maturation: The biofilm develops a mature structure with layered cell clusters embedded in a self-produced matrix. Nutrient and oxygen gradients arise, leading to metabolic heterogeneity (Sharma et al., 2023). Dispersal: Portions of the biofilm disintegrate to release planktonic cells. This dispersal is often driven by agr-controlled factors (PSMs, proteases, etc.) that degrade the matrix and allow cells to escape. These stages ensure that *S. aureus* can colonize surfaces, build protective communities, and later disseminate when conditions permit.

For clarity, the biofilm life cycle can be summarized in four stages:

- Initial Attachment – Single cells adhere via nonspecific forces and MSCRAMMs (e.g., ClfA/B, FnBPs, Cna).
- Accumulation – Attached cells divide, form microcolonies, and begin secreting EPS components (polysaccharides, proteins, DNA).
- Maturation – The biofilm develops a complex structure with oxygen/nutrient gradients, creating metabolic heterogeneity (Sharma et al., 2023).
- Dispersal – Portions of the biofilm disintegrate, often agr-driven (PSMs, proteases, nucleases), releasing planktonic cells to initiate colonization at new sites.

These stages ensure that *S. aureus* can colonize surfaces, build protective communities, and later disseminate when conditions permit.

9.3.2 Role of the EPS

The EPS**** is critical to biofilm integrity and function. *S. aureus* EPS is a heterogeneous mixture comprising polysaccharides, extracellular DNA, proteins, and other host-derived molecules.

TABLE 9.3
Biofilm Matrix Components and Their Functional Roles in *Staphylococcus aureus*

Component	Biological Function	Source
Poly-*N*-acetylglucosamine (PNAG)	Intercellular adhesion, matrix structure	Encoded by icaADBC
Extracellular DNA (eDNA)	Structural scaffold, ion chelation	Released by autolysis
Amyloid proteins	Matrix rigidity and adherence	Secreted proteins like Bap
PSMs	Surfactant activity, biofilm dispersion	Agr-regulated
Teichoic acids	Surface charge modulation	Cell wall-associated
Host-derived fibrin	Adhesion and biofilm integration	Host plasma proteins

Extracellular DNA (eDNA), DNA released from lysed cells, is abundant in many staphylococcal biofilms. eDNA provides a structural scaffold and chelates cations, contributing to biofilm cohesion. Proteins: Several surface and secreted proteins are part of the matrix. This includes adhesive proteins (e.g., SasG, Bap), amyloid fibrils, and peptidoglycan-binding proteins. Notably, PSMs are surfactant-like peptides (agr-regulated) that can integrate into the matrix and form amyloid-like fibers. Enzymes (autolysins, proteases) are also present and can modulate the matrix. Other components, Lipids and host molecules (fibrin, plasma proteins) often become trapped in the matrix, especially on medical devices. Together, these EPS elements perform key functions. The sticky polysaccharide and eDNA glue cells together; matrix proteins and amyloids add strength; and the complex mixture forms a barrier to antibiotics and immune cells (Patel & Rawat, 2023). Table 9.3 categorizes the different matrix constituents that contribute to biofilm integrity and protection.

9.3.3 Characteristics and Properties of *S. aureus* Biofilms

S. aureus biofilms exhibit distinctive traits that challenge therapy **Antibiotic tolerance:** Biofilm-embedded cells show markedly reduced susceptibility to antibiotics. The EPS matrix slows drug penetration, while metabolically inactive subpopulations (slow-growing or dormant cells) further resist killing. Consequently, doses that effectively eradicate planktonic bacteria often only suppress biofilm populations (Yarwood & Schlievert, 2003; Lamret et al., 2020). Clinically, this explains why device-associated and chronic *S. aureus* infections often fail to resolve with standard antibiotic regimens. **Immune evasion:** The biofilm matrix provides a physical barrier against immune effectors. Neutrophils and macrophages struggle to infiltrate mature biofilms, while antibodies and complement factors are hindered from binding. In addition, biofilm-associated cells undergo transcriptional changes that downregulate immunogenic surface proteins, enhancing their ability to evade immune detection. **Altered physiology:** Biofilm growth induces profound metabolic and regulatory shifts. Stress response genes are upregulated, efflux systems enhanced, and persister cells emerge. These phenotypic changes result in a heterogeneous population that behaves very differently from planktonic cells (Sahu & Ruhal, 2025). **Chronicity:** Biofilm-related infections are characteristically persistent and relapse-prone. On implanted devices such as prosthetic joints, catheters, or heart valves, biofilms act as reservoirs that periodically release bacteria into the bloodstream. Cheung et al. (2021) emphasize the central role of biofilms in chronic *S. aureus* diseases, including prosthetic joint infection and infective endocarditis. In summary, *S. aureus* biofilms confer robust protection. The combined effect of a fortified ECM and altered bacterial physiology underlies the persistence and treatment resistance of biofilm-associated infections such as osteomyelitis and device-related infections (Cheung et al., 2021).

9.4 INTERPLAY BETWEEN QS AND BIOFILM FORMATION

9.4.1 QS Regulation of Biofilm-Related Genes

The agr quorum-sensing system negatively regulates biofilm formation while promoting bacterial dispersal in *S. aureus*. When agr is activated at high cell density, it downregulates many biofilm-associated functions. For instance, agr activity represses the ica operon (reducing PNAG production) and suppresses adhesin expression. At the same time, agr upregulates the production of enzymes (proteases, nucleases) and surfactant peptides (PSMs) that degrade the matrix (Lee et al., 2020). Collectively, this shifts the bacteria from a sessile biofilm mode to a planktonic invasion mode. Disruption of agr has the opposite effect. Agr-deficient or agr-inhibited strains form much more robust biofilms. In vitro experiments show that inhibition of agr (e.g., by mutation or inhibitors) leads to increased surface attachment and thicker biofilm biomass (Yarwood & Schlievert, 2003). Agr deficiency causes "abundant biofilm production" and decreases cell lysis. In other words, without agr signaling, the brake on colonization functions is released, and the dispersal mechanism is deactivated, resulting in a tenacious biofilm. Thus, agr QS directly governs biofilm genes: AgrA and RNAIII downregulate adhesion factors (FnBPs, Protein A, capsule) and upregulate dispersal factors (autolysins, PSMs). The net result is that agr serves as a rheostat: when QS is high, biofilm formation is suppressed; when QS is off, biofilm formation is allowed to proceed. This molecular switch helps *S. aureus* balance biofilm persistence with the need to escape and spread when cell density signals it is time to move on.

9.4.2 QS Influence on Biofilm Structure and Function

QS affects not only whether a biofilm forms but also its internal architecture and community behavior. Agr-regulated factors contribute to biofilm structuring: for example, PSMs act as biosurfactants that create channels within the biofilm and can lead to the detachment of clusters. α-Toxin, also agr-regulated, has been implicated in biofilm cell death and turnover. That QS inhibits biofilm formation, allowing cells to spread to new places. Indeed, in agr-active settings, biofilms are looser and have more ragged edges, whereas agr-deficient biofilms are compact and "sticky." Agr also creates variability in biofilm populations. Within a mature biofilm, not all cells are equal: some cells may switch agr "on" and others "off," creating subpopulations. This can produce a division of labor (adhesive cells vs. dispersing cells) that affects biofilm resilience. Moreover, agr induction at high local density can cause microenvironments of detachment; conversely, cells in nutrient-poor niches may not activate agr (Mirzaei et al., 2024). Therefore, QS contributes to the dynamic mosaic of activity in a biofilm. Overall, agr signaling tunes the physical stability of the biofilm. Active QS tends to thin the matrix and promote dispersal, while QS-inactive states allow thick, cohesive biofilms to form. This interplay shapes infection: agr-active *S. aureus* populations may form transient biofilms that release cells, whereas agr-inactive populations form static, drug-resistant communities (Schilcher & Horswill, 2020).

9.4.3 Case Studies: QS–Biofilm Interaction in *S. aureus*

Here are a few case studies where the interplay between QS and biofilm interaction in *S. aureus* shows its multifaceted nature. In vitro, agr blockade often strengthens biofilms, but, *in vivo*, the picture is mixed. These cases underscore a key theme: inhibiting agr (to reduce virulence) can have complex effects on biofilms and vice versa. Targeting QS may thus be a double-edged sword in infections where biofilms play a role. Chronic Infections and Agr Deficiency: Clinical case series reveal that agr-deficient *S. aureus* is often isolated from biofilm-associated infections. For example, isolates from persistent bacteremia and prosthetic joint infections frequently carry agr mutations. These agr-impaired strains form stronger biofilms in vitro and may evade antibiotics better. Conversely, agr-active strains isolated from acute disease may show less biofilm formation. This real-world observation underscores the agr–biofilm trade-off (Howden et al., 2023; Beenken &

TABLE 9.4
Phenotypic Comparison of Agr-Positive and Agr-Defective *Staphylococcus aureus* Strains

Phenotypic Trait	Agr-Positive	Agr-Defective
Toxin production	High	Low
Biofilm thickness	Reduced	Increased
Dispersal ability	Active	Deficient
Persistence in bloodstream	Shorter duration	Prolonged bacteremia
Immune evasion	Moderate	Enhanced via biofilm
Virulence in acute infection	High	Attenuated

Smeltzer, 2025). RIP (RNAIII-inhibiting peptide), RNAIII-inhibiting peptide (RIP), is a small molecule that blocks agr. In a catheter biofilm model, RIP treatment prevented biofilm formation by coagulase-negative staphylococci, but, interestingly, it did not significantly inhibit *S. aureus* biofilms. This suggests species-specific differences: RIP (and other agr inhibitors) may not block established *S. aureus* biofilms easily, highlighting the robustness of *S. aureus* communities (De Oliveira et al., 2021).

Probiotic QS Disruption: A recent study used bioactive peptides from *Lactobacillus* strains to disrupt MRSA QS. These probiotic-derived molecules downregulated agrA and luxS expression, dramatically reducing the production of hemolysins and staphyloxanthin. Importantly, treatment with these peptides altered MRSA biofilm architecture and rendered the bacteria much more sensitive to β-lactam antibiotics (cefoxitin) (Cella et al., 2023). This case illustrates how targeting QS can both attenuate virulence and weaken biofilm defense. Agr Deletion *In Vivo*: In a recent *in vivo* study, researchers deleted agr in a mouse implant infection model. Although agr-null bacteria formed slightly more biofilm in certain in vitro conditions, surprisingly, agr deletion did not increase biofilm formation or disease severity *in vivo* (Jordan et al., 2022). In other words, agr inhibition in the animal did not exacerbate biofilm-associated infection. This suggests that the agr–biofilm relationship is context-dependent: In a complex host environment, other factors can compensate, and preventing QS did not make the implant infection worse. Table 9.4 contrasts the clinical and laboratory features observed in agr+ and agr− *S. aureus* isolates.

9.5 CLINICAL IMPLICATIONS AND THERAPEUTIC STRATEGIES

9.5.1 Clinical Relevance of QS and Biofilms

QS and biofilms are central to the clinical persistence and the pathogenesis of *S. aureus*. Biofilm-associated infections, such as endocarditis, osteomyelitis, and indwelling-device infections, are typically chronic and relapsing because bacteria are shielded from both antibiotics and host immune defenses. Biofilm development confers nonspecific antibiotic resistance in many *S. aureus* infections, particularly those involving prosthetic joints, intravascular catheters, and heart valves. Prosthetic joint infections are especially problematic, as biofilms confer exceptionally high levels of antibiotic tolerance and immune evasion, making eradication extremely difficult. Consequently, patients with biofilm-associated *S. aureus* infections often require prolonged antibiotic therapy and device removal, yet relapse remains common (Kaushik et al., 2024).

The agr QS system further shapes clinical outcomes. Active agr function promotes acute, aggressive infections through toxin production, whereas agr dysfunction is commonly associated with persistent disease. For example, agr-defective bacteremias have been linked to treatment failure and increased mortality. In device-related infections, agr-negative isolates form robust biofilms on implants that resist clearance. Paulander et al. observed that agr-deficient strains are enriched in chronic infections, such as those seen in cystic fibrosis patients. Thus, the agr QS system represents a double-edged sword: activation drives acute, toxin-mediated tissue damage, while inactivation promotes biofilm-mediated persistence. Both phenotypes carry important clinical implications – one for acute virulence, the other for chronicity and relapse (Costerton et al., 2005). Effective management of *S. aureus* infections, therefore, requires strategies that address both aspects.

9.5.2 Therapeutic Strategies Targeting QS and Biofilms

Given the critical roles of QS and biofilms in *S. aureus* pathogenesis, novel therapeutic approaches are being developed to disrupt these systems, particularly for drug-resistant strains.

- **Quorum Quenching.** Targeting the agr QS system can "disarm" *S. aureus* by preventing virulence factor production without directly affecting viability. Investigational compounds include AIP analogs (competitive inhibitors of AgrC), AgrA-targeting molecules, and RNAIII antisense agents. For example, staquorsin, a small molecule that binds AgrA, strongly suppresses agr activity; in a mouse skin infection model, it significantly reduced abscess formation without impairing bacterial growth (Dehbanipour & Ghalavand, 2022; Touati et al., 2025). Similarly, Azan-7, an aza-derivative that binds the AgrA DNA-binding domain, blocks agr signaling, downregulates key MRSA toxin genes (e.g., *hla*, *psmα*), and enhances susceptibility to clindamycin. Importantly, neither staquorsin nor Azan-7 inhibits bacterial growth at effective concentrations, reducing selective pressure for resistance. In addition, natural QS inhibitors – such as plant-derived compounds (e.g., eugenol, flavonoids) and microbial peptides (e.g., *Bacillus*-derived peptides) – have shown promise. Overall, agr-targeted therapies aim to attenuate virulence while synergizing with antibiotics (Touati et al., 2025).
- **Anti-Biofilm Agents.** Direct biofilm disruption is another therapeutic avenue. Enzymes targeting matrix components are under investigation: DNases degrade extracellular DNA (eDNA), dispersin B hydrolyzes PNAG, and proteases such as proteinase K or lysostaphin digest protein adhesins (Donelli et al., 2007). For instance, dispersin B combined with antibiotics has enhanced the clearance of *S. aureus* biofilms in vitro. Antimicrobial peptides and nanoparticles also penetrate biofilms, with nanoparticles serving as carriers for antibiotics or matrix-degrading enzymes to improve drug delivery. Clinically, surface modifications such as anti-adhesive coatings and antibiotic-impregnated catheters are already employed to prevent biofilm formation on medical devices.
- **Combination Therapies.** Because agr inhibition can paradoxically stabilize biofilms, combinatorial strategies are particularly promising. For example, probiotics that downregulate agr signaling increased cefoxitin efficacy against MRSA while modifying biofilm architecture. Azan-7 enhanced clindamycin-mediated killing of MRSA biofilms, and DNase–antibiotic combinations (e.g., DNase with vancomycin) have demonstrated synergy in experimental models (Table 9.5). The overarching rationale is to simultaneously weaken biofilm defenses and suppress virulence while enabling antibiotics to clear infection. Nevertheless, caution is warranted, as indiscriminate agr inhibition may promote biofilm persistence. Still, anti-virulence therapies offer key advantages: by "disarming" rather than killing pathogens, they may impose less selective pressure for resistance (Cella et al., 2023).

9.5.3 Ongoing Research and Future Approaches

In the past five years, researchers have developed numerous candidate therapies targeting QS and biofilms in *S. aureus*, though clinical translation remains limited. High-throughput screening and rational drug design have identified several agr inhibitors. In addition to staquorsin and Azan-7,

TABLE 9.5
Anti-Quorum Sensing and Anti-Biofilm Agents Targeting *Staphylococcus aureus*

Agent	Target/Mechanism	Therapeutic Status
Staquorsin	Inhibits agrA transcriptional activity	*In vivo* tested
Azan-7	Binds agrA DNA-binding domain	Preclinical
RNAIII-inhibiting peptide (RIP)	Blocks RNAIII synthesis	In vitro and *in vivo*
DNase I	Degrades eDNA in biofilms	Adjunct in biofilm treatment
Dispersin B	Hydrolyzes PNAG	In vitro studies
Phage lysins	Disrupt bacterial cell walls and biofilms	Experimental

repurposed drugs such as the antidiabetic sitagliptin (which downregulates *agr*) and synthetic peptides that antagonize AgrC are under evaluation. *In vivo* studies are encouraging: topical *agr* inhibitors improved outcomes in animal models without evident toxicity.

Anti-biofilm approaches are also progressing. Phage therapy is being revisited, as bacteriophages produce depolymerases that degrade the EPS, while engineered phage lysins exhibit potent killing of biofilm-embedded *S. aureus*. Vaccine development is another active area, with strategies targeting biofilm polysaccharide antigens. Notably, antibodies against deacetylated PNAG enhance opsonophagocytosis, making PNAG a strong vaccine candidate. Immunotherapies directed against toxins, such as anti-α-toxin and anti-PVL antibodies, are also under study and may indirectly influence the balance between biofilm persistence and invasive disease (Mobarezi et al., 2025).

Looking forward, personalized combination therapies are likely to emerge. For chronic device-associated infections, combining an agr inhibitor (to disperse biofilm) with high-dose antibiotics or surgical device removal may optimize clearance. In acute toxin-driven infections, rapid agr blockade may mitigate tissue damage. CRISPR-based systems designed to selectively eliminate agr^{+} strains or modulate bacterial communication are also under investigation (Bari et al., 2023).

However, challenges remain. Recent reviews caution that agr inhibition may paradoxically enhance biofilm formation. Therefore, dual-targeting strategies that address both QS and biofilms, coupled with rigorous *in vivo* validation, will be essential for successful clinical translation.

9.6 CONCLUSION

QS and biofilm formation are interconnected aspects of *S. aureus* pathogenicity. The agr QS system functions as a regulatory switch, controlling virulence factor expression while inversely modulating biofilm development. At high cell density, *agr* activation induces toxins and degradative enzymes while repressing adhesins and matrix components, driving acute invasion. Conversely, *agr* inactivity promotes biofilm formation, contributing to chronic infection. Clinical evidence reflects this dichotomy: *agr*-positive strains dominate acute, toxin-mediated infections, whereas *agr*-deficient strains are enriched in chronic biofilm-associated disease.

Understanding this interplay has major clinical implications, as biofilm-associated infections pose significant treatment challenges and *agr*-mediated QS amplifies disease severity. Accordingly, therapies targeting both QS and biofilms are being actively developed. Quorum-quenching drugs, such as AgrA inhibitors, attenuate virulence, while antibiofilm agents – including dispersal enzymes and matrix-degrading compounds – reduce biofilm resilience. Early studies suggest that combining QS inhibitors with antibiotics may synergistically improve bacterial clearance.

Future research should focus on improving the efficacy, safety, and delivery of QS and biofilm inhibitors, identifying novel chemical or biological quenchers, and advancing their clinical translation. Multi-target approaches that simultaneously disrupt QS and biofilm formation appear especially promising. In addition, vaccine and immunotherapy strategies targeting biofilm polysaccharides and toxins provide complementary therapeutic avenues. Further study of the agr–biofilm interplay will clarify how best to manipulate these systems. In conclusion, unraveling the molecular dialogue of QS and biofilm formation in *S. aureus* will guide innovative therapeutic strategies against this adaptable and persistent pathogen.

REFERENCES

Arciola, C. R., Campoccia, D., Ravaioli, S., & Montanaro, L. (2015). Polysaccharide intercellular adhesin in biofilm: structural and regulatory aspects. *Frontiers in Cellular and Infection Microbiology*, *5*. https://doi.org/10.3389/fcimb.2015.00007

Bari, A. K., Belalekar, T. S., Poojary, A., & Rohra, S. (2023). Combination drug strategies for biofilm eradication using synthetic and natural agents in KAPE pathogens. *Frontiers in Cellular and Infection Microbiology*, *13*. https://doi.org/10.3389/fcimb.2023.1155699

Beenken, K. E., & Smeltzer, M. S. (2025). *Staphylococcus aureus* biofilm-associated infections: Have we found a clinically relevant target? *Microorganisms*, *13*(4), 852. https://doi.org/10.3390/microorganisms13040852

Cella, M. A., Coulson, T., MacEachern, S., Badr, S., Ahmadi, A., Tabatabaei, M. S., Labbe, A., & Griffiths, M. W. (2023). Probiotic disruption of quorum sensing reduces virulence and increases cefoxitin sensitivity in methicillin-resistant *Staphylococcus aureus*. *Scientific Reports*, *13*(1). https://doi.org/10.1038/s41598-023-31474-2

Cheung, G. Y. C., Bae, J. S., & Otto, M. (2021). Pathogenicity and virulence of *Staphylococcus aureus*. *Virulence*, *12*(1), 547–569. https://doi.org/10.1080/21505594.2021.1878688

Costerton, J., Montanaro, L., & Arciola, C. (2005). Biofilm in implant infections: Its production and regulation. *The International Journal of Artificial Organs*, *28*(11), 1062–1068. https://doi.org/10.1177/039139880502801103

De Oliveira, A., Pinheiro-Hubinger, L., Pereira, V. C., Riboli, D. F. M., Martins, K. B., Romero, L. C., & De Lourdes Ribeiro De Souza Da Cunha, M. (2021). Staphylococcal biofilm on the surface of catheters: Electron microscopy Evaluation of the inhibition of biofilm growth by RNAIII inhibiting peptide. *Antibiotics*, *10*(7), 879. https://doi.org/10.3390/antibiotics10070879

Dehbanipour, R., & Ghalavand, Z. (2022). Anti-virulence therapeutic strategies against bacterial infections: Recent advances. *GERMS*, *12*(2), 262–275. https://doi.org/10.18683/germs.2022.1328

Donelli, G., Francolini, I., Romoli, D., Guaglianone, E., Piozzi, A., Ragunath, C., & Kaplan, J. B. (2007). Synergistic activity of dispersin B and cefamandole nafate in inhibition of staphylococcal biofilm growth on polyurethanes. *Antimicrobial Agents and Chemotherapy*, *51*(8), 2733–2740. https://doi.org/10.1128/aac.01249-06

Ghasemian, A., Peerayeh, S. N., Bakhshi, B., & Mirzaee, M. (2015). *The microbial surface components recognizing adhesive matrix molecules (MSCRAMMs) genes among clinical isolates of Staphylococcus aureus from hospitalized children*. https://pmc.ncbi.nlm.nih.gov/articles/PMC4539745/

Howden, B. P., Giulieri, S. G., Lung, T. W. F., Baines, S. L., Sharkey, L. K., Lee, J. Y. H., Hachani, A., Monk, I. R., & Stinear, T. P. (2023). *Staphylococcus aureus* host interactions and adaptation. *Nature Reviews Microbiology*, *21*(6), 380–395. https://doi.org/10.1038/s41579-023-00852-y

Jordan, S. C., Hall, P. R., & Daly, S. M. (2022). Nonconformity of biofilm formation in vivo and in vitro based on *Staphylococcus aureus* accessory gene regulator status. *Scientific Reports*, *12*(1). https://doi.org/10.1038/s41598-022-05382-w

Kaushik, A., Kest, H., Sood, M., Steussy, B., Thieman, C., & Gupta, S. (2024). Biofilm producing methicillin-resistant *Staphylococcus aureus* (MRSA) infections in humans: Clinical implications and management. *Pathogens*, *13*(1), 76. https://doi.org/10.3390/pathogens13010076

Lamret, F., Colin, M., Mongaret, C., Gangloff, S. C., & Reffuveille, F. (2020). Antibiotic tolerance of *Staphylococcus aureus* biofilm in periprosthetic joint infections and antibiofilm strategies. *Antibiotics*, *9*(9), 547. https://doi.org/10.3390/antibiotics9090547

Lee, S. O., Lee, S., Lee, J. E., Song, K., Kang, C. K., Wi, Y. M., San-Juan, R., López-Cortés, L. E., Lacoma, A., Prat, C., Jang, H., Kim, E. S., Kim, H. B., & Lee, S. H. (2020). Dysfunctional accessory gene regulator (agr) as a prognostic factor in invasive *Staphylococcus aureus* infection: A systematic review and meta-analysis. *Scientific Reports*, *10*(1). https://doi.org/10.1038/s41598-020-77729-0

Mirzaei, R., Campoccia, D., Ravaioli, S., & Arciola, C. R. (2024). Emerging issues and initial insights into bacterial biofilms: From orthopedic infection to metabolomics. *Antibiotics*, *13*(2), 184. https://doi.org/10.3390/antibiotics13020184

Mobarezi, Z., Esfandiari, A. H., Abolbashari, S., & Meshkat, Z. (2025). Efficacy of phage therapy in controlling staphylococcal biofilms: A systematic review. *European Journal of Medical Research*, *30*(1). https://doi.org/10.1186/s40001-025-02781-3

Patel, H., & Rawat, S. (2023). A genetic regulatory see-saw of biofilm and virulence in MRSA pathogenesis. *Frontiers in Microbiology*, *14*. https://doi.org/10.3389/fmicb.2023.1204428

Paulander, W., Varming, A. N., Bojer, M. S., Friberg, C., Bæk, K., & Ingmer, H. (2018). The agr quorum sensing system in *Staphylococcus aureus* cells mediates death of sub-population. *BMC Research Notes*, *11*(1). https://doi.org/10.1186/s13104-018-3600-6

Peng, Q., Tang, X., Dong, W., Sun, N., & Yuan, W. (2022). A review of biofilm formation of *Staphylococcus aureus* and its regulation mechanism. *Antibiotics*, *12*(1), 12. https://doi.org/10.3390/antibiotics12010012

Podkowik, M., Perault, A. I., Putzel, G., Pountain, A., Kim, J., DuMont, A. L., Zwack, E. E., Ulrich, R. J., Karagounis, T. K., Zhou, C., Haag, A. F., Shenderovich, J., Wasserman, G. A., Kwon, J., Chen, J., Richardson, A. R., Weiser, J. N., Nowosad, C. R., Lun, D. S., . . . Shopsin, B. (2023). Quorum-sensing agr system of *Staphylococcus aureus* primes gene expression for protection from lethal oxidative stress. *eLife*, *12*. https://doi.org/10.7554/elife.89098

Sahu, A., & Ruhal, R. (2025). Immune system dynamics in response to Pseudomonas aeruginosa biofilms. *Npj Biofilms and Microbiomes*, *11*(1). https://doi.org/10.1038/s41522-025-00738-2

Schilcher, K., & Horswill, A. R. (2020). Staphylococcal biofilm development: Structure, regulation, and treatment strategies. *Microbiology and Molecular Biology Reviews*, *84*(3). https://doi.org/10.1128/mmbr.00026-19

Sharma, S., Mohler, J., Mahajan, S. D., Schwartz, S. A., Bruggemann, L., & Aalinkeel, R. (2023). Microbial biofilm: A review on formation, infection, antibiotic resistance, control measures, and innovative treatment. *Microorganisms*, *11*(6), 1614. https://doi.org/10.3390/microorganisms11061614

Thammavongsa, V., Kim, H. K., Missiakas, D., & Schneewind, O. (2015). Staphylococcal manipulation of host immune responses. *Nature Reviews Microbiology*, *13*(9), 529–543. https://doi.org/10.1038/nrmicro3521

Touati, A., Ibrahim, N. A., & Idres, T. (2025). Disarming *Staphylococcus aureus*: Review of strategies combating this resilient pathogen by targeting its virulence. *Pathogens*, *14*(4), 386. https://doi.org/10.3390/pathogens14040386

Turner, N. A., Sharma-Kuinkel, B. K., Maskarinec, S. A., Eichenberger, E. M., Shah, P. P., Carugati, M., Holland, T. L., & Fowler, V. G. (2019). Methicillin-resistant *Staphylococcus aureus*: An overview of basic and clinical research. *Nature Reviews Microbiology*, *17*(4), 203–218. https://doi.org/10.1038/s41579-018-0147-4

Wang, B., & Muir, T. W. (2016). Regulation of virulence in *Staphylococcus aureus*: Molecular mechanisms and remaining puzzles. *Cell Chemical Biology*, *23*(2), 214–224. https://doi.org/10.1016/j.chembiol.2016.01.004

Williams, P., Hill, P., Bonev, B., & Chan, W. C. (2023). Quorum-sensing, intra- and inter-species competition in the staphylococci. *Microbiology*, *169*(8). https://doi.org/10.1099/mic.0.001381

Yarwood, J. M., & Schlievert, P. M. (2003). Quorum sensing in *Staphylococcus* infections. *Journal of Clinical Investigation*, *112*(11), 1620–1625. https://doi.org/10.1172/jci20442

Zhao, A., Bodine, S. P., Xie, Q., Wang, B., Ram, G., Novick, R. P., & Muir, T. W. (2022). Reconstitution of the *S. aureus* agr quorum sensing pathway reveals a direct role for the integral membrane protease MroQ in pheromone biosynthesis. *Proceedings of the National Academy of Sciences*, *119*(33). https://doi.org/10.1073/pnas.2202661119

10 Current Techniques for Detecting Biofilm Formation

P. Gnana Deepu[1], Anmol Kumar[2], Md Rahman uddin[3], Advaith Roy[3], Vislavath Sneha[3], and Bikshu Jaga[4]

[1]Centre for Biotechnology, University College of Engineering Science and Technology, Jawaharlal Nehru Technological University, Hyderabad, India

[2]Department of Biotechnology, Atmiya University, Rajkot, Gujarat, India

[3]Department of Biotechnology, Chaitanya Bharathi Institute of Technology, Hyderabad, Telangana, India

[4]School of Biomedical Sciences, Charles Sturt University, Wagga Wagga, NSW, Australia

ABBREVIATIONS

AFM	atomic force microscopy
C. albicans	*Candida albicans*
CLSM	confocal laser scanning microscopy
CRA	Congo Red Agar method
cryo-EM	cryo-electron microscopy
FTIR	Fourier-transform infrared
PIA	polysaccharide intercellular adhesin
SEM	scanning electron microscopy
SERS	surface-enhanced Raman spectroscopy
ToF-SIMS	time-of-flight secondary ion mass spectrometry

10.1 INTRODUCTION

Biofilm formation is a well-organized microbial strategy that enhances survival and persistence in challenging environments. Among pathogens, *Staphylococcus aureus* is particularly adept at forming biofilms, which is a significant factor in chronic infections and device-related complications. These biofilms consist of microbial cells embedded within an extracellular matrix, providing protection against antibiotics and immune responses, making infections difficult to treat (Archer et al., 2011). Compared to other microorganisms such as *Pseudomonas aeruginosa* or *Candida albicans*, *S. aureus* biofilms are particularly concerning due to their rapid formation, resilience, and ability to cause persistent infections. While *P. aeruginosa* biofilms exhibit structural complexity and antibiotic resistance (Rumbaugh & Sauer, 2020) and *C. albicans* biofilms thrive in mixed-species environments (Köhler et al., 2017), *S. aureus* biofilms uniquely excel at evading host immune defenses and persisting in diverse niches (Otto, 2008).

Detecting biofilm formation is critical for understanding its mechanisms and developing targeted therapeutic strategies. Early and accurate detection can significantly improve treatment outcomes and reduce complications (Percival et al., 2015). Methods such as traditional staining, molecular assays, and advanced imaging techniques are utilized to study and analyze biofilms. This chapter highlights the clinical relevance of *S. aureus* biofilms, compares their impact with biofilms formed

 DOI: 10.1201/9781003647072-13

by other microbes, and reviews current detection methodologies, emphasizing the importance of advancing these tools to combat biofilm-associated infections effectively.

10.1.1 Definition of Biofilms

Biofilms are structured microbial communities embedded in an extracellular polymeric matrix that enhances microbial survival, persistence, and resistance to external stresses. Originating early in Earth's history as a protective adaptation for prokaryotes, biofilms allow microorganisms to thrive in hostile environments by promoting homeostasis and cooperative interactions (Hall-Stoodley et al., 2004). First described systematically by Costerton in 1978, biofilms are now recognized as heterogeneous, three-dimensional consortia that can adhere to both inert surfaces, such as plastic or glass, and biological substrates, including mucosal tissues and skin. These sessile aggregates are primarily composed of extracellular polymeric substances (EPSs), a hydrated matrix of polysaccharides, proteins, lipids, teichoic acids, and nucleic acids that enmesh both viable and nonviable cells (Costerton et al., 1981). Their structural complexity has led to the description of biofilms as "cities for microbes" due to their organized, community-like architecture (Watnick & Kolter, 2000).

Biofilms are not confined to bacteria alone but can also include fungi and protists, forming either surface-attached colonies or free-floating aggregates, as exemplified by *S. aureus* and *P. aeruginosa* (Alhede et al., 2011). These communities play critical roles in clinical and industrial contexts, where they contribute to persistent infections, medical device contamination, and drug resistance challenges, as well as industrial processes such as water treatment and food processing (Costerton et al., 1981). Although a wide range of detection and analytical methods are available, their comparative reliability and translational relevance remain poorly consolidated. This chapter focuses on systematically evaluating current and emerging strategies for biofilm detection and characterization to inform more effective anti-biofilm interventions.

10.1.2 Importance of Studying Biofilm Formation in *S. aureus*

The study of biofilm formation in *S. aureus* is crucial due to its significant impact on public health, clinical outcomes, and industrial settings. Understanding the mechanisms and consequences of biofilm formation in this bacterium can lead to better management of infections, development of targeted therapies, and improved safety protocols in the medical and food industries.

10.1.2.1 Clinical Relevance and Persistent Infections

Biofilm formation in *S. aureus* is a major factor contributing to persistent infections, including endocarditis, osteomyelitis, and infections associated with medical devices such as catheters, prosthetics, and heart valves. Biofilms protect *S. aureus* from the host immune response and antibiotic treatment, leading to chronic infections that are difficult to eradicate (Arciola et al., 2012). According to Otto (2013), biofilms can cause up to 80% of chronic infections, demonstrating their critical role in clinical settings.

10.1.2.2 Antibiotic Resistance

Biofilms contribute significantly to the antibiotic resistance of *S. aureus*. The extracellular matrix of biofilms impedes the penetration of antibiotics, allowing the bacteria to survive in hostile conditions. *S. aureus* biofilms can tolerate antibiotic concentrations 10–1,000 times higher than those needed to inhibit planktonic cells (Costerton, 2007). Research by Høiby et al. (2010) highlights that biofilm-associated *S. aureus* infections often require prolonged and aggressive antibiotic therapy, which increases the risk of resistance development.

10.1.2.3 Mechanisms of Biofilm Formation

Studying the biofilm formation mechanisms in *S. aureus* helps identify potential therapeutic targets. Key components involved in *S. aureus* biofilm formation include surface proteins, extracellular DNA, and polysaccharide intercellular adhesin (PIA). These elements play roles in the initial adhesion, maturation, and maintenance of the biofilm structure (Foster et al., 2013). Research by Schilcher and Horswill (2020) details the genetic and molecular pathways involved in biofilm development, which can be targeted for novel anti-biofilm strategies. Clinical *S. aureus* isolates often rely on proteinaceous scaffolds (FnBP, ClfA/B) for adhesion, whereas laboratory strains default to PIA-driven matrices encoded by the ica operon. This divergence influences susceptibility to dispersin B and proteolytic treatments, highlighting the need to tailor anti-biofilm strategies to strain-specific matrix composition.

10.1.2.4 Implications for Medical Device-Associated Infections

S. aureus is a leading cause of infections related to indwelling medical devices due to its ability to form robust biofilms on various surfaces. These biofilms protect the bacteria from both immune defenses and antimicrobial treatments, posing significant challenges in clinical management (Becker et al., 2014). According to Kiedrowski et al. (2011), the presence of biofilms on medical devices often necessitates surgical intervention and device removal, highlighting the need for better preventive and treatment strategies.

10.1.2.5 Impact on Food Industry and Public Health

Biofilms formed by *S. aureus* can also contaminate food processing environments, leading to foodborne illnesses. The persistence of *S. aureus* biofilms on food contact surfaces poses a public health risk by facilitating the spread of bacteria and associated toxins (Kadariya et al., 2014). A study by Di Ciccio et al. (2015) emphasizes the importance of understanding biofilm dynamics in industrial settings to improve cleaning protocols and reduce contamination.

10.1.2.6 Development of Novel Anti-Biofilm Strategies

Research on *S. aureus* biofilms is essential for developing new therapeutic approaches, including biofilm-disrupting agents, vaccines, and surface coatings that prevent biofilm formation. Targeting specific biofilm components, such as extracellular matrix molecules or regulatory pathways, offers promising strategies to combat biofilm-related infections (Fleming & Rumbaugh, 2017). According to Conlon et al. (2014), understanding biofilm biology is critical for designing effective treatments that can disrupt biofilm integrity and enhance antibiotic efficacy.

10.2 MICROSCOPY TECHNIQUES FOR BIOFILM DETECTION

Microscopy remains central to biofilm research, providing structural and spatial insights that extend beyond biochemical assays. In the case of *S. aureus*, different imaging modalities offer complementary perspectives, ranging from surface morphology to three-dimensional organization and nanoscale mechanics.

10.2.1 Scanning Electron Microscopy

Scanning electron microscopy (SEM) is one of the most widely applied tools for studying biofilm surface morphology. By exploiting electron scattering, SEM generates high-resolution images that reveal bacterial distribution and EPSs within the matrix (Clayborn et al., 2015; Nanakoudis & Blog. phenom-World, 2020). It has been extensively used to assess *S. aureus* colonization under varying environmental conditions and to evaluate anti-biofouling strategies such as nano-silica coatings (Wang et al., 2018; Boguslavsky et al., 2018; Oliveira et al., 2014). However, SEM requires dehydration, vacuum fixation, and coating, all of which can introduce artifacts and make the technique inherently destructive (Doucet et al., 2005). This restricts its use for longitudinal or in situ studies and limits the interpretation of internal biofilm architecture.

10.2.1.1 ToF-SIMS–SEM Hybrid Imaging

Recent advances have attempted to overcome some of these limitations by combining SEM with time-of-flight secondary ion mass spectrometry (ToF-SIMS), where SEM provides high-resolution topography while ToF-SIMS maps chemical signatures such as lipids, proteins, and polysaccharides. Such hybrid approaches preserve biofilm architecture and offer correlative insights into structural and compositional heterogeneity.

10.2.2 Confocal Laser Scanning Microscopy

Confocal laser scanning microscopy (CLSM) has emerged as the method of choice for non-destructive biofilm imaging. By employing fluorescent dyes to label carbohydrates, proteins, and extracellular DNA, CLSM enables three-dimensional reconstructions of biofilms with detailed information on thickness, heterogeneity, and spatial distribution of *S. aureus* cells (Neu & Lawrence, 2014; Dominiak et al., 2011). It can be applied in situ on flat surfaces, which makes it suitable for studying dynamic biofilm processes without fixation or dehydration, although ex situ preparation is often required for samples from complex environments such as bioreactors (Palmer & Sternber, 1999). Despite its advantages, CLSM depends heavily on staining, which may bias data interpretation, and is limited by penetration depth in opaque matrices. Reported resolutions range between 20 and 100 µm depending on instrumentation, making it highly adaptable across different experimental settings.

10.2.3 Atomic Force Microscopy

Atomic force microscopy (AFM) provides a different dimension of analysis by measuring nanoscale topography and mechanical properties of biofilms. Using a scanning probe, AFM can quantify adhesion forces, elasticity, and nanostructural organization of EPS with resolutions between 1 and 10 nm (Chatterjee et al., 2014; Merino et al., 2019; Ozkan et al., 2016). Unlike SEM, AFM can be operated under ambient or liquid conditions, preserving biofilms in a near-native state. The use of tapping mode reduces deformation during imaging, enabling high-resolution analysis of live or hydrated samples (Hannig et al., 2010). Nevertheless, AFM is technically demanding and time-intensive and may still introduce artifacts during thickness measurements, as shown in studies where partial scratching of the biofilm was required (Birarda et al., 2019). Its application is, therefore, more suited to mechanistic investigations of cell surface interactions rather than routine detection.

10.2.4 Emerging Techniques

10.2.4.1 AI-Driven Image Analysis

Artificial intelligence (AI) and deep learning approaches are rapidly transforming biofilm imaging by automating segmentation, classification, and quantitative morphometry from large microscopy datasets. Modern pipelines combine convolutional neural networks and generative models to (1) segment dense, multi-species colonies; (2) classify stages of biofilm development from time-lapse series; and (3) synthesize annotated training data when ground-truth labels are scarce. These advances increase throughput and reduce observer bias, enabling objective, reproducible metrics of biofilm architecture and dynamics; however, model generalizability remains dependent on representative training sets and rigorous validation across modalities and labs. Recent reviews and methodological reports document successful AI workflows for biofilm and bacterial cell image analysis and describe open tools for "segment-anything" style microscopy segmentation and synthetic dataset generation (Li et al., 2025).

10.2.4.2 Cryo-Electron Microscopy (Cryo-EM and Cryo-ET)

Cryo-EM and cryo-electron tomography (cryo-ET) permit ultrastructural visualization of biofilms and their matrix in a near-native, vitrified state, enabling molecular-to-nanometer resolution of polymers, protein complexes, and cell–matrix interfaces. Advances in automated data collection, machine-assisted map interpretation, and AI-based model building have improved throughput and lowered the barrier to interpretable structural models from cryo datasets. Nevertheless, cryo workflows require specialized vitrification hardware, cryo-transfer stages, and expertise in tomography reconstruction and interpretation, which restricts their routine use to well-equipped research centers. For biofilm research, cryo approaches are particularly powerful for mechanistic studies where molecular detail or preservation of hydrated matrix architecture is essential (De La Cruz & Eng, 2023).

10.2.4.3 Correlative Light and Electron Microscopy (CLEM/Cryo-CLEM)

Correlative workflows integrate the molecular specificity and live/immunofluorescent labeling of light microscopy with the ultrastructural resolution of electron microscopy. Cryo-CLEM variants preserve fluorescence localization within vitrified samples and then map that signal to electron tomography, enabling precise identification of labeled matrix components or cells within the native architecture. CLEM reduces interpretive ambiguity (linking molecular markers to ultrastructure) but adds complexity in sample preparation, instrument alignment, and data fusion. Recent methodological reviews highlight improved fiducial-based registration, near-infrared branding, and cryo-CLEM protocols that expand applicability to diverse biological specimens. For biofilm studies, CLEM is uniquely valuable when one must locate specific taxa or matrix molecules within the three-dimensional electron-dense context (Pierson et al., 2024).

10.2.5 Comparative Analysis of Microscopy Techniques

Microscopy techniques are essential tools for studying biofilms, such as those formed by *S. aureus*. Each technique offers unique strengths and limitations, which influence their suitability for various aspects of biofilm research. Table 10.1 shows the comparative analysis of light microscopy, fluorescence microscopy (including CLSM), AFM, and emerging techniques (AI image analysis, Cryo-EM, CLEM).

TABLE 10.1
Comparative Analysis of Microscopy Techniques

Microscopy Technique	Strengths	Limitations
Light microscopy	• Broad accessibility and ease of use • Enables real-time visualization of biofilm formation	• Limited spatial resolution (~200 nm) • Primarily reveals surface-associated structures, with poor penetration into deeper layers
Confocal laser scanning microscopy (CLSM)	• High specificity through fluorescent labeling • Non-destructive imaging suitable for in situ analysis • Allows three-dimensional reconstruction of biofilm structures	• Requires complex sample preparation and fluorescent staining • Restricted to labeled components • Resolution varies (20–101 µm), limiting nanoscale insights
Atomic force microscopy (AFM)	• Nanometer-scale resolution (1–10 nm) • Capable of assessing mechanical and topographical properties (e.g., elasticity, adhesion) • Minimally destructive under optimized conditions	• Demands specialized instrumentation and expertise • Sample preparation can be technically demanding • Potential risk of sample alteration during scanning
Emerging approaches (AI-driven image analysis, cryo-EM, CLEM)	• Provide ultra-high-resolution imaging of biofilm ultrastructure • Enable correlative analysis integrating structural and molecular specificity • AI facilitates automated, quantitative, and high-throughput interpretation	• High operational costs and infrastructure requirements • Significant technical expertise necessary • Currently limited in widespread clinical translation

10.3 BIOCHEMICAL AND MOLECULAR ASSAYS FOR BIOFILM DETECTION

Biochemical and molecular assays form the cornerstone of biofilm detection by providing insights into microbial activity, genetic regulation, and matrix composition. Unlike microscopy, which primarily offers structural visualization, these assays enable scalable, functional, and mechanistic evaluation. Nevertheless, reproducibility, sensitivity, and translational value vary significantly between techniques, making it essential to assess their strengths and limitations (O'Toole et al., 2000; Chavant et al., 2007; Sánchez et al., 2013; Brugger et al., 2014).

10.3.1 Congo Red Agar Method

The Congo Red Agar (CRA) assay is one of the oldest methods used to screen biofilm formation. Colonies are cultured on agar supplemented with Congo red dye, and biofilm-forming strains typically appear black, while non-producers remain red. The presence of a dry crystalline morphology may further indicate biofilm production, although this feature is not consistently observed across all *S. aureus* isolates (Figure 10.1). This approach is appealing because of its simplicity and cost-effectiveness, but its diagnostic value is limited. CRA provides only a binary color change without quantifying biomass or viability. False positives and false negatives are frequent due to strain variability and culture conditions. Although once widely used, CRA is now largely considered outdated and is best suited for basic demonstrations or historical comparisons rather than rigorous scientific investigations (Beenken et al., 2004).

10.3.2 Microtiter Plate Assay

The microtiter plate assay, also referred to as the tissue culture plate (TCP) assay, remains one of the most reliable and widely employed methods for quantifying biofilm biomass. In this approach, bacteria are grown in 96-well plates, stained with crystal violet, and quantified by measuring optical density, usually at 570 nm. This provides a semi-quantitative measurement of total biomass attached to the well surface. For *S. aureus*, assay sensitivity is improved by supplementing culture media with sugars such as glucose or sucrose, which enhance biofilm production (Knobloch et al., 2002). Despite these advantages, the method is prone to variability due to inconsistent washing and

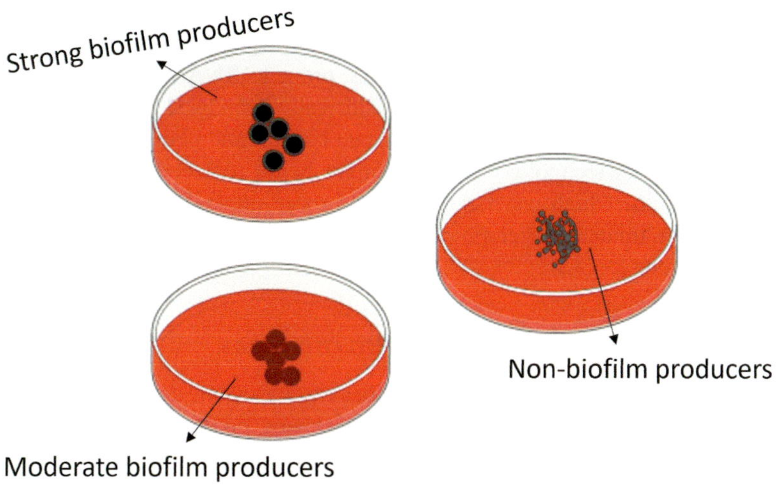

FIGURE 10.1 Biofilm-forming colonies typically appear black on the red agar, whereas non-biofilm-producing colonies maintain a red color. Additionally, the presence of dry, crystalline morphology can indicate biofilm production.

handling, which may influence results. Furthermore, the assay does not capture metabolic activity or biofilm architecture – it merely quantifies the accumulated biomass. Consequently, while the TCP assay is well-suited for large-scale screening of isolates, it should be complemented with additional methods when characterizing clinically relevant biofilms.

10.3.3 Resazurin and Metabolic Assays

Resazurin-based assays extend biofilm detection by measuring bacterial metabolic activity. This redox-sensitive dye is reduced by metabolically active cells, producing a fluorescent or colorimetric signal proportional to bacterial viability (Sánchez et al., 2013). Unlike biomass-only assays, this method provides information on the physiological state of cells within biofilms, thereby differentiating between dormant, active, and potentially persister populations. Despite these strengths, resazurin assays may underrepresent non-dividing persister cells and tend to skew results toward metabolically active subpopulations. As such, they are most valuable when combined with biomass or structural assays, offering a more comprehensive understanding of biofilm physiology.

10.3.4 Quantification of EPSs

EPSs are central to biofilm architecture, stability, and antimicrobial resistance. In *S. aureus*, EPS consists predominantly of polysaccharides such as PNAG/PIA, proteins, extracellular DNA, and lipids. Quantitative assays for EPS components include phenol-sulfuric acid or anthrone-sulfuric acid for polysaccharides, BCA or Lowry assays for proteins, and UV absorbance at 260 nm for DNA. While these assays provide valuable insights into biofilm matrix composition, they often face technical challenges. Cross-interference between different EPS components can complicate interpretation, and results are typically semi-quantitative unless rigorously standardized. Therefore, EPS quantification is highly informative for mechanistic studies but remains unsuitable as a diagnostic tool.

10.3.5 Molecular Biology Techniques

Molecular approaches allow more precise identification of biofilm-related genes and regulatory pathways. PCR is widely applied to detect the presence of biofilm-associated genes such as *icaA*, *icaD*, *fnbA*, and *fnbB*, while qRT-PCR advances this by quantifying transcriptional activity (Nadell et al., 2008). These methods provide detailed insights into the genetic regulation of biofilm formation. However, they are limited to known gene targets, prone to RNA degradation issues, and relatively resource-intensive. DNA microarrays have historically been used to assess global expression profiles, offering insight into regulatory shifts between biofilm and planktonic lifestyles (Beenken et al., 2004). However, this technology is increasingly being replaced by RNA-sequencing and single-cell transcriptomics, which provide deeper and more unbiased genetic insights. Although molecular assays offer unmatched specificity, their cost, complexity, and lack of portability limit direct clinical application.

10.3.6 Gene Expression Analysis of Biofilm-Related Genes

In clinical isolates of *S. aureus*, gene expression studies emphasize the central role of regulatory networks in biofilm development. An investigation of 110 isolates, predominantly MRSA, showed that nearly all produced biofilms to varying degrees, with the *ica* operon (*icaA, icaB, icaC, icaD*) being the most prevalent (Nourbakhsh & Namvar, 2016). High biofilm producers demonstrated marked upregulation of *icaA*, *icaC*, and *icaD*, consistent with their function in PIA synthesis. Adhesion-associated genes such as *fnbA* and *fnbB* were also strongly expressed in certain isolates, whereas *cna* and *ebps* showed lower expression. These observations underline the genetic complexity of biofilm formation and its variation across infection sites.

10.3.6.1 Phenotypic and Genotypic Characterization

Among the 110 isolates, 103 were identified as MRSA. Phenotypic screening revealed universal biofilm formation, which was categorized as high (53.5%), intermediate (33.3%), or low (13.2%). Notably, isolates from decubitus ulcers (85%), wounds (78.2%), and tracheal secretions (65.4%) displayed the highest frequencies of strong biofilm formation (Figure 10.2).

10.3.6.2 Gene Prevalence in Biofilm Formation

Multiplex PCR analysis identified 12 biofilm-associated genes among the MRSA isolates. The prevalence rates were as follows: *icaA* (34.2%), *icaB* (29.7%), *icaC* (69.3%), *icaD* (54.8%), *fnbA* (38.1%), *fnbB* (46.6%), *fib* (39.9%), *clfA* (41.4%), *clfB* (44.1%), *ebps* (26.5%), *cna* (18.3%), and *eno* (29.6%). The *ica* operon was especially prominent, with *icaC* and *icaD* showing the highest prevalence, confirming their pivotal role in PIA synthesis and stable biofilm formation (Figure 10.3).

10.3.6.3 Gene Expression Analysis

The expression levels of these biofilm-related genes were further analyzed in selected isolates. The gene expression profiles revealed that the ica operon genes, particularly icaA, icaC, and icaD, were upregulated in high biofilm-producing isolates. This upregulation is consistent with their role in PIA synthesis and biofilm matrix formation.

- **icaA** and **icaC** were highly expressed in isolates categorized as high biofilm producers, suggesting their critical role in biofilm formation.
- **icaD** also showed elevated expression in these isolates, reinforcing its importance in the biofilm development process.

Other genes, such as **fnbA** and **fnbB**, which are involved in adhesion to host tissues, also demonstrated significant expression levels in high biofilm producers. However, genes like **cna** and **ebps**

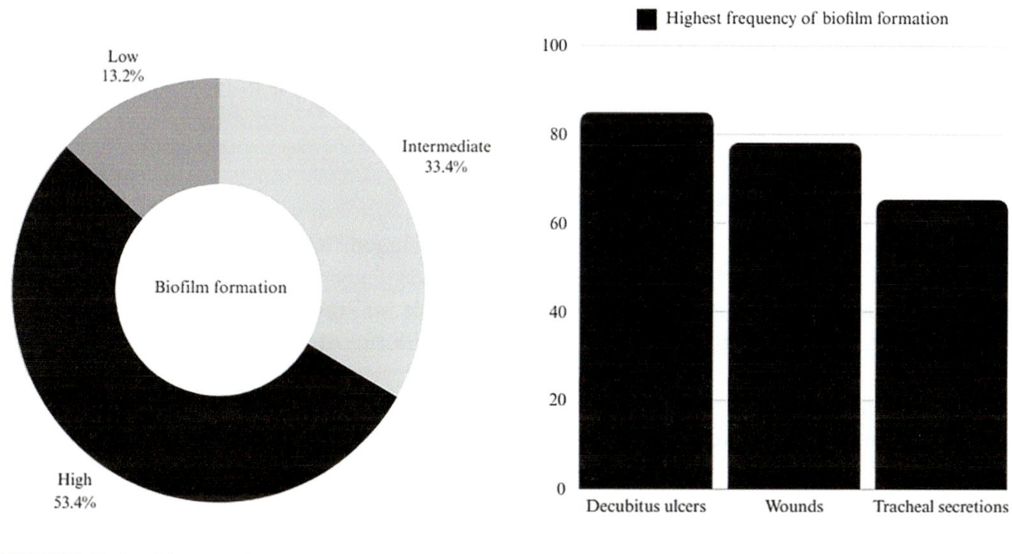

FIGURE 10.2 Phenotypic and genotypic characterization.

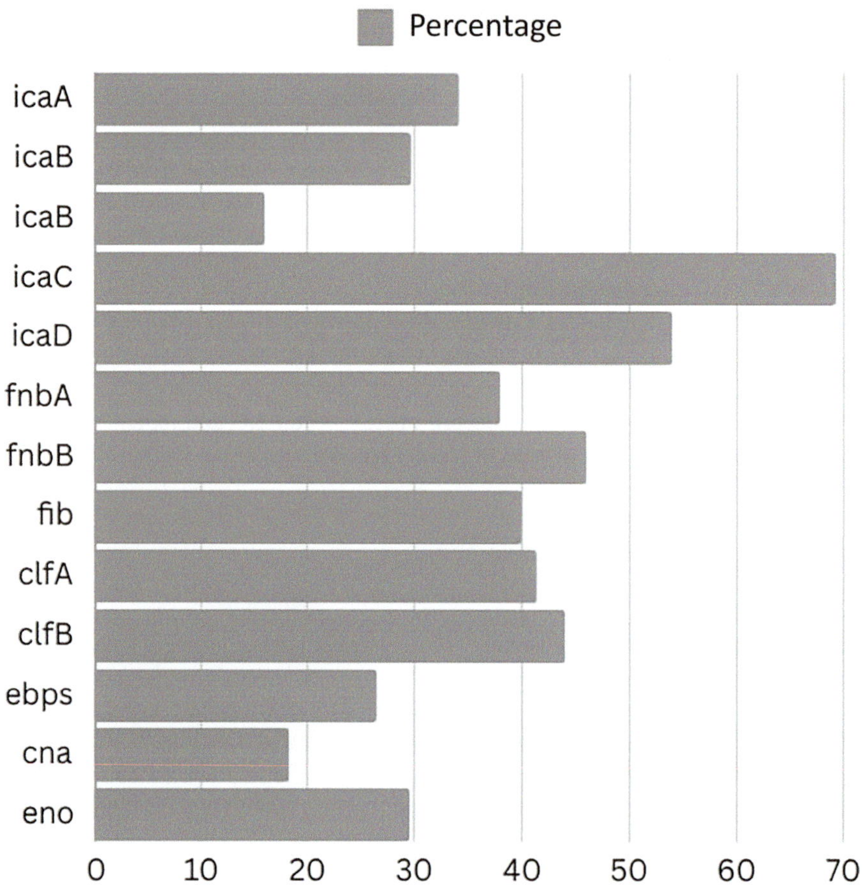

FIGURE 10.3 Gene prevalence in biofilm formation.

exhibited lower expression levels, indicating their potentially lesser role in the biofilm formation under the conditions tested (Nourbakhsh & Namvar, 2016).

10.3.7 Emerging Molecular Approaches

Recent technological advances are redefining biofilm detection. RNA-seq and single-cell transcriptomics offer high-resolution, unbiased views of gene regulatory networks driving biofilm heterogeneity. CRISPR-based fluorescent reporters now enable real-time monitoring of biofilm gene expression in situ. Similarly, biosensor and microfluidic platforms are being developed to integrate genetic and biochemical detection into portable diagnostic devices. These emerging methods carry significant promise for near-patient applications, but their adoption faces hurdles such as high cost, regulatory complexity, and the need for technical standardization. At present, they remain largely experimental but are poised to transform biofilm diagnostics in the future.

10.3.8 Comparative Analysis of the Techniques

Table 10.2 summarizes a comparative analysis of various biochemical and molecular assays.

TABLE 10.2
Comparative Analysis of Biochemical and Molecular Assays for Biofilm Detection

Technique	Advantages	Limitations
Congo Red Agar (CRA)	Simple, low-cost, rapid; allows preliminary screening of biofilm formation	Only qualitative; high rate of false positives/negatives; poor reproducibility
Microtiter plate assay (TCP)	Widely used; semi-quantitative; compatible with high-throughput; sensitive to sugar supplementation	Variability due to washing steps; biomass-only; no viability or structural data
Resazurin/metabolic assays	Provides functional data on metabolic activity; highlights heterogeneity within biofilms	Underrepresents dormant/persister cells; biased to active populations
EPS quantification	Reveals compositional detail (polysaccharides, proteins, DNA); essential for mechanistic insights	Cross-interference between components; semi-quantitative; poor reproducibility
PCR	Specific detection of biofilm-associated genes; reliable for strain identification	Limited to known targets; cannot distinguish gene presence versus expression
qRT-PCR	Quantifies gene expression; provides transcriptional insights; high specificity	Expensive, labor-intensive; RNA instability issues; limited scalability
DNA microarrays	High-throughput analysis of global expression patterns; identifies regulatory shifts	Expensive; require specialized equipment; being replaced by RNA-seq
Emerging tools (RNA-seq, CRISPR, biosensors)	Offer high sensitivity and resolution; integrate detection with portability; enable real-time monitoring	Still experimental; costly; complex; not yet validated for clinical use

10.3.9 Microfluidic Platforms

Microfluidic technology has emerged as a transformative platform for studying biofilm formation under highly controlled, physiologically relevant conditions. Unlike traditional static assays that primarily support surface-attached biofilms, microfluidic devices can simulate complex hydrodynamic environments, including laminar shear, pulsatile flow, and nutrient gradients – features commonly encountered *in vivo*, particularly in vascular and urological systems. These microscale flow chambers typically consist of polydimethylsiloxane channels bonded to glass or plastic substrates. The confined geometry and controllable flow parameters allow for reproducible studies of planktonic aggregation, initial adhesion, microcolony development, and biofilm maturation. *S. aureus*, which forms resilient biofilms in dynamic environments such as blood vessels or indwelling catheters, exhibits unique behaviors under microfluidic conditions. For instance, shear stress modulates expression of adhesion genes and alters the production of extracellular matrix components, including PIA and surface proteins like FnBPs. Advanced microfluidic systems now integrate real-time monitoring tools such as impedance sensors, optical coherence tomography (OCT), and fluorescence imaging modules to track biofilm development dynamically. These setups enable time-resolved data acquisition, allowing researchers to observe biofilm lifecycle transitions, such as detachment events or dispersal under enzymatic or antimicrobial treatment. Furthermore, microfluidic platforms are compatible with live/dead staining, metabolic dyes, or pH-responsive nanoparticles, which can be used to assess cell viability and metabolic heterogeneity within developing *S. aureus* biofilms.

Importantly, microfluidics offers unparalleled resolution for analyzing biofilm behavior in response to environmental perturbations, such as pH shifts, osmotic stress, or nutrient depletion.

These conditions can be precisely modulated using gradient-generating modules or parallel channel designs. Such capabilities are particularly valuable for evaluating anti-biofilm compounds, enabling high-throughput screening of novel antibiotics, dispersal agents, or surface coatings under standardized conditions. Moreover, microfluidic co-culture systems have been developed to study interactions between *S. aureus* and host immune cells, or between polymicrobial species, which often occur in chronic wound or device-related infections. The scalability and modularity of microfluidic systems also make them suitable for incorporation into lab-on-a-chip diagnostic platforms. Recent innovations have combined microfluidic biofilm culture chambers with portable electrochemical sensors and data acquisition systems, laying the groundwork for future point-of-care biofilm diagnostics. These devices could be deployed for early detection of biofilm-related infections on medical implants, enabling timely intervention before full colonization occurs.

10.4 EMERGING AND LABEL-FREE DETECTION TECHNIQUES

Recent advances in biofilm sensing span a diverse range of label-free, real-time methods. For example, OCT has been demonstrated to image thick bacterial biofilms in situ. Gunn et al. (2025) developed an OCT-based approach to noninvasively detect and quantify methicillin-resistant *S. aureus* (MRSA) biofilms on metal surfaces (Demidova et al., 2025). Their technique automatically delineates biofilm thickness, roughness, and pore structure, achieving "unprecedented" resolution of MRSA biofilm architecture while obviating stains or dyes. Similarly, hyperspectral optical imaging has been applied for label-free biofilm detection on surfaces. In one study, fluorescence hyperspectral imaging (HSI) coupled with machine learning distinguished *Escherichia coli* and *Salmonella* biofilms on plastic or steel with >90% accuracy (Lee et al., 2021). Such hyperspectral methods acquire spatial plus spectral data to discriminate biomass signatures without labels. Another optical approach is surface plasmon resonance imaging (SPRi), which provides wide-field, label-free imaging of biofilm growth. SPRi sensors (gold-coated chips with CCD readouts) produce high-resolution images of attached biomass over centimeter-scale areas in real time (Kulshrestha & Gupta, 2024). Kulshrestha and Gupta (2024) note that SPRi "provides label-free, high-resolution images of biomass covering large surfaces . . . in real time, allowing quantitative analysis of biofilm dynamics." Although SPRi has been applied to *E. coli* and *P. aeruginosa*, its label-free mechanism should also apply to gram-positive biofilms like *S. aureus*. Together, these emerging optical modalities (OCT, hyperspectral, SPRi, etc.) offer non-destructive monitoring of biofilm formation and structure with spatial detail, and in the case of OCT can specifically detect *S. aureus* biofilm signals (Demidova et al., 2025).

10.4.1 Real-Time Electrochemical and Label-Free Biosensors

Label-free electrical sensing has become a prominent trend for continuous biofilm monitoring. Electrochemical impedance spectroscopy (EIS) in particular "has emerged as a promising approach for real-time, non-destructive, and rapid monitoring of biofilms," offering high sensitivity and low cost (Ameer et al., 2023). Microfabricated EIS sensors (microelectrode arrays or interdigitated electrodes) can be integrated into flow cells or devices. For instance, McGlennen et al. (2023) reported a microfluidic impedance biosensor that continuously tracked biofilm growth and dispersal under flowing conditions (McGlennen et al., 2023). They observed impedance changes corresponding to *E. coli* colonization and EPS production in real time. These impedimetric sensors require no labels and respond to the dielectric and redox changes induced by accumulating biofilm. Likewise, cyclic voltammetry (CV) has been used: biofilm accumulation on electrodes typically reduces CV peak currents as surface area becomes covered (Kulshrestha & Gupta, 2024). Both CV and EIS can thus report biofilm presence dynamically. While most demonstrations involve *P. aeruginosa* or *E. coli*, the principle extends to *S. aureus*, provided that suitable redox mediators or conductive pathways exist in the biofilm. In all, real-time electrical sensors (EIS/CV) offer label-free, in situ biofilm detection with rapid feedback (Ameer et al., 2023).

Beyond electrochemistry, other label-free biosensors have advanced. Quartz crystal microbalance (QCM) devices track biofilm mass and viscoelastic properties via frequency shifts. As biofilm EPS accumulates, the QCM resonant frequency drops and dissipation rises. For example, Reipa et al. monitored *P. aeruginosa* attachment by QCM, observing decreased frequency and increased damping as the biofilm matured. QCM with dissipation (QCM-D) further quantifies biofilm viscoelasticity ($\Delta D/\Delta f$ ratio). These methods are label-free and real-time, though they often require controlled flow or shear to interpret the data. Surface acoustic wave (SAW) sensors similarly measure biofilm-induced mass/viscosity changes at MHz frequencies. Recent SAW devices even incorporate a bioelectric field (SAW-BE) to both detect and perturb *E. coli* biofilms in situ. In principle, QCM and SAW systems could be adapted for *S. aureus* by using appropriate surface chemistries (Kulshrestha & Gupta, 2024).

10.4.2 Nanotechnology-Enhanced Sensors and Bio Systems

Nanomaterials and hybrid sensor platforms have yielded sensitive new biofilm detectors. For example, Brunetti et al. (2021) designed a dual micro/nanoelectrode photonic device for label-free biofilm monitoring. Their "micro-nano optoelectronic biosensor" uses nanoscale guided-mode resonance structures (Q-factor ~400) in conjunction with electrodes to optically and electrically detect biofilm. It achieved ~0.9 nm resonance shift and nanosecond-scale current changes when a 15 µm-thick biofilm was removed (Brunetti et al., 2021). Similarly, aptamer-based biosensors have emerged for *S. aureus* detection (Chen et al., 2022). These aptasensors often use gold nanoparticles or carbon nanomaterials as signal amplifiers. Recent reviews emphasize that *S. aureus* aptasensors (electrochemical or optical) are highly sensitive and easily miniaturized, with many incorporating nanostructured transducers. For instance, one fiber-optic probe with quantum-dot-labeled antibodies achieved a detection limit of ~10^3 CFU/mL for *S. aureus* in clinical samples. Although this immunofluorescent sensor is not label-free, it highlights how nanoscale fluorescent tags enable rapid *S. aureus* detection (within 2 h) even in viscous secretions (Chen et al., 2022). Future *S. aureus* biofilm sensors may similarly exploit nanotechnology (e.g., magnetic or plasmonic nanoparticles) for enhanced contrast and specificity.

Emerging "nano-plasmonic" devices also show promise. For example, localized surface plasmon resonance (LSPR) sensors – patterned gold nanostructures – allow label-free optical sensing of microbial adhesion. While earlier work used LSPR to monitor *E. coli* biofilm formation, more recent nanoplasmonic biosensors are being developed to detect bacterial analytes in real time (Kulshrestha & Gupta, 2024). In general, any sensor that provides high refractive index contrast (plasmonic, photonic crystals, etc.) can image or quantify biofilm on a surface. Importantly, none of these require traditional stains, making them inherently label-free. Overall, nanomaterial-enabled biosensors – whether electrochemical, optical, or acoustic – are pushing biofilm detection toward higher speed and sensitivity, with *S. aureus* often included as a target.

10.4.3 Vibrational and Spectroscopic Techniques

Advanced spectroscopic methods offer molecular fingerprinting of biofilms without labels. Surface-enhanced Raman spectroscopy (SERS) has been applied to bacterial biofilms. For instance, Peng et al. (2022) collected SERS spectra of nine clinical *S.* strains and used machine learning to distinguish them (Chen et al., 2022). Although their work was on planktonic cultures, the results imply that SERS can identify species such as *S. aureus* by specific Raman peaks. In practice, one can imagine coating a substrate with Ag/Au nanostructures and acquiring SERS spectra of attached cells – enabling label-free detection of biofilm composition. Likewise, Fourier-transform infrared (FTIR) spectroscopy has been used to characterize biofilm matrix chemistry. For example, Kamnev et al. (2023) performed FTIR on the exopolymeric matrix of *Azospirillum* biofilms, correlating infrared absorption bands with polysaccharide and protein composition (Kamnev et al., 2023). This demonstrates that FTIR can reveal biofilm chemical signatures in situ (though *S. aureus*-specific FTIR data are not yet reported).

HSI bridges imaging and spectroscopy. Beyond fluorescence HSI (discussed earlier), HSI in the visible/near-IR can detect multiple species simultaneously by their spectral profiles. This multi-band approach is under active study for biofilms (e.g., in food safety as above; Lee et al., 2021). Together, these spectroscopic tools (Raman, FTIR, HSI) can, in principle, detect *S. aureus* biofilms based on characteristic molecular signatures without any stains. In comparative terms, Raman/SERS provides high chemical specificity but typically requires a plasmonic substrate, whereas FTIR gives broad compositional information for thicker films. All are label-free and can be miniaturized (fiber probes or microspectrometers) for in situ or *in vivo* use.

Many of the previously mentioned methods are generally applicable to biofilms but differ in sensitivity and ease of use for *S. aureus*. For instance, OCT excels at imaging thick, opaque biofilms (like *S. aureus* on implants) but provides lower chemical detail than Raman or FTIR. Conversely, Raman/SERS and FTIR offer biochemical insights but require optical access or specialized optics. Electrical methods (EIS/QCM) are highly sensitive to small amounts of biomass but may need calibration for different species (gram-positive vs. gram-negative). In practice, combining methods (e.g., impedimetric sensing with periodic microscopy) can leverage the strengths of each. Current literature suggests that label-free approaches (OCT, EIS, SPRi, SERS, etc.) are advancing most rapidly, promising automated, real-time biofilm diagnostics with minimal sample processing.

10.5 *IN VIVO* MODELS FOR BIOFILM FORMATION

10.5.1 Animal Models

Animal models are extremely important for understanding the complexities involved in biofilm formation in a host. For instance, *S. aureus* infections have been extensively studied using various kinds of animal models such as rats, pigs, and larger animals such as sheep and pigs (Battaglia & Garrett-Sinha, 2023). These models provide insights into the interaction between the host's immune system and the film-forming bacteria. In many rodents, biofilm infections are simulated using foreign body implants such as catheters (Yin et al., 2021). These models allow researchers to study the chronicity of biofilm infections and the failure of immune clearance. Moreover, large animal models help mimic human-like physiology and provide a closer approximation of how biofilms form in human tissues. Such models are extremely helpful for preclinical testing of novel antimicrobial therapies.

For example:

a. **Mice models**: Mice models are one of the most widely used animal models, mainly due to:
   i. their genetic flexibility
   ii. relatively small size, and
   iii. the ease of availability of transgenic and knockout strains (Hickman et al., 2017).

Scientists can induce biofilm formation by *P. aeruginosa*, *S. aureus*, or other pathogens in different human tissues such as the lungs, skin, urinary tract, and central nervous system (Xuejie Li et al., 2023). One common biofilm infection model uses the mouse's respiratory tract to mimic conditions such as cystic fibrosis or ventilator-associated pneumonia (Duncan et al., 2024).

b. **Other common animal models**: Rabbits, pigs, and other non-human primates are also being used widely in biofilm research when closer similarities to human physiology are required (Esteves et al., 2018; Cauvin et al., 2015). For example, pigs have been used in wound biofilm research due to their skin's resemblance to human skin (Summerfield et al., 2015). Pig models of *S. aureus* and *Pseudomonas* biofilms in burn wounds allow for the study of both bacterial behavior and the immune response in chronic wounds (Maslova et al., 2021).

Rabbit models are often used in studies involving heart valve biofilm formation, particularly in endocarditis and arrhythmia models (Odening et al., 2020). Biofilms forming on heart valves can be challenging to treat and are a leading cause of morbidity and mortality in patients with prosthetic heart valves. Traditional animal models remain indispensable, yet translational gaps persist. While these animal models have their own benefits of providing accurate tissue simulation, they come with the drawbacks of higher ethical and financial costs. Their use is generally restricted to specific applications where larger models fail to provide appropriate conditions for studying human diseases. Recent advances in organ-on-chip and microfluidic technologies offer promising avenues to bridge these gaps, enabling controlled, ethical, and physiologically relevant investigations of biofilm–host interactions. These platforms allow high-resolution analysis of bacterial behavior and immune responses, complementing animal studies and potentially reducing reliance on higher-order species.

10.5.2 Tissue Culture Models

Tissue culture models play a crucial role in understanding the cellular mechanisms of biofilm formation. This allows researchers to observe film formation under controlled conditions. Vertrees et al. (2009). These models provide insight into bacterial adherence, colonization, and the interaction between host cells and bacteria in the early stages of biofilm formation.

1. **Epithelial Cell Cultures**: Biofilms mostly form on mucosal surfaces and epithelial cells, which form the inner lining of organs such as the lungs, intestines, and urinary tract (Motta et al., 2021). They serve as an important in vitro platform for studying biofilm behavior. Epithelial cell cultures can be derived from both human and animal tissues, and they provide an environment conducive to understanding how bacteria adhere to host cells (Hiemstra et al., 2018). For example, in lung epithelial models, researchers studied *P. aeruginosa* biofilms to simulate respiratory infections seen in cystic fibrosis patients. These studies have provided insights into how biofilms resist immune defenses and evade antibiotic treatment, leading to chronic infections (Swart et al., 2024).
2. **Organotypic Models**: Organotypic culture models are advanced tissue culture systems that aim at replicating the complexity of tissues by incorporating multiple cell types, such as epithelial cells, immune cells, and fibroblasts. These models offer a more physiologically relevant environment than simple monolayers, mimicking the three-dimensional structure and cellular interactions found *in vivo* (Hayden et al., 2021). Organotypic skin models are used to study wound biofilms, where bacteria form protective biofilms on the surface of chronic wounds. These models allow for a better understanding of the host's immune response and are useful for testing the efficacy of novel treatments that aim to disrupt biofilms or promote wound healing (Masson-Meyers et al., 2020; Maslova et al., 2021).
3. **Limitations of Tissue Culture Models**: While tissue culture models provide valuable insights into the early stages of biofilm formation, they have significant limitations. These models do not completely replicate the complex interactions between biofilms and the immune system seen in living organisms. Their inability to simulate long-term, chronic infections makes them less suitable for studying persistent biofilm infections. Furthermore, the lack of blood flow and immune components in these systems limits their applicability to real-world infections.

10.5.3 Comparative Analysis of *In Vivo* Models

Each *in vivo* model has its own advantages and disadvantages. While mouse and rat models can provide accurate and valuable insights into the biology, genetics, and molecular pathways of biofilm formation, they may not always replicate the exact condition of biological tissues. Larger models, such as rabbits and pigs, offer better physiological traits but have greater financial and ethical concerns. Tissue culture models, although helpful in studying film formation, lack the complexity of an immune response and do not replicate the chronic nature of a biofilm infection. Thus, a combination of different *in vivo* models may help in eliminating the drawbacks of individual models.

10.5.4 Portable and Field-Deployable Tools

MinION Sequencing: On-site nanopore sequencing identifies *S. aureus* biofilm-associated genes (ica, agr, sarA) within 6 hours. Handheld Raman Spectroscopy: Portable Raman probes detect characteristic EPS peaks for rapid biofilm screening. ATP Bioluminescence Kits: Provide semi-quantitative ATP readouts (<10 minutes) as a surrogate for biofilm metabolic activity on clinical and food surfaces.

10.6 CLINICAL RELEVANCE AND IMPLICATIONS

10.6.1 Role of Biofilms in *S. aureus* Infections

Biofilms are an important factor in the persistence of *S. aureus* infections, especially within the setting of medical devices, which include catheters, prosthetic joints, and heart valves (Peng et al., 2022; Lister & Horswill, 2014). The bacteria in a biofilm are enveloped in a protective extracellular matrix that protects them from the activity of antibiotics and the host immune system. Under these conditions, *S. aureus* may also become dormant, which enhances its resistance to therapy (Parastan et al., 2020; Idrees et al., 2021). These features constitute biofilm-associated *S. aureus* infections as a big clinical challenge due to the ability to evade the immune system, to persist despite antibiotic treatment, and to cause chronic and relapsing infections. Many examples, such as prosthetic joint infections, often need surgical intervention to remove the infected device since antibiotics alone are generally not sufficient for the clearance of biofilm.

10.6.2 Challenges in Treating Biofilm-Associated Infections

One of the main challenges in treating infections associated with biofilms is the poor penetration of antibiotics into the biofilm matrix. This may render bacteria within biofilms resistant to antibiotics by as much as a 1,000-fold increase over their planktonic counterparts. The slow growth or dormancy state of bacteria in the biofilm further complicates treatment, as many antibiotics target actively dividing cells. Apart from this, the immune system cannot clear biofilms easily. The extracellular matrix of biofilms prevents the accumulation of immune cells at the site of infection. In addition, bacteria within the biofilm evade recognition by the immune system. This finally leads to chronic infections that demand long-term therapy with antibiotics, which once more is often not very effective (Pozo & Patel, 2007).

10.6.3 Potential Strategies for Disrupting or Preventing Biofilm Formation

Current research actively explores various strategies to disrupt biofilms, which are known to hinder effective treatment outcomes in biofilm-associated infections. The resilience of biofilms often leads to prolonged and challenging infections, making innovative anti-biofilm methods essential for improving therapeutic efficacy. Promising approaches target the structural integrity of biofilms, aiming to weaken their protective layers and enhance antibiotic access to the encased bacteria, which are typically shielded from immune responses and conventional antibiotics (Yin et al., 2021).

Approaches that hold promise include the following.

10.6.3.1 Anti-Biofilm Agents

These compounds specifically target and degrade the extracellular matrix of biofilms, effectively breaking down the structural defense that protects the bacterial community. For instance, enzymes like DNase and dispersin B have shown potential in degrading extracellular polysaccharides, a primary component of the biofilm matrix. This degradation disrupts the biofilm's cohesion, allowing antimicrobials to penetrate more effectively and target the bacteria directly. Current studies on DNase and dispersin B suggest they may enhance antibiotic action against resilient biofilms and improve treatment outcomes in persistent infections (Kaplan et al., 2024; Li et al., 2024; Wen et al., 2022).

10.6.3.2 Surface Modifications

Altering the surfaces of medical devices to prevent biofilm formation is another promising approach. Coating medical devices, such as catheters, with antimicrobial or anti-adhesive materials can inhibit bacterial attachment, thereby reducing the likelihood of biofilm development. Silver-coated catheters, for example, are effective in lowering biofilm-associated infections in urinary tracts. This method is gaining traction as it minimizes the initial colonization of bacteria on devices, thereby reducing infection risks significantly (Kanti et al., 2022; Rajaramon et al., 2023).

10.6.3.3 Combination Therapies

Employing antibiotics in conjunction with biofilm-disrupting agents or immuno-enhancing drugs is emerging as a highly effective approach. These combination therapies target multiple aspects of the biofilm and the bacteria within, enhancing treatment efficacy. For example, bacteriophages – viruses that infect bacteria exclusively – have shown promise in biofilm eradication. When combined with antibiotics, bacteriophages help break down biofilms, increasing bacterial vulnerability to antibiotics and supporting the host's immune response. Research into combination therapies suggests they may offer a robust approach to overcoming the defenses of biofilms (Szelachowska et al., 2022; Hemmati et al., 2021). While it is well-established that biofilms can increase antibiotic tolerance by up to 1,000-fold, what remains underappreciated is the diagnostic lag. Delayed or inaccurate detection often forces clinicians into empiric, prolonged therapy, which can further accelerate the development of resistance and compromise patient outcomes. Addressing both biofilm tolerance and timely, accurate diagnostics is, therefore, essential for optimizing the effectiveness of combination therapies.

10.6.4 Emerging Trends and Future Directions

Advancements in biofilm detection are driving the development of novel, more efficient techniques that promise enhanced sensitivity, speed, and specificity. Among these, biosensors and nanotechnology-based approaches are emerging as promising tools for the future of biofilm detection. Biosensors, which utilize biological recognition elements, such as antibodies, enzymes, or nucleic acids, in combination with signal transduction systems (e.g., electrochemical or optical), allow for real-time, portable, and highly specific biofilm detection. These sensors can be used for the rapid identification of biofilm-forming bacteria in clinical, environmental, or industrial settings, providing an opportunity for early intervention and monitoring of biofilm-related infections (Santos et al., 2013). Furthermore, nanotechnology enables the development of nanomaterials, such as nanoparticles or nanowires, which can interact with biofilm components at the molecular level. These technologies enhance detection sensitivity by improving the interaction between the sensor surface and biofilm matrix, making them highly effective for detecting biofilms at early stages or in low-biomass environments. Another exciting direction is the integration of multiple detection methods to achieve a more comprehensive understanding of biofilm formation. For example, combining biochemical assays (like crystal violet or resazurin assays) with molecular techniques (such as PCR and qRT-PCR) and advanced imaging methods (like CLSM or AFM) provides multidimensional insights into biofilm dynamics. This integrated approach allows for the simultaneous evaluation of biofilm biomass, metabolic activity, gene expression, and structural integrity. By combining these technologies, researchers can gain a deeper understanding of the biochemical, genetic, and physical properties of biofilms, leading to more precise assessments of biofilm development and resilience.

Further possibilities include the development of point-of-care diagnostic tools, which integrate biosensors and microfluidics for on-site biofilm detection. These tools could revolutionize healthcare by enabling rapid detection of biofilm-related infections, such as those caused by *S. aureus* or *P. aeruginosa*, directly at the patient's bedside. Additionally, lab-on-a-chip devices combining microfluidics with biosensors could be used for high-throughput screening of biofilm formation, providing valuable insights for drug development or antibiotic resistance studies. As these emerging technologies continue to evolve, their integration into biofilm detection protocols holds the potential to not only improve diagnostic and therapeutic strategies but also provide more sustainable and efficient solutions for biofilm management in healthcare, water treatment, and industrial applications (Yang et al., 2011). Despite advances, a critical bottleneck remains the lack of standardized, clinically validated protocols. Biosensor and microfluidic platforms are promising but remain largely confined to proof-of-concept studies. Integrating omics approaches with AI-driven predictive models offers a disruptive opportunity for real-time biofilm detection across both healthcare and industrial pipelines. With ongoing innovations in molecular biology, nanotechnology, and bioengineering, the future of biofilm detection looks highly promising, offering new avenues to tackle the challenges posed by biofilm-related infections and biofilm-associated issues in diverse industries.

10.7 CONCLUSION

Biofilm formation by *S. aureus* continues to pose serious complications across medical, industrial, and environmental sectors. The ability of this pathogen to form robust, adhesive communities within an extracellular matrix contributes to its persistence, enhanced resistance to antibiotics, and difficulty in eradication. Such resilience underpins chronic infections, particularly those associated with implanted medical devices and recurring wounds, as well as contamination in food processing and water systems. Historically, techniques such as CLSM, SEM, and biochemical assays like crystal violet staining have provided insight into biofilm architecture and biomass. Molecular assays, including PCR and gene expression profiling, have shed light on the regulatory mechanisms driving biofilm development. While these methods have significantly advanced the field, they often come with drawbacks – such as destructive sample preparation, endpoint analysis, or reliance on labeling agents – that restrict their application in real-time or clinical settings.

In the last few years, notable innovations have emerged that address many of these limitations. Technologies like OCT, HSI, and SPRi now allow non-invasive, label-free visualization of biofilm morphology in situ. These techniques are particularly advantageous for observing *S. aureus* biofilms on medically relevant surfaces, offering high spatial resolution and three-dimensional mapping capabilities without disrupting the sample. Electrochemical tools such as EIS and CV have made it possible to continuously monitor biofilm growth and metabolic changes, offering real-time data with high sensitivity. Acoustic and mass-sensitive platforms like QCM and SAW devices add to this capability by tracking shifts in viscoelastic properties and biomass accumulation, which are hallmarks of biofilm progression. Furthermore, nanotechnology is driving the next wave of innovation. Aptamer-based biosensors, plasmonic nanostructures, and quantum dot probes have improved the specificity and speed of *S. aureus* detection. These platforms are particularly promising for point-of-care diagnostics, with potential applications in implantable sensors and portable biosensor kits. Complementary spectroscopic methods – such as SERS and FTIR spectroscopy – also enable detailed molecular analysis of biofilm components without the need for dyes or labels, contributing to the molecular fingerprinting of biofilms.

As a collective trend, the biofilm detection field is shifting from traditional, isolated methodologies to integrated, real-time systems capable of high-throughput, label-free, and in situ monitoring. These next-generation tools are essential for capturing the dynamic and heterogeneous nature of biofilm communities, particularly in environments where early detection and intervention are critical.

Nevertheless, translating these advanced techniques into standard clinical and industrial workflows remains a challenge. Cost, complexity, and the need for specialized expertise can limit widespread adoption. Future work should focus on refining these technologies for ease of use, ensuring reproducibility across diverse settings, and integrating multi-modal platforms that combine structural, biochemical, and genetic analyses. To effectively combat *S. aureus* biofilms and reduce their impact on human health and infrastructure, continued research must bridge disciplines – from microbiology and nanoscience to systems engineering and clinical diagnostics. The integration of emerging detection strategies with preventive and therapeutic approaches will be key to overcoming the persistent challenge of biofilm-related infections.

REFERENCES

Alhede, M., Kragh, K. N., Qvortrup, K., Allesen-Holm, M., van Gennip, M., Christensen, L. D., . . . Bjarnsholt, T. (2011). Phenotypes of non-attached Pseudomonas aeruginosa aggregates resemble surface attached biofilm. *PloS One*, *6*(11), e27943.

Ameer, S., Ibrahim, H., Yaseen, M. U., Kulsoom, F., Cinti, S., & Sher, M. (2023). electrochemical impedance spectroscopy-based sensing of biofilms: A comprehensive review. *Biosensors*, *13*(8), 777. https://doi.org/10.3390/bios13080777

Archer, N. K., Mazaitis, M. J., Costerton, J. W., Leid, J. G., Powers, M. E., & Shirtliff, M. E. (2011). *Staphylococcus aureus* biofilms: Properties, regulation, and roles in human disease. *Virulence*, *2*(5), 445–459.

Arciola, C. R., Campoccia, D., Speziale, P., Montanaro, L., & Costerton, J. W. (2012). Biofilm formation in *Staphylococcus* implant infections. A review of molecular mechanisms and implications for biofilm-resistant materials. *Biomaterials*, *33*(26), 5967–5982.

Battaglia, M., & Garrett-Sinha, L. A. (2023). *Staphylococcus xylosus* and *Staphylococcus aureus* as commensals and pathogens on murine skin. *Laboratory Animal Research*, *39*(1), 18.

Becker, K., Heilmann, C., & Peters, G. (2014). Coagulase-Negative staphylococci. *Clinical Microbiology Reviews*, *27*(4), 870–926. https://doi.org/10.1128/cmr.00109-13

Beenken, K. E., Blevins, J. S., & Smeltzer, M. S. (2004). Mutation of the *S. aureus* sarA global regulator reduces biofilm formation and increases the bacterial clearance in a rat model of foreign-body infection. *Journal of Infectious Diseases*, *190*(5), 893–899.

Birarda, G., Delneri, A., Lagatolla, C., Parisse, P., Cescutti, P., Vaccari, L., Rizzo, R. (2019). Multi-technique microscopy investigation on bacterial biofilm matrices: A study on Klesbsiella pneumoniae clinical strains. *Analytical and Bioanalytical Chemistry*, *411*, 7315–7325.

Boguslavsky, Y., Shemesh, M., Friedlander, A., Rutenberg, R., Filossof, A M., Buslovich, A., & Poverenov, E. (2018). Eliminating the need for biocidal agents in anti-biofouling polymers by applying grafted nano-silica instead. *ACS Omega*, *3*, 12437–12445.

Brunetti, G., Conteduca, D., Armenise, M., & Ciminelli, C. (2021). Novel Micro-Nano Optoelectronic biosensor for label-free real-time biofilm monitoring. *Biosensors*, *11*(10), 361. https://doi.org/10.3390/bios11100361

Brugger, S. D., Mühlemann, K., & Böttger, E. C. (2014). Molecular techniques for detecting biofilm formation in *Staphylococcus aureus*. *Journal of Clinical Microbiology*, *52*(8), 2842–2849.

Cauvin, A. J., Peters, C., & Brennan, F. (2015). Advantages and limitations of commonly used nonhuman primate species in research and development of biopharmaceuticals. *The Nonhuman Primate in Nonclinical Drug Development and Safety Assessment*, 379–395.

Chatterjee, S., Biswas, N., Datta, A., Dey, R. & Maiti, P. (2014). Atomic force microscopy in biofilm study. *Microscopy*, *63*, 269–278.

Chavant, P., Maillard, M. B., & Loret, J. F. (2007). Biofilm ring test: A simple and rapid method for detecting biofilm formation. *Journal of Microbiological Methods*, *69*(1), 45–51.

Chen, W., Lai, Q., Zhang, Y., & Liu, Z. (2022). Recent advances in APTAsensors for rapid and sensitive detection of *Staphylococcus aureus*. *Frontiers in Bioengineering and Biotechnology*, *10*. https://doi.org/10.3389/fbioe.2022.889431

Clayborn, J.; Adams, J.; Baker, C.; Ricke, S. (2015). Assessment of Salmonella spp. attachment to reusable plastic containers based on scanning electron microscopy and BAX PCR. *Journal of Food Research*, *4*, 166.

Conlon, B. P., Geoghegan, J. A., Waters, E. M., McCarthy, H., Rowe, S. E., Davies, J. R., . . . O'Gara, J. P. (2014). Role for the A domain of unprocessed accumulation-associated protein (Aap) in the attachment phase of the *Staphylococcus* epidermidis biofilm phenotype. *Journal of bacteriology*, *196*(24), 4268–4275.

Costerton, J. W. (2007). The Biofilm Primer. In *Springer series on biofilms*. https://doi.org/10.1007/b136878

Costerton, J. W., Irvin, R. T., Cheng, K. J., & Sutherland, I. W. (1981). The role of bacterial surface structures in pathogenesis. *CRC Critical Reviews in Microbiology*, *8*(4), 303–338.

De La Cruz, M. J., & Eng, E. T. (2023). Scaling up cryo-EM for biology and chemistry: The journey from niche technology to mainstream method. *Structure*, *31*(12), 1487–1498. https://doi.org/10.1016/j.str.2023.09.009

De Oliveira, D. F. J. A., Fernandes, A., Kaneno, R., Silva, M. G., Araujo, J. P., Silva, N. C. C., Rall, V., L., M. (2014). Ability of *Salmonella* spp. to produce biofilm is dependent on temperature and surface material. *Foodborne Pathogens and Disease*, *11*, 478–843.

Del Pozo, J. L., & Patel, R. (2007). The challenge of treating biofilm-associated bacterial infections. *Clinical Pharmacology and Therapeutics*, *82*(2), 204–209.

Demidova, N., Gunn, J. R., Gitajn, I. L., Vitkin, I. A., Elliott, J. T., & Demidov, V. V. (2025). Optical coherence tomography for label-free detection and characterization of methicillin-resistant *S. aureus* biofilms. *Journal of Biomedical Optics*, *30*(04). https://doi.org/10.1117/1.jbo.30.4.046003

Di Ciccio, P. A., Ghidini, S., Zanardi, E., Anna, R. F., Paludi, D., Alberto, D. B., . . . Ianieri, A. (2015). Biofilm-forming ability of *Staphylococcus aureus* from food environment. In *Atti società italiana delle scienze veterinarie-sis vet* (pp. 287–287). Società Italiana Delle Scienze Veterinarie-Sis Vet.

Dominiak, D.; Nielsen, J.; Nielsen, P. (2011) Extracellular DNA is abundant and important for microcolony strength in mixed microbial biofilms. *Environmental Microbiology.*, *13*, 710–721.

Doucet, F. J., Lead, J. R., Maguire, L., Achterberg, E. P., Millward, G. E. (2005). Visualisation of natural aquatic colloids and particles – A comparison of conventional high vacuum and environmental scanning electron microscopy. *Journal of Environmental Monitoring*, *7*, 115–121.

Duncan, R. P., Moustafa, D. A., Lewin, G. R., Diggle, F. L., Bomberger, J. M., Whiteley, M., & Goldberg, J. B. (2024). Improvement of a mouse infection model to capture *Pseudomonas aeruginosa* chronic physiology in cystic fibrosis. *Proceedings of the National Academy of Sciences of the United States of America*, *121*(33), e2406234121.

Esteves, P. J., Abrantes, J., Baldauf, H., BenMohamed, L., Chen, Y., Christensen, N., González-Gallego, J., Giacani, L., Hu, J., Kaplan, G., Keppler, O. T., Knight, K. L., Kong, X., Lanning, D. K., Pendu, J. L., De Matos, A. L., Liu, J., Liu, S., Lopes, A. M., . . . Mage, R. (2018). The wide utility of rabbits as models of human diseases. *Experimental & Molecular Medicine*, *50*(5), 1–10.

Fleming, D., & Rumbaugh, K. P. (2017). Approaches to dispersing medical biofilms. *Microorganisms*, 5(2), 15. https://doi.org/10.3390/microorganisms5020015. PMID: 28368320; PMCID: PMC5488086.

Foster, T. J., Geoghegan, J. A., Ganesh, V. K., & Höök, M. (2013). Adhesion, invasion and evasion: the many functions of the surface proteins of *Staphylococcus aureus*. *Nature Reviews Microbiology*, *12*(1), 49–62. https://doi.org/10.1038/nrmicro3161

Hall-Stoodley, L., Costerton, J. W., & Stoodley, P. (2004). Bacterial biofilms: From the natural environment to infectious diseases. *Nature Reviews Microbiology*, *2*(2), 95–108.

Hannig, C.; Follo, M.; Hellwig, E.; Al-Ahmad, A. (2010). Visualization of adherent micro-organisms using different techniques. *Journal of Environmental Monitoring*, *59*, 1–7.

Hayden, P. J., & Harbell, J. W. (2021). Special review series on 3D organotypic culture models: Introduction and historical perspective. *In Vitro Cellular & Developmental Biology. Animal*, *57*(2), 95–103.

Hemmati, F., Rezaee, M. A., Ebrahimzadeh, S., Yousefi, L., Nouri, R., Kafil, H. S., & Gholizadeh, P. (2021). Novel strategies to combat bacterial biofilms. *Molecular Biotechnology*, *63*(7), 569–586.

Hickman, D. L., Johnson, J., Vemulapalli, T. H., Crisler, J. R., & Shepherd, R. (2017). Commonly used animal models. *Principles of Animal Research for Graduate and Undergraduate Students*, 117–175.

Hiemstra, P. S., Grootaers, G., van der Does, A. M., Krul, C. A. M., & Kooter, I. M. (2018). Human lung epithelial cell cultures for analysis of inhaled toxicants: Lessons learned and future directions. *Toxicology in Vitro*, *47*, 137–146.

Høiby, N., Bjarnsholt, T., Givskov, M., Molin, S., & Ciofu, O. (2010). Antibiotic resistance of bacterial biofilms. *International Journal of Antimicrobial Agents*, *35*(4), 322–332. https://doi.org/10.1016/j.ijantimicag.2009.12.011

Idrees, M., Sawant, S., Karodia, N., & Rahman, A. (2021). *Staphylococcus aureus* Biofilm: Morphology, genetics, pathogenesis and treatment strategies. *International Journal of Environmental Research and Public Health*, *18*(14), 7602.

Kadariya, J., Smith, T. C., & Thapaliya, D. (2014). *Staphylococcus aureus* and staphylococcal food-borne disease: An ongoing challenge in public health. *BioMed Research International*, *2014*(1), 827965.

Kamnev, A. A., Dyatlova, Y. A., Kenzhegulov, O. A., Fedonenko, Y. P., Evstigneeva, S. S., & Tugarova, A. V. (2023). Fourier transform infrared (FTIR) spectroscopic study of biofilms formed by the rhizobacterium

Azospirillum baldaniorum Sp245: Aspects of methodology and matrix composition. *Molecules*, *28*(4), 1949. https://doi.org/10.3390/molecules28041949
Kanti, S. P. Y., Csóka, I., Jójárt-Laczkovich, O., & Adalbert, L. (2022). Recent advances in antimicrobial coatings and material modification strategies for preventing urinary catheter-associated complications. *Biomedicines*, *10*(10), 2580. https://doi.org/10.3390/biomedicines10102580
Kaplan, J. B., Sukhishvili, S. A., Sailer, M., Kridin, K., & Ramasubbu, N. (2024). Aggregatibacter actinomycetemcomitans Dispersin B: The Quintessential Antibiofilm Enzyme. *Pathogens*, *13*(8), 668.
Kiedrowski, M. R., Kavanaugh, J. S., Malone, C. L., Mootz, J. M., Voyich, J. M., Smeltzer, M. S., Bayles, K. W., & Horswill, A. R. (2011). Nuclease modulates biofilm formation in community-associated methicillin-resistant *Staphylococcus aureus*. *PLoS One*, *6*(11), e26714. doi:10.1371/journal.pone.0026714. Epub 2011 Nov 11. PMID: 22096493; PMCID: PMC3214024.
Knobloch, J. K. M., Horstkotte, M. A., Rohde, H., & Mack, D. (2002). Evaluation of different detection methods of biofilm formation in *Staphylococcus aureus*. *Medical Microbiology and Immunology*, *191*, 101–106.
Köhler, J. R., Casadevall, A., & Perfect, J. (2017). The spectrum of fungi that infect humans. *Cold Spring Harbor Perspectives in Medicine*, *5*(1), a019273.
Kulshrestha, A., & Gupta, P. (2024). Real-time biofilm detection techniques: Advances and applications. *Future Microbiology*, *19*(11), 1003–1016. https://doi.org/10.1080/17460913.2024.2350285
Lee, A., Park, S., Yoo, J., Kang, J., Lim, J., Seo, Y., Kim, B., & Kim, G. (2021). Detecting bacterial biofilms using fluorescence hyperspectral imaging and various discriminant analyses. Sensors, 21(6), 2213. https://doi.org/10.3390/s21062213
Li, X., Gu, N., Huang, T. Y., Zhong, F., & Peng, G. (2023). *Pseudomonas aeruginosa*: A typical biofilm forming pathogen and an emerging but underestimated pathogen in food processing. *Frontiers in Microbiology*, *13*, 1114199.
Li, X., Lin, S., Wang, Y., Chen, Y., Zhang, W., Shu, G., . . . Fu, H. (2024). Application of biofilm dispersion-based nanoparticles in cutting off reinfection. *Applied Microbiology and Biotechnology*, *108*(1), 386.
Lister, J. L., & Horswill, A. R. (2014). *Staphylococcus aureus* biofilms: Recent developments in biofilm dispersal. *Frontiers in Cellular and Infection Microbiology*, *4*, 178.
Łusiak-Szelachowska, M., Międzybrodzki, R., Drulis-Kawa, Z., Cater, K., Knežević, P., Winogradow, C., . . . Górski, A. (2022). Bacteriophages and antibiotic interactions in clinical practice: What we have learned so far. *Journal of Biomedical Science*, *29*(1), 23.
Maslova, E., Eisaiankhongi, L., Sjöberg, F., & McCarthy, R. R. (2021). Burns and biofilms: Priority pathogens and in vivo models. *NPJ Biofilms and Microbiomes*, *7*(1), 73.
Masson-Meyers, D. S., Andrade, T. A. M., Caetano, G. F., Guimaraes, F. R., Leite, M. N., Leite, S. N., & Frade, M. A. C. (2020). Experimental models and methods for cutaneous wound healing assessment. *International Journal of Experimental Pathology*, *101*(1–2), 21–37.
McGlennen, M., Dieser, M., Foreman, C. M., & Warnat, S. (2023). Monitoring biofilm growth and dispersal in real-time with impedance biosensors. *Journal of Industrial Microbiology & Biotechnology*, *50*(1). https://doi.org/10.1093/jimb/kuad022
Merino, L., Procura, F., Trejo, F. M., Bueno, D. J., Golowczyc, M. A. (2019) Biofilm formation by *Salmonella* sp. in de poultry industry: Detection, control and iradication strategies. *Food Research International*, *119*, 530–540.
Motta, J.-P., Wallace, J. L., Buret, A. G., Deraison, C., & Vergnolle, N. (2021). Gastrointestinal biofilms in health and disease. *Nature Reviews Gastroenterology & Hepatology*, *18*(5), 314–334.
Nadell, C. D., Drescher, K., & Foster, K. R. (2008). Biofilm formation and dispersal in *Pseudomonas aeruginosa*: A molecular perspective. *Environmental Microbiology*, *10*(5), 1627–1637.
Nanakoudis, A. & Blog.phenom-World. 2019. Available online: https://blog.phenomworld.com/sem-tem-difference (accessed on 2 January 2020).
Neu, T., & Lawrence, J. (2014). Investigation of microbial biofilm structure by laser scanning microscopy. In *Productive biofilms*; Springer: Cham, Switzerland, pp. 1–51.
Nourbakhsh, F., & Namvar, A. E. (2016). Detection of genes involved in biofilm formation in *Staphylococcus aureus* isolates. GMS Hygiene and infection control, 11.
Odening, K. E., Baczko, I., & Brunner, M. (2020). Animals in cardiovascular research: Important role of rabbit models in cardiac electrophysiology. *European Heart Journal*, *41*(21), 2036 2036.
O'Toole, G. A., Kaplan, H. B., & Kolter, R. (2000). Biofilm formation as microbial development. *Annual Review of Microbiology*, *54*(1), 49–79.
Otto, M. (2008). Staphylococcal biofilms. *Current Topics in Microbiology and Immunology*, *322*, 207–228.
Otto, M. (2013). Staphylococcal infections: Mechanisms of biofilm maturation and detachment as critical determinants of pathogenicity. *Annual Review of Medicine*, *64*(1), 175–188.

Ozkan, A. D., Topal, A. E., Dana, A., Gruler, M. O., Tekinay, A. B. (2016). Atomic force microscopy for the investigation of molecular and cellular behaviour. *Micron*, *89*, 60–76.

Palmer Jr, R., Sternber, C. (1999). Modern microscopy in biofilm research: Confocal microscopy and other approaches. *Current Opinion in Biotechnology*, *10*, 263–268.

Parastan, R., Kargar, M., Solhjoo, K., & Kafilzadeh, F. (2020). *Staphylococcus aureus* biofilms: Structures, antibiotic resistance, inhibition, and vaccines. *Gene Reports*, *20*, 100739.

Peng, Q., Tang, X., Dong, W., Sun, N., & Yuan, W. (2022). A review of biofilm formation of *Staphylococcus aureus* and its regulation mechanism. *Antibiotics (Basel, Switzerland)*, *12*(1), 12.

Percival, S. L., Suleman, L., Vuotto, C., & Donelli, G. (2015). Healthcare-associated infections, medical devices and biofilms: Risk, tolerance and control. *Journal of Medical Microbiology*, *64*(4), 323–334.

Pierson, J. A., Yang, J. E., & Wright, E. R. (2024). Recent advances in correlative cryo-light and electron microscopy. *Current Opinion in Structural Biology*, *89*, 102934. https://doi.org/10.1016/j.sbi.2024.102934

Rajaramon, S., Shanmugam, K., Dandela, R., & Solomon, A. P. (2023). Emerging evidence-based innovative approaches to control catheter-associated urinary tract infection: A review. *Frontiers in Cellular and Infection Microbiology*, *13*, 1134433.

Rumbaugh, K. P., & Sauer, K. (2020). Biofilm dispersion. *Nature Reviews Microbiology*, *18*(10), 571–586.

Sánchez, C. J., Salazar, J. K., & Cantey, J. R. (2013). Resazurin-based assay for bacterial biofilm formation. *Journal of Microbiological Methods*, *95*(2), 334–336.

Santos, A. M., Soares, F., & Rodrigues, L. (2013). Biosensors for detection of biofilms and microbial pathogens: A review. *Sensors and Actuators B: Chemical*, *176*, 137–151.

Schilcher, K., & Horswill, A. R. (2020). Staphylococcal biofilm development: Structure, regulation, and treatment strategies. *Microbiology and Molecular Biology Reviews*, *84*(3), e00026-19. https://doi.org/10.1128/MMBR.00026-19. PMID: 32792334; PMCID: PMC7430342.

Summerfield, A., Meurens, F., & Ricklin, M. E. (2015). The immunology of the porcine skin and its value as a model for human skin. *Molecular Immunology*, *66*(1), 14–21.

Swart, A. L., Laventie, B.-J., Sütterlin, R., Junne, T., Lauer, L., Manfredi, P., . . . Jenal, U. (2024). Pseudomonas aeruginosa breaches respiratory epithelia through goblet cell invasion in a microtissue model. *Nature Microbiology*, *9*(7), 1725–1737.

Vertrees, R. A., Jordan, J. M., Solley, T., & Goodwin, T. J. (2009). Tissue Culture Models. *Basic Concepts of Molecular Pathology*, *2*, 159–182.

Wang, J., Ren, H., Li, X., Li, J., Ding, L., Geng, J., Xu, K., Huang, H., & Hu, H. (2018). In situ monitoring of wastewater biofilm formation process via ultrasonic time domain reflectometry (UTDR). *Chemical Engineering Journal 334*, 2134–2141.

Watnick, P., & Kolter, R. (2000). Biofilm, city of microbes. *Journal of Bacteriology*, *182*(10), 2675–2679.

Wen, J., Wang, Z., Du, X., Liu, R., & Wang, J. (2022). Antibioflm effects of extracellular matrix degradative agents on the biofilm of different strains of multi-drug-resistant Corynebacterium striatum. *Annals of Clinical Microbiology and Antimicrobials*, *21*(1), 53.

Yang, L., Xie, J., & Yang, Z. (2011). Integration of multiple biofilm detection techniques: Current advances and future perspectives. *Journal of Microbiological Methods*, *87*(3), 330–340.

Yin, W., Xu, S., Wang, Y., Zhang, Y., Chou, S. H., Galperin, M. Y., & He, J. (2021). Ways to control harmful biofilms: Prevention, inhibition, and eradication. *Critical Reviews in Microbiology*, *47*(1), 57–78.

11 Harnessing Quorum-Sensing Inhibitors for Biofilm Control in *Staphylococcus aureus*

Himanshu Narayan Singh[1], *Bangaru Naidu Thaddi*[2], *P. Gnana Deepu*[3], *Bhagyashree Mohakhud*[3], *Patel Shaifali Ben*[3], and *Aleti Akila*[3]

[1]Memorial Sloan Kettering Cancer Centre, New York, USA

[2]Department of Biotechnology, Anil Neerukonda Institute of Technology & Sciences (Autonomous), Sangivalasa, Visakhapatnam, Andhra Pradesh, India

[3]Center for Biotechnology, University College of Engineering Science and Technology, Hyderabad, JNTUH, Hyderabad, Telangana, India

ABBREVIATIONS

AGR	accessory gene regulator
AHLs	acyl-homoserine lactones
EPS	extracellular polymeric substances
FPLAM	fluorescent photo-activated localization microscopy
MRSA	methicillin-resistant *S. aureus*
MSCRAMMs	microbial surface components recognizing adhesive matrix molecules
PDT	photodynamic treatment
PIA	polysaccharide intercellular adhesin
PLAM	photo-activated localization microscopy
QSIs	quorum-sensing inhibitors
S. aureus	*Staphylococcus aureus*
STORM	stochastic optical reconstruction microscopy

11.1 INTRODUCTION

11.1.1 Background on *Staphylococcus aureus* and Biofilm Formation

11.1.1.1 Overview of *S. aureus*

S. aureus is a gram-positive coccus, typically measuring between 0.5 and 1.5 μm in diameter (Bitrus et al., 2018; Gnanamani et al., 2017; Harris et al., 2002). It is characterized by its non-motile, non-spore-forming nature (Harris et al., 2002). When subjected to Gram staining, *S. aureus* exhibits a distinct bluish-purple coloration, as shown in Figure 11.1 (Gnanamani et al., 2017; Taylor & Unakal, 2018). Microscopically, the organism may appear as individual cells, pairs, or arranged in grape-like clusters (Gnanamani et al., 2017). As a facultative anaerobe, it is capable of generating energy through either aerobic respiration or fermentation pathways (Harris et al., 2002). The bacterium reproduces through binary fission, and its optimal growth occurs within a temperature range of 18°C to 40°C (Taylor & Unakal, 2018).

DOI: 10.1201/9781003647072-14

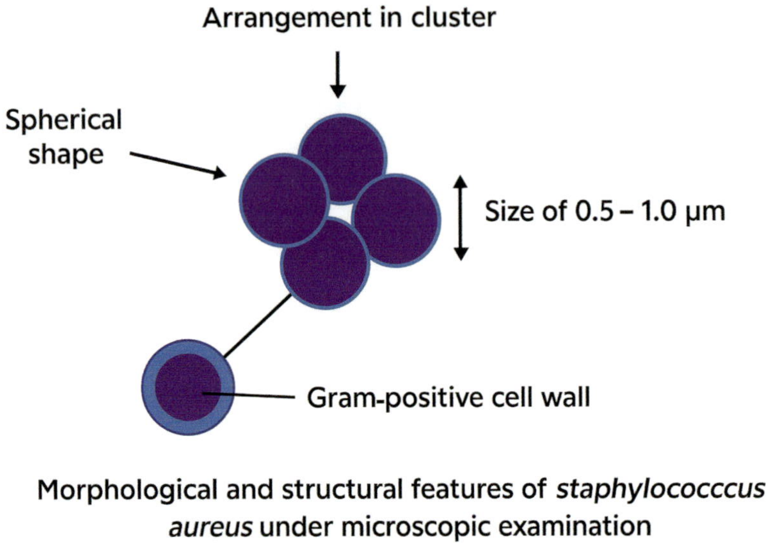

FIGURE 11.1 Morphological and structural features of *Staphylococcus aureus* under microscopic examination.

S. aureus exists as part of the commensal flora on human skin and mucosal surfaces (Boucher & Corey, 2008). *S. aureus* is an opportunistic pathogen that also causes a wide range of diseases in both humans and animals, making it a serious pathogen. *S. aureus* causes a wide range of illnesses in humans, from simple infections of the skin and soft tissues to serious, life-threatening illnesses (Chessa et al., 2015). Clinically, it is regarded as the most harmful member of the *Staphylococcus* genus, producing a number of diseases, including food poisoning, bacteremia, endocarditis, necrotic pneumonia in infants, and superficial skin abscesses (Shaw et al., 2004). *S. aureus* is linked to dermatitis in dogs, septicemia and arthritis in chickens, mastitis in cows, and botryomycosis in horses (Zunita et al., 2008; Luzzago et al., 2014).

11.1.1.2 Dual Nature of *S. aureus*: Virulence and Adaptability

S. aureus exhibits a remarkable ability to produce multiple virulence determinants that play a crucial role in promoting its pathogenic nature. These include the toxins produced by *S. aureus*, such as hemolysins, enterotoxins, and toxic shock syndrome toxin, which are important in interfering with host cellular processes and immunological responses (Foster et al., 2014; Lowy, 1998). The bacterium secretes proteases and coagulases, among other enzymes, which aid in immune evasion and tissue invasion (Otto, 2013). One of the key factors contributing to the adaptability of *S. aureus* is its ability to develop biofilms. These structured microbial communities enable the organism to endure unfavorable environmental conditions, reduce its susceptibility to antibiotics, and resist clearance by the host immune system (Archer et al., 2011). Furthermore, biofilm formation plays a major role in persistent and recurring infections, posing considerable difficulties in clinical management (Boles et al., 2005).

11.1.1.3 Biofilm Development in *S. aureus*

S. aureus can form biofilms, which are complex bacterial communities enclosed within a matrix secreted by the cells themselves. This matrix, often referred to as extracellular polymeric substances (EPSs), allows the biofilm to anchor onto various surfaces, both living tissues and inert materials. The EPS structure is a composite of biomolecules such as structural proteins, polysaccharides, lipids, and strands of extracellular DNA, which together provide protection and cohesion to the biofilm (Flemming et al., 2016).

The formation of a biofilm by *S. aureus* unfolds in a series of well-defined stages. The initial contact stage is characterized by bacterial cells beginning to adhere to a surface. Specific adhesion proteins, including microbial surface components recognizing adhesive matrix molecules (MSCRAMMs), play a central role by binding to host tissue or artificial substrates (Foster et al., 2014). Structural development is a process where, once attached, the cells multiply and begin constructing a layered, three-dimensional structure. These layers create localized environments within the biofilm that support survival under nutrient limitations and stress conditions. Cell dispersal: As the biofilm matures, some bacterial cells break away and move to establish new sites of colonization. This release phase helps *S. aureus* persist within the host and contributes to the recurrence of infections, especially in medical settings (Boles et al., 2005).

11.1.1.4 Contribution of Biofilms to *S. aureus* Infections

The ability of *S. aureus* to form biofilms significantly enhances its survival in unfavorable conditions and contributes to its virulence. Biofilms act as a protective mode of growth, offering several functional advantages that aid the organism during infection. One key benefit of biofilm formation is the reduction in antibiotic efficacy. The extracellular matrix (ECM) surrounding the bacterial community acts as a diffusion barrier, limiting drug penetration. In addition, dormant cells known as persisters embedded within the biofilm display a high level of tolerance to antimicrobial agents, making treatment more difficult (Høiby et al., 2010).

Biofilms also play a critical role in evading the host immune system. The matrix shields bacterial cells from immune mechanisms such as phagocytosis and complement-mediated lysis, enabling prolonged survival within the host. Moreover, the resilience of *S. aureus* to environmental stressors is greatly enhanced when it exists in a biofilm. Conditions such as oxidative stress, nutrient depletion, and desiccation are better tolerated due to the structural and metabolic adaptations provided by the biofilm mode of growth (Otto, 2008). Because of these factors, *S. aureus* biofilms are frequently associated with chronic and recurrent infections, especially in clinical scenarios involving indwelling medical devices or compromised tissue. This makes biofilm-associated infections particularly challenging to manage in therapeutic settings.

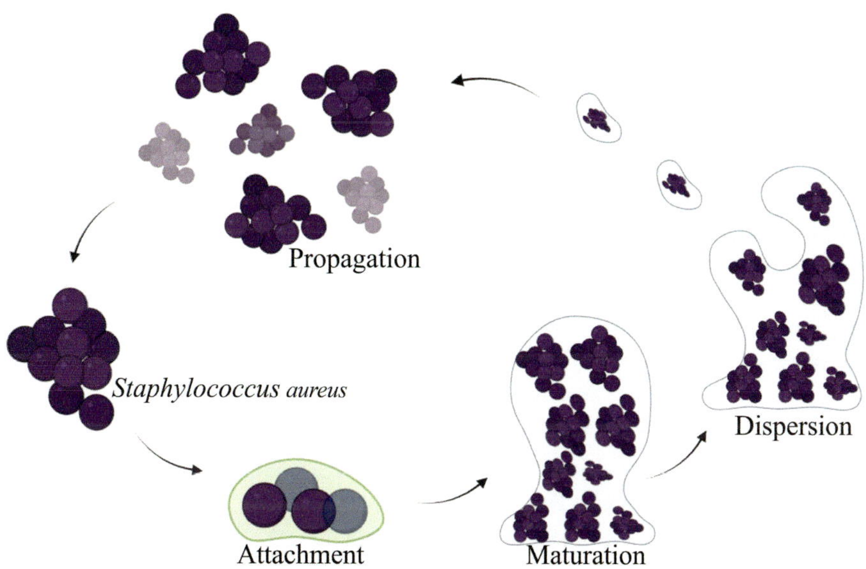

FIGURE 11.2 Stages of biofilm formation in *Staphylococcus aureus*.

11.1.1.5 Clinical Conditions Involving *S. aureus* Biofilms

S. aureus biofilms are a significant factor in a range of persistent and device-associated infections. According to Zimmerli et al. (2004), these biofilms are commonly implicated in infections involving indwelling medical devices such as heart valves, prosthetic joints, and catheters. In the context of non-healing wounds, *S. aureus* biofilm formation has been closely associated with the chronicity of conditions like diabetic foot ulcers and venous leg ulcers, where they impede tissue repair and sustain inflammatory responses (Bjarnsholt, 2013). Moreover, in cases of infective endocarditis, the bacterium's colonization of heart valves frequently results in biofilm establishment, thereby complicating therapeutic interventions due to enhanced resistance to antibiotics and the ability to evade host immune defenses (Fowler et al., 2005).

11.1.1.6 Significance of Investigating Biofilm Formation in *S. aureus*

Understanding how *S. aureus* develops biofilms is crucial for improving treatment approaches, especially in cases where standard antibiotic therapies are ineffective. Biofilm-associated infections often show resistance to conventional drugs due to the protective nature of the biofilm structure and the altered metabolic states of the bacterial cells within. This has led researchers to explore alternative therapeutic options, such as targeting bacterial communication systems (quorum sensing (QS)) and employing agents that promote biofilm disassembly (Rabin et al., 2015). Studying these mechanisms not only advances our knowledge of *S. aureus* pathogenicity but also plays a key role in designing more effective clinical interventions.

11.1.2 Significance of Biofilm Formation in Bacterial Infections

Biofilm formation plays a pivotal role in the pathogenesis of bacterial infections, particularly those caused by *S. aureus*. These structured microbial communities confer several advantages to the bacteria, enhancing their survival and persistence within the host. One significant advantage is the increased resistance to antibiotics. Bacteria residing within biofilms can exhibit up to 1,000 times greater resistance (Mah and O'Toole, 2001) to antimicrobial agents compared to their planktonic counterparts. This heightened resistance is attributed to multiple factors, including the presence of persister cells, reduced metabolic activity of cells within the biofilm, and the limited penetration of antibiotics through the EPS matrix (Foster et al., 2014).

Biofilms also facilitate evasion of the host immune system. The biofilm matrix acts as a physical barrier, impeding the access of immune cells and molecules to the bacterial cells. Additionally, *S. aureus* within biofilms can modulate immune responses by releasing toxins and enzymes that interfere with host defense mechanisms. The ability of *S. aureus* to form biofilms is closely associated with recurrent and chronic infections. Biofilms enable the bacteria to persist in the host, leading to infections that are difficult to eradicate with conventional treatments (Bjarnsholt, 2013). This persistence is a common feature in various chronic conditions, such as diabetic foot ulcers and infections associated with medical implants.

The impact of biofilm-associated infections extends to global healthcare systems. These infections contribute to increased morbidity, prolonged hospital stays, and elevated medical costs. The situation is further exacerbated by the emergence of antibiotic-resistant strains of *S. aureus*, such as methicillin-resistant *S. aureus* (Tong et al., 2015), which pose significant challenges in clinical management. The contrasting characteristics between planktonic and biofilm-associated bacteria are summarized in Table 11.1, highlighting differences in antibiotic resistance, immune evasion, and infection chronicity.

11.1.3 QS and Its Relevance in Biofilm Development

QS is a cell-density–dependent regulatory mechanism that allows bacterial communities to coordinate gene expression collectively. This process operates through the synthesis and diffusion of small signaling compounds called autoinducers (AIs). As the bacterial population expands, the

TABLE 11.1
Comparison of Planktonic and Biofilm-Associated Bacterial Characteristics

Characteristic	Planktonic Bacteria	Biofilm-Embedded Bacteria
Antibiotic resistance	Typically susceptible to antibiotics	Up to 1,000 times more resistant due to factors like limited antibiotic penetration, presence of persister cells, and altered metabolic states
Immune evasion	Readily targeted and cleared by immune cells	Evade immune responses through physical barriers (extracellular matrix), impaired phagocytosis, and modulation of host immune signaling
Infection chronicity	Often cause acute, transient infections	Associated with chronic, recurrent infections due to persistent colonization and resistance to treatment

concentration of these molecules increases in the surrounding environment. Upon reaching a threshold level, they activate a synchronized genetic response that governs key physiological behaviors such as virulence factor production, motility, and biofilm formation (Miller and Bassler, 2001).

11.1.3.1 QS System in *S. aureus*

In *S. aureus*, cell-to-cell communication is primarily mediated by a genetic network referred to as the accessory gene regulator (Agr) system. This system responds to the presence of specific signaling molecules called autoinducing peptides (AIPs), which are secreted and sensed by the bacterial cells. When AIP concentration reaches a sufficient level, the Agr system becomes activated, resulting in a shift in gene expression. This modulation influences the production of toxins, enzymes, and other virulence-related components. Additionally, it plays a critical role in determining whether the bacteria remain in a biofilm-associated state or initiate detachment and spread (Otto, 2008).

11.1.3.2 Functional Role of QS in Biofilm Development

QS plays a pivotal role in the biofilm lifecycle, influencing both its formation and dispersal. During the initial stages, QS regulates the production of adhesive molecules, thereby promoting bacterial attachment and biofilm development. Conversely, under specific environmental conditions, QS signaling can trigger the disassembly of biofilms, enabling bacterial cells to disperse and colonize new niches (Boles et al., 2005). In *S. aureus*, the accessory gene regulator (Agr) system is a key QS pathway that modulates both biofilm dynamics and virulence. Due to its central role in biofilm regulation and pathogenicity, the Agr system presents a promising target for therapeutic strategies aimed at mitigating *S. aureus* infections (Novick and Geisinger, 2008).

11.2 QS IN *S. AUREUS*

11.2.1 QS Mechanisms in *S. aureus*

QS in bacteria refers to a cell-to-cell signaling process that enables coordinated regulation of gene expression in response to population density. In *S. aureus*, this regulatory system plays a vital role in modulating a wide range of physiological behaviors. These include the expression of virulence factors, the ability to form and maintain biofilms, and the capacity to adapt to fluctuating environmental conditions (Atkinson and Williams, 2009).

11.2.1.1 The Agr System

The principal QS pathway in *S. aureus* is governed by the accessory gene regulator (Agr) system. This system is composed of four key genes – agrA, agrB, agrC, and agrD – which are organized into two distinct operons and function together to mediate cell-density-dependent gene regulation (Novick and Geisinger, 2008). The Agr system in *S. aureus* functions via a classical two-component signal transduction mechanism that regulates the switch between biofilm formation and virulence. Initially, the *agrD* gene encodes a precursor peptide that is processed and exported by the membrane protein AgrB, leading to the extracellular accumulation of an AIP. When AIP concentration reaches a critical threshold, it binds to the membrane-bound sensor kinase AgrC, initiating its autophosphorylation and subsequent phosphorylation of the response regulator AgrA. Activated AgrA then drives the transcription of RNAIII, the primary effector molecule of the Agr system. RNAIII modulates gene expression post-transcriptionally by repressing surface adhesion factors and enhancing the production of secreted toxins and enzymes. This regulatory cascade facilitates the bacterial transition from a sessile, biofilm-associated state to an invasive, pathogenic phenotype (Yarwood and Schlievert, 2003).

11.2.1.2 Other Regulatory Pathways

Apart from the agr system, some additional mechanisms/systems can impact S. *aureus* QS, such as the following: the two-component SaeRS system, which frequently functions in tandem with agr, controls virulence factors such as hemolysins and coagulase (Otto, 2013); the ArlRS system: under some circumstances, this mechanism, which is involved in adhesion and biofilm formation, can inhibit agr-mediated dispersal (Novick and Geisinger, 2008); the SigB regulon is a stress response regulator that, especially in the presence of environmental stress, regulates the production of adhesins, toxins, and genes linked to biofilms (Zhang et al., 2002).

11.2.2 QS and Gene Regulation in Biofilm Formation

Bacterial populations that are organized and covered in EPSs are called biofilms. QS controls the growth and spread of biofilms in *S. aureus* (Atkinson and Williams, 2009). First attachment: Surface proteins such as clumping factors and fibronectin-binding proteins mediate adhesion to surfaces. When QS activity is minimal during the early stages of biofilm formation, it is increased (Novick and Geisinger, 2008). As biofilms progress through their maturation stage, the ica operon plays a central role in the synthesis of the EPS, primarily consisting of polysaccharide intercellular adhesin (PIA). This matrix not only stabilizes the biofilm architecture but also contributes to resistance against environmental stress. QS mechanisms influence this phase by modulating metabolic activity, nutrient acquisition, and stress response pathways (Otto, 2013). In the dispersal stage, activation of the Agr QS system triggers the expression of degradative enzymes such as proteases and nucleases. These enzymes target and break down the structural elements of the biofilm, including proteins and extracellular DNA, thereby weakening the matrix. As a result, bacterial cells are released from the biofilm, facilitating their transition from a sessile to a motile and invasive form (Yarwood and Schlievert, 2003).

11.3 QS INHIBITORS

11.3.1 Overview of QS Inhibitors

Bacterial populations coordinate group behaviors – such as biofilm development, virulence expression, and antibiotic resistance – through a signaling process known as QS. In *S. aureus*, these functions are regulated by the accessory gene regulator (agr) system. Interrupting this signaling network through quorum-sensing inhibitors (QSIs) has emerged as a potential strategy for mitigating bacterial infections (Kalia, 2013; Whiteley et al., 2017).

11.3.1.1 Disruption of Signal Production

This approach involves inhibiting enzymes responsible for synthesizing signaling molecules. In gram-positive organisms like *S. aureus*, this includes interference with AIP synthesis. Compounds such as farnesol have been shown to suppress signal generation in gram-positive species (Njoroge and Sperandio, 2009).

11.3.1.2 Enzymatic Breakdown of Signaling Molecules

Certain enzymes are capable of deactivating QS signals by catalyzing their degradation. Enzymes like lactonases and acylases are effective in this regard. For example, acyl-homoserine lactone (AHL)-lactonase hydrolyzes AHLs, rendering them ineffective in gram-negative QS systems (Dong et al., 2001).

11.3.1.3 Blocking Signal Reception

Some molecules can interfere with the interaction between signaling peptides and their receptors. In *S. aureus*, a well-studied example is the RNAIII-inhibiting peptide (RIP), which prevents the binding of AIP to the AgrC sensor kinase, effectively silencing the agr system (Balaban et al., 1998).

11.3.1.4 Structural Analogs of QS Signals

Molecules that closely resemble natural QS signals in structure can act as competitive inhibitors by binding to signal receptors and blocking their normal function. A well-known example is the group of halogenated furanones synthesized by the marine red alga *Delisea pulchra*. These compounds imitate the structure of AHLs used in gram-negative bacterial communication. Upon binding to LuxR-type receptors, the furanones interfere with receptor stability, often triggering their degradation. As a result, genes typically activated by QS – including those involved in biofilm formation – fail to express properly, disrupting bacterial coordination and reducing pathogenic potential (Hentzer et al., 2002).

11.3.2 Rationale for Using QSIs to Control Biofilm Formation

Antibiotic resistance, immunological evasion, and chronic infections are all significantly influenced by *S. aureus* biofilm development. By focusing on QS-regulated processes, QSIs offer a unique solution to these problems.

Inhibiting Biofilm Maturation: QSIs increase the susceptibility of bacteria to treatment by interfering with the expression of genes necessary for biofilm maturation (Kong et al., 2006; Otto, 2013). Reducing the Production of Virulence Factors: QSIs such as RIP impair the stability of biofilms by reducing the production of toxins and enzymes controlled by the agr system (Balaban et al., 1998). Increasing Antibiotic Sensitivity: QSIs increase the effectiveness of antibiotics against bacteria that are trapped in biofilms by interfering with QS. Avoiding Chronic Infections: QSIs inhibit bacterial communication, which lowers virulence and biofilm formation and, in turn, avoids chronic infections (Kalia, 2013).

11.3.3 QS Inhibition: Strategies and Applications

The QS process can be disrupted through several strategic mechanisms: (1) attenuation of the functional activity of the AHL synthase enzyme or its corresponding receptor protein, (2) inhibition of the biosynthesis of QS signaling molecules, (3) enzymatic degradation of AHLs, and (4) molecular mimicry through synthetic analogs that imitate native signaling compounds (Naga et al., 2023). Among these approaches, enzymatic degradation of AHLs has emerged as the most extensively explored and operationally leveraged strategy. Additionally, the use of antibodies and decoy receptors to neutralize QS signals has been proposed as an innovative paradigm for anti-infective therapeutics (Rajkhowa et al., 2024). Both natural and synthetic compounds have demonstrated QSI activity. Naturally derived QSIs include polyphenolic extracts from sources such as beans, carrots, chamomile, and *Trigonella stellata*. On the synthetic front, compounds such as aspirin, β-lactam antibiotics, azithromycin, and benzothiazole derivatives have shown promising QSI potential (Naga et al., 2023).

11.3.3.1 Bacterial Reporters to Detect QS Signals and Screening QSI

One of the most essential requirements for avoiding the pathogenicity problems associated with QS is its identification. Because of the growing diversity in the QS signals generated by various species, it was recognized that there was a need for biosensors to be developed. Several researchers created AHL and AI-2 reporters throughout time to demonstrate the link between cause and effect. The biomonitor strains allow sensitive, quantitative, and real-time detection of QS signals such as AHLs (Toure et al., 2019).

11.3.3.2 Criteria for Selecting QSIs

To effectively identify and deploy QSIs, certain critical criteria have been proposed. These include (1) the compound should be a small molecule capable of significantly downregulating genes governed by QS systems, (2) it must exhibit high specificity toward the target QS regulator without eliciting detrimental effects on either the host or the bacterial population, (3) the compound should demonstrate chemical stability and exhibit resistance to metabolic degradation by host systems, and (4) ideally, the molecule should possess a structural length exceeding that of native AHLs (Hetta et al., 2024). Adhering to these selection parameters ensures minimal risk of resistance development by pathogenic bacteria, facilitates selective targeting without disrupting the host's beneficial microbiota, and accelerates drug discovery efforts. Furthermore, owing to their low molecular weight and consequent lack of antigenicity, such compounds present promising candidates in the development of next-generation anti-infective therapies (Kalia et al., 2013).

11.3.3.2.1 Natural

In natural ecosystems, microbial communities exhibit complex interspecies interactions, characterized by both cooperative and competitive dynamics. Quorum quenching (QQ) capabilities offer a competitive edge to certain bacteria by suppressing QS-regulated activities in rival species, thereby enhancing survival and ecological fitness (Castillo-Juárez et al., 2015).

11.3.3.2.2 Prokaryotic QSI

The diversity of organisms with abilities to produce QQ enzymes ranges quite a bit: (1) Firmicutes – *Bacillus*, *Arthrobacter*, and *Oceanobacillus*; (2) Proteobacteria – *Acinetobacter*, *Comomonas*, *Alteromonas*, *Pseudomonas aeruginosa*, *Halomonas*, *Ralstonia*, *Klebsiella pneumoniae*, *Hyphomonas*, *Stappia*, *A. tumefaciens*, and *Variovoraxparadaoxus*; (3) cyanobacteria – *Anabaena*, (4) Actinobacteria – *Rhodococcus* and *Streptomyces*; (5) Bacteroidetes – *Tenacibaculum (*Hemmati et al., 2020). Additionally, certain bacteria synthesize small molecules with QQ activity such as non-cognate AHLs, intermediates of the AHL biosynthetic pathway, and dicyclic peptides. Enzymatically, four distinct classes exhibit QS signal degradation functionality: AHL-lactonases and decarboxylases cleave the lactone ring, whereas AHL-acylases and deaminases target the acyl side chain (Hetta et al., 2024).

11.3.3.2.3 Animal-Based QSIs

Host–pathogen interactions in eukaryotic systems, particularly in the presence of bacterial QS signals, elicit a range of host defense mechanisms. For instance, plants respond to AHLs by synthesizing QS-mimicking molecules, stimulating root development, absorbing QS molecules systemically, and modifying stomatal conductance. In the animal kingdom, various QQ enzymes have been identified in species such as mice, rats, and zebrafish. Specifically, porcine kidney-derived acylase I has demonstrated inactivation capabilities against C6HSL and 3OC12HSL, though it is ineffective against C4HSL (Majdura et al., 2023). Moderate inhibition of biofilm formation by *A. hydrophila* and *P. putida* has been attributed to acylase I. Human epithelial cells can degrade AHLs from *P. aeruginosa*, with activity being influenced by the acyl chain length – 3OC12HSL is degraded, while shorter chains are not. Moreover, mammalian serum from bovine, goat, horse, mouse, and rabbit sources has shown efficacy in neutralizing 3OC12HSL (Lade et al., 2014). From a nutritional perspective, various animal-derived foods – including turkey patties, chicken breast, cheeses, and beef products – have demonstrated significant AI-2 inhibitory activity (84.4–99.8%). Long-chain fatty acids in poultry showed 25–99% inhibition of AI-2 activity in *Vibrio harveyi* BB170. Ground beef extracts were also shown to suppress bioluminescence and virulence gene expression (e.g., *yadK*, *hha*) in *V. harveyi* by over 90% (Nychas et al., 2009).

11.3.3.2.4 Fungus-Based QSI

Fungi are prolific producers of bioactive secondary metabolites, including antibiotics such as penicillin from *Penicillium* spp., which revolutionized antibacterial therapy. Approximately 33 *Penicillium* species have been documented to produce QSIs, including compounds like patulin and penicillic acid (Sułkowska-Ziaja et al., 2023). *In vivo* studies have demonstrated the therapeutic efficacy of patulin in significantly reducing *P. aeruginosa* infections in murine lung infection models. Furthermore, fungi from the phyla Ascomycota and Basidiomycota, particularly those associated with the plant rhizosphere, possess lactonase enzymes capable of degrading AHLs such as C6HSL and 3OC6HSL. Notably, natural pigments synthesized by *Auricularia auricula* have been identified to function as QSIs by suppressing violacein production in *Chromobacterium violaceum* (Vaillancourt et al., 2024).

11.3.3.2.5 Synthetic QSI

It has been known that many organisms produce QSIs naturally. But being natural has some issues, as they are produced in small quantities and have some toxicity in some cases. These problems can be overcome by chemically synthesizing them. Studies have shown these chemically synthesized compounds to be very effective. Synthetic QSIs are man-made compounds designed to specifically target and inhibit bacterial QS systems (Chen et al., 2024). These compounds are often engineered to interfere with bacterial communication by disrupting the signaling molecules or their receptors. Synthetic QSIs can be developed for targeted therapeutic use, especially to treat infections involving biofilms.

11.3.3.2.6 Enzyme-Based Synthetic QSIs

Synthetic versions of enzymes like AHL-lactonases are designed to break down AHL molecules, thereby preventing bacterial communication and biofilm formation. These enzymes are often engineered for specific pathogens, such as *P. aeruginosa* or *V. harveyi* (Rémy et al., 2020). AHL-acylases are synthetic enzymes that cleave the acyl chain from AHL molecules, rendering them inactive and unable to bind to bacterial QS receptors.

11.3.3.2.7 Small-Molecule QSIs

Several synthetic QSIs have demonstrated significant potential in disrupting bacterial communication and biofilm formation. Brominated furanones, derived from natural compounds produced by marine algae, are chemically modified to enhance their inhibitory activity by preventing AHL binding to QS receptors in gram-negative bacteria such as *P. aeruginosa* (Muñoz-Cázares et al., 2022). Quercetin, a flavonoid, and its synthetic derivatives also exhibit strong QSI activity by interfering with AHL-mediated signaling, thereby inhibiting biofilm formation and virulence factor production in both *P. aeruginosa* and *S. aureus* (Ouyang et al., 2020). Additionally, S-adenosylmethionine analogs have been engineered to disrupt AHL biosynthesis by targeting and blocking specific enzymes involved in QS signal production in gram-negative bacteria (Papenfort& Bassler, 2016).

11.3.3.2.8 Receptor Antagonists

Synthetic QS receptor antagonists are small molecules that specifically bind to bacterial QS receptors, blocking their activation by the natural signaling molecules. For example, PQS (*Pseudomonas*

TABLE 11.2
Classes of Quorum-Sensing Inhibitors (QSIs) and Their Modes of Action

S. No.	Class of QSI	Mode of Action	Examples
1.	**Natural compounds**	Interfere with QS pathways through various mechanisms, often derived from plants or microbes	Furanones, garlic extract (ajoene)
2.	**Peptide-based inhibitors**	Synthetic peptides designed to disrupt QS signaling or receptor binding	RNAIII-inhibiting peptide derivatives
3.	**Receptor antagonists**	Compete with natural ligands for binding to QS receptors, inhibiting signal transduction	Solonamide B, savirin
4.	**Signal analogs**	Mimic natural QS signals to disrupt normal signaling pathways	Synthetic AIP analogs
5.	**Signal degradation enzymes**	Degrade or modify QS signaling molecules, preventing signal accumulation	AHL-lactonases, AHL-acylases
6.	**Signal synthesis inhibitors**	Block the production of autoinducing peptides (AIPs) essential for QS activation	RNAIII-inhibiting peptide (RIP), ambuic acid

quinolone signal) antagonists have been developed to prevent the activation of QS receptors in *P. aeruginosa*, which are critical for biofilm formation and virulence factor production (Mattmann and Blackwell, 2010).

11.3.3.2.9 Peptide-Based QSIs

Synthetic peptides that mimic the native signaling peptides involved in QS can be designed to interfere with bacterial communication. These peptides often act as antagonists, binding to QS receptors and preventing the binding of the natural QS signals.

11.3.3.2.10 Hydrophobic Small Molecules

Synthetic cyclic peptides or organic molecules can be designed to disrupt the binding of QS signals to bacterial receptors or inhibit the enzymes involved in signal production. These compounds can be tailored for use against specific bacterial species, like *P. aeruginosa* or *V. cholera* (Rutherford and Bassler, 2012). Table 11.2 summarizes the various classes of QSIs, detailing their specific modes of action and representative examples, which are pivotal in disrupting QS-mediated biofilm formation in *S. aureus*.

11.4 INHIBITION OF BIOFILM FORMATION WITH QSIS

Bacteria, like all higher organisms, communicate using their own language. Bacteria behave as single cells at low cell densities, but they may convert to a "multicellular" mode of action when they detect that their population density exceeds a particular threshold. They can express genes for a variety of traits, including those that regulate their pathogenic behavior. At this moment, they communicate via microscopic signaling molecules. Opportunistic infections, such as *P. aeruginosa*, prefer to remain "dormant" and not exhibit a virulent phenotype until their population has grown to the point where the host's defenses are overwhelmed. This phenomenon, termed as bacterial QS, operates through a wide range of signals such as (1) methyl dodecanoic acid, (2) *N*-AHLs, (3) oligopeptides (5–10 amino acid cyclic thiolactone), (4) furanosyl borate (AI-2, AI-2), and (5) hydroxylpalmitic acid methyl ester (Jiang et al., 2019).

The two principal QS signal classes that have garnered substantial scientific attention are (1) AHLs – synthesized by over 70 gram-negative bacterial species – which freely diffuse across cellular membranes and subsequently interact with intracellular transcriptional regulators, and

(2) peptide-based AIs – prevalent in gram-positive bacteria – which initiate signal transduction via membrane-associated histidine kinase receptors (Khalid et al., 2021). QS is a sophisticated intercellular communication mechanism that enables bacterial populations to monitor their own density and modulate gene expression accordingly. Upon reaching a critical cell density threshold, QS activates coordinated behavioral responses, including virulence factor production, siderophore biosynthesis, extracellular enzyme secretion, and biofilm development. This communication relies on the biosynthesis and detection of low-molecular-weight signaling compounds known as AIs, whose accumulation is tightly correlated with population density. In gram-positive organisms such as *S. aureus*, AIPs serve as QS activators. Conversely, in gram-negative species including *Pseudomonas* spp., *Burkholderia* spp., and *Acinetobacter* spp., QS is primarily mediated by AHLs. Beyond AHLs and AIPs, other chemical classes of AIs have also been characterized. For instance, ketone-based signals are employed by species such as *Vibrio cholerae* and *Legionella* spp., while fatty acid-derived AIs function in *Burkholderia* spp. and *Xanthomonas* spp. Importantly, AI-2 represents a chemically distinct signal that is recognized and utilized by both gram-negative and gram-positive bacteria, underscoring its role as a universal QS molecule (Rutherford and Bassler, 2012). An overview of these QS signal types – including their chemical nature, primary bacterial producers, receptor systems, and representative examples – is encapsulated in Table 11.3, highlighting the mechanistic diversity of QS across bacterial taxa.

11.4.1 Effects of QSIs on Biofilm Development Stages

11.4.1.1 Initial Reversible Attachment

The initial phase of biofilm formation is characterized by reversible adhesion, wherein individual planktonic bacterial cells transiently attach to surfaces immersed in aqueous environments. This stage marks the onset of biofilm development, during which a conditioning layer – a thin organic film – is rapidly established on the substrate. This layer serves as a preliminary interface facilitating microbial attachment. Adhesion is widely recognized as a critical determinant in the biofilm formation process, as the initial colonization of a surface by bacteria enhances the recruitment of additional planktonic microorganisms. This leads to coaggregation and the eventual development of multilayered biofilm structures. The conditioning layer, forming within minutes of exposure to water or other aqueous media, remains stable for several hours and plays a pivotal role in determining the structural integrity and cohesion of the nascent biofilm (Sharma et al., 2023).

TABLE 11.3
Major Quorum-Sensing Signals and Their Characteristics

Signal Type	Chemical Nature	Predominant In	Receptor Type	Examples of Producing Bacteria
Acyl-homoserine lactones (AHLs)	Small molecules with a lactone ring	Gram-negative bacteria	Intracellular LuxR-type proteins	*Pseudomonas* spp., *Burkholderia* spp.
Autoinducing peptides (AIPs)	Short peptides	Gram-positive bacteria	Membrane-bound histidine kinases	*Staphylococcus aureus*, *Enterococcus* spp.
Autoinducer-2 (AI-2)	Furanosyl borate diester	Both gram-positive and gram-negative bacteria	LuxP-type receptors	*Escherichia coli*, *Salmonella* spp.
Ketone-based signals	Ketone compounds	Various bacteria	Diverse receptor types	*Vibrio cholerae*, *Legionella* spp.
Fatty acid signals	Fatty acid derivatives	Various bacteria	Diverse receptor types	*Burkholderia* spp., *Xanthomonas* spp.

The conditioning film influences cellular or microbial adhesion and is dependent on the concentration of organic molecules in the fluids in contact with the surface. Organic molecules cling to surfaces based on their physical characteristics, such as charge and roughness. Due to their larger surface area, rougher surfaces promote biofilm formation and microbial colonization. Planktonic microorganisms in aquatic media connect to the conditioning layer once it is created. Gravity, diffusion, and fluid dynamics in the surrounding liquid phase all have an impact on how microbes and planktonic cells adhere (Donlan, 2002). Bacterial appendages such as pili, flagella, and fimbriae can cross physical barriers to reach the conditioning layer, triggering chemical reactions and enhancing the bacteria–surface connection. Nutrient availability can influence adhesion. Following the first connection between bacterial cells and the substrate, reversible adhesion can be modified by a variety of factors, including electrostatic, van der Waals attraction, and hydrophobic interactions. Fluid shear forces can also easily remove bacterial cells at this point. Negatively charged bacteria and inert surfaces generate electrostatic forces that encourage repulsion (Sharma et al., 2023).

11.4.1.2 Irreversible Attachment

During this stage, bacteria form an irreversible adhesion to the surface and produce EPS. Bacteria secrete EPS in a specific amount, causing considerable contact with the surface. Planktonic microbes can stick to one another or other surface-bound species, producing aggregation on the substratum. This adhesion is irreversible without physical or chemical intervention, leaving bacterial cells securely attached to the surface (Saharan et al., 2024).

11.4.1.3 Biofilm Architecture Development

After an initial adaptation period, a significant rise in cell population is observed during the exponential growth phase, which is largely influenced by the prevailing physical and chemical conditions. Permanently adhered bacterial cells begin to proliferate, forming microcolonies by utilizing nutrients present in both the conditioning film and the surrounding fluid. These cells produce EPSs, which serve to anchor them more securely to the surface and shield the developing community from environmental stresses. While adhesion is initially governed by physical and chemical factors, biological mechanisms increasingly dominate as the biofilm matures (Dang and Lovell, 2015). PIA molecules, along with divalent cations, establish durable connections. Additionally, sessile cells downregulate the production of surface appendages due to reduced motility requirements, while simultaneously upregulating genes responsible for surface proteins and secreted factors.

11.4.1.3.1 Maturation

In this stage, the initially attached microcolonies continue to reproduce, accumulating metabolic byproducts and incorporating additional planktonic bacteria from the surrounding environment. This leads to the development of a fully mature biofilm, characterized by its complex three-dimensional structure. EPS becomes the primary mediator of both intercellular and surface adhesion. As the bacterial density increases, QS systems – utilizing diverse signaling molecules – are activated. These QS mechanisms are critical for biofilm maturation, allowing the microbial community to sense population thresholds and coordinate collective behavior (Guzmán-Soto et al., 2021). While traditionally associated with intraspecies communication, QS is also shown to facilitate interspecies interactions. Gram-positive and gram-negative bacteria employ distinct QS pathways involving the production and detection of AIs: peptides in gram-positives and acylated homoserine lactones in gram-negatives (Jin et al., 2024). Through these signals, bacteria modulate gene expression in response to population density. Upon maturation, the biofilm consists of three distinct structural layers: a basal connecting film that anchors it to the substrate, a dense bacterial layer at the core, and a superficial layer from which planktonic cells can detach and disseminate.

11.4.1.3.2 Dispersal

Dispersal represents the terminal phase of biofilm development, during which previously attached cells separate and migrate to colonize new environments. This process can occur through the active shedding of daughter cells from proliferating bacteria or through detachment triggered by factors such as nutrient limitation, hydrodynamic forces, secreted enzymes, and catabolite repression (Kaplan, 2010). The detachment process includes mechanisms such as sloughing, erosion, and abrasion. *Erosion* refers to the gradual removal of individual cells or small clusters from the biofilm surface. *Sloughing* involves the detachment of large biomass fragments, typically due to nutrient or oxygen depletion at the biofilm base or sudden nutrient influx in the surrounding fluid. *Abrasion* describes the mechanical loss of biofilm due to contact with moving particles. Once released, these cells may relocate and initiate biofilm formation in new niches. Importantly, biofilm dispersal is recognized as a major contributor to the propagation of infectious diseases (Rumbaugh and Sauer, 2020).

11.4.2 Synergistic Effects with Other Antimicrobial Agents

Antibiotic resistance is a rising global issue requiring novel therapeutic techniques and antibiotic administration strategies. A "synergistic effect" in biofilm formation occurs when combining two or more antimicrobial agents results in a significantly greater inhibition of biofilm development compared to using each agent alone, meaning the combined effect is more potent than the sum of their individual effects; this is often achieved by targeting different aspects of biofilm formation, like cell adhesion, matrix production, or bacterial communication pathways. To address the resistance problem, a novel strategy called combination therapy is being investigated (Bonincontro et al., 2023). In combination therapy, two or more medications are combined to treat an infection. Innovative techniques that improve the efficiency of existing antibiotics while also preventing the development of resistance are critical for dealing with this rising challenge. Synergistic therapy techniques that combine antibiotics have demonstrated potential in fighting drug-resistant illnesses (Morales-Durán et al., 2024). Here are a few ways antimicrobial agents can show synergistic effects in biofilm inhibition or eradication:

11.4.2.1 Use of Antimicrobial Peptides and Antibiotics

Antimicrobial peptides (AMPs) are the natural defense system that is found in many multicellular organisms, also in humans, that exhibit a broad spectrum of antimicrobial activity. AMPs attach to target cells' membranes and create pore-like openings across the membrane. Unlike traditional antibiotics, which target particular bacterial enzymes, proteins, or growth/proliferation pathways. AMPs are known for their quick bactericidal activity, low risk of resistance development, immunomodulatory effects, and ability to improve wound healing. In recent decades, several *in vivo*, in vitro, and preclinical investigations have investigated the use of AMPs in combination with conventional antibiotics against a variety of infections (Zhang et al., 2021). While the majority of published research suggests synergistic effects that improve antimicrobial activity and may reduce resistance, there have also been reports of combinations that do not synergize. Combining conventional antibiotics and AMPs can often result in synergistic benefits against biofilms, as AMPs can break the biofilm matrix, allowing antibiotics to penetrate more effectively.

(i) **Plant Compounds and Antibiotics**: Certain plant extracts like thymol or curcumin can exhibit synergistic activity with antibiotics against biofilms by interfering with biofilm formation mechanisms.
(ii) **Enzymes and Antibiotics**: Enzymes that degrade biofilm components like polysaccharides can be combined with antibiotics to enhance biofilm disruption.

11.5 CHALLENGES AND LIMITATIONS

11.5.1 Potential Challenges in Using QSIs

Biofilm formation presents significant challenges and limitations, especially in the context of medical treatments, industrial processes, and environmental management. Biofilms are complex communities of microorganisms (bacteria, fungi, etc.) that adhere to surfaces and are encased in an ECM. They are difficult to treat, and the formation of biofilms can lead to persistent infections and other issues. Here are some of the key challenges and limitations associated with biofilm formation:

11.5.1.1 Antimicrobial Resistance

Biofilms significantly contribute to antimicrobial resistance. The ECM, composed ofpolysaccharides, proteins, and DNA, acts as a barrier, limiting the penetration of antimicrobial agents into the biofilm. Within biofilms, microorganisms often exhibit reduced susceptibility to antibiotics due to factors such as nutrient gradients that create a nutrient-poor environment in deeper layers, leading to less metabolically active cells that are less responsive to treatments (Singh et al., 2021). Additionally, the presence of persister cells – dormant or slow-growing bacteria – further enhances resistance. Biofilm formation can also lead to altered gene expression in bacteria, enhancing their resistance mechanisms, such as increased efflux pump activity or modification of antibiotic target sites (Shree et al., 2023).

11.5.1.2 Difficulty in Diagnosis and Detection

Detecting biofilms poses significant challenges, especially in chronic infections or those associated with indwelling medical devices. Traditional diagnostic methods may not accurately reflect biofilm formation, leading to delayed or inappropriate treatment. Biofilm-related infections often present as chronic, low-grade infections that are hard to diagnose until they become severe, potentially mimicking other conditions and leading to misdiagnosis or underestimation of their severity (Mendhe et al., 2023).

11.5.1.3 Persistence of Infections

Biofilms are responsible for persistent infections, such as those found in chronic wounds, sinus infections, periodontal diseases, and infections associated with medical implants like catheters and prosthetic joints. Once established, biofilms can act as reservoirs for infection, leading to recurrent episodes even after apparent clinical resolution (Sahoo and Meshram, 2024).

11.5.1.4 Difficulty in Treatment and Eradication

Standard antimicrobial therapies are often ineffective against biofilm-associated infections due to the physical barrier presented by the biofilm matrix and the altered physiology of the embedded bacteria. Antibiotics may have reduced efficacy because they are typically designed to target actively growing bacteria, whereas many bacteria in biofilms are in a dormant or slow-growing state. Treating biofilm-associated infections often requires higher doses of antibiotics or prolonged treatment durations, increasing the risk of side effects and encouraging resistance development (Pai et al., 2023).

11.5.1.5 Formation and Growth of Medical Devices

Biofilms are a major concern in the context of medical implants and long-term indwelling devices, including catheters, artificial heart valves, and orthopedic prosthetics. Once bacterial cells adhere to the surfaces of these devices, they can initiate the formation of complex biofilm structures. These microbial communities are notoriously difficult to eliminate and are often resistant to conventional antibiotic therapy. The persistence of biofilms can impair device function, provoke chronic immune responses, and cause localized tissue injury. In severe cases, such complications may require the complete removal or replacement of the affected device to manage the infection effectively (Khatoon et al. 2018).

11.5.1.6 Environmental and Industrial Impact

Biofilms can form in water pipes, filtration systems, and cooling towers, leading to the buildup of pathogens, scaling, and corrosion, which compromise water quality and increase maintenance costs. In industries such as food production, oil and gas, and pharmaceuticals, biofilms can cause contamination, equipment malfunction, product spoilage, and pose health risks. Additionally, biofilms contribute to microbial-induced corrosion, deteriorating metals and other materials used in pipelines, reactors, and storage tanks (Waqas et al., 2023).

11.5.1.7 Complexity of Biofilm Microbial Communities

Biofilms are often polymicrobial, containing a mix of bacteria, fungi, and sometimes viruses or archaea. This diversity complicates treatment, as different microorganisms may respond differently to various antimicrobial agents. Interactions between species within the biofilm can promote resistance, mutualistic relationships, or shared resistance mechanisms like horizontal gene transfer, making it challenging to target a single species effectively (Anju et al., 2022).

11.5.1.8 Challenges in Developing Effective Therapies

There is a lack of effective biofilm-targeting drugs. While research into biofilm-targeting agents is ongoing, current therapeutic options are limited in their ability to effectively target and eliminate biofilms without causing toxicity or resistance. Alternative strategies, such as enzyme-based therapies (e.g., DNases or proteases) and QS inhibitors, are under investigation but face challenges regarding safety, efficacy, and cost-effectiveness (Koo et al., 2017).

11.5.1.9 Host Immune Evasion

Biofilms can evade the host immune system by limiting immune cell access to the bacteria, creating an immunologically protected environment. This contributes to the persistence of infection and chronic inflammation. The biofilm matrix can also prevent immune cells, like macrophages, from effectively engulfing and destroying bacteria within the biofilm (Cangui-Panchi et al., 2023).

11.5.2 Strategies to Overcome Challenges

Biofilm production increases bacterial pathogenicity and resistance to antibiotics, demanding specific measures to address the issue. Foreign bodies are a leading source of biofilm infections, according to recent reports. To treat biofilm infections, it's important to remove infected medical equipment and replace it with fresh, uninfected equipment. Additionally, antibiotics should be administered aggressively. Furthermore, the implant removal should be timed correctly so that the new or replacement implant does not get contaminated when implanted into the patient's body. To prevent biofilm growth, antibiotics should be administered long-term if removal is not possible (Kang et al., 2023).

Previous study indicates that antibiotics are more effective against premature biofilms than mature biofilms. Inadequate detection of early biofilms in the body contributes significantly to clinical problems caused by mature biofilms. Antibiotics should be chosen based on their sensitivity as well as their penetration into the biofilm matrix to treat them successfully. Studies have demonstrated that bacteria within biofilms exhibit significantly greater resistance to antibiotics compared to their planktonic counterparts. Consequently, combination therapies are often more effective than single-agent antibiotic treatments. This synergistic approach is advantageous because different agents can target distinct bacterial states – such as one acting on dormant cells and another on actively growing populations (Abebe, 2020).

Optimal therapeutic outcomes also depend on precise antibiotic administration, including appropriate dosing and treatment duration. To mitigate biofilm formation, the development of antifouling or antimicrobial surfaces is a strategic measure. Hydrophilic polymer coatings, such as polyethylene glycol, are frequently employed to reduce microbial adhesion. Meanwhile, antimicrobial surfaces are typically engineered by incorporating antibiotics or disinfectants into materials like polyurethane polymers. Additionally, nanoparticle-based coatings – such as those containing silver or antioxidant nanoparticles – offer a promising solution for inhibiting biofilm formation. However, the coating method has faced challenges due to the surface. Biofilms can emerge as a result of fast erosion (Asma et al., 2022). Photodynamic treatment might help reduce wound biofilm infections. A photoactive dye is applied, followed by irradiation with oxygen to kill microorganisms. During therapy, ensure the patient's eyes are not exposed to laser light. When using photosensitizers or photochemical reactions, it's important to properly stain and destroy bacterial cells without harming the patient's surrounding tissues. Using anti-biofilm compounds or biofilm-dissolving chemicals is another successful emerging method (Allamyradov et al., 2024).

Anti-biofilm chemicals disrupt bacterial signaling pathways in both gram-positive and gram-negative bacteria. Anti-biofilm compounds may include enzymes, peptides, antibiotics, and polyphenols. This review focuses on anti-biofilm compounds identified to combat various bacterial diseases. We have also discussed the mechanism of action of certain anti-biofilm compounds. This review will aid researchers in discovering novel antibiotics for gram-negative bacteria by better understanding the targets for anti-biofilm compounds (Asma et al., 2022).

11.6 CONCLUSIONS AND FUTURE PROSPECTS

There has been a lot of research and understanding concerning bacterial biofilm development. The rise of severe biofilm infections and their resistance to antimicrobial therapy has created a significant concern in the medical industry. Resistant bacteria are widely found in humans, farm animals, fruits, vegetables, dairy products, seafood, and poultry. As a result, it is vital to investigate effective techniques for addressing this issue and developing an alternative to antibiotics. Biofilm imaging methods have advanced dramatically to meet today's enormous challenge. Super-resolution microscopy techniques include photo-activated localization microscopy, fluorescent photo-activated localization microscopy, and stochastic optical reconstruction microscopy. They employ fluorescent proteins or probes to provide pictures with substantially better resolution than CSLM. As a result, these approaches will be increasingly commonly utilized to research biofilm. The current study reveals the mechanisms of action employed by several small compounds with anti-biofilm characteristics.

All anti-biofilm compounds have several mechanisms of action. For example, EGCG inhibits the AHL-mediated QS system or degrades peptidoglycan, causing membrane rupture. Understanding the mechanism of action of biofilms can lead to the development of novel and effective therapeutic compounds targeting previously identified targets. This can boost the efficacy of already recognized medications. This can be achieved either by combinatorial therapy or by making modifications. According to the current research, the most significant anti-biofilm drugs with high clinical relevance are EPS273, mannosides, and pilicides. More *in vivo* testing is required, followed by clinical trials. Natural antimicrobials are more biochemically and structurally diverse than produced drugs, making them excellent for developing alternative therapies and in silico pharmacological approaches. Natural products are very complex, allowing for more specific binding to targets. The use of organically generated anti-biofilm chemicals has limitations, including high cost, limited sustainability, time-consuming extraction, and variable effectiveness.

Synthetic medications, on the other hand, are less expensive and take less time to administer, but many have negative side effects. Furthermore, because of their huge size, many peptides and polysaccharides become inappropriate from a pharmacological perspective. For combating biofilm, it's necessary to ensure optimal distribution and dose of the substance in an *in vivo* system. However, discovering novel medications for new targets may be a lengthy and challenging process. Modifying already authorized medications for non-biofilm targets might be a more effective and beneficial approach. Bioinformatic technologies may be used to screen and improve drug libraries, making a valuable contribution. Further study is needed to identify tiny active pieces of these peptides and polysaccharides that can efficiently bind to newly found targets.

REFERENCES

Abebe, G. M. (2020). The role of bacterial biofilm in antibiotic resistance and food contamination. *International Journal of Microbiology*, 1–10. https://doi.org/10.1155/2020/1705814

Allamyradov, Y., Yosef, J. B., Annamuradov, B., Ateyeh, M., Street, C., Whipple, H., & Er, A. O. (2024). Photodynamic therapy review: Past, present, future, opportunities and challenges. *Photochem*, *4*(4), 434–461. https://doi.org/10.3390/photochem4040027

Anju, V. T., Busi, S., Imchen, M., Kumavath, R., Mohan, M. S., Salim, S. A., . . . Dyavaiah, M. (2022). Polymicrobial infections and biofilms: Clinical significance and eradication strategies. *Antibiotics*, *11*(12), 1731. https://doi.org/10.3390/antibiotics11121731

Archer, N. K., Mazaitis, M. J., Costerton, J. W., Leid, J. G., Powers, M. E., & Shirtliff, M. E. (2011). *Staphylococcus aureus* biofilms: Properties, regulation, and roles in human disease. *Virulence*, *2*(5), 445–459. https://doi.org/10.4161/viru.2.5.17724

Asma, S. T., Imre, K., Morar, A., Herman, V., Acaroz, U., Mukhtar, H., . . . Gerlach, R. (2022). An overview of biofilm formation–combating strategies and mechanisms of action of antibiofilm agents. *Life*, *12*(8), 1110. https://doi.org/10.3390/life12081110

Atkinson, S., & Williams, P. (2009). Quorum sensing and social networking in the microbial world. *Trends in Microbiology*, *17*(12), 594–600. https://doi.org/10.1016/j.tim.2009.09.009

Balaban, N., Goldkorn, T., Nhan, R. T., Dang, L. B., Scott, S., Ridgley, R. M., . . . Rasmussen, H. (1998). Autoinducer of virulence as a target for vaccine and therapy against *Staphylococcus aureus*. *Science*, *280*(5362), 438–440. https://doi.org/10.1126/science.280.5362.438

Bitrus, A. A., Peter, O. M., Abbas, M. A., & Goni, M. D. (2018). *Staphylococcus aureus*: A review of antimicrobial resistance mechanisms. *Veterinary Sciences: Research and Reviews*, *4*(2), 43–54. http://dx.doi.org/10.17582/journal.vsrr/2018/4.2.43.54

Bjarnsholt, T. (2013). The role of bacterial biofilms in chronic infections. *APMIS Supplementum*, *136*(1), 1–51. https://doi.org/10.1111/apm.12099

Boles, B. R., Thoendel, M., & Singh, P. K. (2005). Self-generated diversity produces "insurance effects" in biofilm communities. *Proceedings of the National Academy of Sciences*, *102*(30), 11558–11563. https://doi.org/10.1073/pnas.0504923102

Bonincontro, G., Scuderi, S. A., Marino, A., & Simonetti, G. (2023). Synergistic effect of plant compounds in combination with conventional antimicrobials against biofilm of *Staphylococcus aureus*, *Pseudomonas aeruginosa*, and *Candida* spp. *Pharmaceuticals*, *16*(11), 1531. https://doi.org/10.3390/ph16111531

Boucher, H. W., & Corey, G. R. (2008). Epidemiology of methicillin-resistant *Staphylococcus aureus*. *Clinical Infectious Diseases*, *46*(Supplement_5), S344-S349. https://doi.org/10.1086/533590

Cangui-Panchi, S. P., Ñacato-Toapanta, A. L., Enríquez-Martínez, L. J., Salinas-Delgado, G. A., Reyes, J., Garzon-Chavez, D., & Machado, A. (2023). Battle royale: Immune response on biofilms – host-pathogen interactions. *Current Research in Immunology*, *4*, 100057. https://doi.org/10.1016/j.crimmu.2023.100057

Castillo-Juárez, I., Maeda, T., Mandujano-Tinoco, E. A., Tomás, M., Pérez-Eretza, B., García-Contreras, S. J., . . . García-Contreras, R. (2015). Role of quorum sensing in bacterial infections. *World Journal of Clinical Cases*, *3*(7), 575. https://doi.org/10.12998/wjcc.v3.i7.575

Chen, A., Long, Z., Xiao, Y., Feng, Y., Zhou, Y., Yang, S., . . . Yang, S. (2024). Application of natural product-based quorum sensing inhibitors in plant pathogen control: A review. *Arabian Journal of Chemistry*, *18*(1), 106050. https://doi.org/10.1016/j.arabjc.2024.106050

Chessa, D., Ganau, G., & Mazzarello, V. (2015). An overview of *Staphylococcus* epidermidis and *Staphylococcus aureus* with a focus on developing countries. *The Journal of Infection in Developing Countries*, *9*(06), 547–550. https://doi.org/10.3855/jidc.6923

Dang, H., & Lovell, C. R. (2015). Microbial surface colonization and biofilm development in marine environments. *Microbiology and Molecular Biology Reviews*, *80*(1), 91–138. https://doi.org/10.1128/mmbr.00037-15

Dong, Y. H., Wang, L. H., & Zhang, L. H. (2001). Quenching quorum-sensing-dependent bacterial infection by an N-acyl homoserine lactonase. *Nature*, *411*(6839), 813–817. https://doi.org/10.1038/35081101

Donlan, R. M. (2002). Biofilms: Microbial life on surfaces. *Emerging Infectious Diseases*, *8*(9), 881–890. https://doi.org/10.3201/eid0809.020063

Flemming, H. C., Wingender, J., Szewzyk, U., Steinberg, P., Rice, S. A., & Kjelleberg, S. (2016). Biofilms: An emergent form of bacterial life. *Nature Reviews Microbiology*, *14*(9), 563–575. https://doi.org/10.1038/nrmicro.2016.94

Foster, T. J., Geoghegan, J. A., Ganesh, V. K., & Höök, M. (2014). Adhesion, invasion, and evasion: The many functions of the surface proteins of *Staphylococcus aureus*. *Nature Reviews Microbiology*, *12*(1), 49–62. https://doi.org/10.1038/nrmicro3161

Fowler, V. G., Miro, J. M., Hoen, B., Cabell, C. H., Abrutyn, E., Rubinstein, E., . . . Corey, G. R. (2005). *Staphylococcus aureus* endocarditis: A consequence of medical progress. *JAMA*, *293*(24), 3012–3021. https://doi.org/10.1001/jama.293.24.3012

Gnanamani, A., Hariharan, P., & Paul-Satyaseela, M. (2017). *Staphylococcus aureus*: Overview of bacteriology, clinical diseases, epidemiology, antibiotic resistance and therapeutic approach. *Frontiers in Staphylococcus aureus*, *4*(28), 10–5772.

Guzmán-Soto, I., McTiernan, C., Gonzalez-Gomez, M., Ross, A., Gupta, K., Suuronen, E. J., . . . Alarcon, E. I. (2021). Mimicking biofilm formation and development: Recent progress in in vitro and in vivo biofilm models. *iScience*, *24*(5), 102443. https://doi.org/10.1016/j.isci.2021.102443

Harris, L. G., Foster, S. J., & Richards, R. G. (2002). An introduction to *Staphylococcus aureus*, and techniques for identifying and quantifying *S. aureus* adhesins in relation to adhesion to biomaterials: Review. *European Cells & Materials*, *4*(3), 100–120.

Hemmati, F., Salehi, R., Ghotaslou, R., Kafil, H. S., Hasani, A., Gholizadeh, P., . . . Rezaee, M. A. (2020). Quorum quenching: A potential target for antipseudomonal therapy. *Infection and Drug Resistance*, *13*, 2989–3005. https://doi.org/10.2147/idr.s263196

Hentzer, M., Riedel, K., Rasmussen, T. B., Heydorn, A., Andersen, J. B., Parsek, M. R., . . . Givskov, M. (2002). Inhibition of quorum sensing in *Pseudomonas aeruginosa* biofilm bacteria by a halogenated furanone compound. *Microbiology*, *148*(1), 87–102. https://doi.org/10.1099/00221287-148-1-87

Hetta, H. F., Ramadan, Y. N., Rashed, Z. I., Alharbi, A. A., Alsharef, S., Alkindy, T. T., . . . Donadu, M. G. (2024). Quorum sensing inhibitors: An alternative strategy to win the battle against multidrug-resistant (MDR) *Bacteria. Molecules*, *29*(15), 3466. https://doi.org/10.3390/molecules29153466

Høiby, N., Ciofu, O., & Bjarnsholt, T. (2010). Pseudomonas aeruginosa biofilms in cystic fibrosis. *Future Microbiology*, *5*(11), 1663–1674. https://doi.org/10.2217/fmb.10.125

Jiang, Q., Chen, J., Yang, C., Yin, Y., & Yao, K. (2019). Quorum sensing: A prospective therapeutic target for bacterial diseases. *BioMed Research International*, 1–15. https://doi.org/10.1155/2019/2015978

Jin, Y., Chen, W., Hu, J., Wang, J., & Ren, H. (2024). Constructions of quorum sensing signaling network for activated sludge microbial community. *ISME Communications*, *4*(1). https://doi.org/10.1093/ismeco/ycae018

Kalia, V. C. (2013). Quorum sensing inhibitors: An overview. *Biotechnology Advances*, *31*(2), 224–245. https://doi.org/10.1016/j.biotechadv.2012.10.004

Kalia, V. C., Wood, T. K., & Kumar, P. (2013). Evolution of resistance to Quorum-Sensing inhibitors. *Microbial Ecology*, *68*(1), 13–23. https://doi.org/10.1007/s00248-013-0316-y

Kang, X., Yang, X., He, Y., Guo, C., Li, Y., Ji, H., . . . Wu, L. (2023). Strategies and materials for the prevention and treatment of biofilms. *Materials Today Bio*, *23*, 100827. https://doi.org/10.1016/j.mtbio.2023.100827

Kaplan, J. (2010). Biofilm dispersal: Mechanisms, clinical implications, and potential therapeutic uses. *Journal of Dental Research*, *89*(3), 205–218. https://doi.org/10.1177/0022034509359403

Khalid, S. J., Ain, Q., Khan, S. J., Jalil, A., Siddiqui, M. F., Ahmad, T., . . . Adnan, F. (2021). Targeting acyl homoserine lactones (AHLs) by the quorum quenching bacterial strains to control biofilm formation in *Pseudomonas aeruginosa*. *Saudi Journal of Biological Sciences*, *29*(3), 1673–1682. https://doi.org/10.1016/j.sjbs.2021.10.064

Khatoon, Z., McTiernan, C. D., Suuronen, E. J., Mah, T., & Alarcon, E. I. (2018). Bacterial biofilm formation on implantable devices and approaches to its treatment and prevention. *Heliyon*, *4*(12), e01067. https://doi.org/10.1016/j.heliyon.2018.e01067

Kong, K. F., Vuong, C., & Otto, M. (2006). *Staphylococcus* quorum sensing in biofilm formation and infection. *International Journal of Medical Microbiology*, *296*(2–3), 133–139. https://doi.org/10.1016/j.ijmm.2006.01.042

Koo, H., Allan, R. N., Howlin, R. P., Stoodley, P., & Hall-Stoodley, L. (2017). Targeting microbial biofilms: Current and prospective therapeutic strategies. *Nature Reviews Microbiology*, *15*(12), 740–755. https://doi.org/10.1038/nrmicro.2017.99

Lade, H., Paul, D., & Kweon, J. H. (2014). N-acyl homoserine lactone-mediated quorum sensing with special reference to use of quorum quenching bacteria in membrane biofouling control. *BioMed Research International*, 1–25. https://doi.org/10.1155/2014/162584

Lowy, F. D. (1998). *Staphylococcus aureus* infections. *The New England Journal of Medicine*, *339*(8), 520–532. https://doi.org/10.1056/NEJM199808203390806

Luzzago, C., Locatelli, C., Franco, A., Scaccabarozzi, L., Gualdi, V., Viganò, R., . . . Cremonesi, P. (2014). Clonal diversity, virulence-associated genes and antimicrobial resistance profile of *Staphylococcus*

aureus isolates from nasal cavities and soft tissue infections in wild ruminants in Italian Alps. *Veterinary Microbiology*, *170*(1), 157–161. https://doi.org/10.1016/j.vetmic.2014.01.016
Mah, T.-F. C., & O'Toole, G. A. (2001). Mechanisms of biofilm resistance to antimicrobial agents. *Trends in Microbiology*, *9*(1), 34–39. https://doi.org/10.1016/S0966-842X(00)01913-2
Majdura, J., Jankiewicz, U., Gałązka, A., & Orzechowski, S. (2023). The role of quorum sensing molecules in Bacterial–Plant interactions. *Metabolites*, *13*(1), 114. https://doi.org/10.3390/metabo13010114
Mattmann, M. E., & Blackwell, H. E. (2010). Small molecules that modulate quorum sensing and control virulence in *Pseudomonas aeruginosa*. *The Journal of Organic Chemistry*, *75*(20), 6737–6746. https://doi.org/10.1021/jo101237e
Mendhe, S., Badge, A., Ugemuge, S., & Chandi, D. (2023). Impact of biofilms on chronic infections and medical challenges. Cureus. https://doi.org/10.7759/cureus.48204
Miller, M. B., & Bassler, B. L. (2001). Quorum sensing in bacteria. *Annual Review of Microbiology*, *55*(1), 165–199. https://doi.org/10.1146/annurev.micro.55.1.165
Morales-Durán, N., León-Buitimea, A., & Morones-Ramírez, J. R. (2024). Unraveling resistance mechanisms in combination therapy: A comprehensive review of recent advances and future directions. *Heliyon*, *10*(6), e27984. https://doi.org/10.1016/j.heliyon.2024.e27984
Muñoz-Cázares, N., Castillo-Juárez, I., García-Contreras, R., Castro-Torres, V. A., Díaz-Guerrero, M., Rodríguez-Zavala, J. S., . . . Martínez-Vázquez, M. (2022). A brominated furanone inhibits *Pseudomonas aeruginosa* quorum sensing and type III secretion, attenuating its virulence in a murine cutaneous abscess model. *Biomedicines*, *10*(8), 1847. https://doi.org/10.3390/biomedicines10081847
Naga, N. G., El-Badan, D. E., Ghanem, K. M., & Shaaban, M. I. (2023). It is the time for quorum sensing inhibition as alternative strategy of antimicrobial therapy. *Cell Communication and Signaling*, *21*(1). https://doi.org/10.1186/s12964-023-01154-9
Njoroge, J., & Sperandio, V. (2009). Jamming bacterial communication: New approaches for the treatment of infectious diseases. *EMBO Molecular Medicine*, *1*(4), 201–210. https://doi.org/10.1002/emmm.200900032
Novick, R. P., & Geisinger, E. (2008). Quorum sensing in staphylococci. *Annual Review of Genetics*, *42*, 541–564. https://doi.org/10.1146/annurev.genet.42.110807.091640
Nychas, G., Dourou, D., Skandamis, P., Koutsoumanis, K., Baranyi, J., & Sofos, J. (2009). Effect of microbial cell-free meat extract on the growth of spoilage bacteria. *Journal of Applied Microbiology*, *107*(6), 1819–1829. https://doi.org/10.1111/j.1365-2672.2009.04377.x
Otto, M. (2008). Staphylococcal biofilms. *Current Topics in Microbiology and Immunology*, *322*, 207–228. https://doi.org/10.1007/978-3-540-75418-3_10
Otto, M. (2013). Community-associated MRSA: What makes them special? *International Journal of Medical Microbiology*, *303*(6–7), 324–330. https://doi.org/10.1016/j.ijmm.2013.02.007
Ouyang, J., Feng, W., Lai, X., Chen, Y., Zhang, X., Rong, L., . . . Chen, Y. (2020). Quercetin inhibits *Pseudomonas aeruginosa* biofilm formation via the vfr-mediated lasIR system. *Microbial Pathogenesis*, *149*, 104291. https://doi.org/10.1016/j.micpath.2020.104291
Pai, L., Patil, S., Liu, S., & Wen, F. (2023). A growing battlefield in the war against biofilm-induced antimicrobial resistance: Insights from reviews on antibiotic resistance. *Frontiers in Cellular and Infection Microbiology*, *13*. https://doi.org/10.3389/fcimb.2023.1327069
Papenfort, K., & Bassler, B. L. (2016). Quorum sensing signal–response systems in Gram-negative bacteria. *Nature Reviews Microbiology*, *14*(9), 576–588. https://doi.org/10.1038/nrmicro.2016.89
Rabin, N., Zheng, Y., Opoku-Temeng, C., Du, Y., Bonsu, E., & Sintim, H. O. (2015). Biofilm formation mechanisms and targets for developing antibiofilm agents. *Future Medicinal Chemistry*, *7*(4), 493–512. https://doi.org/10.4155/fmc.15.6
Rajkhowa, S., Hussain, S. Z., Agarwal, M., Zaheen, A., Al-Hussain, S. A., & Zaki, M. E. A. (2024). Advancing antibiotic-resistant microbe combat: Nanocarrier-based systems in combination therapy targeting quorum sensing. *Pharmaceutics*, *16*(9), 1160. https://doi.org/10.3390/pharmaceutics16091160
Rémy, B., Plener, L., Decloquement, P., Armstrong, N., Elias, M., Daudé, D., & Chabrière, É. (2020). Lactonase specificity is key to quorum quenching in *Pseudomonas aeruginosa*. *Frontiers in Microbiology*, *11*. https://doi.org/10.3389/fmicb.2020.00762
Rumbaugh, K. P., & Sauer, K. (2020). Biofilm dispersion. *Nature Reviews Microbiology*, *18*(10), 571–586. https://doi.org/10.1038/s41579-020-0385-0
Rutherford, S. T., & Bassler, B. L. (2012). Bacterial quorum sensing: Its role in virulence and possibilities for its control. *Cold Spring Harbor Perspectives in Medicine*, *2*(11), a012427. https://doi.org/10.1101/cshperspect.a012427
Saharan, B. S., Beniwal, N., & Duhan, J. S. (2024). From formulation to function: A detailed review of microbial biofilms and their polymer-based extracellular substances. *The Microbe*, 100194. https://doi.org/10.1016/j.microb.2024.100194

Sahoo, K., & Meshram, S. (2024). Biofilm formation in chronic infections: A comprehensive review of pathogenesis, clinical implications, and novel therapeutic approaches. *Cureus*. https://doi.org/10.7759/cureus.70629

Sharma, S., Mohler, J., Mahajan, S. D., Schwartz, S. A., Bruggemann, L., & Aalinkeel, R. (2023). Microbial Biofilm: A review on formation, infection, antibiotic resistance, control measures, and innovative treatment. *Microorganisms*, *11*(6), 1614. https://doi.org/10.3390/microorganisms11061614

Shaw, L., Golonka, E., Potempa, J. & Foster, S. J. (2004). The role and regulation of the extracellular proteases of *Staphylococcus aureus*. *Microbiology*, *150*(1), 217–228. https://doi.org/10.1099/mic.0.26634-0

Shree, P., Singh, C. K., Sodhi, K. K., Surya, J. N., & Singh, D. K. (2023). Biofilms: Understanding the structure and contribution towards bacterial resistance in antibiotics. *Medicine in Microecology*, *16*, 100084. https://doi.org/10.1016/j.medmic.2023.100084

Singh, S., Datta, S., Narayanan, K. B., & Rajnish, K. N. (2021). Bacterial exo-polysaccharides in biofilms: Role in antimicrobial resistance and treatments. *Journal of Genetic Engineering and Biotechnology*, *19*(1), 140. https://doi.org/10.1186/s43141-021-00242-y

Sułkowska-Ziaja, K., Trepa, M., Olechowska-Jarząb, A., Nowak, P., Ziaja, M., Kała, K., & Muszyńska, B. (2023). Natural compounds of fungal origin with antimicrobial activity – potential cosmetics applications. *Pharmaceuticals*, *16*(9), 1200. https://doi.org/10.3390/ph16091200

Taylor, T. A., & Unakal, C. G. (2018). *Staphylococcus aureus*. Treasure Island, FL: Stat Pearls.

Tong, S. Y., Davis, J. S., Eichenberger, E., Holland, T. L., & Fowler, V. G. (2015). *Staphylococcus aureus* infections: Epidemiology, pathophysiology, clinical manifestations, and management. *Clinical Microbiology Reviews*, *28*(3), 603–661. https://doi.org/10.1128/CMR.00134-14

Toure, S. L., Lou, Z., Mamadi, A., Cliff, B., & AL-Ansi, W. (2019). Quorum sensing inhibitor and its inhibition mechanism: Mushroom derived agents. *International Journal of Agriculture Innovations and Research*, *7*(5). https://www.ijair.com/administrator/components/com_jresearch/files/publications/IJAIR_3002_FINAL.pdf

Vaillancourt, M., Aguilar, D., Fernandes, S. E., & Jorth, P. A. (2024). A chronic *Pseudomonas aeruginosa-mouse* lung infection modeling the pathophysiology and inflammation of human cystic fibrosis. bioRxiv (Cold Spring Harbor Laboratory). https://doi.org/10.1101/2024.10.07.617039

Waqas, U., Farhan, A., Haider, A., Qumar, U., & Raza, A. (2023). Advancements in biofilm formation and control in potable water distribution systems: A comprehensive review and analysis of chloramine decay in water systems. *Journal of Environmental Chemical Engineering*, *11*(6), 111377. https://doi.org/10.1016/j.jece.2023.111377

Whiteley, M., Diggle, S. P., & Greenberg, E. P. (2017). Progress in and promise of bacterial quorum sensing research. *Nature*, *551*(7680), 313–320. https://doi.org/10.1038/nature24624

Yarwood, J. M., & Schlievert, P. M. (2003). Quorum sensing in *Staphylococcus* infections. *Journal of Clinical Investigation*, *112*(11), 1620–1625. https://doi.org/10.1172/JCI20442

Zhang, L., Gray, L., Novick, R. P., & Ji, G. (2002). Transmembrane topology of AgrB, the protein involved in the post-translational modification of AgrD in *Staphylococcus aureus*. *Journal of Biological Chemistry*, *277*(37), 34736–34742. https://doi.org/10.1074/jbc.M205209200

Zhang, Q., Yan, Z., Meng, Y., Hong, X., Shao, G., Ma, J., . . . Fu, C. (2021). Antimicrobial peptides: Mechanism of action, activity, and clinical potential. *Military Medical Research*, *8*(1). https://doi.org/10.1186/s40779-021-00343-2

Zimmerli, W., Trampuz, A., & Ochsner, P. E. (2004). Prosthetic-joint infections. The New *England Journal of Medicine*, *351*(16), 1645–1654. https://doi.org/10.1056/NEJMra040181

Zunita, Z., Bashir, A., & Hafizal, A. (2008). Occurrence of multidrug resistant *Staphylococcus aureus* in horses in Malaysia. *Veterinary World*, *1*(6), 165–167.

Section IV

Innovative Therapeutic Strategies and Future Research Directions

12 Advanced Approaches to Identify, Characterize, and Track the Spread of *Staphylococcus aureus*

Moulina Bose[1], *Kondakindi Venkateswar Reddy*[1], *Kasoju Sravya*[2], *Jogipeta Harihara*[1], *Surisetty Navyatha*[1], and *Mohd Arif Uddin*[3]

[1]Centre for Biotechnology, University College of Engineering Science and Technology, Jawaharlal Nehru Technological University Hyderabad, Kukatpally, Telangana, India

[2]MSN R&D Centre, MSN Laboratories Ltd, Pashamylaram, Hyderabad, Telangana, India

[3]Department of Clinical Research and Experimental Medicine, Malla Reddy University, Kompally, Hyderabad, Telangana, India

ABBREVIATIONS

AI	artificial intelligence
AST	antibiotic susceptibility testing
CA-MRSA	community-acquired methicillin-resistant *S. aureus*
CARD	comprehensive antibiotic resistance database
CNS	coagulase-negative staphylococci
ECDC	European Centre for Disease Prevention and Control
HA-MRSA	hospital-acquired methicillin-resistant *S. aureus*
MALDI-TOF MS	matrix-assisted laser desorption/ionization–time-of-flight mass spectrometry
MIC	minimum inhibitory concentration
MLST	multi-locus sequence typing
MRSA	methicillin-resistant *Staphylococcus aureus*
PCR	polymerase chain reaction
S. aureus	*Staphylococcus aureus*
SCCmec	staphylococcal cassette chromosome mec
SIRU	staphylococcal interspersed repeat unit
spa	staphylococcal protein A (gene) typing
VNTR	variable-number tandem-repeat
WGS	whole genome sequencing

12.1 INTRODUCTION

Staphylococcus aureus is an incredibly resilient bacterium that holds a singular duality in human existence: as a normal commensal found on skin and mucous membranes and as a powerful pathogen

DOI: 10.1201/9781003647072-16

that causes a broad range of diseases. In healthy individuals, it rarely causes harm, but under certain conditions, compromised immunity, wounds, or medical interventions, it can shift gears, causing anything from minor skin infections to severe illnesses like sepsis, osteomyelitis, endocarditis, and pneumonia. The clinical significance of the species is compounded by its propensity to rapidly develop antibiotic resistance and forever change its genetic content to evade both therapeutic and immune attacks (Otto et al., 2014; Turner et al., 2019).

The evolution of *S. aureus* as a human pathogen is intertwined with milestones in 20th-century medicine: advances in laboratory diagnostics allowed for faster detection, while successful antibiotics kept deadly outbreaks at bay for decades. But the advent of methicillin-resistant *S. aureus* (MRSA) since the 1960s highlighted how quickly bacteria can outsmart our pharmacologic defenses. Now, MRSA is a known threat: linked to hospital (HA-MRSA) and community-acquired (CA-MRSA) infections, its persistence continues to present ongoing clinical and economic challenges (Turner et al., 2019; Tong et al., 2015).

The toll of *S. aureus* infections is staggering is one of the most frequent causes of healthcare-associated infection, taking advantage of opportunities for transmission by growing on hospital surfaces and enduring drying intervals (Tong et al., 2015). It is the leading cause of surgical-site infection, device-associated illness (e.g., with catheters and prostheses), and ventilator-associated pneumonia. Delayed or incorrect identification usually results in greater morbidity, mortality, and healthcare expense (Tong et al., 2015). Beyond hospitals, the bacterium's adaptability has enabled it to establish itself in the community: outbreaks of CA-MRSA have occurred in athletic teams, military barracks, and correctional facilities worldwide. The rise of such strains demonstrates that antibiotic resistance is not confined to healthcare settings alone and that public health strategies must be equally vigilant in both community and hospital environments (Turner et al., 2019).

The repercussions extend beyond medical care. Healthcare costs due to prolonged hospitalizations, complicated therapies, and intractable outbreaks are substantial. Additionally, there are emotional, social, and logistical consequences for affected families and communities.

Initial identification in the early 20th century relied almost exclusively on traditional microbiological methods: culturing on selective media, observing characteristic colony morphology, and basic biochemical reactions (Tong et al., 2015). While these culture-based methods remain foundational, technological advancements have dramatically enhanced the precision and speed of diagnostics. Clinical laboratories now often employ polymerase chain reaction (PCR), matrix-assisted laser desorption/ionization–time-of-flight (MALDI-TOF) mass spectrometry, and whole genome sequencing (WGS). These modern approaches yield actionable information in hours, a significant improvement over previous investigational timelines (Turner et al., 2019). Rapid, accurate identification is vital: timely diagnosis can be decisive in initiating appropriate therapy, improving clinical outcomes, and limiting transmission. Genotypic methods also enable surveillance by allowing the recognition of hypervirulent or highly resistant clones (Turner et al., 2019). Detection is insufficient; *S. aureus* needs to be fully characterized to direct hospital infection control and treatment. Characterization entails antibiotic susceptibility testing (AST), virulence determinant profiling, and genetic typing of strains (Otto et al., 2013). AST informs clinicians of the most effective treatments and prevents the use of futile therapies. Identification of certain virulence factors gives information regarding the pathogen's clinical potential and probable course of infection (Otto et al., 2013).

Genetic typing, for example, by multi-locus sequence typing (MLST) or WGS, facilitates surveillance, outbreak detection, and epidemiological mapping. It is essential in identifying the source of infection, determining transmission patterns, and restricting pathogenic clone spread (Turner et al., 2019). Successful surveillance supports public health interventions for *S. aureus*. Various challenges remain, including high asymptomatic rates of colonization, considerable survival on inanimate surfaces, and easy progression from colonization to infection (Tong et al., 2015). Surveillance programs can employ a mix of routine screening, molecular genotyping, and data-sharing tools to quickly identify outbreaks and trigger containment (Turner et al., 2019).

Globalization of travel, extensive food distribution, and frequent transfer of patients across hospitals create new channels of transmission, making it more challenging to conduct surveillance. Comparative genomics and phylogenetics have shed an unprecedented level of insight into tracking transmission events, allowing detailed control of outbreaks (Turner et al., 2019).

Scientific innovation keeps abreast of research on *S. aureus*. Metagenomics, CRISPR-Cas-based diagnostics, and machine learning algorithms widen the range of both the pace and depth of pathogen recognition and characterization (Otto et al., 2013). Implementation of these technologies not only elucidates infection dynamics but also provides possibilities for novel therapeutic and preventive interventions.

The success of new technologies ultimately depends on integration into clinical practice. Achieving this requires collaboration among laboratory professionals, clinicians, epidemiologists, and public health practitioners. Education of both healthcare providers and patients is crucial to preventing the spread of resistant strains and ensuring compliance with recommended therapies (Tong et al., 2015).

12.2 IDENTIFICATION OF *S. AUREUS*

Recognizing *S. aureus* swiftly and accurately is critical in both clinical and public health contexts. Diagnostic tools have evolved from classic phenotype-based techniques to include rapid molecular and proteomic technologies, providing increasingly sophisticated options to tackle this adaptable pathogen. Many laboratories begin with time-honored, culture-based approaches. *S. aureus* typically forms golden-yellow, creamy colonies with beta-hemolysis on blood agar and turns Mannitol Salt Agar yellow due to mannitol fermentation (Moraes et al., 2021). Under the microscope, this bacterium appears as gram-positive cocci in grape-like clusters, not chain formations. The catalase test (bubble formation with hydrogen peroxide) distinguishes staphylococci from catalase-negative genera, while the coagulase test is highly specific – only *S. aureus* reliably clots plasma (Boerlin et al., 2003).

Other auxiliary techniques, like the DNase test and latex agglutination for protein A, further increase diagnostic confidence. However, classic methods can require 24–48 hours and may yield ambiguous results in the case of atypical or mixed flora, prompting a need for molecular confirmation (Karmakar et al., 2016).

12.2.1 Molecular Techniques: Sensitivity and Speed Reimagined

Advances in molecular biology have revolutionized *S. aureus* diagnostics. PCR-based assays target unique gene sequences such as nuc (thermostable nuclease), coa (coagulase), and mecA (methicillin resistance) – allowing for species confirmation and resistance profiling within hours (Karimzadeh & Ghassab, 2022). Real-time and multiplex PCR assays can identify multiple genetic targets simultaneously, streamlining comprehensive diagnostics and providing vital information during outbreak investigations (Sanchini, 2022).

Techniques like 16S rRNA gene sequencing and next-generation sequencing offer even greater resolution for strain identification and epidemiological surveillance, especially during complex outbreak scenarios (Tenover et al., 1994).

Recent developments in isothermal amplification – such as the MIRA assay – have shown rapid turnaround and high accuracy, by facilitating nucleic acid detection without a thermocycler (Heng et al., 2022). A comparative overview of the key identification methods, highlighting their principles, advantages, limitations, and supporting references, is provided in Table 12.1. This summary illustrates both the evolution of diagnostic workflows and the factors that may influence method selection in clinical laboratories. Other innovative approaches like SEA and SSEA amplify target DNA at constant temperature, bypassing limitations of classic PCR (Zhang et al., 2023).

TABLE 12.1
Comparative Table of *Staphylococcus aureus* Identification Techniques

Method	Principle	Time to Result	Strengths	Limitations	References
Blood agar culture	Morphology and hemolysis	24–48 h	Inexpensive, widely available	Slow, can be ambiguous	Moraes et al. (2021); Boerlin et al. (2003)
Mannitol salt agar	Mannitol fermentation	18–24 h	Selective for Staphylococci	Occasional false positives	Moraes et al. (2021)
Catalase/coagulase test	Enzymatic activity	Minutes–1 h	Specific for *S. aureus* (coagulase test)	May confuse with CNS species	Boerlin et al. (2003)
DNase/latex agglutination	DNase activity, protein A	<1 h	Improves specificity of ID	Not always definitive	Karmakar et al. (2016)
PCR and isothermal amplifications	Gene amplification	<2 h	Rapid, sensitive and specific	Requires equipment; inhibitors	Heng et al. (2022); Zhang et al. (2023)
MALDI-TOF MS	Protein profiling (mass spec)	<1 h	Fast, high-throughput	High cost, needs database	Sanchini (2022)

Proteomic profiling, particularly with MALDI-TOF mass spectrometry, has dramatically increased the speed and specificity of microbial identification (Seng et al., 2010). This technology generates a unique protein spectrum for each isolate, which is rapidly compared to comprehensive databases, often enabling reliable species or even strain differentiation within minutes. However, pure isolates and high-quality databases remain prerequisites for accuracy.

Emerging diagnostic platforms are pushing the boundaries of accessibility and speed. Isothermal amplification methods such as LAMP and MIRA, along with CRISPR-based diagnostics, are showing great promise for use at the bedside or in resource-limited environments (Cheng et al., 2024). Biosensor-based tools, sometimes utilizing magnetic bead sample processing, bring additional rapidity and sensitivity to detecting *S. aureus* in complex matrices, sometimes achieving results in under an hour (Duan & Wang, 2022). Automated, multiplex systems such as BioFire FilmArray and Accelerate Pheno™ provide quick and comprehensive analysis of positive cultures – identifying *S. aureus* and its resistance determinants in a single workflow (Sanchini, 2022). The integration of traditional, molecular, and proteomic diagnostics allows for tailored, rapid, and accurate identification of *S. aureus*. Advances in isothermal amplification, mass spectrometry, and CRISPR promise more accessible tools for diverse settings, enhancing clinical outcomes and infection control globally (Tenover et al., 1994).

12.3 CHARACTERIZATION OF *S. AUREUS*

Thorough characterization of *S. aureus* isolates provides crucial information for clinical management, epidemiological tracking, and outbreak containment. This process integrates phenotypic and genotypic approaches, with each category offering distinct insights into antibiotic susceptibility, virulence profile, and genetic lineage.

12.3.1 Phenotypic Characterization

Phenotypic characterization focuses on assessing antibiotic susceptibility, virulence attributes, and metabolic capabilities.

12.3.1.1 Antibiotic Susceptibility Testing

AST is essential due to the global rise of multidrug-resistant strains – particularly MRSA. Disk Diffusion (Kirby–Bauer) Antibiotic disks are placed on inoculated agar; zones of inhibition are measured to determine susceptibility. This method is standardized and widely used but does not provide exact MIC values. Broth Microdilution Strains are tested in serially diluted broth containing antibiotics; the MIC is defined as the lowest concentration without visible growth. This method is quantitative and supports critical patient care. E-test (Gradient Diffusion): An antibiotic-impregnated strip with a gradient is applied to the plate; the MIC is read at the intersection of growth and strip. Automated Systems (e.g., VITEK, BD Phoenix): These systems streamline testing and generate rapid results, especially valuable in high-throughput settings (Rahmani et al., 2024; Sanchini, 2022).

Identifying resistance early guides therapy and control measures. For example, when MRSA is detected, clinicians avoid beta-lactams and employ alternative agents, impacting recovery and hospital resource use (Reddy et al., 2017).

12.3.1.2 Virulence Factor Assessment

Key phenotypic assays highlight pathogenic potential. Hemolysin and Coagulase Production Blood agar detects hemolysin activity; slide and tube coagulase tests confirm plasma clotting. Cell culture or immunoassays quantify production of toxins such as PVL and alpha-toxin. Biofilm Formation Microtiter plate staining with crystal violet estimates a strain's ability to form biofilms, relevant for persistent infections. Adhesion and Invasion Assays. Microplate-based and cell culture assays measure how well bacteria stick to extracellular matrix proteins or invade host cells, providing information on the potential for severe disease. Elevated biofilm formation or toxin production may signal higher risks of chronic infection or tissue invasion and influence therapy duration and device management (Rahmani et al., 2024; Lacoma et al., 2021).

12.3.2 Genotypic Characterization

Genotypic approaches offer high-resolution information, especially valuable for strain tracking and understanding resistance evolution. WGS deciphers the entire genome, revealing resistance loci, virulence genes, and evolutionary relationships with unmatched detail. MLST uses sequences from several housekeeping genes to assign isolates to specific sequence types (STs) (Azarian et al., 2021; Yang et al., 2025). To aid comparison and guide method selection, Table 12.2 presents an integrated view of commonly used phenotypic and genotypic characterization techniques, the type of data each provides, their typical use cases, and representative references. By sequencing the variable X region of the *spa* gene, this method quickly classifies strains for epidemiological investigations (Frénay et al., 1996). PFGE uses restriction enzymes to cut DNA into fragments, which are separated to create a unique banding fingerprint (Meemken et al., 2012). SCCmec typing categorizes MRSA based on the genetic structure of the mec element, offering clues to origin (hospital, community, animal) and resistance potential (Abbasian et al., 2018).

Other genotypic methods like agr typing differentiate between accessory gene regulator (agr) groups with distinct virulence profiles. VNTR/SIRU typing variable-number tandem-repeat and staphylococcal interspersed repeat unit methods offer fine-scale discrimination among strains (Sabat et al., 2003). DNA microarrays detect a wide range of resistance and virulence genes and automate clonal complex or ST assignment (Sabat et al., 2006).

12.3.3 Integrating Phenotypic and Genotypic Data

Merging phenotypic and genotypic insights provides a complete picture. Rapid PCR or WGS-based detection of resistance genes supports early intervention, which is confirmed with phenotypic AST before clinical decisions. Combining virulence profiles with sequence data identifies clones responsible for outbreaks or severe disease (Yang et al., 2025). Integrated approaches clarify patterns of nosocomial or community transmission during epidemiological surveillance.

TABLE 12.2
Phenotypic versus Genotypic Methods for Characterization of *S. aureus*

Category	Method	Data Provided	Use Cases	References
Phenotypic	Disk diffusion	Antibiotic susceptibility (zones)	Routine AST	Rahmani et al. (2024); Sanchini (2022)
Phenotypic	Broth microdilution	MIC values (quantitative)	Critical care, research	Rahmani et al. (2024)
Phenotypic	E-test	MIC via gradient	Difficult-to-test drugs	Rahmani et al. (2024)
Phenotypic	Toxin/biofilm assays	Toxin/biofilm phenotype	Virulence estimation	Lacoma et al. (2021)
Genotypic	WGS	Genome, resistance and virulence genes	Outbreaks, research studies	Azarian et al. (2021); Yang et al. (2025)
Genotypic	MLST	Sequence type (ST)	Epidemiology, clonal tracking	Azarian et al. (2021)
Genotypic	spa typing	spa gene repeats (typing)	Quick strain comparisons	Frénay et al. (1996)
Genotypic	SCCmec typing	MRSA genetic element structure	Determining MRSA origin	Abbasian et al. (2018)

12.4 TRACKING THE SPREAD OF *S. AUREUS*

Tracking the epidemiological spread of *S. aureus* – particularly multidrug-resistant clones – has emerged as a priority for infection control, hospital surveillance, and public health policy. Modern epidemiology integrates surveillance networks, molecular typing, real-time data analytics, and global collaboration to understand transmission dynamics, identify sources and outbreaks, and limit the pathogen's dissemination.

12.4.1 Epidemiological Surveillance

Hospital-Associated and Community Surveillance. Continuous surveillance in healthcare settings underpins early detection of outbreaks and guides infection control strategies. Surveillance typically involves routine screening of at-risk patients (e.g., ICU, surgical wards) for nasal, wound, and device-related colonization; monitoring of surgical-site, catheter-associated, and ventilator-associated infections, with culture-based and rapid molecular assays (Donker et al., 2012; Sanchini, 2022); and aggregation and reporting of infection and resistance rates to hospital epidemiologists and national databases (ECDC, 2023). In community settings, tracking focuses on outbreaks among groups with close contact, such as athletes or people in correctional institutions (Uhlemann et al., 2014). Sentinel laboratories and public health authorities coordinate this network of surveillance to detect and investigate clusters quickly.

Outbreak responses include case finding and contact tracing, often using electronic health records and laboratory data. Environmental sampling from suspected sources (e.g., surfaces, healthcare workers' hands, medical devices) (Ben-David et al., 2008). Temporal and spatial mapping of cases to identify common exposures or super-spreader events. Epidemiological data must be rapidly linked to laboratory characterization (genotyping, phenotyping), enabling precise source attribution and assessment of ongoing transmission risk (Bihan et al., 2017).

12.4.2 Molecular Epidemiology and Typing in Surveillance

Molecular methods have revolutionized our understanding of how *S. aureus* spreads within hospitals, communities, and even globally.

WGS is now increasingly used to reconstruct transmission chains. WGS can pinpoint strain relationships with unprecedented accuracy, distinguishing between persistent carriage and recent transmission events. Single-nucleotide polymorphism (SNP) analysis highlights microevolution within outbreaks, identifying "who infected whom" in real time (Bihan et al., 2017; Harris et al., 2012). Global surveillance of clonal lineages, MLST, spa, and SCCmec typing, enables researchers to track movement of high-risk clones (e.g., USA300, ST239) across regions and continents (Uhlemann et al., 2014). Databases such as PubMLST and GenBank support global epidemiology, with real-time sharing of STs and resistance profiles (Sanchini, 2022). In neonatal intensive care units, WGS-based investigations unmasked hidden chains of MRSA transmission, allowing hospitals to target interventions accurately (Bihan et al., 2017). Community outbreaks have been mapped through spa and MLST typing, revealing community-to-hospital spread, or vice versa, of dominant clones (Uhlemann et al., 2014).

12.4.3 Computational Tools and Advanced Analytics

The volume of genomic and surveillance data now requires advanced informatics approaches. Bioinformatics pipelines automatically align, compare, and annotate genome sequences, producing phylogenetic trees and SNP matrices for epidemiological inference (Harris et al., 2012). Integrative surveillance platforms combine laboratory, clinical, and movement data with spatial mapping, enabling hospital infection control teams to visualize and interrupt transmission networks efficiently (ECDC, 2024). Machine learning models are increasingly used to predict outbreaks, resistance trends, and risk factors by analyzing large-scale surveillance data (Sanchini, 2022). Despite remarkable progress, challenges remain.

- **Asymptomatic carriage**. Many carriers spread *S. aureus* silently, complicating detection unless active screening is implemented (Ben-David et al., 2008).
- **Data sharing and privacy**. Integrating sequencing and patient movement data requires secure, standardized systems and careful attention to patient confidentiality.
- **Resource limitations**. Routine WGS and surveillance may not be feasible in low-resource settings, highlighting the need for scalable, affordable tools (Donker et al., 2012).
- **Emerging trends** include on-site rapid sequencing, real-time analytics dashboards for clinicians, and integration of laboratory, surveillance, and antimicrobial stewardship programs.

12.5 ADVANCED TECHNIQUES IN *S. AUREUS* RESEARCH

Biotechnology's acceleration has introduced strong technologies that are revolutionizing the way scientists and clinicians investigate, diagnose, and treat *S. aureus*. New sophisticated strategies are refining our knowledge of bacterial biology, resistance, and infection dynamics while facilitating the creation of accurate diagnostics and targeted treatments.

12.5.1 Metabolomics and Proteomics

Metabolomics of the small-molecule metabolites in bacterial cells charts metabolic pathways and uncovers responses to environmental stress (Ahmadi et al., 2024). Metabolomic research in *S. aureus* is elucidating how nutrient uptake, stress response, and metabolic adaptation determine survival and pathogenicity. For example, metabolic profiling can detect biomarkers for antibiotic resistance and virulence or expose vulnerabilities to new drug targets. Clinical applications involve metabolomics-guided diagnostics, which may allow for the investigation of infection state or antibiotic resistance from patient samples with quick identification through monitoring of pathogen- or resistance-associated metabolic signatures (Ahmadi et al., 2024).

Proteomic methods, mostly mass spectrometry (e.g., LC-MS/MS, MALDI-TOF), enable global protein analysis of *S. aureus*, including those associated with virulence, stress resistance, and drug resistance. Proteomics reveals the ways in which *S. aureus* reorganizes protein production in response to host defenses or antimicrobial therapy, revealing mechanisms such as biofilm formation, toxin expression, and immune evasion. Protein biomarker identification contributes to improving diagnostics and can result in new vaccines or therapies aimed at essential virulence factors. MALDI-TOF MS, now common in clinical laboratories, performs rapid bacterial identification within minutes by comparing protein spectra with a library, even resolving some antibiotic resistance phenotypes (Bizzini & Greub, 2010). CRISPR-Cas systems have revolutionized genome editing in bacteria. In *S. aureus*, CRISPR tools are being applied to knock out or alter genes, revealing functions of antibiotic resistance elements, regulatory circuits (e.g., agr system), and virulence loci (Penewit & Salipante, 2019). These approaches help pinpoint genetic determinants of pathogenicity and uncover new drug targets. CRISPR diagnostics, such as SHERLOCK and DETECTR systems, use guide RNAs to detect specific *S. aureus* DNA or resistance genes with exceptional sensitivity and speed. These platforms can provide results directly from clinical samples, bringing point-of-care precision to settings with limited infrastructure (Kostyusheva et al., 2021). Several studies report CRISPR-based assays that sensitively identify MRSA or toxin genes, sometimes in under an hour, supporting their potential for both clinical and field deployment (Wang et al., 2023).

12.5.2 Bioinformatics, Databases, and Computational Tools

With the expansion of high-throughput sequencing and omics data, bioinformatics is now central to *S. aureus* research and diagnostic strategy.

- **Public Databases**. NCBI GenBank, PubMLST, and the Comprehensive Antibiotic Resistance Database (CARD) provide expansive genomic, proteomic, and resistance information.
- **Software Platforms**. Tools such as MEGA for phylogenetic analysis, BLAST for sequence comparison, Prokka for genome annotation, and Roary for pan-genome studies are widely used (Seemann, 2014; Page et al., 2015).
- **Clinical Microbiology Integration**. Automated computational pipelines interpret WGS data in real time, linking genotype with resistance phenotypes, informing infection control, and supporting surveillance. Artificial intelligence now aids in pattern recognition from large datasets, predicting outbreaks or drug resistance trends based on genomic and clinical metadata (Qu et al., 2019).
- **Single-Cell Sequencing**. Emerging single-cell genomics and transcriptomics are revealing within-host diversity, adaptation during infection, and real-time resistance development (Bakeer et al., 2024). Portable sequencing devices like Oxford Nanopore's MinION enable rapid field sequencing of *S. aureus* for outbreak investigation or low-resource clinical environments (Quick et al., 2016).
- **Metagenomics**. Direct sequencing of clinical or environmental samples uncovers the composition of complex microbial communities, including unculturable strains and detailed resistome profiles (Loman et al., 2012).

Certainly! Here's the next major section, "Case Studies and Future Directions," crafted with integrated references throughout the text (Figure 12.1). I have drawn citations from the full set of references used in previous sections to ensure a well-supported and scholarly narrative with in-text citations in et al. format and DOIs included in the reference list.

12.6 FUTURE DIRECTIONS

The integration of CRISPR-based diagnostics promises ultra-rapid identification and resistance detection even in low-resource settings, as proof-of-concept studies show sub-hour results for MRSA marker genes directly from clinical samples (Kostyusheva et al., 2021).

Portable, real-time sequencing tools, such as Oxford Nanopore's MinION, are enabling rapid outbreak resolution in field settings, including resource-limited hospitals and during global health emergencies (Quick et al., 2016; Loman et al., 2012).

The deluge of genomic and clinical data requires advanced computational methods; machine learning models are now being trained to predict outbreaks, resistance trends, and even patient outcomes from integrated clinical and laboratory data (Harris et al., 2012).

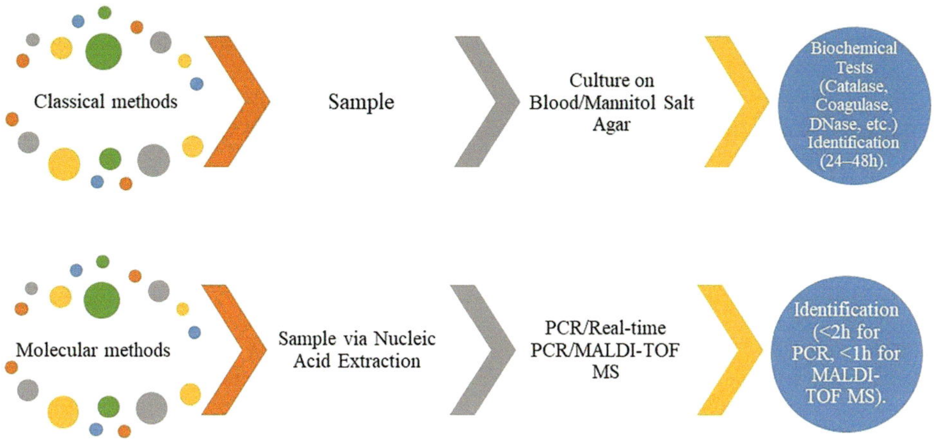

FIGURE 12.1 Comparison of classical microbiological and advanced molecular workflows for identifying *Staphylococcus aureus*.

Public repositories like GenBank, PubMLST, and the CARD database are being leveraged to cross-reference local findings with global trends, supporting the detection of new high-risk clones and the emergence of novel resistance mechanisms (Jolley & Maiden, 2010; Sanchini, 2022).

A major challenge is the harmonization of data from hospitals, regional health networks, and international consortia (European Centre for Disease Prevention and Control, 2024). Ongoing initiatives focus on expanding access to WGS, facilitating real-time data sharing, and improving infection prevention resources in low-to-middle-income countries.

Increasing omics data are informing the development of novel therapeutics, such as monoclonal antibodies against toxin components and vaccines targeting conserved surface proteins. Single-cell genomics is shedding light on within-host adaptation, guiding precision interventions (Sanchini, 2022).

Asymptomatic carriage and environmental reservoirs remain barriers to full. Data privacy and infrastructure needs are limiting the adoption of real-time analytics in some regions (Donker et al., 2012; European Centre for Disease Prevention and Control, 2024). Resource disparities continue to challenge global equity in diagnostics and outbreak response.

REFERENCES

Abbasian, S., Farahani, N., Mir, Z., Alinejad, F., Haeili, M., Dahmardehei, M., Mirzaii, M., Khoramrooz, S., Nasiri, M., & Darban-Sarokhalil, D. (2018). Genotypic characterization of *Staphylococcus aureus* isolated from a burn centre by using agr, spa and SCCmec typing methods. *New Microbes and New Infections*, *26*, 15–19. https://doi.org/10.1016/j.nmni.2018.08.001

Ahmadi, S., Sedaghat, F. R., Memar, M. Y., & Yekani, M. (2024). Metabolomics in the diagnosis of bacterial infections. *Clinica Chimica Acta*, *565*, 120020. https://doi.org/10.1016/j.cca.2024.120020

Azarian, T., Cella, E., Baines, S. L., Shumaker, M. J., Samel, C., Jubair, M., Pegues, D. A., & David, M. Z. (2021). Genomic epidemiology and global population structure of exfoliative toxin A-producing *Staphylococcus aureus* strains associated with staphylococcal scalded skin syndrome. *Frontiers in Microbiology*, *12*. https://doi.org/10.3389/fmicb.2021.663831

Bakeer, W., Gaafar, M., El-Gendy, A. O., Badry, M. A. E., Alblwi, N. A. N., Selim, H. M. R. M., Alharbi, N. K., & Bendary, M. M. (2024). Exploring phenotypic and genotypic diversity among methicillin-resistant, vancomycin-resistant, and sensitive *Staphylococcus aureus*. *Medicine*, *103*(52), e41051. https://doi.org/10.1097/md.0000000000041051

Ben-David, D., Mermel, L. A., & Parenteau, S. (2008). Methicillin-resistant *Staphylococcus aureus* transmission: The possible importance of unrecognized health care worker carriage. *American Journal of Infection Control*, *36*(2), 93–97. https://doi.org/10.1016/j.ajic.2007.05.013

Bihan, C. L., Zahar, J., & Timsit, J. (2017). *Staphylococcus aureus* transmission in the intensive care unit: The potential role of the healthcare worker carriage. *Annals of Infection*, *1*, 1. https://doi.org/10.21037/aoi.2017.08.01

Bizzini, A., & Greub, G. (2010). Matrix-assisted laser desorption ionization time-of-flight mass spectrometry, a revolution in clinical microbial identification. *Clinical Microbiology and Infection*, *16*(11), 1614–1619. https://doi.org/10.1111/j.1469-0691.2010.03311.x

Boerlin, P., Kuhnert, P., HüSsy, D., & Schaellibaum, M. (2003). Methods for identification of *Staphylococcus aureus* Isolates in cases of Bovine mastitis. *Journal of Clinical Microbiology*, *41*(2), 767–771. https://doi.org/10.1128/jcm.41.2.767-771.2003

Cheng, S., Tu, Z., Zheng, S., Khan, A., Yang, P., Shen, H., & Gu, B. (2024). Development of a Magnetically-Assisted SERS biosensor for rapid bacterial detection. *International Journal of Nanomedicine*, *19*, 389–401. https://doi.org/10.2147/ijn.s433316

Duan, R., & Wang, P. (2022). Rapid and simple approaches for diagnosis of *Staphylococcus aureus* in bloodstream infections. *Polish Journal of Microbiology*, *71*(4), 481–489. https://doi.org/10.33073/pjm-2022-050

Donker, T., Wallinga, J., Slack, R., & Grundmann, H. (2012). Hospital networks and the dispersal of hospital-acquired pathogens by patient transfer. *PLoS ONE*, *7*(4), e35002. https://doi.org/10.1371/journal.pone.0035002

European Centre for Disease Prevention and Control. (2024). Antimicrobial resistance in the EU/EEA – AER 2019 surveillance report. In *Antimicrobial resistance in the EU/EEA (EARS-Net) – Annual epidemiological report 2023*. ECDC. https://www.ecdc.europa.eu/sites/default/files/documents/antimicrobial-resistance-annual-epidemiological-report-EARS-Net-2023.pdf

Frénay, H. M. E., Bunschoten, A. E., Schouls, L. M., Leeuwen, W. J., Vandenbroucke-Grauls, C. M. J. E., Verhoef, J., & Mooi, F. R. (1996). Molecular typing of methicillin-resistant *Staphylococcus aureus* on the basis of protein A gene polymorphism. *European Journal of Clinical Microbiology & Infectious Diseases*, *15*(1), 60–64. https://doi.org/10.1007/bf01586186

Harris, S. R., Cartwright, E. J., Török, M. E., Holden, M. T., Brown, N. M., Ogilvy-Stuart, A. L., Ellington, M. J., Quail, M. A., Bentley, S. D., Parkhill, J., & Peacock, S. J. (2012). Whole-genome sequencing for analysis of an outbreak of meticillin-resistant *Staphylococcus aureus*: A descriptive study. *The Lancet Infectious Diseases*, *13*(2), 130–136. https://doi.org/10.1016/s1473-3099(12)70268-2

Heng, P., Liu, J., Song, Z., Wu, C., Yu, X., & He, Y. (2022). Rapid detection of *Staphylococcus aureus* using a novel multienzyme isothermal rapid amplification technique. *Frontiers in Microbiology*, *13*. https://doi.org/10.3389/fmicb.2022.1027785

Jolley, K. A., & Maiden, M. C. (2010). BIGSdb: Scalable analysis of bacterial genome variation at the population level. *BMC Bioinformatics*, *11*(1). https://doi.org/10.1186/1471-2105-11-595

Karimzadeh, R., & Ghassab, R. K. (2022). Identification of nuc nuclease and sea enterotoxin genes in *Staphylococcus aureus* isolates from nasal mucosa of burn hospital staff: A cross-sectional study. *New Microbes and New Infections*, *47*, 100992. https://doi.org/10.1016/j.nmni.2022.100992

Karmakar, A., Dua, P., & Ghosh, C. (2016). Biochemical and molecular analysis of *Staphylococcus aureus* clinical isolates from hospitalized patients. *Canadian Journal of Infectious Diseases and Medical Microbiology*, *2016*, 1–7. https://doi.org/10.1155/2016/9041636

Kostyusheva, A., Brezgin, S., Babin, Y., Vasilyeva, I., Glebe, D., Kostyushev, D., & Chulanov, V. (2021). CRISPR-Cas systems for diagnosing infectious diseases. *Methods*, *203*, 431–446. https://doi.org/10.1016/j.ymeth.2021.04.007

Lacoma, A., Laabei, M., Sánchez-Herrero, J. F., Young, B., Godoy-Tena, G., Gomes-Fernandes, M., Sumoy, L., Plans, O., Arméstar, F., & Prat, C. (2021). Genotypic and phenotypic characterization of *Staphylococcus aureus* isolates from the respiratory tract in mechanically-ventilated patients. *Toxins*, *13*(2), 122. https://doi.org/10.3390/toxins13020122

Loman, N. J., Constantinidou, C., Chan, J. Z. M., Halachev, M., Sergeant, M., Penn, C. W., Robinson, E. R., & Pallen, M. J. (2012). High-throughput bacterial genome sequencing: An embarrassment of choice, a world of opportunity. *Nature Reviews Microbiology*, *10*(9), 599–606. https://doi.org/10.1038/nrmicro2850

Meemken, D., Blaha, T., Hotzel, H., Strommenger, B., Klein, G., Ehricht, R., Monecke, S., & Kehrenberg, C. (2012). Genotypic and phenotypic characterization of *Staphylococcus aureus* isolates from wild boars. *Applied and Environmental Microbiology*, *79*(5), 1739–1742. https://doi.org/10.1128/aem.03189-12

Moraes, G. F. Q., Cordeiro, L. V., & De Andrade Júnior, F. P. (2021). Main laboratory methods used for the isolation and identification of *Staphylococcus* spp. *Revista Colombiana De Ciencias Químico Farmacéuticas*, *50*(1). https://doi.org/10.15446/rcciquifa.v50n1.95444

Otto, M. (2013). *Staphylococcus aureus* toxins. *Current Opinion in Microbiology*, *17*, 32–37. https://doi.org/10.1016/j.mib.2013.11.004

Page, A. J., Cummins, C. A., Hunt, M., Wong, V. K., Reuter, S., Holden, M. T., Fookes, M., Falush, D., Keane, J. A., & Parkhill, J. (2015). Roary: Rapid large-scale prokaryote pan genome analysis. *Bioinformatics, 31*(22), 3691–3693. https://doi.org/10.1093/bioinformatics/btv421

Penewit, K., & Salipante, S. J. (2019). Genome editing in *Staphylococcus aureus* by conditional recombineering and CRISPR/Cas9-mediated counterselection. *Methods in Molecular Biology*, 127–143. https://doi.org/10.1007/978-1-4939-9740-4_14

Qu, K., Guo, F., Liu, X., Lin, Y., & Zou, Q. (2019). Application of machine learning in Microbiology. *Frontiers in Microbiology, 10*. https://doi.org/10.3389/fmicb.2019.00827

Quick, J., Loman, N. J., Duraffour, S., Simpson, J. T., Severi, E., Cowley, L., Bore, J. A., Koundouno, R., Dudas, G., Mikhail, A., Ouédraogo, N., Afrough, B., Bah, A., Baum, J. H. J., Becker-Ziaja, B., Boettcher, J. P., Cabeza-Cabrerizo, M., Camino-Sánchez, Á., Carter, L. L., . . . Carroll, M. W. (2016). Real-time, portable genome sequencing for Ebola surveillance. *Nature, 530*(7589), 228–232. https://doi.org/10.1038/nature16996

Rahmani, Z., Hosseini, S. S., Bagheri, P., Dadashi, M., Haghighi, M., & Goudarzi, M. (2024). Phenotypic and genotypic characterization of *Staphylococcus aureus* strains isolated from otitis externa: Emergence of CC30-spa t019-SCCmec IV carrying PVL as major genotype. *Heliyon, 10*(11), e32002. https://doi.org/10.1016/j.heliyon.2024.e32002

Reddy, P. N., Srirama, K., & Dirisala, V. R. (2017). An update on clinical burden, diagnostic tools, and therapeutic options of *Staphylococcus aureus*. *Infectious Diseases Research and Treatment, 10*. https://doi.org/10.1177/1179916117703999

Sabat, A., Krzyszton-Russjan, J., Strzalka, W., Filipek, R., Kosowska, K., Hryniewicz, W., Travis, J., & Potempa, J. (2003). New method for typing *Staphylococcus aureus* strains: Multiple-locus variable-number tandem repeat analysis of polymorphism and genetic relationships of clinical isolates. *Journal of Clinical Microbiology, 41*(4), 1801–1804. https://doi.org/10.1128/jcm.41.4.1801-1804.2003

Sabat, A., Malachowa, N., Miedzobrodzki, J., & Hryniewicz, W. (2006). Comparison of PCR-based methods for typing *Staphylococcus aureus* isolates. *Journal of Clinical Microbiology, 44*(10), 3804–3807. https://doi.org/10.1128/jcm.00395-06

Sanchini, A. (2022). Recent developments in phenotypic and molecular diagnostic methods for antimicrobial resistance detection in *Staphylococcus aureus*: A narrative review. *Diagnostics, 12*(1), 208. https://doi.org/10.3390/diagnostics12010208

Seemann, T. (2014). Prokka: Rapid prokaryotic genome annotation. *Bioinformatics, 30*(14), 2068–2069. https://doi.org/10.1093/bioinformatics/btu153

Seng, P., Rolain, J., Fournier, P. E., La Scola, B., Drancourt, M., & Raoult, D. (2010). MALDI-TOF-mass spectrometry applications in clinical microbiology. *Future Microbiology, 5*(11), 1733–1754. https://doi.org/10.2217/fmb.10.127

Tenover, F. C., Arbeit, R., Archer, G., Biddle, J., Byrne, S., Goering, R., Hancock, G., Hébert, G. A., Hill, B., & Hollis, R. (1994). Comparison of traditional and molecular methods of typing isolates of *Staphylococcus aureus*. *Journal of Clinical Microbiology, 32*(2), 407–415. https://doi.org/10.1128/jcm.32.2.407-415.1994

Uhlemann, A., Dordel, J., Knox, J. R., Raven, K. E., Parkhill, J., Holden, M. T. G., Peacock, S. J., & Lowy, F. D. (2014). Molecular tracing of the emergence, diversification, and transmission of *S. aureus* sequence type 8 in a New York community. *Proceedings of the National Academy of Sciences, 111*(18), 6738–6743. https://doi.org/10.1073/pnas.1401006111

Wang, Y., Yang, T., Liu, G., Xie, L., Guo, J., & Xiong, W. (2023). Application of CRISPR/Cas12a in the rapid detection of pathogens. *Clinica Chimica Acta, 548*, 117520. https://doi.org/10.1016/j.cca.2023.117520

Yang, C., Chen, X., Li, M., Yuan, W., Li, S., Han, D., Feng, J., Luo, H., Zheng, M., Liang, J., Chen, C., Qu, P., & Li, S. (2025). Genomic epidemiology and phenotypic characterization of *Staphylococcus aureus* isolated from atopic dermatitis patients in South China. *Scientific Reports, 15*(1). https://doi.org/10.1038/s41598-025-87317-9

Turner, N. A., Sharma-Kuinkel, B. K., Maskarinec, S. A., Eichenberger, E. M., Shah, P. P., Carugati, M., Holland, T. L., & Fowler, V. G. (2019). Methicillin-resistant *Staphylococcus aureus*: An overview of basic and clinical research. *Nature Reviews Microbiology, 17*(4), 203–218. https://doi.org/10.1038/s41579-018-0147-4

Tong, S. Y. C., Davis, J. S., Eichenberger, E., Holland, T. L., & Fowler, V. G. (2015). *Staphylococcus aureus* infections: Epidemiology, pathophysiology, clinical manifestations, and management. *Clinical Microbiology Reviews, 28*(3), 603–661. https://doi.org/10.1128/cmr.00134-14.

Zhang, J., Han, X., Wang, Y., Mu, X., Shi, C., Li, Y., & Ma, C. (2023). Rapid and sensitive detection of *Staphylococcus aureus* via an all-in-one staggered strand exchange amplification platform. *The Analyst, 148*(9), 1970–1977. https://doi.org/10.1039/d2an02092k

13 AI Applications in Detecting, Treating, and Researching *Staphylococcus aureus*

Prafulla Kumar[1], *Abhay Kumar*[2], *Ankit Agrawal*[3], *Ravindra Kumar*[4], *Neelesh Kapoor*[4], *and L.K. Gangwar*[5]

[1]Department of Biotechnology, UIEMT, Guru Nanak University, Hyderabad, India

[2]PG Department of Botany, M.S. College Motihari, BRABU University, Muzzafarpur, Bihar, India

[3]School of Biotechnology, IFTM University, Moradabad, India

[4]Division of Plant Biotechnology, College of Biotechnology, Sardar Vallabhbhai Patel University of Agriculture and Technology, Meerut, Uttar Pradesh, India

[5]Department of Genetics and Plant Breeding, Sardar Vallabhbhai Patel University of Agriculture and Technology, Meerut, Uttar Pradesh, India

ABBREVIATIONS

AFB	acid-fast bacilli
AI	artificial intelligence
AMR	antimicrobial resistance
APAS	automated plate assessment system
AST	antimicrobial susceptibility testing
CRKP	carbapenem-resistant *Klebsiella pneumoniae*
CT	computed tomography
DBN	dynamic Bayesian networks
DL	deep learning
DNA	deoxyribonucleic acid
DNN	deep neural network
EHR	electronic health record
GIS	geographic information system
IDD	infectious disease dynamics
K-mer	short DNA sequence of length *k* used in bioinformatics analysis
MALDI-TOF MS	matrix-assisted laser desorption/ionization–time-of-flight mass spectrometry
ML	machine learning
MRSA	methicillin-resistant *Staphylococcus aureus*
MS	mass spectrometry
MSSA	methicillin-sensitive *Staphylococcus aureus*
NLP	natural language processing
PaPrBaG	pathogen profile bagging (machine learning-based pathogen classification tool)
PCR	polymerase chain reaction
PHISTO	pathogen–host interaction search tool

DOI: 10.1201/9781003647072-17

RNA	ribonucleic acid
RV	reverse vaccinology
T-SPOT	T cell ELISPOT assay for detecting TB infection
UTI	urinary tract infection
VRE	vancomycin-resistant **Enterococcus**
WASPLab	walk-away specimen processor laboratory

13.1 INTRODUCTION

Bacterial infections continue to pose a significant threat to global public health, accounting for millions of deaths each year, particularly in low- and middle-income countries. According to a 2022 comprehensive study published in *The Lancet*, bacterial infections were responsible for 7.7 million deaths globally in 2019, making them the second leading cause of death after ischemic heart disease (Murray et al., 2022). The burden is further exacerbated by the increasing prevalence of antimicrobial resistance (AMR), which is projected to cause up to 10 million deaths annually by 2050 if left unaddressed (O'Neill, 2016). Early and accurate diagnosis, along with rapid antimicrobial susceptibility testing (AST), is critical for initiating targeted therapy, improving patient outcomes, and curbing the spread of resistant strains. However, conventional diagnostic approaches – including culture-based methods and phenotypic AST – are time-consuming, often taking 24–72 hours or more, which can lead to delayed or inappropriate treatment (Ventola, 2015). Molecular diagnostics, while faster, often remain inaccessible in resource-limited settings due to high costs and technical complexity. Recent advances in artificial intelligence (AI) offer transformative potential in addressing these challenges. AI, through its subfields such as machine learning (ML), deep learning (DL), natural language processing (NLP), and computer vision, simulates human cognitive functions and can analyze large-scale biomedical data with high accuracy and speed (Esteva et al., 2019). In the context of bacterial infections, AI applications range from automated interpretation of diagnostic imaging and electronic health records to real-time prediction of pathogen identity, drug resistance patterns, and clinical outcomes (Topol, 2019). Moreover, AI-powered platforms have shown promise in streamlining epidemiological surveillance, facilitating contact tracing, and forecasting disease outbreaks, which are particularly relevant for controlling nosocomial and community-acquired infections (Ching et al., 2018). In drug discovery, AI accelerates the identification of novel antimicrobial agents by mining vast chemical libraries and predicting bioactivity profiles (Stokes et al., 2020). AI has also been employed in vaccine design by analyzing pathogen genomes and host–pathogen interactions to predict immunogenic targets (Nguyen et al., 2021). Despite these advancements, several challenges hinder the widespread adoption of AI in clinical microbiology. These include issues related to data privacy, interoperability, algorithm transparency, and the need for large, curated datasets for training AI models. Furthermore, clinical validation, regulatory approval, and integration into existing healthcare systems remain critical hurdles. This study aims to explore the multifaceted applications of AI in diagnosing and managing bacterial infections, focusing on how AI enhances diagnostic precision, facilitates personalized treatment, and contributes to global efforts against AMR. By examining current developments, challenges, and future directions, this chapter seeks to support healthcare professionals and researchers in harnessing AI for improved patient care and public health outcomes.

13.2 APPLICATION OF AI IN EPIDEMIOLOGICAL SURVEILLANCE OF BACTERIAL INFECTIOUS DISEASES

AI and big data are transforming infectious disease epidemiology, improving research and public health emergency management. Models like infectious disease dynamics and dynamic Bayesian networks enhance disease forecasting and outbreak analysis (Gao & Wang, 2022). AI, via cloud computing, enables real-time data processing for monitoring outbreaks, despite long model training

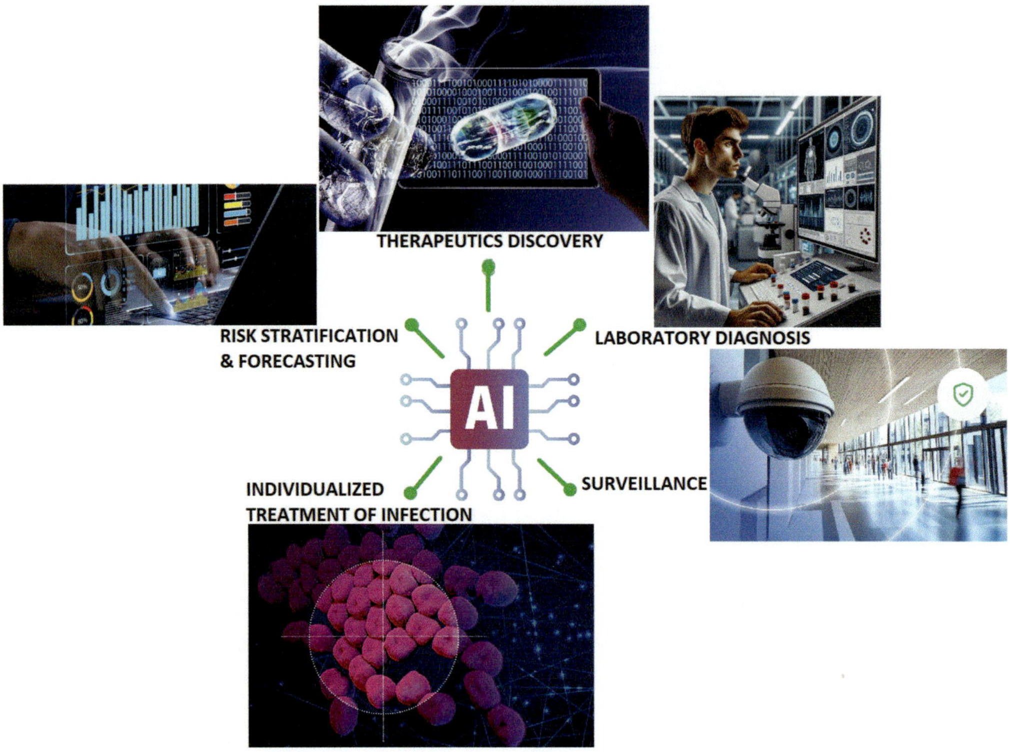

FIGURE 13.1 Application of AI against bacterial diseases.

times (Li et al., 2023). Geographic information systems (GISs) and tools like ToxPi*GIS Toolkit optimize public health data integration (Wells et al., 2021; Fleming et al., 2022). Additionally, cloud storage and Internet search data, such as Google Flu Trends, boost real-time disease surveillance (Pfeiffer & Stevens, 2015). Advanced tools like machine learning and real-time locator systems, currently used in viral epidemiology, are expanding in bacterial epidemiology. For instance, machine learning can predict *Clostridioides difficile* infection risks in hospitals, enabling proactive preventive measures (Oh et al., 2018). Real-time locator systems improve contact tracing in emergency departments, enhancing efficiency and resource use (Hellmich et al., 2017). Studies have also shown AI's value in controlling hospital-acquired infections, such as methicillin-resistant *Staphylococcus aureus* (MRSA), through simulations and cross-hospital transmission models (Lesosky et al., 2011; Ciccolini et al., 2014). AI is creating new opportunities to predict and prevent bacterial infections. By integrating and analyzing large volumes of complex data, AI enables early detection and precise forecasting of bacterial outbreaks. This enhances prevention and control efforts, informs public health decisions, and contributes to the global fight against infectious diseases with innovative solutions (Figure 13.1).

13.3 REVOLUTION OF AI TO STUDY THE MECHANISM OF BACTERIAL INFECTION

Studying the pathogenesis of bacterial infections is key to understanding their nature, involving both pathogen–host interactions and the host's immune response. Animal models have traditionally been crucial for observing these processes, but they are costly, time-consuming, and raise ethical concerns. With AI advancements, machine learning models now allow researchers to simulate pathogen–host interactions without animal experiments. Tools like PHISTO and techniques such as 3D microscopy and fluorescence microscopy enable deeper insights into infection mechanisms and microbial behavior, aiding drug and vaccine development. Machine learning models offer a flexible approach to predict and simulate the complex mechanisms of pathogen–host interactions, speeding up research and lowering costs. While AI models cannot fully replace all animal studies, they open up new possibilities for exploring previously unexplored areas.

13.4 APPLICATIONS OF AI IN THE DIAGNOSIS OF BACTERIAL INFECTIONS

In traditional bacterial infection diagnosis, laboratory technicians use microbiological and biochemical tests to identify pathogens, including bacterial cultures, morphological observations, biochemical reactions, and serological techniques (Váradi et al., 2017). Molecular biology methods, such as polymerase chain reaction (PCR), are also widely used for bacterial DNA identification (Deusenbery et al., 2021). While PCR is more advanced than traditional methods, it is time-consuming. The integration of AI technology not only optimizes traditional detection and management processes but also has the potential to revolutionize them completely (Ho et al., 2019; Wang et al., 2020; Paquin et al., 2022; Howard et al., 2024)

13.5 AI IMPROVES THE EFFICIENCY AND ACCURACY OF IDENTIFICATION OF *S. AUREUS*

AI technology enables rapid and accurate bacterial infection diagnosis. For example, MALDI-TOF MS with ClinProTools software achieved 100% accuracy in identifying *S. aureus* subspecies through genetic analysis (Pérez-Sancho et al., 2018). Findaureus, an open-source Python app, automates bacterial detection in tissue sections, improving efficiency over manual methods (Mandal et al., 2024). The PhenoMatrix Colorimetric Detection Module uses digital imaging to detect Group B *Streptococcus* with high sensitivity, increasing productivity and reducing human error. DNA microarray technology with the DendrisChips machine learning algorithm identifies 11 bacteria linked to respiratory infections in under 4 hours with over 95% accuracy (Senescau et al., 2018). A neural network-based sensor can detect 16 bacterial species and their Gram-staining properties with >90% accuracy, requiring much less dye and sample compared to traditional sensors, and is stable for up to 6 months (Laliwala et al., 2022). This method offers an innovative, efficient diagnostic tool, ideal for resource-limited settings. Traditional microscopy methods for diagnosing diseases like tuberculosis are slow and have limited sensitivity. AI advancements, such as Metasystems' automated AFB smear scanning and deep learning-based analysis (Neon Metafer), enhance speed and accuracy in recognizing smear-negative slides (Desruisseaux et al., 2024). A deep neural network combined with automated scanning reduces analysis time to about 10 seconds per slide (Horvath et al., 2020). Additionally, a system combining T-SPOT with DL-based CT analysis improves classification accuracy for mycobacterial lung diseases (Ying et al., 2022). AI tools, like neural networks, improve pathogen detection speed, accuracy, and reliability, reducing human error in diagnostics.

13.6 AI OPTIMIZES ANTIMICROBIAL SUSCEPTIBILITY TESTING AGAINST *S. AUREUS*

In clinical labs, pathogen identification and AST often rely on culturing, with standard methods like disk diffusion and microbroth dilution taking 2–3 days or more (Abu-Aqil et al., 2023). To prevent infection spread, clinicians often use broad-spectrum antibiotics empirically, but this can lead to drug-resistant strains. There is an urgent need for faster and more accurate AST technologies to guide diagnosis and treatment. AI is enhancing bacterial AST with efficient, rapid methods (Figure 13.2). For example, Raman spectroscopy with image stitching enables single-cell detection

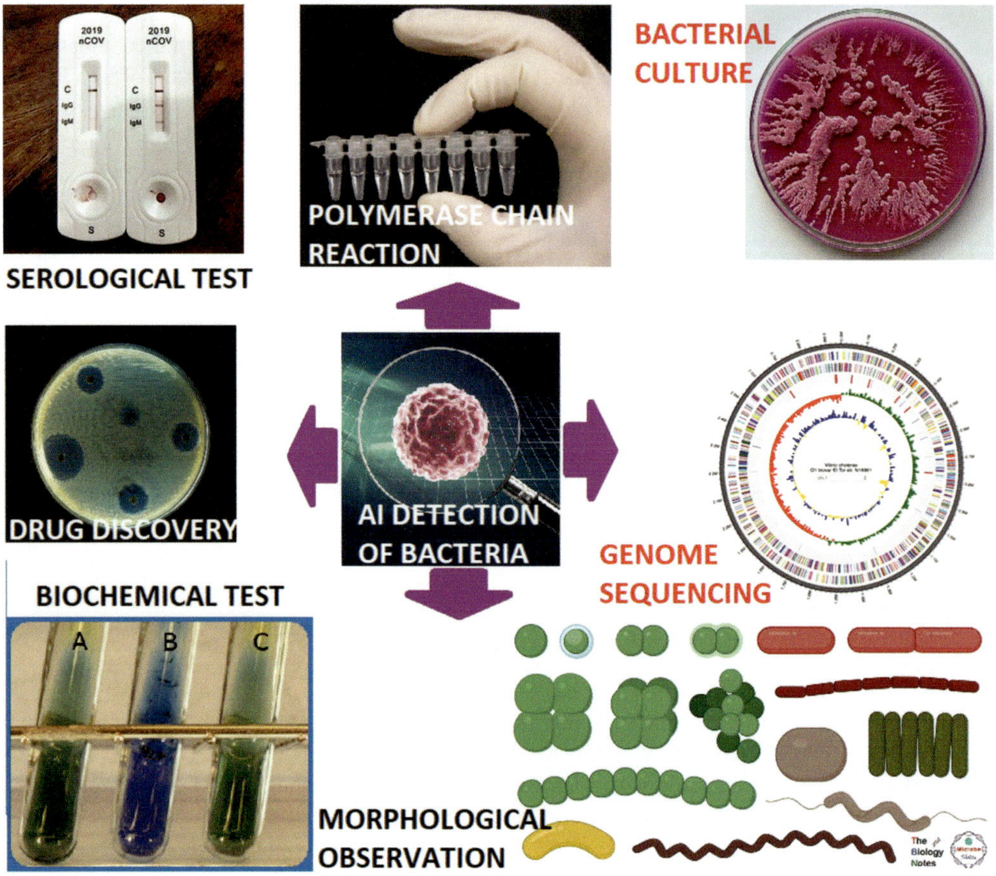

FIGURE 13.2 Different AI-based technologies used to diagnose disease caused by *Staphylococcus* species.

of drug-resistant bacteria (Nakar et al., 2022). Combining machine learning and infrared spectroscopy speeds up urinary tract infection diagnosis and drug resistance identification, reducing time from 48 hours to 40 minutes. The SlipChip microfluidic device extracts and enriches bacteria from blood cultures, providing AST results in 3–8 hours for faster, targeted treatment (Yi et al., 2019). Automation technologies, such as MALDI-TOF MS, are highly efficient in detecting drug-resistant bacteria like MRSA and CRKP (Zhang et al., 2023). The ML-based MALDI-TOF MS method enables rapid identification of MRSA and CRKP from blood cultures in under 1 hour (Yu et al., 2023). Computer analysis of MALDI-TOF MS data offers insights into resistant versus sensitive isolates (Wang et al., 2021). The WASPLab system speeds up VRE recognition (Cherkaoui et al., 2019), and the APAS Independence system improves laboratory productivity by automatically classifying MRSA and MSSA cultures (Gammel et al., 2021).

13.7 IMPROVEMENT OF BACTERIAL GENOME SEQUENCING THROUGH AI

Genome sequencing technologies, including whole genome and next-generation sequencing, have accelerated the identification of infectious agents, tracking transmission in healthcare settings, and analyzing microbial communities' impact on human health (d'Humières et al., 2021). They also offer a powerful tool for monitoring and addressing global AMR (Waddington et al., 2022; Sherry et al., 2023). Current genetic testing mainly relies on sequence similarity but struggles with identifying new species without closely related genomes in reference databases. The machine learning-based PaPrBaG method offers a reliable prediction approach, even with low genome coverage (Deneke et al., 2017). Additionally, combining machine learning with metagenomic sequencing enhances diagnostic accuracy for hard-to-diagnose diseases like tuberculous meningitis (Ramachandran et al., 2022). As sequencing costs decrease and throughput increases, rapidly interpreting high-dimensional genomic data becomes a challenge. Machine learning helps process large genomic datasets, such as predicting health impacts of Shiga toxin-producing *E. coli* infections (Njage et al., 2019). Additionally, Bayesian neural networks use nonparametric algorithms to accelerate genetic association studies and accurately identify variant strains of infection. Combining machine learning with genomics improves pathogen-resistance prediction, outperforming existing methods. For example, machine learning has been used to create a knowledge map of AMR in *E. coli*, revealing unknown resistance genes (Youn et al., 2022). Using XGBoost and convolutional neural networks, researchers accurately predicted *Klebsiella pneumoniae* resistance and identified high-resistance or virulent strains (Liu et al., 2021). Additionally, the Treesist-TB decision tree method surpasses TB-Profiler in identifying mutant strains and predicting drug resistance (Deelder et al., 2022).

13.8 APPLICATION OF AI IN THE TREATMENT OF *STAPHYLOCOCCUS* INFECTIONS

AMR is a major challenge in treating bacterial infections. Declared a global concern at the 2016 UN General Assembly, AMR led to nearly 5 million deaths in 2019 (Antimicrobial Resistance Collaborators, 2022). Projections suggest 10 million deaths annually by 2050 (Walsh et al., 2023). Bacteria evolve resistance through natural selection, complicating treatment and vaccine development. This highlights the urgent need for new antimicrobial strategies and therapies (Stracy et al., 2022). In this context, AI technology effectively simulates the complex interactions between pathogens, hosts, and drugs, uncovering microbial infection patterns and enhancing drug and vaccine development. Additionally, AI's use in phage therapy offers new hope in the battle against bacterial resistance.

13.9 REVOLUTION OF AI FOR DRUG DISCOVERY AND DEVELOPMENT

AI is revolutionizing drug research by overcoming drug resistance. For example, combining high-throughput biophysical analysis with machine learning has helped identify bioactive targets of antimicrobial drugs (Santa Maria et al., 2017). Additionally, combining fragment-based drug design with structure–activity modeling highlights the potential of artificial neural networks in drug discovery (Kleandrova and Speck-Planche, 2020). Data-driven techniques, such as machine learning and molecular pair analysis, have identified key chemical features that influence bacterial activity, expanding the chemical space for broad-spectrum antimicrobials (Gurvic et al., 2022). A support vector machine approach analyzed genomics and other data from *Pseudomonas aeruginosa*, uncovering a molecular mechanism distinguishing pathogenic from non-pathogenic strains, offering valuable targets for new therapeutics. AI has made significant strides in anti-tuberculosis drug development. Machine learning and neural networks have identified inhibitors targeting LeuRS and MetRS in *Mycobacterium tuberculosis* (Volynets et al., 2022) and small-molecule inhibitors for topoisomerase I (Ekins et al., 2017), offering new strategies to combat drug resistance. Additionally, combining public Mtb data with machine learning boosts drug discovery efficiency and provides valuable resources for future research (Lane et al., 2022). These advanced technologies not only speed up the drug research and development process but also increase the likelihood of discovering new therapeutic options. They are transforming how researchers understand and work with complex biological systems, ushering in a new era of smarter and more precise pharmaceutical development.

13.10 AI BRINGS BREAKTHROUGHS IN VACCINE DEVELOPMENT FOR *STAPHYLOCOCCUS* INFECTIONS

Significant progress has been made in vaccine development for viral diseases, especially in response to emerging epidemics like COVID-19, where computer-aided design played a key role (Abbasi et al., 2022). However, developing vaccines for bacterial diseases is more challenging due to bacterial variability, evolving drug resistance, and complex host interactions. To overcome these obstacles, leveraging AI, computer-aided design, and advanced immunological techniques is crucial for accelerating vaccine development. Vaccine design faces the challenge of identifying key antigens that trigger lasting immune memory and broad immune responses. Reverse vaccinology (RV) has been widely used, with Bexsero for *Neisseria meningitidis* B being a successful example. Computational tools like deep learning, RV, and immunoinformatics support antigen identification, accelerating vaccine development and reducing reliance on traditional methods. These approaches show promise in addressing drug-resistant bacteria, with machine learning models offering higher precision in predicting *M. tuberculosis* aspects (Khanna and Rana, 2019). Machine learning technology enhances the vaccine development process by improving efficiency, reducing dependence on traditional experiments and animal testing, and providing robust scientific and technological support to address the evolving challenges of bacterial infections.

13.11 AI DRIVES INNOVATIVE APPLICATIONS OF PHAGE THERAPY FOR *S. AUREUS*

Phage therapy shows promise in fighting drug-resistant bacterial infections (Kulshrestha et al., 2024), but predicting phage-pathogen-host interactions is challenging (Cisek et al., 2017). AI models, like machine learning-based K-mer strategies, are helping to overcome this issue by accurately predicting these interactions (Qiu et al., 2024). Machine learning aids in designing clinical phage therapy for multidrug-resistant *E. coli* urinary tract infections (Keith et al., 2024). Additionally, HostPhinder accurately predicts phage-host genus and species with 81% and 74% accuracy, respectively, highlighting its potential for identifying therapeutic targets. The use of phages, either alone or combined with antimicrobial agents, offers a promising alternative for treating infections caused by resistant pathogens (Tagliaferri et al., 2019). AI technology boosts the effectiveness of phage therapy by accurately predicting complex pathogen–phage interactions, aiding in the design of personalized treatments. This accelerates phage therapy development and improves treatment success.

13.12 USES OF AI FOR PERSONALIZED MEDICAL DEVELOPMENT

AI analyzes vast patient data, including genetics, lifestyle, and health history, to enable accurate diagnoses and personalized treatment plans. In cancer treatment, it helps select the best drug combinations, reduce side effects, and improve outcomes, while also predicting efficacy, complications, and creating tailored health management strategies (Bilgin et al., 2024; Elemento, 2024). The CombiANT method rapidly quantifies antimicrobial synergy through automated analysis, improving personalized infection treatment (Fatsis-Kavalopoulos et al., 2020). Kuo-Wei Hsu developed a portable antimicrobial susceptibility test for urinary tract infections, completing in 4.5–9 hours, offering potential for personalized medicine (Hsu et al., 2021). ChatGPT-3 demonstrated >90% success in generating correct diagnoses from ten differential options (Hirosawa et al., 2023). Future research will focus on complex cases and developing AI chatbots for more accurate, personalized diagnoses (Figure 13.3).

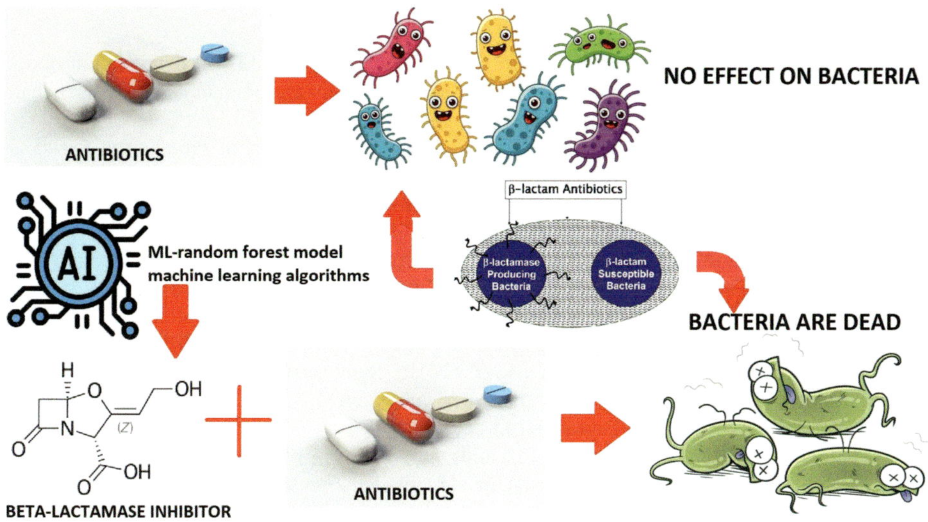

FIGURE 13.3 Artificial intelligence facilitates the diagnosis of bacterial infectious diseases.

13.13 CHALLENGES OF AI IN THE MEDICAL FIELD

AI in bacterial infections holds great potential but faces several challenges. These include issues with data quantity and quality due to privacy and standardization concerns, limiting AI model efficiency (Baowaly et al., 2019). Additionally, deep learning's "black box" problem affects accuracy and trust (Schwartz et al., 2024). Most AI research is still in experimental stages, with limited integration into clinical care (Alami et al., 2020). Moreover, complex bacterial behavior and rapid mutations make predictions difficult, and developing AI models requires interdisciplinary expertise, posing challenges for resource-limited research teams. Currently, there is no established legal system or authoritative standards in the field of AI. As technology advances and its applications expand, it is crucial to develop and update relevant regulations, which will be an ongoing and evolving process (Rees and Müller, 2022).

13.14 CONCLUSION

AI technology offers a new approach to tackling bacterial infections. With machine learning and deep learning, AI enhances pathogen detection, antimicrobial susceptibility analysis, genomic data interpretation, and personalized treatment development. By optimizing algorithms, AI improves pathogen identification speed and accuracy, predicts antibiotic resistance, and supports clinical decisions. In epidemiology, AI strengthens real-time monitoring and early warning systems, helping public health decision-making through data analysis. Despite AI's potential in treating bacterial infections, challenges remain, including ensuring transparency, interpretability, and maintaining patient safety. To overcome these hurdles and enable broader clinical use, interdisciplinary collaboration, innovation, and policy support are essential. AI technology will revolutionize the diagnosis and treatment of bacterial infections. As AI improves in pathogen identification, drug susceptibility testing, and genomic analysis, it will support clinicians in providing more precise, efficient, and personalized care, ultimately enhancing patient outcomes.

REFERENCES

Abbasi, B. A., Saraf, D., Sharma, T., Sinha, R., Singh, S., Sood, S., et al. (2022). Identification of vaccine targets and design of vaccine against SARS-CoV-2 coronavirus using computational and deep learning-based approaches. *PeerJ*, *10*, e13380. https://doi.org/10.7717/peerj.13380

Abu-Aqil, G., Lapidot, I., Salman, A., & Huleihel, M. (2023). Quick detection of *Proteus* and *Pseudomonas* in patients' urine and assessing their antibiotic susceptibility using infrared spectroscopy and machine learning. *Sensors (Basel)*, *23*(8132). https://doi.org/10.3390/s23198132

Alami, H., Lehoux, P., Denis, J.-L., Motulsky, A., Petitgand, C., Savoldelli, M., et al. (2020). Organizational readiness for artificial intelligence in health care: Insights for decision-making and practice. *Journal of Health Organization and Management*, *35*, 106–114. https://doi.org/10.1108/JHOM-03-2020-0074

Antimicrobial Resistance Collaborators. (2022). Global burden of bacterial antimicrobial resistance in 2019: A systematic analysis. *The Lancet*, *399*, 629–655. https://doi.org/10.1016/S0140-6736(21)02724-0

Baowaly, M. K., Lin, C.-C., Liu, C.-L., & Chen, K.-T. (2019). Synthesizing electronic health records using improved generative adversarial networks. *Journal of the American Medical Informatics Association*, *26*, 228–241. https://doi.org/10.1093/jamia/ocy142

Bilgin, G. B., Bilgin, C., Burkett, B. J., Orme, J. J., Childs, D. S., Thorpe, M. P., et al. (2024). Theranostics and artificial intelligence: New frontiers in personalized medicine. *Theranostics*, *14*, 2367–2378. https://doi.org/10.7150/thno.94788

Cherkaoui, A., Renzi, G., Charretier, Y., Blanc, D. S., Vuilleumier, N., & Schrenzel, J. (2019). Automated incubation and digital image analysis of chromogenic media using Copan WASPLab enables rapid detection of vancomycin-resistant *Enterococcus*. *Frontiers in Cellular and Infection Microbiology*, *9*, 379. https://doi.org/10.3389/fcimb.2019.00379

Ching, T., Himmelstein, D. S., Beaulieu-Jones, B. K., Kalinin, A. A., Do, B. T., Way, G. P., et al. (2018). Opportunities and obstacles for deep learning in biology and medicine. *Journal of the Royal Society Interface*, *15*(141), 20170387. https://doi.org/10.1098/rsif.2017.0387

Ciccolini, M., Donker, T., Grundmann, H., Bonten, M. J. M., & Woolhouse, M. E. J. (2014). Efficient surveillance for healthcare-associated infections spreading between hospitals. *Proceedings of the National Academy of Sciences of the United States of America*, *111*, 2271–2276. https://doi.org/10.1073/pnas.1308062111

Cisek, A. A., Dąbrowska, I., Gregorczyk, K. P., & Wyżewski, Z. (2017). Phage therapy in bacterial infections treatment: One hundred years after the discovery of bacteriophages. *Current Microbiology*, *74*, 277–283. https://doi.org/10.1007/s00284-016-1166-x

d'Humières, C., Salmona, M., Dellière, S., Leo, S., Rodriguez, C., Angebault, C., et al. (2021). The potential role of clinical metagenomics in infectious diseases: Therapeutic perspectives. *Drugs*, *81*, 1453–1466. https://doi.org/10.1007/s40265-021-01572-4

Deelder, W., Napier, G., Campino, S., Palla, L., Phelan, J., & Clark, T. G. (2022). A modified decision tree approach to improve the prediction and mutation discovery for drug resistance in *Mycobacterium tuberculosis*. *BMC Genomics*, *23*, 46. https://doi.org/10.1186/s12864-022-08291-4

Deneke, C., Rentzsch, R., & Renard, B. Y. (2017). PaPrBaG: A machine learning approach for the detection of novel pathogens from NGS data. *Scientific Reports*, *7*, 39194. https://doi.org/10.1038/srep39194

Desruisseaux, C., Broderick, C., Lavergne, V., Sy, K., Garcia, D. J., Barot, G., & Charles, M. K. (2024). Retrospective validation of MetaSystems' deep-learning-based digital microscopy platform with assistance compared to manual fluorescence microscopy for detection of mycobacteria. *Journal of Clinical Microbiology*, *62*(3), e01069-23. https://doi.org/10.1128/jcm.01069-23

Deusenbery, C., Wang, Y., & Shukla, A. (2021). Recent innovations in bacterial infection detection and treatment. *ACS Infectious Diseases*, *7*, 695–720. https://doi.org/10.1021/acsinfecdis.0c00890

Ekins, S., Godbole, A. A., Kéri, G., Orfi, L., Pato, J., Bhat, R. S., et al. (2017). Machine learning and docking models for *Mycobacterium tuberculosis* topoisomerase I. *Tuberculosis (Edinburgh)*, *103*, 52–60. https://doi.org/10.1016/j.tube.2017.01.005

Elemento, O. (2024). How artificial intelligence unravels the complex web of cancer drug response. *Cancer Research*, *84*, 1745–1746. https://doi.org/10.1158/0008-5472.CAN-24-1123

Esteva, A., Robicquet, A., Ramsundar, B., Kuleshov, V., DePristo, M., Chou, K., et al. (2019). A guide to deep learning in healthcare. *Nature Medicine*, *25*(1), 24–29. https://doi.org/10.1038/s41591-018-0316-z

Fatsis-Kavalopoulos, N., Roemhild, R., Tang, P.-C., Kreuger, J., & Andersson, D. I. (2020). CombiANT: Antibiotic interaction testing made easy. *PLoS Biology*, *18*, e3000856. https://doi.org/10.1371/journal.pbio.3000856

Fleming, J., Marvel, S. W., Supak, S., Motsinger-Reif, A. A., & Reif, D. M. (2022). ToxPi*GIS toolkit: Creating, viewing, and sharing integrative visualizations for geospatial data using ArcGIS. *Journal of Exposure Science & Environmental Epidemiology*, *32*, 900–907. https://doi.org/10.1038/s41370-022-00433-w

Gammel, N., Ross, T. L., Lewis, S., Olson, M., Henciak, S., Harris, R., et al. (2021). Comparison of an automated plate assessment system (APAS Independence) and artificial intelligence (AI) to manual plate reading of methicillin-resistant and methicillin-susceptible *Staphylococcus aureus* CHROMagar surveillance cultures. *Journal of Clinical Microbiology*, *59*, e00971.21. https://doi.org/10.1128/JCM.00971-21

Gao, S., & Wang, H. (2022). Scenario prediction of public health emergencies using infectious disease dynamics model and dynamic Bayes. *Future Generation Computer Systems*, *127*, 334–346.

Gurvic, D., Leach, A. G., & Zachariae, U. (2022). Data-driven derivation of molecular substructures that enhance drug activity in gram-negative bacteria. *Journal of Medicinal Chemistry*, *65*, 6088–6099. https://doi.org/10.1021/acs.jmedchem.1c01984

Hellmich, T. R., Clements, C. M., El-Sherif, N., Pasupathy, K. S., Nestler, D. M., Boggust, A., et al. (2017). Contact tracing with a real-time location system: A case study of increasing relative effectiveness in an emergency department. *American Journal of Infection Control*, *45*, 1308–1311. https://doi.org/10.1016/j.ajic.2017.08.014

Hirosawa, T., Harada, Y., Yokose, M., Sakamoto, T., Kawamura, R., & Shimizu, T. (2023). Diagnostic accuracy of differential-diagnosis lists generated by generative pretrained transformer 3 chatbot for clinical vignettes with common chief complaints: A pilot study. *International Journal of Environmental Research and Public Health*, *20*, 3378. https://doi.org/10.3390/ijerph20043378

Ho, C.-S., Jean, N., Hogan, C. A., Blackmon, L., Jeffrey, S. S., Holodniy, M., et al. (2019). Rapid identification of pathogenic bacteria using Raman spectroscopy and deep learning. *Nature Communications*, *10*, 4927. https://doi.org/10.1038/s41467-019-12898-9

Horvath, L., Hänselmann, S., Mannsperger, H., Degenhardt, S., Last, K., Zimmermann, S., et al. (2020). Machine-assisted interpretation of auramine stains substantially increases throughput and sensitivity of microscopic tuberculosis diagnosis. *Tuberculosis (Edinburgh)*, *125*, 101993. https://doi.org/10.1016/j.tube.2020.101993

Howard, A., Aston, S., Gerada, A., Reza, N., Bincalar, J., Mwandumba, H., et al. (2024). Antimicrobial learning systems: An implementation blueprint for artificial intelligence to tackle antimicrobial resistance. *Lancet Digital Health*, *6*, e79–e86. https://doi.org/10.1016/S2589-7500(23)00221-2

Hsu, K.-W., Lee, W.-B., You, H.-L., Lee, M. S., & Lee, G.-B. (2021). An automated and portable antimicrobial susceptibility testing system for urinary tract infections. *Lab on a Chip*, *21*, 755–763. https://doi.org/10.1039/d0lc01315c

Keith, M., Park de la Torriente, A., Chalka, A., Vallejo-Trujillo, A., McAteer, S. P., Paterson, G. K., et al. (2024). Predictive phage therapy for *Escherichia coli* urinary tract infections: Cocktail selection for therapy based on machine learning models. *Proceedings of the National Academy of Sciences of the United States of America*, *121*, e2313574121. https://doi.org/10.1073/pnas.2313574121

Khanna, D., & Rana, P. S. (2019). Ensemble technique for prediction of T-cell *Mycobacterium tuberculosis* epitopes. *Interdisciplinary Sciences: Computational Life Sciences*, *11*, 611–627. https://doi.org/10.1007/s12539-018-0309-0

Kleandrova, V. V., & Speck-Planche, A. (2020). The QSAR paradigm in fragment-based drug discovery: From the virtual generation of target inhibitors to multi-scale modeling. *Mini-Reviews in Medicinal Chemistry*, *20*, 1357–1374. https://doi.org/10.2174/1389557520666200204123156

Kulshrestha, M., Tiwari, M., & Tiwari, V. (2024). Bacteriophage therapy against ESKAPE bacterial pathogens: Current status, strategies, challenges, and future scope. *Microbial Pathogenesis*, *186*, 106467. https://doi.org/10.1016/j.micpath.2023.106467

Laliwala, A., Svechkarev, D., Sadykov, M. R., Endres, J., Bayles, K. W., & Mohs, A. M. (2022). Simpler procedure and improved performance for pathogenic bacteria analysis with a paper-based ratiometric fluorescent sensor array. *Analytical Chemistry*, *94*, 2615–2624. https://doi.org/10.1021/acs.analchem.1c05021

Lane, T. R., Urbina, F., Rank, L., Gerlach, J., Riabova, O., Lepioshkin, A., et al. (2022). Machine learning models for *Mycobacterium tuberculosis* in vitro activity: Prediction and target visualization. *Molecular Pharmaceutics*, *19*, 674–689. https://doi.org/10.1021/acs.molpharmaceut.1c00791

Lesosky, M., McGeer, A., Simor, A., Green, K., Low, D. E., & Raboud, J. (2011). Effect of patterns of transferring patients among healthcare institutions on rates of nosocomial methicillin-resistant *Staphylococcus aureus* transmission: A Monte Carlo simulation. *Infection Control & Hospital Epidemiology*, *32*(2), 136–147. https://doi.org/10.1086/657945

Li, R., Shen, M., Liu, H., Bai, L., & Zhang, L. (2023). Do infrared thermometers hold promise for an effective early warning system for emerging respiratory infectious diseases? *JMIR Formative Research*, *7*, e42548. https://doi.org/10.2196/42548

Liu, W., Ying, N., Mo, Q., Li, S., Shao, M., Sun, L., et al. (2021). Machine learning for identifying resistance features of *Klebsiella pneumoniae* using whole-genome sequence single nucleotide polymorphisms. *Journal of Medical Microbiology*, *70*. https://doi.org/10.1099/jmm.0.001474

Mandal, S., Tannert, A., Löffler, B., Neugebauer, U., & Silva, L. B. (2024). Findaureus: An open-source application for locating *Staphylococcus aureus* in fluorescence-labelled infected bone tissue slices. *PLOS ONE*, *19*, e0296854. https://doi.org/10.1371/journal.pone.0296854

Murray, C. J. L., Ikuta, K. S., Sharara, F., Swetschinski, L., Robles Aguilar, G., Gray, A., et al. (2022). Global burden of bacterial antimicrobial resistance in 2019: A systematic analysis. *The Lancet*, *399*(10325), 629–655. https://doi.org/10.1016/S0140-6736(21)02724-0

Nakar, A., Pistiki, A., Ryabchykov, O., Bocklitz, T., Rösch, P., & Popp, J. (2022). Detection of multi-resistant clinical strains of *E. coli* with Raman spectroscopy. *Analytical and Bioanalytical Chemistry*, *414*, 1481–1492. https://doi.org/10.1007/s00216-021-03800-y

Nguyen, N. T., Chinn, J., Nahmias, J., Yuen, S., Kirby, K. A., Hohmann, S., & Amin, A. (2021). Outcomes and mortality among adults hospitalized with COVID-19 at US medical centers. *JAMA Network Open*, *4*(3), e210417–e210417. https://doi.org/10.1001/jamanetworkopen.2021.0417

Njage, P. M. K., Leekitcharoenphon, P., & Hald, T. (2019). Improving hazard characterization in microbial risk assessment using next generation sequencing data and machine learning: Predicting clinical outcomes in shigatoxigenic *Escherichia coli*. *International Journal of Food Microbiology*, *292*, 72–82. https://doi.org/10.1016/j.ijfoodmicro.2018.11.016

Oh, J., Makar, M., Fusco, C., McCaffrey, R., Rao, K., Ryan, E. E., et al. (2018). A generalizable, data-driven approach to predict daily risk of *Clostridium difficile* infection at two large academic health centers. *Infection Control & Hospital Epidemiology*, *39*(4), 425–433. https://doi.org/10.1017/ice.2018.16

O'Neill, J. (2016). *Tackling drug-resistant infections globally: Final report and recommendations*. Review on Antimicrobial Resistance. https://amr-review.org

Paquin, P., Durmort, C., Paulus, C., Vernet, T., Marcoux, P. R., & Morales, S. (2022). Spatio-temporal based deep learning for rapid detection and identification of bacterial colonies through lens-free microscopy time-lapses. *PLOS Digital Health*, *1*, e0000122. https://doi.org/10.1371/journal.pdig.0000122

Pérez-Sancho, M., Vela, A. I., Horcajo, P., Ugarte-Ruiz, M., Domínguez, L., Fernández-Garayzábal, J. F., et al. (2018). Rapid differentiation of *Staphylococcus aureus* subspecies based on MALDI-TOF MS profiles. *Journal of Veterinary Diagnostic Investigation*, *30*(6), 813–820. https://doi.org/10.1177/1040638718805537

Pfeiffer, D. U., & Stevens, K. B. (2015). Spatial and temporal epidemiological analysis in the big data era. *Preventive Veterinary Medicine*, *122*(1–2), 213–220. https://doi.org/10.1016/j.prevetmed.2015.05.012

Qiu, J., Nie, W., Ding, H., Dai, J., Wei, Y., Li, D., et al. (2024). PB-LKS: A python package for predicting phage-bacteria interaction through local K-mer strategy. *Briefings in Bioinformatics*, *25*, bbae010. https://doi.org/10.1093/bib/bbae010

Ramachandran, P. S., Ramesh, A., Creswell, F. V., Wapniarski, A., Narendra, R., Quinn, C. M., et al. (2022). Integrating central nervous system metagenomics and host response for diagnosis of tuberculosis meningitis and its mimics. *Nature Communications*, *13*, 1675. https://doi.org/10.1038/s41467-022-29353-x

Rees, C., & Müller, B. (2022). All that glitters is not gold: Trustworthy and ethical AI principles. *AI and Ethics*, *3*, 1241–1254. https://doi.org/10.1007/s43681-022-00232-x

Santa Maria, J. P., Park, Y., Yang, L., Murgolo, N., Altman, M. D., Zuck, P., et al. (2017). Linking high-throughput screens to identify MoAs and novel inhibitors of *Mycobacterium tuberculosis* dihydrofolate reductase. *ACS Chemical Biology*, *12*(9), 2448–2456. https://doi.org/10.1021/acschembio.7b00468

Schwartz, I. S., Link, K. E., Daneshjou, R., & Cortés-Penfield, N. (2024). Black box warning: Large language models and the future of infectious diseases consultation. *Clinical Infectious Diseases*, *78*(4), 860–866. https://doi.org/10.1093/cid/ciad633

Senescau, A., Kempowsky, T., Bernard, E., Messier, S., Besse, P., Fabre, R., et al. (2018). Innovative DendrisChips® technology for a syndromic approach of *in vitro* diagnosis: Application to the respiratory infectious diseases. *Diagnostics*, *8*(4), 77. https://doi.org/10.3390/diagnostics8040077

Sherry, N. L., Horan, K. A., Ballard, S. A., Gonçalves da Silva, A., Gorrie, C. L., Schultz, M. B., et al. (2023). An ISO-certified genomics workflow for identification and surveillance of antimicrobial resistance. *Nature Communications*, *14*, 60. https://doi.org/10.1038/s41467-022-35713-4

Stokes, J. M., Yang, K., Swanson, K., Jin, W., Cubillos-Ruiz, A., Donghia, N. M., et al. (2020). A deep learning approach to antibiotic discovery. *Cell*, *180*(4), 688–702.e13. https://doi.org/10.1016/j.cell.2020.01.021

Stracy, M., Snitser, O., Yelin, I., Amer, Y., Parizade, M., Katz, R., et al. (2022). Minimizing treatment-induced emergence of antibiotic resistance in bacterial infections. *Science*, *375*(6585), 889–894. https://doi.org/10.1126/science.abg9868

Tagliaferri, T. L., Jansen, M., & Horz, H.-P. (2019). Fighting pathogenic bacteria on two fronts: Phages and antibiotics as combined strategy. *Frontiers in Cellular and Infection Microbiology*, *9*, 22. https://doi.org/10.3389/fcimb.2019.00022

Topol, E. J. (2019). High-performance medicine: The convergence of human and artificial intelligence. *Nature Medicine*, *25*(1), 44–56. https://doi.org/10.1038/s41591-018-0300-7

Váradi, L., Luo, J. L., Hibbs, D. E., Perry, J. D., Anderson, R. J., Orenga, S., et al. (2017). Methods for the detection and identification of pathogenic bacteria: Past, present, and future. *Chemical Society Reviews*, *46*(16), 4818–4832. https://doi.org/10.1039/c6cs00693k

Ventola, C. L. (2015). The antibiotic resistance crisis: Part 1: Causes and threats. *Pharmacy and Therapeutics*, *40*(4), 277–283.

Volynets, G. P., Usenko, M. O., Gudzera, O. I., Starosyla, S. A., Balanda, A. O., Syniugin, A. R., et al. (2022). Identification of dual-targeted *Mycobacterium tuberculosis* aminoacyl-tRNA synthetase inhibitors using machine learning. *Future Medicinal Chemistry*, *14*(15), 1223–1237. https://doi.org/10.4155/fmc-2022-0085

Waddington, C., Carey, M. E., Boinett, C. J., Higginson, E., Veeraraghavan, B., & Baker, S. (2022). Exploiting genomics to mitigate the public health impact of antimicrobial resistance. *Genome Medicine*, *14*, 15. https://doi.org/10.1186/s13073-022-01020-2

Walsh, T. R., Gales, A. C., Laxminarayan, R., & Dodd, P. C. (2023). Antimicrobial resistance: Addressing a global threat to humanity. *PLOS Medicine*, *20*(2), e1004264. https://doi.org/10.1371/journal.pmed.1004264

Wang, H., Ceylan Koydemir, H., Qiu, Y., Bai, B., Zhang, Y., Jin, Y., et al. (2020). Early detection and classification of live bacteria using time-lapse coherent imaging and deep learning. *Light: Science & Applications*, *9*, 118. https://doi.org/10.1038/s41377-020-00358-9

Wang, H.-Y., Chung, C.-R., Wang, Z., Li, S., Chu, B.-Y., Horng, J.-T., et al. (2021). A large-scale investigation and identification of methicillin-resistant *Staphylococcus aureus* based on peaks binning of matrix-assisted laser desorption ionization-time of flight MS spectra. *Briefings in Bioinformatics*, *22*, bbaa138. https://doi.org/10.1093/bib/bbaa138

Wells, J., Grant, R., Chang, J., & Kayyali, R. (2021). Evaluating the usability and acceptability of a geographical information system (GIS) prototype to visualise socio-economic and public health data. *BMC Public Health*, *21*, 2151. https://doi.org/10.1186/s12889-021-12072-1

Yi, Q., Cai, D., Xiao, M., Nie, M., Cui, Q., Cheng, J., et al. (2019). Direct antimicrobial susceptibility testing of bloodstream infection on SlipChip. *Biosensors and Bioelectronics*, *135*, 200–207. https://doi.org/10.1016/j.bios.2019.04.003

Ying, C., Li, X., Lv, S., Du, P., Chen, Y., Fu, H., et al. (2022). T-SPOT with CT image analysis based on deep learning for early differential diagnosis of nontuberculous mycobacteria pulmonary disease and pulmonary tuberculosis. *International Journal of Infectious Diseases*, *125*, 42–50. https://doi.org/10.1016/j.ijid.2022.09.031

Youn, J., Rai, N., & Tagkopoulos, I. (2022). Knowledge integration and decision support for accelerated discovery of antibiotic resistance genes. *Nature Communications*, *13*, 2360. https://doi.org/10.1038/s41467-022-29993-z

Yu, J., Lin, H.-H., Tseng, K.-H., Lin, Y.-T., Chen, W.-C., Tien, N., et al. (2023a). Prediction of methicillin-resistant *Staphylococcus aureus* and carbapenem-resistant *Klebsiella pneumoniae* from flagged blood cultures by combining rapid Sepsityper MALDI-TOF mass spectrometry with machine learning. *International Journal of Antimicrobial Agents*, *62*, 106994. https://doi.org/10.1016/j.ijantimicag.2023.106994

Zhang, Y.-M., Tsao, M.-F., Chang, C.-Y., Lin, K.-T., Keller, J. J., & Lin, H.-C. (2023). Rapid identification of carbapenem-resistant *Klebsiella pneumoniae* based on matrix-assisted laser desorption ionization time-of-flight mass spectrometry and an artificial neural network model. *Journal of Biomedical Science*, *30*, 25. https://doi.org/10.1186/s12929-023-00918-2

14 Applications of Systems Biology Tools to Understand *Staphylococcus aureus* Behavior

Jogipeta Harihara[1], *Vasanthu Sai Jahnavi*[2], *Jogipeta Pruthviraj*[3], *Betha Srinu*[4], *and Jhansi Lakshmi Vendra*[5]

[1]Centre for Biotechnology, University College of Engineering Science and Technology, Jawaharlal Nehru Technological University Hyderabad, Kukatpally, Telangana, India

[2]Allele Tech Private Limited, Jeedimetla, Hyderabad, Telangana, India

[3]ICAR – Central Institute of Agricultural Engineering, Bhopal, Madhya Pradesh, India

[4]MSN R&D Centre, MSN Laboratories Ltd, Pashamylaram, Hyderabad, Telangana, India

[5]University Institute of Emerging Technologies, Guru Nanak University, Ibrahimpatnam, Hyderabad, Telangana, India

ABBREVIATIONS

2D-DIGE	two-dimensional difference gel electrophoresis
AI	artificial intelligence
AP-MS	affinity purification–mass spectrometry
cDNA	complementary DNA
Co-IP	Co-immunoprecipitation
FBA	flux balance analysis
GRN	gene regulatory network
LC-MS/MS	liquid chromatography with tandem mass spectrometry
MALDI-TOF/TOF	matrix-assisted laser desorption ionization–time-of-flight/time-of-flight mass spectrometry
MGE(s)	mobile genetic element(s)
ML	machine learning
MRSA	methicillin-resistant *Staphylococcus aureus*
NGS	next-generation sequencing
ODEs	ordinary differential equations
PPIN	protein–protein interaction network
PPIs	protein–protein interactions
RNA-seq	RNA sequencing
rRNA	ribosomal RNA
S. aureus	*Staphylococcus aureus*
SMRT	single-molecule real-time (sequencing)
SNP	single-nucleotide polymorphism
WGS	whole genome sequencing
XL-MS	crosslinked mass spectrometry

DOI: 10.1201/9781003647072-18

14.1 INTRODUCTION

Staphylococcus aureus is an extremely adaptive opportunistic pathogen capable of producing both superficial and life-threatening systemic infections (sepsis and endocarditis). The incessant appearance of antibiotic-resistant strains, the best known of which is the methicillin-resistant *S. aureus* (MRSA), has given further weight to the necessity of an attentive understanding of their pathology and pathogenic mechanisms. Conventional microbiological studies have yielded consequential information; unfortunately, this is often insufficient to interpret the complex and multifaceted nature of *S. aureus* physiology. This gap is filled through systems biology, which, in the integration of several omics technologies through multiple omics technologies, offers simultaneous analysis of a pathogen at the molecular level/cellular level/population level (Ideker et al., 2001; Kitano, 2002).

Genomic research is the primary element of such integrative approach, and, as whole genome sequencing (WGS) is the most promising attribute of this type of research, it is likely to provide the details of the way the genetic inheritance of *S. aureus* is set. Comparative genomic research demonstrates a high variability of strains, which, in addition to the presence of mobile genetic elements, or pathogenicity islands and pathogen-resistance genes, each contributed to hold a significant impact on virulence and adaptability (Holden et al., 2004; Lindsay, 2013). Using these data, transcriptomic profiling can describe the manner in which environmental alterations or immune requirements cause the expression of gene alterations, which characterize regulatory networks that must be incurred throughout infection and persistence (Wang et al., 2008).

S. aureus biochemistry has become understood through system-wide developments. Proteomic studies identify the entire repertoire of expressed proteins, including those associated with stress, pathogenicity, and resistance. A combination of high-resolution mass spectrometry (MS) permits decoding of the protein–protein network and post-translational alterations and, therefore, offers a deeper look at the functions (Aebersold & Mann, 2003). At the same time, other metabolomic tests examine the chemical imprint and metabolomics pathways utilized by *S. aureus* and disclose the way the organism re-utilizes its biochemistry to evade the host defense and both external and internal pressures (Lei et al., 2019).

Combining these datasets using computational modeling and network biology, it is feasible to reverse engineer complex gene regulation and metabolic networks and predict the behavior of bacteria in certain conditions, as well as predict important molecular nodes which would be good therapeutic targets (Taiwo et al., 2024). This type of integrative study is especially effective in explaining complicated phenomena such as virulence modulation, host–pathogen interaction, and antimicrobial resistance (Figure 14.1).

Over the past few years, systems biology has made a significant venture in the clinical field, especially in multidrug-resistant infections and individual attention toward therapeutic patient care. Cross-omics has contributed to a new, better portrait of the pathogenesis of MRSA, therefore, leading to narrower diagnostics, even more successful therapy, and more efficient infection control procedures (Bosi et al., 2016; Westermann et al., 2016). Some technological innovations that must have appeared in the future to further shape our knowledge about *S. aureus* and prepare us to deal with the problematic pathogen even more successfully are single-cell omics, spatial transcriptomics, and artificial-intelligence-based data integration, just to name only a few (Imdahl et al., 2020; Angermueller et al., 2016).

14.2 GENOMICS

Genomics in the world of systems biology is central to discovery, providing the precise, quantitative framework for the complete genetic makeup of living things. This lens has led to far more than strands of nucleotides; it has enabled us to read the genetic blueprints onto which *S. aureus* builds its virulent nature, resilience, and knack in evading antibiotics. The next ten years brought huge advancements in sequencing tech as it became much faster, more accurate, and much less expensive.

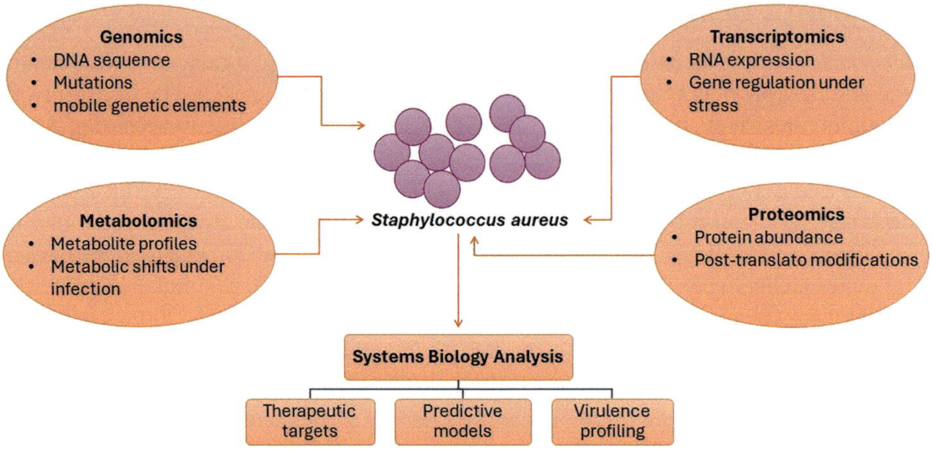

FIGURE 14.1 Multi-omics systems biology workflow to study *Staphylococcus aureus* behavior. The figure explains that a multi-omics approach with *S. aureus* deconstructs physiology and virulence. Transcriptomics, genomics, proteomics, and metabolomics collaborate in regulating the mapping of the variation of DNA, in gauging the activity of a gene, in evaluating the protein, and tracking the changes associated with metabolic activities. These methods can be combined, leading to the identification of previously unknown regulatory and metabolic pathways, the identification of virulence factors, and the determination of novel targets of therapy, advancing diagnostic and therapeutic approaches to drug-resistant strains.

An enormous task, WGS can now work like no other parallel, making it possible to see the molecular landscape of the pathogen in a manner that has never been done before. It is at this scale that researchers are going beyond establishing an enhanced understanding of *S. aureus* biology but developing bespoke interventions that may turn the tide in the battle against microbe-based illness.

14.2.1 WGS Techniques and Technologies

The technique of WGS provides a comprehensive picture of the genetic profile of a living organism at that point in time, with the whole length of the DNA being sequenced. In *S. aureus*, this has most commonly been done by next-generation sequencing platforms, especially Illumina, which produce massive amounts of very accurate short-read information. To overcome the short-read drawbacks, the long-read sequencing technologies are utilized, for example, Oxford Nanopore and PacBio Single-Molecule Real-Time sequencing. The techniques have performed well in resolving complex genomic regions such as repetitive sequences and mobile genetic material. The harnessing of both short- and long-read data enables researchers to assemble near-complete or fully closed genome sequences, with a resulting massive increase in accuracy. The degree of accuracy allows the identification of single-nucleotide polymorphisms, insertions, deletions, and other structural variations, which provide the phenotypic diversity in *S. aureus* strains (Koren et al., 2013).

The WGS research has found that the nature of *S. aureus* strains is that although the core genome remains relatively stable, there is the existence of accessory genome characterized by a substantial degree of variation in which most of the variation is the result of horizontal gene transfer. MGEs such as plasmids, transposons, and bacteriophages are also vehicles carrying antibiotic-resistance genes and virulence determinants, and they speed up the evolutionary flexibility of the organism. In addition to its usefulness in fundamental research, WGS has gained prominence as a pillar of epidemiological surveillance, where the source of outbreaks, routes of transmission, and also lineages of epidemic organisms such as MRSA have been tracked. The provided genomic information gives vital instructions to the infection control and antimicrobial stewardship, and it is, ultimately, beneficial to the clinical practice of *S. aureus* in the clinic.

14.2.2 Comparative Genomics

A systematic alignment of genome sequences of different strains of *S. aureus* – analysis of similarities and differences within and between the genomes – is referred to as comparative genomics. This approach has detected strain-specific gene sets such as lineage-related virulence determinants, the presence of unique metabolic capability, and antibiotic resistance genes. Analyzing the large set of genomes enables researchers to determine phylogenetic relations and chart evolutionary stains and thus one can differentiate hospital-associated, community-associated, and livestock-associated strains (Holden et al., 2004).

Among the key observations made as a result of such comparisons are pathogenicity islands, which are groups of genes that have frequently been acquired via horizontal gene transfer, encoding toxins, adhesins, and proteins that allow them to evade the immune system. Among this group, the *Staphylococcal aureus* pathogenicity islands (SaPIs) are important in virulence enhancement. They are transferred, often via helper bacteriophages, and this genetic variability among bacterial populations facilitates the flexibility and evolutionary change (Frost et al., 2005).

Comparative genomics has also thrown light on the genomic context of the major genes of antibiotic resistance, especially mecA, which causes methicillin resistance and is hosted by the staphylococcal cassette chromosome mec (SCCmec). Characterizing the pattern of these mobile genetic elements along with their distribution and structure is necessary because it is through this mapping that the mechanisms of how the resistance traits spread and how *S. aureus* adapts to the pressure of antimicrobials can be known (Malachowa & DeLeo, 2010). This knowledge finds its key element in the formulation of strategies that will focus on containing and eventually curbing the effects of the antibiotic-resistant strains.

14.3 TRANSCRIPTOMICS

Transcriptomics involves an attempt to measure the entire extent of the RNA transcripts generated by a genome in a specific environment, making it akin to a snapshot of what is happening in real time with regard to the regulation and expression of those genes. In *S. aureus*, transcriptomic profiling has played a critical role in describing pathogen adaptation to various environmental conditions, its stress tolerance via exposure to antibiotics, and its means of orchestrating its virulence within the context of an infection. RNA sequencing (RNA-Seq) has come to be the technique of choice to study the complete transcriptome. That is initiated by the purification of the total RNA of *S. aureus* grown under clearly defined experimental conditions. Since ribosomal RNA (rRNA) is abundant in bacteria, rRNA selection is commonly incorporated into workflows as a method of enriching the messenger RNA (mRNA) and other less-abundant non-coding RNAs of interest. The resulting enriched RNA is converted back to complementary DNA (cDNA) and read on next-generation platforms like Illumina.

The raw sequencing read submits to quality control filters after which they are aligned to a reference *S. aureus* genome by way of applications such as Bowtie2 or HISAT2. Statistical packages like DESeq2 or edgeR are used to quantify the levels of expression, and they allow the researcher to figure out genes that are substantially up- or downregulated at altered conditions (Wang et al., 2008). Such datasets can show condition-specific patterns of transcription, showing how *S. aureus* alters its metabolism, stress responses, and virulence responses to survive in unfavorable conditions (Beenken et al., 2004). Through such analyses we can compare antibiotic-treated vs. untreated cells or free-floating (planktonic) bacteria and biofilm-forming communities. Such comparisons of *S. aureus* have revealed important virulence determinants in this organism: toxins, such as alpha-hemolysin; adhesion factors, such as fibronectin-binding proteins; and global

TABLE 14.1
Transcriptomics Workflow in *S. aureus* Research

Step	Description	Tools/Software	References
RNA enrichment	Isolation of total RNA and removal of rRNA	RNA extraction kits, rRNA-depletion	Wang et al. (2008)
cDNA library construction	Conversion of RNA to cDNA and library preparation	Illumina TruSeq kits	
Sequencing	High-throughput sequencing	Illumina platform	
Quality filtering	Removal of low-quality reads	FastQC, Trimmomatic	
Alignment	Mapping reads to the reference genome	Bowtie2, HISAT2	
Quantification and DE analysis	Identification of differentially expressed genes	DESeq2, edgeR	

regulators, such as the agr quorum-sensing system, which play key roles in the infection processes (Novick & Geisinger, 2008).

The effect of transcriptomics is also seen through the response of antibiotic resistance genes to environmental pressures. Cell wall biosynthesis can be altered, and efflux pump expression and expression of beta-lactamase can go both ways. On the same note, metabolic pathways are rewired to generate energy even under nutrient shifts in the host.

A dramatic example of transcriptomics is the investigation of a biofilm. Biofilms also place *S. aureus* at an advantageous position since it safeguards it against the strength of immunity systems and antibiotic penetration. Transcriptomic studies have found that during biofilm formation genes involved in the extracellular matrix (including those responsible for the synthesis of the polysaccharide intercellular adhesin) and stress tolerance and regulatory systems were upregulated between the biofilm and planktonic cells (Resch et al., 2006). These findings will lead the way to a possible treatment approach to interfere with biofilms and effectively address chronic infections.

This Table 14.1 outlines the standard transcriptomics workflow used to analyze gene expression in *S. aureus*. It begins with RNA extraction and enrichment, followed by the construction of cDNA libraries suitable for sequencing. Subsequent steps include sequencing on high-throughput platforms, quality filtering of raw data, alignment to a reference genome, and data analysis to identify differentially expressed genes under various conditions. This pipeline enables researchers to decode transcriptional responses and regulatory mechanisms driving *S. aureus* physiology and pathogenicity.

14.4 PROTEOMICS

A thorough knowledge of the biology of *S. aureus*, a highly pathogenic and exceptionally flexible microorganism that can cause a spectrum of infections, requires the systems-level analysis of the proteome of the microbe under varying physiological conditions. In this context, proteomics, notably MS-based approaches, and functional proteomic concepts are vital in discovering the complex system of the bacterium's regulation and adaptability.

14.4.1 MS-Based Proteomics

MS-based proteomics has transformed *S. aureus* research, allowing the identification and quantification of thousands of proteins at a single time. Shotgun proteomics, using liquid chromatography coupled with tandem MS (LC-MS/MS), is used extensively for global protein profiling (Mann et al., 2013).

Label-free quantification and stable isotope labeling methods like SILAC and iTRAQ also enable the precise measurement of protein abundance (Bantscheff et al., 2012) Read quantification was conducted using software such as HTSeq or featureCounts, followed by differential expression analysis. In *S. aureus*, these methods have shown differential protein expression due to antibiotic stress, host immune pressure, and nutritional limitation. Label-free LC–MS/MS was used to investigate the effect of oxacillin on *S. aureus*, revealing induction of proteins associated with cell wall synthesis and repression of metabolic enzymes (Singh et al., 2001).

Environmental stresses – iron starvation, oxidative stress, and antibiotic exposure – cause dynamic changes in the *S. aureus* proteome. Proteomic profiling enables the identification of stress-specific virulence factors and potential biomarkers. Two-dimensional difference gel electrophoresis (2D-DIGE) was also used with MALDI-TOF/TOF MS to study *S. aureus* in an iron-starved condition and identified proteins that are important in the process of iron acquisition. On the same note, comparison of the proteome of biofilm-forming cells with non-biofilm cells has led to the significance of proteins like Atl, SarA, and various extracellular matrix proteins, all of which are key to strong biofilm growth (Resch et al., 2006).

14.4.2 Functional Proteomics

Determination of protein–protein interactions (PPIs) is a direct path toward how *S. aureus* coordinates its signaling cascades and regulates virulence. Technological breakthroughs like functional proteomics, that is, systems based on co-immunoprecipitation (Co-IP), affinity purification/MS (AP-MS), and crosslinked MS (XL-MS) have revealed detailed systems of regulation working within the bacterium.

These studies keep pointing out a collection of master regulators, such as AgrA, SarA, and Rot. Behaving as molecular conductors, they modify the level at which virulence genes are expressed in response to the new stimuli. Fine-tuning proteomic studies have discovered, too, that these molecules are not binary on/off regulators but that their activity can be controlled using post-translational modifications, including phosphorylation and acetylation (Burckhardt et al., 2019).

A prominent display can be seen in phosphoproteomics that has been used to clarify the role of the bacterial kinase Stk1. This enzyme is involved in a number of crucial actions – cell division, biofilm formation, and even the acquisition of antibiotic resistance, which proves the significance of this enzyme to the adaptability and persistence of *S. aureus*.

14.5 METABOLOMICS AND FLUXOMICS

The full inventory of the metabolism of a biological system – the subject of metabolomics – yields a potent tool through which to investigate the way *S. aureus* re-engineers its small-molecule metabolism in response to various environmental stresses and host interactions. Mapping changes in metabolism allows characterizing the physiological state of the bacterium in real-time, since a snapshot is obtained of the adaptive mechanisms that are evident without necessarily being evident in the genetic or protein compartment. When the metabolomic data are combined with discoveries of genomics, transcriptomics, and proteomics, the image widens into the actual systems biology view. This multifaceted study progress enables a synthetic cartography of the pathophysiology of *S. aureus* linking metabolic rewiring to gene expression, protein functioning, and virulence expression.

14.5.1 Metabolic Profiling

When the researcher examines the metabolome of the *S. aureus*, it provides insight into the way this pathogen modifies its chemistry by being able to survive. A big part of this work is accomplished with the help of such strong analytical instruments as nuclear magnetic resonance (NMR) spectroscopy and liquid chromatography-linked MS (LC-MS). NMR is best known because of its exact quantification and capacity to identify molecular structures with minimum preparation, but LC-MS is better in terms of expanded sensitivity and comprehensive coronary stitching of the metabolome (Dettmer et al., 2006).

Application of NMR-based metabolomics by the researchers has enabled them to map out the central carbon metabolism of *S. aureus*, through which it turns out that the intermediates of glycolysis and the tricarboxylic acid (TCA) cycle form the core of the adaptive strategies employed by the bacterium. In its turn, the LC-MS has contributed to developing the perception of how *S. aureus* alters its resource distribution and metabolic pathway under the stress of antibiotics (Troitzsch et al., 2021). The findings narrate an impressively accommodating being. A low level of oxygen triggers *S. aureus* to enter fermentative energetics and to enhance lactate and acetate production during low nutrient circumstances (Somerville & Proctor, 2009). It usually resorts to amino acids as a source of energy in the host environment. The arginine deiminase pathway is especially notable among them, as, in the environment of acidosis and oxygen shortage, it is turned on and ATP is produced, which is then used to maintain the internal pH in the bacterium (Freiberg et al., 2024). Such metabolic transitions do not exist solely as a trade-off against survival but also confer an advantage on *S. aureus* in disease evasion of the immune system. That is, its virulence is inextricably linked to its metabolism as well.

14.5.2 Fluxomics

While metabolomics provides a snapshot of cellular metabolites at a given time, fluxomics quantifies the dynamic rates of metabolic processes by monitoring labeled substrates like ^{13}C-glucose. It helps to understand the real flow of metabolites through pathways under different conditions (Wiechert, 2001). *S. aureus* flux analysis has shown rerouting of metabolism during biofilm development and exposure to antibiotics. For example, ^{13}C flux measurements reported decreased activity of the TCA cycle and enhanced fermentation in SCVs (Small Colony Variant), a phenotype that is linked with chronic infections (Koren et al., 2013).

Fluxomics is especially strong when combined with transcriptomics and proteomics to allow for the construction of genome-scale metabolic models (GEMs). These models permit the simulation of *S. aureus* growth in a range of environmental conditions and the prediction of gene essentiality. Systems biology tools like the COBRA toolbox have been employed to create context-specific models that connect metabolic activity to regulatory networks (Emwas et al., 2022). Such combined analyses are able to reveal possible drug targets by highlighting bottlenecks or weaknesses in bacterial metabolism.

14.6 INTEGRATIVE OMICS APPROACHES

The intricacies of *S. aureus* behavior under various environmental and host-conditioned stresses cannot be comprehended through a solitary approach of omics. Integrative omics integrates genomics, transcriptomics, proteomics, and metabolomics to give a systems-level view of bacterial physiology and pathogenesis. Holistic frameworks of this nature allow predictive models of bacterial response to be built and molecular mechanisms of virulence, resistance, and metabolic plasticity to be revealed.

14.6.1 Multi-Omics Data Integration

Methods for Integrating Genomics, Transcriptomics, Proteomics, and Metabolomics. Multi-omics integration is usually performed by combining statistical, machine learning (ML), and model-based methods. Multivariate analysis (e.g., PCA, PLS-DA), Bayesian networks, and network-based integration are frequently used. More sophisticated platforms such as iCluster, mixOmics, and multi-omics factor analysis (MOFA) are progressively utilized to uncover intricate interactions between

TABLE 14.2
Integrative Omics Applications

Integrated Data	Application	Example Outcomes	References
Genomic + transcriptomic + proteomic + metabolomic	Build systems-level models of *Staphylococcus aureus* physiology	Linked genetic variation to transcript/protein/metabolite changes during nasal colonization	Holtfreter et al. (2016)
Transcriptomics + metabolic modeling (FBA)	Predict metabolic vulnerabilities	Identified essential metabolic genes in host-like conditions	Baumler and Fang (2013)
Dual RNA-seq + proteomics	Study host–pathogen interactions	Dynamic regulation of virulence factors in response to host immune signals	Westermann et al. (2016)

layers of biological information. In *S. aureus*, such integration has permitted the association of genetic differences with variation in transcript and protein amounts, as well as metabolite levels. Integrate transcriptomic and proteomic information to examine *S. aureus* during nasal epithelium colonization in humans, identifying adaptive features in energy metabolism as well as iron acquisition processes (Holtfreter et al., 2016).

This Table 14.2 compares important applications where a combination of various omics layers – genomic, transcriptomic, proteomic, and metabolomic data – can allow systems-level knowledge concerning the biology of *S. aureus*. These multi-dimensional strategies of integrating data and taking computational modeling into consideration have allowed complex gene–environment interactions and detection of metabolic weaknesses and provided insight into host–pathogen interactions. The examples provide illustrations of how multi-omics integration can improve predictions of key genes, capacity to regulate virulence, and use of adaptive mechanisms that have proven relevant to colonization and infection.

14.6.2 Systems-Level Understanding of *S. aureus* Physiology

Omics is integrative and enables building of system biology models where it is possible to predict phenotypic effects of the interaction at the molecular level. One such class of models is known as GEMs, where the model is generated on the basis of genomic data and used to interrogate transcriptomic, proteomic, and metabolomic data. These models can replicate bacterial growth in all kinds of settings and predict possible targets of drugs (Bosi et al., 2016).

These strategies have facilitated deciphering metabolic transitions in the process of biofilm formation, antibiotic resistance, and immune evasion to the host in *S. aureus*. Integration also reveals energy trade-offs between allocation and virulence factor production – information needed to develop anti-virulence therapeutics.

14.6.3 Network Biology

Graph Theory. Graph theory (in particular, graph theory) has been used to study biological processes where nodes are molecules (genes, proteins, metabolites), and the edges are interactions. The ways in which genes are controlled (gene regulatory networks or GRN), the interaction between proteins (protein–protein interaction networks or PPIN), and the metabolic networks bring forth. The *S. aureus* regulates its physical activities. GRNs are often derived based on transcriptomics and ChIP, and metabolic networks are normally constructed upon designed hand-curated data sets of pathways (e.g., KEGG, BioCyc). This has seen the placement of regulator proteins of transcription, such as SarA, Rot, and AgrA, as master regulators in GRNs regulating the expression of virulence genes (Cheung et al., 2003).

The nodes making up a hub in networks represent important genes or proteins that are essential to survival or pathogenicity. These have centrality measures such as degree, betweenness, and closeness through which these major nodes are identified. At the system level, it has been identified in *S. aureus* that the RNAIII (regulatory RNA), as well as the two-component system Agr, is central to virulence regulation and quorum sensing (Novick & Geisinger, 2008).

Moreover, independent network motifs determine further connectivity, including feedback and feed-forward loops, which provide data on regulation mechanisms in bacteria, such as the amplification of signal, stability, and decision-making in bacterial regulatory networks.

14.7 COMPUTATIONAL MODELING

The systems biology uses dynamic modeling approaches to understand the temporal behavior of biological systems, simulate and predict it is computational modeling. In the case of *S. aureus*, mathematical models may allow researchers to integrate experimental data with methods of introducing mechanisms of pathogenesis, resistance, and adaptation in silico. This enables the possibility to test hypotheses faster, to optimize experimental design, and to uncover important regulatory factors to target them therapeutically. Model designs for *S. aureus* include predictive models, as well as computational and mathematical models. The *S. aureus* mathematical modeling is usually done by building ordinary differential equations (ODEs) or Boolean networks, which model the existing biological processes. To capture the dynamic regulation of gene expression and regulatory feedback and metabolic flow, they are founded on omics data to enable modeling; predictive models have been used to model the *S. aureus* quorum-sensing system (Agr system) to show how external conditions control the synthesis of virulence factors via autoinducing peptides (Jabbari et al., 2009). Also, genome-scale metabolic models like iYS854 and iSB619 enable in silico growth predictions under different nutrient environments or under antibiotic stress (Renz & Dräger, 2021).

Simulations can determine key control points in regulatory networks, make predictions about phenotypic behavior under gene knockouts, and analyze the effect of external perturbations (e.g., antibiotics, immune pressure). Systems dynamics investigates how biological networks change over time and react to internal or external perturbations. Stability analysis – frequently with Jacobian matrices, bifurcation theory, or Lyapunov methods – assists in making predictions about whether the network returns to a steady state upon disturbance or transitions into other states (e.g., persistence, dormancy, or virulence). *S. aureus* stability analysis has been applied to test the stability of the Agr quorum-sensing system, identifying conditions under which the feedback is bistable, resulting in heterogeneous population response (Shukla et al., 2020). Bistability is considered to be crucial for survival against immune pressure and antibiotic treatment. Computational modeling is becoming a more prevalent tool to forecast how *S. aureus* responds to environmental stress like nutrient starvation, oxidative pressure, or host immune attack. These models are able to mimic therapeutic intervention (e.g., drug delivery, metabolic inhibitions) and make predictions about bacterial response, which is crucial for rational antimicrobial discovery. For instance, flux balance analysis (FBA) in conjunction with transcriptomics has been applied to forecast essential host-like conditions metabolic genes, facilitating the determination of metabolic vulnerabilities (Baumler & Fang, 2013). Perturbation models can, in addition, model how drug or gene knockout impacts global regulatory networks and fitness landscapes.

14.8 APPLICATIONS IN DRUG RESISTANCE AND PATHOGENESIS

The rise of multidrug-resistant *S. aureus* (particularly MRSA) poses a global health challenge. Systems biology provides a holistic framework to dissect the molecular underpinnings of drug resistance and pathogenesis, facilitating the discovery of new therapeutic targets and strategies. By integrating genomics, transcriptomics, proteomics, and metabolomics, systems-level approaches enable deeper insights into antibiotic resistance mechanisms, virulence regulation, and host–pathogen interactions.

14.8.1 Insights from Systems Biology into Resistance Mechanisms

Conventional strategies have used resistance-conferring genes (e.g., mecA, blaZ, efflux pumps); however, systems biology shows a far broader capacity to reveal global regulatory networks and metabolic reprogramming linked to resistance. Recent high-throughput transcriptomic and proteomic analyses under antibiotic stress showed *S. aureus* undergoes significant reprogramming, including the upregulation of stress response genes, rewiring of metabolism, and modulation of cell wall biosynthesis (Sulaiman et al., 2022). For instance, genome-scale metabolic modeling of MRSA describes a reprogramming of metabolism as an adaptive response to resistance, and metabolic gene expression changes can be used to predict adaptive responses in additional phenotypes. Integrating transcriptomic data and fluxomic data can allow us to observe pathways compensating for resistance deployed during resistance development.

Systems-level approaches have enabled the discovery of new drug targets by revealing unexplored essential genes and pathways under relevant conditions for infection. Tools such as genome-scale metabolic models (i.e., iSB619, iYS854) have facilitated in silico gene knockouts to discover essential nodes within the bacterial systems under study, and studies have discovered metabolic chokepoints and non-redundant enzymes (e.g., peptidoglycan remodeling or nucleotide biosynthesis) as novel antibiotic targets (Bosi et al., 2016).

Therapeutic strategies designed to perturb regulatory nodes regulating resistance and virulence (e.g., two-component systems such as WalKR or the accessory gene regulator Agr) produce high-value approaches for targeting bacterial virulence using strategies designed to attenuate or eliminate virulence without killing the host.

14.8.2 Systems Biology Approaches to Study Host–Pathogen Dynamics

Systems-level delineation of the interaction between host and pathogen is needed to understand how *S. aureus* has developed to survive in the host. Bacterial and host gene expression have also been monitored simultaneously during infection by dual RNA-seq and proteomics. The focus of such studies is on the dynamic regulation of the virulence factors (hemolysins, adhesions, and immune evasion proteins) expressed by the *S. aureus* after sensing host immune signals (Westermann et al., 2016).

In addition, systems biology can facilitate predictive modeling of infection kinetics, immune evasion tactics, and tissue tropism through a combination of omics data with computational simulations of infection (e.g., agent-based models or ODE systems).

Transcriptomics and network inference are frequently used to study regulatory networks controlling virulence, the Agr, SarA, and SaeRS systems, among others. Tools of systems biology have illuminated the way in which *S. aureus* switches from the commensal to pathogenic under environmental and host-induced signals. The identification of RNAIII as a significant post-transcriptional regulator aligns with the production of exotoxins and adhesion inhibition through the use of perturbation studies and modeling of GRNs (Novick & Geisinger, 2008). These results highlight possible therapeutic options to consider virulence regulation control that may not cause selection pressure for drug resistance.

14.8.3 Applications in Personalized Medicine and Treatment Strategies

Precision medicine will hold the key to the future of infectious disease care. Individualized therapeutic choices can be made on the basis of the systems biology of *S. aureus* clinical isolates.

As an illustration, the host–pathogen dual RNA-qp allows the immune response and pathogenic gene expression of the host to be mapped individually during infection (Westermann et al., 2016). Further, it has been shown that the metabolic modeling of clinical isolates can predict how changes in metabolic flux distributions point toward which antibiotic treatments are most likely to fail or succeed. These evidence-based methods allow keeping the treatment regimen optimal and reducing the development of resistance.

14.9 FUTURE DIRECTIONS

Studies on *S. aureus* are on the frontiers of infectious diseases research with the development of the discipline of systems biology. New instruments, cross-disciplinary, and data-based modeling are poised to change *S. aureus* pathogenicity, resistance to antibiotics, and treatment. Key areas of potential research, diagnosis, and treatment enabled as a result of next-generation systems biology technologies are outlined in this chapter.

14.9.1 Advances in Single-Cell Omics and Spatial Transcriptomics

In contrast to bulk population averages, which form the basis of the majority of conventional omics applications, single-cell omics has revealed the inherent variability of natural bacterial populations. The single-cell RNA sequencing (scRNA-seq) and mass cytometry can be utilized to study bacterial cells and allow scientists to capture variations in gene expression, stress response, and virulence at a single-cell resolution (Imdahl et al., 2020).

A more traditionally used field of spatial transcriptomics involves host tissue profiling, but the technique is also increasingly used in studies of host–pathogen interactions. These tools can monitor the location and specificity of *S. aureus* expressing individual genes during colonization and infection in host tissues, which provides a greater understanding of tissue-specific virulence mechanisms.

14.9.2 Potential of Artificial Intelligence and ML in Systems Biology

Artificial intelligence (AI) and ML are revolutionizing the analysis of biological data. ML algorithms have the capability to detect patterns and make predictions of resistance phenotypes, virulence levels, and metabolic capacities from genomic and transcriptomic data (Angermueller et al., 2016). Deep learning paradigms have been applied to simulate host–pathogen interactions and rank candidate drug targets. For instance, convolutional neural networks that learn from multi-omics data have been shown to attain high accuracy in MRSA resistance pattern prediction. ML also helps to study high-dimensional imaging data from infection models and surgical wound tissue.

14.9.3 Translational Research

One of the key objectives of systems biology is to translate multiscalar molecular intelligence into practical clinical applications. Omics-based diagnostics are already being tested in clinical microbiology for predicting resistance, typing strains, and assessing infection risk. Real-time PCR with transcriptomic biomarkers can predict the severity of disease and direct the choice of antimicrobial treatment. Moreover, it aims to combine the information on a patient's immune response with pathogen genomics toward developing personalized treatment plans. The tool will be able to improve treatment outcomes and reduce the misuse of antibiotics. The i4Health project in the European Union aims to integrate multi-omics data and clinical variables for developing decision-support systems for infectious disease management.

Smart diagnostics and adaptive treatment protocols are being developed to address anticipated diseases. Rapid multi-omics-based diagnostics for real-time pathogen identification and antimicrobial resistance profiling. Synthetic biology approaches to design bacteriophages and antimicrobial peptides targeting S. networks in *S. aureus*, a systems biology approach. AI-driven surveillance techniques to prevent infections in hospitals by predicting and analyzing strain genomics and outbreak history.

14.10 CONCLUSION

The biology of *S. aureus* has greatly benefited from systems biology approaches that provide a systems-level overview of the complex organism. The early insights from WGS revealed the extensive genetic diversity and evolutionary processes that allow *S. aureus* to grow in diverse environments and evolve resistance to antibiotics. The potential to follow the dissemination of virulence and drug resistance in and between bacterial populations has been valuable with the identification of genetic characteristics, such as pathogenicity islands and resistance genes. Other than it being a fixed genome, comparative analysis of how these organisms respond to alluffling in the environment, such as the exposure to antibiotics and host immunity, has given responses on what enables adaptation of organisms to environmental inputs. The transcriptome characterization of various time points and at various conditions has brought to the fore crucial genes that play roles in virulence, metabolism, and biofilm formation, thus giving a diverse aspect of how the bacteria adapt. These genes were discovered, and this explains why the regulation of genes is essential in the characterization of the possible pathogenicity and survival mechanisms of the bacterium. The knowledge supplied by protein and metabolite studies also introduces functional implications of genetic and transcriptomic adaptations. Protein interaction and protein expression analysis demonstrate regulator bands of proteins which regulate significant activities and functions within the cells. Metabolomics has the potential to identify how *S. aureus* optimizes the performance of different biochemical pathways to regulate growth and survive a stressful environment; thus it provides new therapeutic manipulation targets. The combination of these multi-omic systems with computer models enables scientists to create complex cellular networks and predict how the bacteria will react to specific conditions. The accumulation of systems-level knowledge is of decisive significance regarding the control of antibiotic resistance, the enhancement of diagnostics, and the establishment of treatment interventions to manage an infectious disease. With new advances in emerging technology, such as single-cell analyses and AI still in development, there will be more opportunities to enhance the knowledge of *S. aureus*, which will only make the intended application of these findings in the clinic even more appealing.

REFERENCES

Aebersold, R., & Mann, M. (2003). Mass spectrometry-based proteomics. *Nature*, *422*(6928), 198–207. https://doi.org/10.1038/nature01511.

Angermueller, C., Pärnamaa, T., Parts, L., & Stegle, O. (2016). Deep learning for computational biology. *Molecular Systems Biology*, *12*(7). https://doi.org/10.15252/msb.20156651

Bantscheff, M., Lemeer, S., Savitski, M. M., & Kuster, B. (2012). Quantitative mass spectrometry in proteomics: Critical review update from 2007 to the present. *Analytical and Bioanalytical Chemistry*, *404*(4), 939–965. https://doi.org/10.1007/s00216-012-6203-4

Baumler, A., & Fang, F. C. (2013). Host specificity of bacterial pathogens. *Cold Spring Harbor Perspectives in Medicine*, *3*(12), a010041. https://doi.org/10.1101/cshperspect.a010041

Beenken, K. E., Dunman, P. M., McAleese, F., Macapagal, D., Murphy, E., Projan, S. J., . . . Smeltzer, M. S. (2004). Global gene expression in *Staphylococcus aureus* biofilms. *Journal of Bacteriology*, *186*(14), 4665–4684. https://doi.org/10.1128/jb.186.14.4665-4684.2004

Bosi, E., Monk, J. M., Aziz, R. K., Fondi, M., Nizet, V., & Palsson, B. Ø. (2016). Comparative genome-scale modelling of *Staphylococcus aureus* strains identifies strain-specific metabolic capabilities linked to pathogenicity. *Proceedings of the National Academy of Sciences*, *113*(26). https://doi.org/10.1073/pnas.1523199113

Burckhardt, R. M., Buckner, B. A., & Escalante-Semerena, J. C. (2019). *Staphylococcus aureus* modulates the activity of acetyl-coenzyme A synthetase (Acs) by sirtuin-dependent reversible lysine acetylation. *Molecular Microbiology*, *112*(2), 588–604. https://doi.org/10.1111/mmi.14276

Cheung, A. L., Bayer, A. S., Zhang, G., Gresham, H., & Xiong, Y. (2003). Regulation of virulence determinants in vitro and in vivo in *Staphylococcus aureus*. *FEMS Immunology & Medical Microbiology*, *40*(1), 1–9. https://doi.org/10.1016/s0928-8244(03)00309-2

Dettmer, K., Aronov, P. A., & Hammock, B. D. (2006). Mass spectrometry-based metabolomics. *Mass Spectrometry Reviews*, *26*(1), 51–78. https://doi.org/10.1002/mas.20108

Emwas, A., Szczepski, K., Al-Younis, I., Lachowicz, J. I., & Jaremko, M. (2022). Fluxomics – new metabolomics approaches to monitor metabolic pathways. *Frontiers in Pharmacology*, *13*. https://doi.org/10.3389/fphar.2022.805782

Freiberg, J. A., Ruiz, V. M. R., Gimza, B. D., Murdoch, C. C., Green, E. R., . . . Skaar, E. P. (2024). Restriction of arginine induces antibiotic tolerance in *Staphylococcus aureus*. *Nature Communications*, *15*(1). https://doi.org/10.1038/s41467-024-51144-9

Frost, L. S., Leplae, R., Summers, A. O., & Toussaint, A. (2005). Mobile genetic elements: The agents of open source evolution. *Nature Reviews Microbiology*, *3*(9), 722–732. https://doi.org/10.1038/nrmicro1235

Holden, M. T. G., Feil, E. J., Lindsay, J. A., Peacock, S. J., Day, N. P., Enright, M. C., . . . Parkhill, J. (2004). Complete genomes of two clinical *Staphylococcus aureus* strains: Evidence for the rapid evolution of virulence and drug resistance. *PNAS*, *101*(26), 9786–9791.

Holtfreter, S., Kolata, J., Stentzel, S., Bauerfeind, S., Schmidt, F., Sundaramoorthy, N., & Bröker, B. (2016). Omics approaches for the study of adaptive immunity to *Staphylococcus aureus* and the selection of vaccine candidates. *Proteomes*, *4*(1), 11. https://doi.org/10.3390/proteomes4010011

Ideker, T., Galitski, T., & Hood, L. (2001). A new approach to decoding life: Systems biology. *Annual Review of Genomics and Human Genetics*, *2*(1), 343–372. https://doi.org/10.1146/annurev.genom.2.1.343

Imdahl, F., Vafadarnejad, E., Homberger, C., Saliba, A., & Vogel, J. (2020). Single-cell RNA-sequencing reports growth-condition-specific global transcriptomes of individual bacteria. *Nature Microbiology*, *5*(10), 1202–1206. https://doi.org/10.1038/s41564-020-0774-1

Jabbari, S., King, J. R., Koerber, A. J., & Williams, P. (2009). Mathematical modelling of the agr operon in *Staphylococcus aureus*. *Journal of Mathematical Biology*, *61*(1), 17–54. https://doi.org/10.1007/s00285-009-0291-6

Kitano, H. (2002). Systems biology: A brief overview. *Science*, *295*(5560), 1662–1664. https://doi.org/10.1126/science.1069492

Koren, S., Harhay, G. P., Smith, T. P., Bono, J. L., Harhay, D. M., Mcvey, S. D., . . . Phillippy, A. M. (2013). Reducing assembly complexity of microbial genomes with single-molecule sequencing. *Genome Biology*, *14*(9). https://doi.org/10.1186/gb-2013-14-9-r101

Lei, T., Mao, Q., Chen, C., & Ji, Y. (2019). Metabolomic profiling of *Staphylococcus aureus*. *Methods in Molecular Biology*, 177–186. https://doi.org/10.1007/978-1-4939-9849-4_13

Lindsay, J. A. (2013). *Staphylococcus aureus* genomics and the impact of horizontal gene transfer. *International Journal of Medical Microbiology*, *304*(2), 103–109. https://doi.org/10.1016/j.ijmm.2013.11.010

Malachowa, N., & DeLeo, F. R. (2010). Mobile genetic elements of *Staphylococcus aureus*. *Cellular and Molecular Life Sciences*, *67*(18), 3057–3071. https://doi.org/10.1007/s00018-010-0389-4

Mann, M., Kulak, N. A., Nagaraj, N., & Cox, J. (2013). The coming age of complete, accurate, and ubiquitous proteomes. *Molecular Cell*, *49*(4), 583–590. https://doi.org/10.1016/j.molcel.2013.01.029

Novick, R. P., & Geisinger, E. (2008). Quorum Sensing in Staphylococci. *Annual Review of Genetics*, *42*(1), 541–564. https://doi.org/10.1146/annurev.genet.42.110807.091640

Renz, A., & Dräger, A. (2021). Curating and comparing 114 strain-specific genome-scale metabolic models of *Staphylococcus aureus*. *Npj Systems Biology and Applications*, *7*(1). https://doi.org/10.1038/s41540-021-00188-4

Resch, A., Leicht, S., Saric, M., Pásztor, L., Jakob, A., Götz, F., & Nordheim, A. (2006). Comparative proteome analysis of *Staphylococcus aureus* biofilm and planktonic cells and correlation with transcriptome profiling. *PROTEOMICS*, *6*(6), 1867–1877. https://doi.org/10.1002/pmic.200500531

Shukla, S. K., Carter, T. C., Ye, Z., Pantrangi, M., & Rose, W. E. (2020). Modeling of effective antimicrobials to reduce *Staphylococcus aureus* virulence gene expression using a two-compartment hollow fiber infection model. *Toxins*, *12*(2), 69. https://doi.org/10.3390/toxins12020069

Singh, V. K., Jayaswal, R., & Wilkinson, B. J. (2001). Cell wall-active antibiotic induced proteins of *Staphylococcus aureus* identified using a proteomic approach1. *FEMS Microbiology Letters*, *199*(1), 79–84. https://doi.org/10.1111/j.1574-6968.2001.tb10654.x

Somerville, G. A., & Proctor, R. A. (2009). At the crossroads of bacterial metabolism and virulence factor synthesis in staphylococci. *Microbiology and Molecular Biology Reviews*, *73*(2), 233–248. https://doi.org/10.1128/mmbr.00005-09

Sulaiman, J. E., Long, L., Qian, P., & Lam, H. (2022). Proteome profiling of evolved methicillin-resistant *Staphylococcus aureus* strains with distinct daptomycin tolerance and resistance phenotypes. *Frontiers in Microbiology*, *13*. https://doi.org/10.3389/fmicb.2022.970146

Taiwo, O. R., Onyeaka, H., Oladipo, E. K., Oloke, J. K., & Chukwugozie, D. C. (2024). Advancements in Predictive Microbiology: Integrating new technologies for efficient food safety models. *International Journal of Microbiology*, *2024*, 1–21. https://doi.org/10.1155/2024/6612162

Troitzsch, A., Van Loi, V., Methling, K., Zühlke, D., Lalk, M., Riedel, K., Bernhardt, J., Elsayed, E. M., Bange, G., Antelmann, H., & Pané-Farré, J. (2021). Carbon source-dependent reprogramming of anaerobic metabolism in *Staphylococcus aureus*. *Journal of Bacteriology*, *203*(8). https://doi.org/10.1128/jb.00639-20

Wang, Z., Gerstein, M., & Snyder, M. (2008). RNA-Seq: A revolutionary tool for transcriptomics. *Nature Reviews Genetics*, *10*(1), 57–63. https://doi.org/10.1038/nrg2484

Westermann, A. J., Förstner, K. U., Amman, F., Barquist, L., Chao, Y., Schulte, L. N., . . . Vogel, J. (2016). Dual RNA-seq unveils noncoding RNA functions in host–pathogen interactions. *Nature*, *529*(7587), 496–501. https://doi.org/10.1038/nature16547

Wiechert, W. (2001). 13C metabolic flux analysis. *Metabolic Engineering*, *3*(3), 195–206. https://doi.org/10.1006/mben.2001.0187

15 Advancements in *Staphylococcus aureus* Vaccine Development

Recent Insights

Kalal Akhila[1], *Teetla Shivani*[1], *and P. Jayasree*[2]

[1]Department of Immunology & Molecular Biology, Mahavir Hospital & Research Centre, Lakdikapul, Hyderabad, Telangana, India

[2]Department of Pharmacology, School of Allied and Healthcare Sciences, Malla Reddy University, Maisammaguda, Hyderabad, Telangana, India

ABBREVIATIONS

BSIs	bloodstream infections
Can	collagen-binding protein
ClfA	committee clumping factor A
CnBP	collagen-binding protein
CTL	cytotoxic T lymphocyte
EAC	event adjudication committee
EVs	extracellular vesicles
FnBP	fibronectin-binding protein
FnBPA and FnBPB	fibronectin-binding proteins A and B
GMT	geometric mean titers
Hla	α-hemolysin
IDMC	independent data monitoring
IsdA	iron-regulated surface determinant A
LMICs	low- and middle-income countries
MEVs	multi-epitope vaccines
MntC	manganese transport protein C
MRSA	methicillin-resistant form
mSEC	mutant of staphylococcal enterotoxin C
OPA	opsonophagocytic activity
OPK	opsonophagocytic killing
PFT	pore-forming toxin
PMNs	polymorphonuclear leukocytes
PNAG	poly-*N*-acetyl glucosamine
PS/A	polysaccharide/adhesin
SA4Ag	*Staphylococcus aureus* 4-antigen vaccine
SAEs	serious adverse events
SEA	staphylococcal enterotoxin A
SEB	staphylococcal enterotoxin B

DOI: 10.1201/9781003647072-19

SSIs	surgical-site infections
SSTIs	skin/soft tissue infections
Tdap	tetanus–diphtheria–pertussis

15.1 INTRODUCTION

Recent progress in the development of *Staphylococcus aureus* vaccines highlights the considerable scientific and clinical challenges in implementing an effective vaccination strategy against this highly adaptable, infectious agent. While decades of research have produced several vaccine candidates that have reached preclinical testing models and early phases of clinical trials, few have resulted in usable human vaccines. A major factor in these successful uses is *S. aureus*'s great success at evading and responding to altered immune responses, regulating its surface antigens to induce immune tolerance, and rapidly changing its upper pole at the epidermal and or protective levels within various host environments (Ansari et al., 2019). The propensity of this pathogen to move through various bodily compartments inventoried as both immunogen and infection is further complicated by its status as a ubiquitous component of human microbiota and found often asymptomatically on the skin and within the nasal passages, for instance. Moreover, the presence of SA could be considered a commensal as it can exist within the human body, presence-conditioning the immune system, which may greatly alter the response of protective or immunogenic antigen identification (Proctor 2015).

Conventional vaccine approaches that rely primarily on the induction of antibody-mediated immunity were inadequate in combating *S. aureus*, as this organism uses many immune evasion strategies, including: secretion of proteins to prevent opsonization and phagocytosis, modification of surface molecules to escape detection, and reprogramming host immune signaling. As a result, recent studies have put more emphasis on the inclusion of T cell-mediated immunity, specifically Th1 and Th17 activation, which are critical in inducing effective immune clearance of intracellular *S. aureus* and associated biofilms. Complicating the vaccine initiative are the bacteria's virulence factors that are utilized to suppress immunity and secure nutritional elements, primarily iron. *S. aureus* possesses the ability to produce siderophores and heme acquisition systems, allowing it to proliferate in human blood and other environments where iron is limiting. These virulence factors are also being evaluated as possible vaccine targets, manipulating the capacity of the bacteria to survive and cause disease (DeDent et al., 2012) (Daum & Spellberg, 2012).

Due to the wide array of diseases caused by *S. aureus* – from mild skin infection to life-threatening conditions such as bacteremia, endocarditis, and pneumonia – a "one-size-fits-all" vaccine strategy is probably not going to be feasible; thus, a multi-antigen strategy targeting multiple conserved proteins and/or immunologically relevant domains or virulence determinants is growing in popularity. These strategies aim for broad-spectrum efficacy by simultaneously interrupting multiple pathogenic pathways along with stimulating allusions of adaptive immunity. Fortunately, technological advances in genomics and proteomics, as well as systems biology approaches, are facilitating not only a greater selection of vaccine targets but also a greater understanding of the host–pathogen interface at the molecular level. Personalized medicine will also improve upon vaccine designs by providing a snapshot of the host's genetic variability, immune status, and microbiome. To develop a successful *S. aureus* vaccine, it is likely that a balance of basic immunology, molecular microbiology, and personalized therapeutic strategies will be needed to address the multifaceted nature of this formidable pathogen (Wong Fok Lung et al., 2022).

15.2 RECENT CLINICAL TRIAL OUTCOMES AND LESSONS LEARNED

15.2.1 SA4Ag Vaccine Phase IIb Failure Analysis

The SA4Ag vaccine was developed to be preventive against *S. aureus* infections. SA4Ag was a multi-antigenic vaccine developed to provide a better immune response against this highly virulent and adaptable pathogen. The vaccine entered clinical evaluation as part of a phase IIb trial called STRIVE, but this was stopped early despite a rational development phase and potential. There were a number of important factors that led to this outcome.

The most important reason for stopping the trial early was related to a recommendation from the Independent Data Monitoring Committee (IDMC) following a scheduled interim analysis, which indicated that the interim results were related to futility. IDMCs are independent panels of experts that provide oversight of the safety of participants during trial-related events as well as overall treatment efficacy. A recommendation to stop the STRIVE trial indicates that the interim results suggested futility – that is, the chance of achieving a significant (statistically) benefit at the end of the trial. This conclusion often results from the trial's inability to meet pre-established efficacy thresholds, suggesting that the vaccine offered no real clinical benefits over background controls. In these cases, it is concluded that closing the study is the right move to conserve resources and protect participants from unnecessary risks. Another key factor leading to the trial's ultimate failure can also be reflected in the perceived lack of efficacy. Although detailed efficacy outcomes were ultimately never disclosed to the public, the decision to close the trial indicated that the SA4Ag vaccine failed to meet its primary endpoints, which likely included being able to reduce the numbers of *S. aureus* infections among vaccinated participants (Creech et al., 2020). The vaccine may not have elicited sufficient immune response to offer protective benefit against key types of infections, including bloodstream infections (BSIs) and surgical-site infections (SSIs). In vaccine trials, efficacy performance is usually established through a series of clear endpoints, which are often expressed as infection rates or seroconversion rates. The fact that there was no statistically significant improvement over placebo or standard of care indicates that the vaccine had very limited protective performance.

The Vaccine antigenic structure was another important factor. SA4Ag was designed with four antigens (to increase immune coverage). This includes two capsular polysaccharide (CP) serotypes 5 and 8 (CP5 and CP8), which are the dominant serotypes responsible for *S. aureus* infections, conjugated to CRM197, which is a non-toxic variant of diphtheria toxin and is used to enhance T cell dependent immune responses. The SA4Ag vaccine also included two recombinant proteins, ClfA (Clumping factor A) and IsdA (Iron-regulated surface determinant A), as shown in Figure 15.1. These are important for bacterial adhesion and iron acquisition, both of which are critical virulence pathways (Inoue et al., 2018). Although the four virulence pathogen-targeting strategy was intended to simultaneously target several *S. aureus* virulence pathway signatures, complexity sometimes has the potential for creating immunological interference. Specifically, the influence of immunological interference is most often observed when the immune response to one antigen inhibits or diminishes the immune response to another, with the potential for inducing differential or suboptimal protection from all pathogens. Additionally, it is not unlikely that the strength of recognition between human immune recognition and the respective antigens is unequal, resulting in inconsistent immunogenicity across the studied population. The outcome of the trial may have also been vulnerable to modulating conditions within the study population characteristics or trial design. The Phase IIb STRIVE trial enrolled a broad, diverse sample of healthy adults aged 18 to 85, including a sample of healthy Japanese adults (Creech et al., 2020; Inoue et al., 2018). While this diversity is useful to measure performance of the vaccine across relative populations, it also brings in many independent variables, including different baseline immunity, genetic makeup, environmental exposures, or lifestyle factors that can have a robust effect on immune responsiveness, results in terms of effectiveness of the vaccine might have been influenced by the wide age range of participants as older adults typically experience immunosenescence or age-related immune decline, which can affect efficacy of a vaccine. It is also possible that certain aspects of trial design, such as dosing regimens, timing of booster doses, and clinical endpoints, may not have aligned with the immunological profile of the vaccine, resulting in inconclusive or poor results.

Challenges with immunogenicity further complicated the development of SA4Ag. Immunogenicity, the ability of the vaccine to induce a strong, specific, and lasting immune

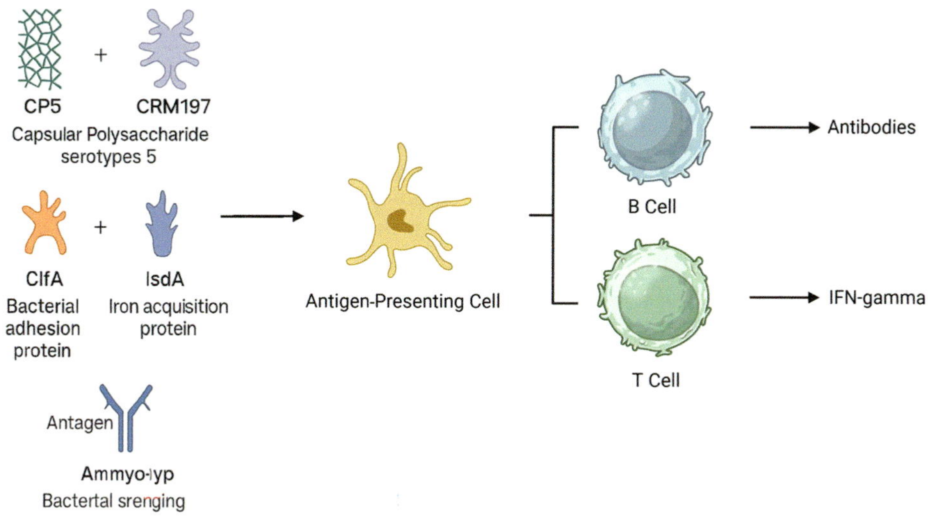

FIGURE 15.1 SA4Ag Vaccine: antigen components and immunological response. Schematic representation of the SA4Ag vaccine composition and its intended immunological action. The vaccine includes four antigens: capsular polysaccharides CP5 and CP8 conjugated to CRM197, and two recombinant proteins – ClfA (clumping factor A) and IsdA (iron surface determinant A). These components aim to stimulate a robust antibody and T cell response targeting adhesion, iron acquisition, and immune evasion mechanisms of *Staphylococcus aureus*.

response, is a fundamental measure of a vaccine's success, as the immune response is the measure of success for an immune health product. Despite including multiple well-characterized antigens, SA4Ag may not have achieved adequate antibody titers or T cell responses to provide protection (Inoue et al., 2018). One of the fundamental challenges is that *S. aureus* is a uniquely agile pathogen, with many immune evasion mechanisms, such as antibody neutralization by protein A, biofilm formation, and direct manipulation of host immunity. Complicating this further is the fact that we do not yet have well-defined correlates of protection for *S. aureus*; thus, there are no widely accepted immune correlates of vaccine-mediated protection to guide findings. This also complicates how we interpret the immunogenicity results, as well as presents fundamental barriers in selecting and optimizing antigen combinations. On the positive side, the SA4Ag vaccine exhibited an acceptable safety profile, and follow-up extension studies have shown that there haven't been any serious adverse events (SAEs) related to vaccination reported up to 36 months (Creech et al., 2020). This suggests that the vaccine was reasonably well tolerated; however, safety is only one aspect of vaccine success. Even the presence of mild side effects like local injection-site effects, fever, or fatigue can impact participant willingness to accept subsequent doses or remain in long-term studies. Additionally, in cases where efficacy is absent, even modest tolerability concerns can change the perceived benefit-to-risk ratio to the negative, supporting a decision to discontinue future developments.

In summary, the failure of SA4Ag in the Phase IIb STRIVE trial was a multifaceted event, influenced by a weak efficacy signal, immunological complexity of a multi-antigen vaccine, the diverse and heterogeneous population of the study, immunogenicity challenges with *S. aureus* as a target, and limited predictive markers for protection. Although it was not safety issues that prompted the termination of the trial, the overall lack of clinical benefit highlighted the need to reconsider vaccine strategies against this difficult pathogen. (Creech et al., 2020; Inoue et al., 2018).

15.2.2 Efficacy Endpoints and Futility Criteria in Spinal Surgery Patients

The SA4Ag vaccine, designed to prevent *S. aureus* infections in patients undergoing surgery, was evaluated in a Phase IIb trial that enrolled patients undergoing elective spinal surgery to assess the initiation of *S. aureus* post-operative infections. Post-operative infections are significant and common complications of surgery. The primary efficacy endpoint was the prevention of *S. aureus* BSIs and/or the occurrence of deep incisional or organ/space SSIs at any time within 90 days of surgery, with all events independently adjudicated by an Event Adjudication Committee (EAC) to ensure objective assessment. Secondary endpoints extended the observation period to 180 days, with a focus on *S. aureus* BSIs; deep SSIs; superficial SSIs; and infections of other types that occurred at any time within 90 or 180 days to capture delayed or less severe infections that were potentially important for recovery. Pre-established interim analysis showed equal totals for primary endpoint events for both vaccine and placebo groups (12 events each) with a conditional power of 6.09%, much lower than the defined futility threshold. We recommended that we halt enrollment. The final analysis confirmed the interim analysis findings, with 14 infections in each group and a vaccine efficacy of 0.0%. This zero efficacy result was particularly surprising, especially since earlier studies in healthy populations demonstrated strong immunogenicity with good antibody responses to all antigens. These results reaffirm the difficulty of assessing vaccine efficacy in surgical patients, as there are a number of confounding factors and patient variability that can influence the clinical effect of even immunogenic vaccines, including: (1) patient variability; (2) variability in surgical practices among surgeons (i.e., procedures provided vary per provider); and (3) a lack of established immune correlates of protection (Hassanzadeh et al., 2023).

15.3 CURRENT VACCINE CANDIDATES IN CLINICAL TRIALS

15.3.1 GSK's SA-5Ag Adjuvanted Vaccine

Pre-specified interim analysis showed equal numbers of primary endpoint events for the vaccine and placebo groups (12 events each), with conditional power of 6.09% – substantially below the futility threshold. We recommended that the month-long enrollment stop. The final analysis also showed equal numbers of infections in both groups (14 events each), resulting in an efficacy estimate of 0.0%. The zero efficacy result was especially surprising, especially because earlier studies in healthy populations showed strong immunogenicity, with good antibody responses to all antigens. These findings reinforce the very tenuous ability to measure vaccine efficacy when assessing surgical patients, due to multiple confounding factors and patient variability, which can alter the clinical effect of even immunogenic vaccines, including variability between patients, variability in surgical practice (i.e., procedures provided differ per provider), and lack of established immune correlates of protection (Clegg et al., 2021)

15.3.2 Novadigm's NDV-3A Cross-Reactive Vaccine

The vaccine candidate NDV-3A from Novadigm is a cross-protective vaccine candidate designed to protect against the infection with *S. aureus* infection (most clinically relevant in the methicillin-resistant form (MRSA)) as well as *Candida albicans* infections. This dual purpose is particularly important because both pathogens have an increased clinical burden. NDV-3A induces a variety of immune responses, including IgG production and Th17-derived cytokines IL-17A and IL-22, which help recruit immune cells and contribute to pathogen elimination. Preclinical studies show that NDV-3A significantly decreases the severity of MRSA-induced skin infections in mice, demonstrated by smaller abscesses and less tissue damage. Additionally, NDV-3A induced strong T cell responses, which provided protection from both local and systemic infections. Overall, the highlight of NDV-3A is the potential for avoiding serious infections, especially in at-risk populations (Yeaman et al., 2014).

15.3.3 Olymvax's rFSAV Multi-Antigen Formulation

Olymvax is developing a fully recombinant multi-antigen vaccine referred to as rFSAV, which is currently in Phase II clinical assessment for efficacy against *S. aureus*. The vaccine incorporates five different antigens that have been chosen to induce robust immune responses against the organism, which is a common cause of virulent infections, particularly in people who are immunocompromised. Researchers used antiserum from immune volunteers and overlapping peptides to identify the B cell immunodominant epitopes from four of the five antigens, detailing eight epitopes, five of which from the vaccine platform described here had not previously been identified. After defining the epitopes, they produced ten mAbs, which they screened in a septic mouse model. All ten showed partial protection, and four mAbs specified in a cocktail had a protective effect against MRSA *in vivo* that reduced both bacterial load and pathology. These findings provide credence to the specific rFSAV formulation and associated monoclonal antibodies, as speculative vaccine and therapeutic agents (Zeng et al., 2021).

15.3.4 Integrated Biotherapeutics' IBT-V02 and STEBVax

Integrated Biotherapeutics has developed IBT-V02, a novel multi-component toxoid vaccine against *S. aureus*, an organism responsible for many infections such as skin/soft tissue infections, bacteremia, and pneumonia. IBT-V02 has shown broad efficacy against numerous *S. aureus* strains in high-income countries. To help overcome cold-chain issues experienced in low- and middle-income countries (LMICs), we implemented a lyophilized (freeze-dried) formulation and a spray freeze-dried formulation for IBT-V02. These formulations retained the biophysical and toxin-neutralizing properties of the liquid formulation. For efficacy testing of the lyophilized IBT-V02 form in animal models, we included multiple *S. aureus* isolates from LMICs. The lyophilized form exhibited strong protective efficacy against the *S. aureus* isolates, providing a significant global vaccine candidate for improving vaccine accessibility (Wang et al., 2024). STEBVax is a different candidate from Integrated Biotherapeutics, consisting of a recombinant staphylococcal enterotoxin B (rSEB) protein, which has been designed with three mutations in the MHC class II binding site to provide additional safety and efficacy. It is administered parenterally with alum (Alhydrogel) as an adjuvant. The vaccine was tested as a first-in-human, open-label, dose-escalation study in healthy volunteers aged 23 to 38 years, which graded doses from 0.01 μg to 20 μg. The vaccine was tolerable in all doses administered, and there were no SAEs. Of note, approximately 93% of subjects receiving doses of 2.5 μg to 20 μg seroconverted and developed SEB-neutralizing antibodies. Additionally, a significant relationship between serum IgG levels and neutralization of interferon-gamma production in vitro was observed, supporting the vaccine's immunogenicity and safety (Chen et al., 2016)

15.4 VACCINE DESIGN STRATEGIES AND ANTIGEN SELECTION

15.4.1 Multi-Component Approaches (4-Antigen versus 5-Antigen Formulations)

A new multi-component vaccination strategy employing bacterial outer membrane vesicles (OMVs) was designed to co-express four virulence factors from *S. aureus*: ClfA^Y338A, LukE, SpA^KKAA, and Hla^H35L. The combination of these four components is referred to as CLSH-OMVsΔ60. The purpose of this tetravalent vaccine formulation is to provide enhanced protective immunity by targeting three antigens simultaneously – something that is well-suited to the antigenic diversity associated with *S. aureus* and its ability to circumvent host immune responses. CLSH-OMVsΔ60 elicited strong antigen-specific antibody responses associated with opsonophagocytosis, inhibition of hemolysis, and blocking leukocytic killing, suggesting strong functional immune activation. Additionally, in several murine models, including skin infection, sepsis, and kidney abscess models, CLSH-OMVsΔ60 provided significant protection against *S. aureus* infections. Additionally, this strategy allows for a simpler and potentially less expensive method of vaccine design from serologically distinct pathogens with antigenic diversity, demonstrating the capacity to extend this approach to other bacterial vaccine developments (König et al., 2021).

A new multi-component vaccination strategy employing bacterial OMVs was designed to co-express four virulence factors from *S. aureus*: ClfA^Y338A, LukE, SpA^KKAA, and Hla^H35L. The combination of these four components is referred to as CLSH-OMVsΔ60. The purpose of this tetravalent vaccine formulation is to provide enhanced protective immunity by targeting three antigens simultaneously – something that is well-suited to the antigenic diversity associated with *S. aureus* and its ability to circumvent host immune responses. CLSH-OMVsΔ60 elicited strong antigen-specific antibody responses associated with opsonophagocytosis, inhibition of hemolysis, and blocking leukocytic killing, suggesting strong functional immune activation. Additionally, in several murine models, including skin infection, sepsis, and kidney abscess models, CLSH-OMVsΔ60 provided significant protection against *S. aureus* infections. Additionally, this strategy allows for a simpler and potentially less expensive method of vaccine design from serologically distinct pathogens with antigenic diversity, demonstrating the capacity to extend this approach to other bacterial vaccine developments (Deng et al., 2019).

15.4.2 Protein-Based Vaccines (ClfA, MntC, CPs)

Driven by the rise of antibiotic resistance in *S. aureus*, there has been a greater focus on developing effective vaccines. Some of the most promising candidates are protein-based vaccines that will target two essential virulence factors: Clumping factor A (ClfA), manganese transport protein C (MntC), and CPs. These components help promote appropriate immune responses and induce protective immunity against staphylococcal infections by targeting different parts critical for pathogenesis and immune evasion. ClfA (clumping factor A) is a surface protein of *S. aureus* that promotes the adherence of staphylococci to host tissues, enabling colonization and the initial step in the infection process. Due to its fundamental role in pathogenesis, ClfA has been widely studied as a vaccine target. One study investigated the combination of ClfA in a vaccine formulation, combined with CPs CP5 or CP8. Though the addition of ClfA showed a slight reduction of bacterial loads from infected rat bones when compared to the capsular-only vaccine group, this decrease in burden from ClfA was not statistically significant. Not surprisingly, the change of ClfA for detoxified alpha-toxin (dHla) in the formulation showed a much larger reduction in the burden and severity of infection. These data suggest that even though ClfA does show some potential in these data, its importance in efficacy is limited either alone or in specific combinations. As well, previous clinical trials involving ClfA-based passive immunotherapy were not successful, showing the difficulty of considering ClfA as an independent antigen (Lattar et al., 2014). Nonetheless, other data lend credence to the idea that ClfA has great potential in a multi-component vaccine approach. For example, in rabbit models of endocarditis, rabbits immunized with recombinant ClfA peptides showed significant reductions in bacterial counts in blood and tissue after challenge compared to controls, indicating that vaccines targeting ClfA could provide at least some protection against severe infection, particularly when administered amid other antigens in high-risk populations (Lee, 2001).

Manganese transport protein C (MntC) is another attractive protein antigen that contributes to manganese uptake, an important process for both bacterial survival and virulence. Vaccines that specifically target MntC may inhibit this acquisition pathway for an essential metal cation, thereby attenuating the pathogen and allowing for immune-mediated clearance. MntC may not have been the main target in the studies mentioned in the cited work, but because of its essential role in *S. aureus* physiology and infection, it should be included in multivalent vaccine formulations (Pier & Maira-Litran, 2021). The SA2412 polypeptide of *S. aureus* has also gained importance as an immunogen for vaccines in the last few years. These types of vaccines, which include SA2412 and its analog-derived sub-epitopes, are showing promise to generate immune responses that may ultimately prevent staphylococcal infections. Research implies that combining SA2412 with other antigens (e.g., IsdB) could improve vaccine performance, mirroring a broader shift toward multivalent approaches. In addition, passive immunization with antibodies against SA2412, or its fragments, has been posited as an auxiliary approach to vaccination (McNeely et al., 2013). Capsular polysaccharides (CPs), which include CP5, CP8, and the CP/adhesin (PS/A), are an important immune-evasive part of the bacterial capsule, allowing the pathogen to avoid host defenses. Vaccines containing these PSs can induce active and passive immunity. The CPs minor components may be purified, or modified to enhance immunogenicity. The targeting of capsular structures has reduced bacterial virulence and improved vaccine protection in preclinical studies (McNeely et al., 2013; Pier et al., 2021).

In conclusion, an integrated vaccine development approach targeting ClfA, MntC, SA2412, and the CPs represents an innovative approach to treating *S. aureus* infections. Multi-component formulations consider multiple aspects of bacterial pathogenesis, such as adhesion, nutrient acquisition, immune evasion, and toxin excretion, contributing toward the induced immune response and overall enhanced immune response. In consideration of the complexity and adaptability of *S. aureus*, especially in the emergence of increasing antibiotic resistance, these types of approaches are more relevant than ever (McNeely et al., 2013; Pier et al., 2021).

15.4.3 Toxin-Based Approaches (α-hemolysin, Enterotoxins)

α-Hemolysin (Hla) is a clinically significant pore-forming toxin (PFT) produced by *S. aureus*, and it plays an important role in the pathogenesis of *S. aureus* by lysing host cells, particularly RBCs, thereby promoting tissue damage. α-Hemolysin is mainly found in human isolates; however, it has been implicated in strains isolated from animals as well. A modified, non-toxic form of α-hemolysin (HlaH35L toxoid) was developed in previous studies toward vaccine development, and in a series of experiments, HlaH35L toxoid provided self-protection to rabbit RBCs, indicating it was a reasonable objective as a vaccine component. The rest of the studies evaluated a combination of the six primary staphylococcal PFTs and found that a combination of HlgA, LukS, HlaH35L, and LukAE323AB gave the most protection from RBC lysis, which supports the proposed strategy of targeting several toxins to provoke a broader immune response and protection. Enterotoxins were not the main focus of this study; however, the high immunogenicity and known virulence contribution will likely enhance vaccine formulations. In immunization experiments with α-hemolysin toxoid (campaigns of vaccinations), repeated α-hemolysin toxoid vaccinations resulted in high antibody responses in rabbits (titers of 4–8 international units per milliliter of serum). The skin lesions from α-hemolysin in the vaccinated animals were consistently less severe compared to controls, confirming effective protection. Together, the work shows that α-hemolysin and perhaps other enterotoxins as part of a multi-antigen vaccine may offer strong and broad protection against *S. aureus* infections (Zhang et al., 2021; Wleklinski et al., 1982)

15.4.4 Cross-Kingdom Antigen Strategies

Cross-kingdom antigen strategies repurpose proteins from one type of organism, such as fungi, to protect against infections caused by another type of organism, such as bacteria. This unique strategy uses the structural and functional similarities in antigens across a large variety of pathogens to develop cross-protective immune responses. One excellent example is the recombinant protein rAls3, made from the fungus *C. albicans*, which bears significant similarity to surface proteins of *S. aureus*, specifically clumping factor A (ClfA) and collagen-binding protein (Cna). These common characteristics allow the immune system to successfully recognize and respond to both fungal and bacterial pathogens. Studies done on mice in a preclinical setting have proven that immunization with rAls3 can offer considerable protection against *S. aureus* infections, suggesting that it could elicit a strong immune response. Building on these findings, the rAls3-based vaccine NDV-3 has now progressed to human clinical trials and demonstrates an acceptable safety profile and induces a strong immune response, including the activation of humoral (antibody) response and cell-mediated (T cell) immunity. The success of rAls3 is significant in showing the role cross-kingdom antigens can play in developing broad-spectrum vaccines, and this route for vaccine design is especially important as antibiotic resistance continues to expand. Overall, the cross-kingdom antigen approach exemplified by rAls3 is a promising pathway for the development of effective vaccines against *S. aureus* and potentially other pathogens (Yeaman, Filler, Chaili et al., 2014).

15.5 SAFETY AND IMMUNOGENICITY PROFILES

15.5.1 Local and Systemic Reaction Patterns

The most frequently reported local reactions after vaccination were pain, redness, and swelling at the injection site. Pain was typically mild (grade 1) following the first and second doses; however, a small number of subjects reported grade 3 pain following the first and third doses. Mild redness and swelling were also noted in addition to grade 3 reactions, following various previous doses in various vaccine groups. These local symptoms almost always resolved within 1–3 days following vaccination. Systemic reactions included generally observed symptoms such as fatigue and headache. Fatigue was reported as the most commonly reported systemic event, which impacted as many as 93.3% of subjects following the first dose, with incidence decreasing for subsequent doses. Headaches were reported by up to 71.4% of participants following the first dose, decreasing again with subsequent doses. Fever was generally mild, with mostly recorded temperatures <38.5°C; however, there were a few cases of grade 2 fever (Levy et al., 2015).

15.5.2 Functional Immune Response Measurements

Various *S. aureus* vaccine candidates have been assessed for functional immune responses using diverse immunological assays. In the 4-antigen vaccine (SA4Ag) evaluation, there were opsonophagocytic activity (OPA) killing assays for the CP serotypes (CP5 & CP8), a fibrinogen-binding inhibition (FBI) assay for the clumping factor A (ClfA), and a Luminex assay for ClfA and recombinant manganese transporter protein C (rMntC). It was demonstrated that 24 and 36 months post-vaccination, recipients of the SA4Ag consistently exhibited significantly greater geometric mean titers (GMTs) than baseline GMT for CP5, CP8, and ClfA. The GMT fold rises ranged from 2.7 to 8.1. In the case of rMntC GMT's shows decaying antibody responses by 36 months of follow-up, although GMFRs for younger and older cohorts remained above baseline at 1.8 and 1.5, respectively (Creech et al., 2020). These responses are important as they hold relevance to assessing not only the presence of antibody but also their functionality. OPA assays reflect the abilities of antibodies to opsonize the bacteria for neutrophil-mediated killing, while FBI assays reflect the ability of immune sera to block bacterial adherence to host tissues. Using diverse clinical isolates of *S. aureus* in such assays reflects vaccine efficacy across genetically diverse *S. aureus* strains (Scully et al., 2014).

A further example of the importance of functional responses comes from the immunization study using a mutated staphylococcal protein A (SpAmut) plus AS01. In this study, the authors found the following results: when the subject mice were infected with *S. aureus*, the initial infection was poorly protected in the vaccinated group. However, when secondarily infected with *S. aureus*, the vaccinated mice exhibited broader and functional immune responses, as evidenced by the ability to produce heightened levels of antibody, increased phagocytic activity by macrophages, and a more substantial humoral response against the antigens which SpA usually covers. These factors helped prevent systemic bacterial spread (Mandelli et al., 2024). In this study, the use of functional responses was able to grade the distinction between initial infection and subsequent re-infection. The authors also illustrated an increase in opsonizing antibodies (IgG1, IgG2a, IgG2b) to MntC, along with increased opsonophagocytosis (opsonophagocytic killing [OPK]) by macrophages upon immunization in a peritonitis mouse model using manganese transport protein C (MntC). The IgG not only allowed but also was responsible for the protection against re-infection with *S. aureus* (in this case, it did not prevent its initial infection with the bacteria). Both the elevated humoral antibody response and the humoral activity further confirmed the functional quality of the antibody responses that occurred as a result of active immunization with MntC. The cellular analytical studies showed increased CD4+ T cell proliferation and cytokine production, including IL-17 and IFN-γ, which likely describe the strength of a Th17 response. Passive immunization using sera from mice vaccinated against MntC exhibited decreased colonization of bacteria and an increase in survival. Cytokine neutralization studies confirmed the critical role of IL-17 was responsible for protection (Yu et al., 2018).

In another research project, it was reported that mice vaccinated with a non-toxic mutant of staphylococcal enterotoxin C (mSEC) provoked a mostly Th2-skewed immune response characterized by elevated IL-10/IL-4 levels and decreased IFN-γ levels. Additionally, it was noted that there was a shift toward a Th2 dominance, since the vaccine sera were able to inhibit IFN-γ and TNF-α production in vitro, the researchers hinted that a protective environment is favorable for limiting *S. aureus* infection (Hu et al., 2007). All future studies will need to ensure both humoral and cell-mediated components of vaccine-induced immunity are measured, but also need to do so with functional measures that are linked to real-world protection against *S. aureus*.

15.5.3 Duration of Antibody Responses

The duration and pattern of antibody responses are critically important for determining the duration of protection from *S. aureus* vaccines. We monitored serum antibody responses longitudinally in 15 patients with *S. aureus* bacteremia for a median of 29 days (range: 9–74 days) from the time of diagnosis to hospital discharge. Antibody levels of IgG, IgA, and IgM were measured, with IgG exhibiting the most common and sustained increases important for long-term immunity. In 12 of the 13 adult patients, the anti-IsdA IgG level increased, with peak response occurring between 7 and 37 days post-diagnosis. The median fold increase in IgG level from baseline to peak was 1.71, indicating a response from the adaptive immune system. There was patient-level variation: some patients, for example, exhibited a decline in antibody levels at first, but rebounded, suggesting that immune profiles can vary widely in people. A unique case was the 5-day-old neonate who exhibited a continual decline in antigen-specific IgG levels, as it is likely that the waning of maternal antibodies outpaced any endogenous antibody production. Our findings suggest that the immune response to *S. aureus* is highly individualized and that the immune response can be statistically inferred to target a variety of antigens. Multivalent formulations would likely confer broad and durable immune protection against *S. aureus* (Verkaik et al., 2010).

In addition, a long-term study utilizing the *S. aureus* 4-antigen vaccine (SA4Ag) has made some important observations about the longevity of the antibody response. There was good functional immunity against all four of its target antigens, CP5, CP8, ClfA, and rMntC, out to 12 months post-vaccination. Furthermore, immunity against three of these antigens (CP5, CP8, and ClfA) was sustained to 36 months post-vaccination. At both 24 and 36 months post-vaccination, the GMT for CP5, CP8, and ClfA was still significantly higher than baseline, with GMFRs ranging from 2.7 to 8.1. Although antibody levels dropped off from the maximum value for rMntC, the antibody levels were still elevated over baseline, with GMFRs of 1.8 in younger adults and 1.5 in older adults. Taken together, these data suggest a moderate but persistent antibody response. It should be emphasized that the study cohort consisted of participants between the ages of 18 and 85 years, and while age can affect immunogenicity, participants aged 65 and older still had considerable antibody levels that persisted over time. Lastly, there were no serious vaccine-related adverse events through the 36 months of the study, showing both long-term safety and immune persistence (Creech et al., 2020).

15.5.4 OPK Assays

OPK assays are an important in vitro procedure for measuring the functional capacity of vaccine-induced antibodies against *S. aureus*. These assays measure the ability of antibodies to promote opsonophagocytosis, where immune cells, mainly polymorphonuclear leukocytes (PMNs), detect and ingest bacteria which have been opsonized, or bound, by antibodies. OPK assays can also assess whether the antibodies allow for effective complement fixation, which also contributes to immune-mediated clearance of the bacteria. During a seminal study, two separate OPK assay protocols were described, one using human PMNs and another using differentiated HL-60 cells, which are neutrophil-like cells. Researchers can compare the efficiency of vaccine-induced antibodies to promote bacterial opsonization and phagocytic killing to determine their functional activity against *S. aureus*. OPK assays are important because *S. aureus* infections have a broad disease spectrum, from minor skin infections to serious diseases such as pneumonia or sepsis, and OPK assays provide a mechanism for identifying and assessing potential vaccine candidates. Therefore, OPK assays were described as a critical correlate of protection in the development of *S. aureus* vaccines (Dwyer & Gadjeva, 2013).

To further highlight the usefulness of OPK assays, the same group explored the bactericidal efficacy of human PMNs against a range of *S. aureus*, including highly encapsulated, type 1, and microencapsulated types 5 and 8. In their results, they found that over 85% of the microencapsulated strains (types 5 and 8) were killed by PMNs in the presence of fresh normal serum with active complement, indicating susceptibility to OPK. In contrast, the highly encapsulated type 1 strain was resistant to phagocytic killing in serum with active complement, which indicates a role for the bacterial capsule as a means for immunological escape. Interestingly, they found that the degree to which heat-inactivated sera from rabbits immunized with killed *S. aureus* was sufficient for opsonizing both the encapsulated and microencapsulated strains to achieve uptake by phagocytes. This implies that antibodies toward either the capsule or other cell wall-associated antigens may be sufficient for opsonization and killing in the absence of an intact capsule. Most importantly, they concluded the need for the production of vaccine-induced antibodies, which would opsonize as many *S. aureus* strains, especially partially encapsulated strains, as possible, to ensure effective immunologic clearance. Overall, OPK assays should be obligatory for assessing the protective capacity of *S. aureus* vaccine candidates (Xu et al., 1992).

15.6 EMERGING VACCINE DEVELOPMENT APPROACHES

Antibiotic-resistant *S. aureus* has become a major global health concern, prompting the development of advanced vaccine strategies. Among these, computationally designed multi-epitope vaccines (MEVs) show strong potential by targeting key virulence factors and stimulating both humoral and cellular immunity. Other promising approaches include glycoconjugate vaccines, extracellular vesicles (EVs), live-attenuated strains, and nucleic acid-based vaccines. Innovations in adjuvants, such as TLR-based formulations and nanoemulsions, further enhance vaccine efficacy, as shown in Figure 15.2. These novel strategies offer hope for effective prevention of *S. aureus* infections.

15.6.1 Computational MEV Design

Antibiotic-resistant *S. aureus* is becoming a growing concern; efforts are underway to create new vaccine strategies, such as computationally designed MEVs to induce a robust, broad-spectrum immune response. By presenting multiple antigenic epitopes, MEVs have the potential to elicit both humoral and cell-mediated immunity. One recent study sought to design an MEV of hemolysin proteins (important virulence factors of *S. aureus*) that consisted of four B cell, four cytotoxic T lymphocyte (CTL), and four helper T lymphocyte (HTL) epitopes. The MEV construct also contained the ribosomal protein L7/L12 and pan-HLA DR-binding epitope adjuvants to increase immunogenicity. In their MEP, safety assessments suggested that the vaccine would be hypoallergenic and non-toxic, and immune simulation showed strong interactions with Toll-like receptors (TLR2, TLR4, TLR6). Furthermore, the MEV had good antigenicity, stability, and 94.4% global population coverage, suggesting broad applicability for preventing *S. aureus* infections (Garrido-Palazuelos et al., 2024).

In a different computational effort, the collagen-binding protein of *S. aureus* was chosen as the target antigen. Epitopes were analyzed with an immunoinformatics pipeline based on antigenicity, allergenicity, toxicity, and cytokine induction. The epitopes were fused with appropriate linkers and a phenol-soluble modulin α4 adjuvant to form the MEV construct. Molecular docking and dynamics simulations showed a strong and stable interaction with TLR2, further validating the vaccine. Moreover, the T cell epitope set showed an astonishing global population coverage at 99.14% (Mahapatra et al., 2023). Another study used fibronectin-binding proteins A and B (FnBPA and FnBPB), which have critical roles in bacterial adhesion and invasion. The authors predicted three linear B cell epitopes, five CTL epitopes, and two HTL epitopes to be highly immunogenic, non-allergenic, and non-toxic. The finalized MEV construct was combined with suitable adjuvants and evaluated by molecular docking and a 100-nanosecond molecular dynamics simulation, confirming a stable complex with TLR2. The vaccine also had an incredible estimated population coverage, at 99.51%, supporting its potential development as a globally relevant prophylactic vaccine (Chatterjee et al., 2023).

The fourth study used reverse vaccinology to design an MEV targeting three important exotoxins – staphylococcal enterotoxin A (SEA), enterotoxin B (SEB), and toxic shock syndrome toxin-1 (TSST-1). Epitope predictions were made for B cell, CTL, and HTL responses, which we then concatenated using KK, AAY, and GPGPG linkers, respectively. Human β-defensin-3 (hBD-3) was used as an adjuvant to enhance immunogenicity. It was also predicted to be antigenic, non-allergenic, and to be suitable for expression in *E. coli* in a pET28a(+) vector. The in silico immune modeling suggested the potential to elicit a robust immune response to the vaccine and demonstrated the possibilities of this MEV as a vaccine candidate against *S. aureus* (Rahman et al., 2023).

Altogether, these computational studies have demonstrated the potential of immunoinformatics in rational vaccine development. They provide valuable templates for multi-epitope constructs that have potential for experimental validation and consideration for addressing the urgent need for effective strategies to combat *S. aureus* infections.

15.6.2 Novel Vaccine Formulations for Enhanced Immune Responses

A major approach to making *S. aureus* vaccines is to use recombinant proteins or polysaccharide antigens to elicit specific host immune responses. The components of these vaccines primarily focus on the surface antigens and secreted toxins of *S. aureus*, as these factors are important for opsonophagocytosis: when antibodies bind to pathogens and tag them for destruction by phagocytic immune cells. Learning from successful conjugate vaccines against other related bacterial pathogens, researchers are exploring glycoconjugation (the conjugation of protein antigens to polysaccharides) of protein antigens, which are designed to elicit protein-specific immune responses, to maximize immunogenicity and the success of subunit vaccines. Another new approach is to use EVs, which are naturally secreted by *S. aureus* and contain immunogenic proteins that are capable of generating a host immune response. Studies have shown that vaccination with *S. aureus* EVs will provide long-lasting protection in mouse animal models. These EVs are important vaccine candidates and platforms because of their inbuilt antigenic neutralizing ability and their unique ability to mimic the process of natural infection (Clegg et al., 2021).

An alternative strategy in vaccine development is live-attenuated or whole-cell strains of *S. aureus* that promote the exposure of the immune system to a wider variety of antigens. This strategy has shown some success in preclinical settings, and has been applied to veterinary uses, but some safety concerns – mainly associated with adverse effects – have created major barriers to the use of whole-cell vaccines with humans. Nucleic acid vaccines, toward which there are both DNA and RNA platforms, are another area of *S. aureus* vaccine development. In nucleic acid vaccines, genetic material encoding *S. aureus* antigens is delivered directly into host cells, where the cells develop the antigens and create an immune response against the antigens. Human clinical assessments for nucleic acid vaccines are still very in the early stages, but preclinical work suggests nucleic acid vaccines may have future development potential (Clegg et al., 2021).

Adjuvants are essential components of modern-day vaccines. They can improve the quality and duration of immune responses. While some adjuvants, such as aluminum salts, have been used for many years, they often fail to facilitate strong cell-mediated immunity. New adjuvant formulations are being evaluated/assessed (Clegg et al., 2021). One recently tested vaccine candidate called the Combo vaccine is comprised of HlaH35L (a non-toxic mutant of the α-toxin), EsxAB (a fusion protein of EsxA and EsxB), and two surface proteins, FhuD2 and Sur2. The Combo vaccine has shown promise in both preclinical models in mice and in phase I clinical studies in humans. An adjuvant (alum/SMIP–7.10 [alum-S-7]) was added to the Combo vaccine. SMIP–7.10 is an adjuvant that contains aluminum salts and a Toll-like receptor 7 agonist. The Combo vaccine with alum-S-7 outperformed the Combo vaccine with alum (survival: 75% versus 40%). The Combo vaccine with alum-S-7 induced significant antibody responses and could drive a Th1/Th17 CD4+ T cell response with IFN-γ and IL-17 production. Furthermore, CD4+ T cell depletion in mice reduced effectiveness by 20%, and passive serum transfer experiments demonstrated that antibodies had protective potential, demonstrating the joint importance of humoral and cellular immunity (Mancini, 2014).

Another area of development regarding formulation approaches is using nanoemulsion-based adjuvants. Nanoemulsions represent dispersions that contain nanoscale droplets, and for vaccines, the delivery to immune cells and immune activation can be enhanced by nanoemulsions transmitting the vaccine antigens. A recent study illustrated three ways to incorporate antigens into nanoemulsions: encapsulation (where the antigen is encapsulated into the nanoemulsion), mixture (the antigen mixed into the nanoemulsion), and combination (where both processes are completed). Encapsulation resulted in the strongest immune response, as encapsulating the vaccine antigen in nanoemulsions significantly improved the protective efficacy of humoral and cellular immunity. Mice vaccinated with encapsulated nanoemulsion formulations demonstrated higher antibody titers, robust T cell responses, lower bacterial loads in target organs, and a greater survival following MRSA challenge, demonstrating superior protective efficacy (Yang et al., 2018).

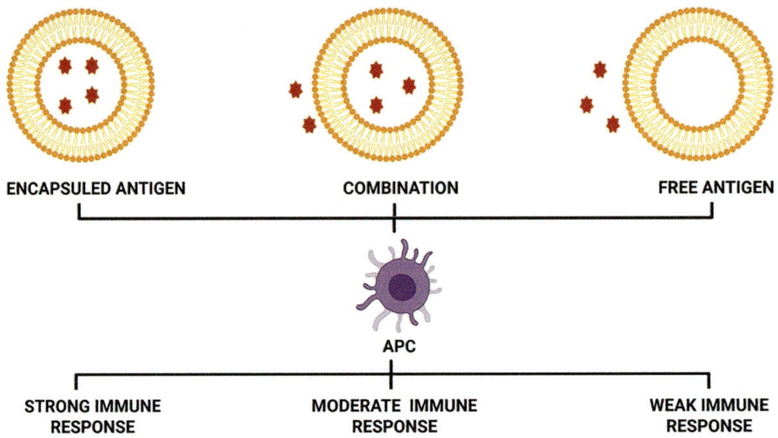

FIGURE 15.2 Comparison of antigen delivery techniques using nanoemulsions. Illustration of three nanoemulsion-based antigen delivery strategies: (1) encapsulation, where antigens are enclosed within nanodroplets for enhanced uptake and immune activation; (2) mixture, with antigens mixed externally with nanoemulsions resulting in weaker responses; and (3) combination, where partial encapsulation improves immune stimulation. Encapsulation leads to the strongest humoral and cellular immunity, as evidenced by antibody titers and T cell activation.

One of the main strategies for creating *S. aureus* vaccines is using recombinant proteins or polysaccharide antigens that generate a unique immune reaction. The components of many vaccines focus on the surface antigens and secreted toxins of *S. aureus*, as these are important in initiating opsonophagocytosis – the process where antibodies mark pathogens for destruction by phagocytic immune cells. If we look at other successful conjugate vaccines against bacterial pathogens, we are now seeing glycoconjugation of protein antigens to enhance overall immunogenicity and the result of using subunit vaccines. An additional interesting and newer strategy is to use EVs, which are routinely secreted by *S. aureus* and have immunogenic proteins that could activate immune responses, since the vaccine platform from *S. aureus*-derived EVs has been shown to produce lasting protection in mice. EVs present an attractive vaccination approach, as they represent the actual antigenic palette of the parent bacteria and their ability to emulate natural infection may allow the host to mount an effective broad immune response (Clegg et al., 2021).

Another approach to vaccine development is using whole-cell or live-attenuated strains of *S. aureus*, which can stimulate the immune system with many different antigens. This approach has been proven effective in preclinical models and has had some utility for veterinary medicine. However, significant barriers related to safety, especially concerning adverse reactions, continue to limit the widespread availability of whole-cell vaccines to humans. The area of nucleic acid vaccines, which includes DNA and RNA vaccines, is another avenue for *S. aureus* vaccine development. This type of vaccine works by delivering DNA or RNA that encodes *S. aureus* antigens into host cells. Once the host cells acquire the vaccines, they will produce the antigens and generate an immune response. At this time, they have not been evaluated clinically in humans, but preclinical studies suggest the potential for nucleic acid vaccine development (Clegg et al., 2021).

Adjuvants are critical to modern vaccine formulations by improving both the magnitude and duration of immune responses. While traditional adjuvants, such as aluminum salts, have been used for many years and are effective in providing a humoral immune response, they do not elicit robust cellular immune responses. Newer adjuvant formulations are under investigation for this purpose (Clegg et al., 2021). One example of a vaccine candidate, the Combo vaccine, consists of four components: HlaH35L (alpha-toxin non-toxic mutant), EsxAB (a fusion protein of EsxA and EsxB), and two surface proteins, FhuD2 and Sur2. In preclinical mouse models and a phase I clinical trial in humans, the Combo vaccine showed promise. The addition of a new adjuvant, alum/SMIP–7.10 (alum-S-7), which is a combination of aluminum salts and a Toll-like receptor 7 agonist, greatly improved the vaccine performance. Mice that received the Combo with alum-S-7 had a 75% survival rate compared to 40% with the Combo formulated with alum only. The Combo with the alum-S-7 formulation produced robust antibody responses and provided a Th1/Th17 CD4+ T cell phenotype that produced IFN-γ and IL–17. Further, depleting CD4+ T cells reduced efficacy by 20%, and passive serum transfer studies demonstrated that antibodies provided protection, showing the importance of both humoral and cellular immunity (Mancini, 2014).

Advanced formulation strategies utilizing nanoemulsion-based adjuvants constitute another innovation worth discussing. Nanoemulsions (which contain nanoscale droplets) can enhance antigen delivery and immune activation because they have effectively been shown to move vaccine antigens to immune cells. In a new study, researchers were able to establish three ways of incorporating the antigen in nanoemulsions: encapsulation (where the antigen is found within the nanoemulsion), mixture (where the antigen is mixed with the nanoemulsion), and combination (which allows for both). The encapsulation method had the greatest immune response, improving both humoral and cellular immunity. Mice that received the encapsulated nanoemulsion formulations had higher antibody titers, greater T cell responses, lower bacterial loads of target organs, and improved survival after the MRSA challenge, showing better protective efficacy (Yang et al., 2018)

15.6.3 Preclinical Vaccine Candidates (4C-Staph)

The 4C-Staph vaccine is an exciting protein-based candidate still in preclinical work and has the potential to provide broad protection against *S. aureus*, the pathogen responsible for a considerable range of serious infections, including MRSA. One of the key challenges in developing a vaccine against *S. aureus* is its antigenic diversity and ability to evade the immune system. The 4C-Staph formulation incorporates five antigens chosen because of their importance to bacterial virulence and pathogenesis, including two surface-associated lipoproteins associated with virulence: FhuD2, which is involved in iron acquisition; and Csa1A, a protein associated with surface adhesion, as well as three secreted virulence factors (α-hemolysin (Hla), EsxA, and EsxB). The α-hemolysin is a PFT that damages host cell populations, and a genetically detoxified (HlaH35L) version was used in the vaccine to ensure safety and immunogenicity. EsxA and EsxB, which are recognized as secreted immune-modulating proteins, were included as a fusion protein (EsxAB) to promote maximum immunogenic potential (Monaci et al., 2015; Redi et al., 2018)

The 4C-Staph vaccine has been formulated with adjuvants (MF59 and alum) to increase immune responses. MF59 is an oil-in-water emulsion that enhances antibody and T cell-mediated immune responses by increasing antigen uptake and activating antigen-presenting cells. While alum is capable of enhancing antibody responses, it is not as effective at inducing strong T cell-mediated immune responses. The preclinical studies indicated that the MF59 adjuvanted formulation of 4C-Staph produced stronger and longer-lasting immune responses in mice than the alum adjuvanted formulation. In particular, the mice vaccinated with 4C-Staph with MF59 developed high levels of IgG and IgM antibodies, as well as strong $CD4^+$ T cell responses necessary for long-term protective immunity against infection (Monaci et al., 2015). The immunological mechanisms of the 4C-Staph vaccine utilize both humoral immunity and cell-mediated immunity. Specific antibodies can be generated that mediate opsonophagocytosis, a crucial process in which immune cells are able to recognize and engulf pathogens. The vaccine-mediated activation of $CD4^+$ T cells, like those that activate B cells to produce antibodies, produces cytokines and helps amplify the immune response and antibody production. The importance of T cells is particularly important in situations with relatively low antibody responses and emphasizes the need for both forms of immunity for effective control of pathogens. Interestingly, the vaccine had protective efficacy in neutropenic mice with low neutrophils, suggesting it activated other immune mechanisms that can compensate for innate immune deficiencies and underscores the relevance of this for immunocompromised individuals (Torre et al., 2015)

The outcome measures of efficacy have been extensively studied by the 4C-Staph vaccine in a wide range of preclinical mouse studies. The preclinical studies especially came from peritonitis and systemic infection challenge models. In these experiments, the mice that received the vaccine lived significantly longer than those that didn't, and they also had reduced bacterial loads in the target organs as compared to controls. The typical immunization protocol used was three separate doses of 10 μg of each of the vaccine antigen from the 4C-Staph vaccine. The vaccine was given by the intramuscular route on days 0, 14, and 28. The vaccine indicated long-term protection since immunity and immune memory could be tested up to four months post-vaccination. The described immunity indicates a prophylactic and durable immunity against a pathogen that is known for both frequent repeat infections and persistence (Monaci et al., 2015; Redi et al., 2018; Torre et al., 2015).

15.7 PASSIVE IMMUNIZATION STRATEGIES

15.7.1 Monoclonal and Polyclonal Antibody Development

Monoclonal and polyclonal antibodies are being developed as a prominent means to design a safe and effective vaccine against *S. aureus*, a pathogen with a large spectrum of infections (including methicillin-resistant strains such as MRSA). There are two major targets for antibody development, namely CP and poly-*N*-acetyl glucosamine (PNAG), which are highly important antigens in terms of a pathogen's ability to escape the host immune response. It has been demonstrated that vaccination with either CP or PNAG antigens can provide polyclonal antisera from animals that can make use of a strong opsonic killing (OPK) activity in human blood. Polyclonal antibodies have distinct advantages because they can recognize several epitopes on the antigens, thus facilitating a broad and vigorous immune response, and monoclonal antibodies (MAbs) directed against the same antigens, which have consistent specificity and strong OPK activity, also offer advantages. However, an important finding from these studies was that mixes of polyclonal and monoclonal antibodies against CP and PNAG interfere with the OPK activity of each other. This antagonism was demonstrated by diminished antibody binding on the bacterial surface and diminished complement deposition, both of which are needed for effective opsonization and bacterial clearance. These findings illustrate the careful design of vaccine formulations and the selection of antibodies when combining (Skurnik et al., 2010).

In addition to surface polysaccharides, antibody development has been directed toward virulence factors, including the fibronectin-binding protein (FnBP), a major player in the adherence of these pathogens to the host tissue. Researchers have taken it to the next step and created synthetic peptides from amino acid sequences critical for FnBP binding to basic component fibronectin and have made polyclonal and monoclonal antibodies. The rabbits were immunized with peptide conjugates and produced very high-titered polyclonal antibodies that inhibited the binding of fibronectin. The second part of their work utilized mice and immunized them with synthetic peptide D3 (20–33 D) and produced a monoclonal antibody that was very effective, so 9C3 exhibited very potent blocking of fibronectin binding, even without the presence of fibronectin. The authors were careful to note that monoclonal antibodies demonstrated specificity – MAb 9C3 recognized a unique linear epitope (SVDFEED) and amino acids, which were important to FnBP biological function. This specificity is important for effective inhibition of bacterial adhesion and other off-target effects. The data yielded from this study indicate that monoclonal antibodies directed toward conserved, functionally important regions of virulence proteins may provide a novel aspect of vaccines aimed at inhibiting *S. aureus* (Huesca et al., 2000).

Another strategy is to find polypeptide epitopes that could induce protective immune responses in humans. The epitopes can be recognized by monoclonal antibodies, which could serve as both diagnostic elements and the basis of vaccine development. Because monoclonal antibodies bind to their respective epitopes with very high-affinity, monoclonal antibodies are optimal for use as therapeutics and prophylactics. Including these in vaccine formulations could greatly enhance immunogenicity and provide protection from *S. aureus* infections, including MRSA infection. This is an example of prioritizing both epitope-based design and antibody engineering to develop vaccines for complex bacterial pathogens (Klimka et al., 2023).

15.7.2 Passive Protection Approaches

Passive immunization provides an important part of the defense against *S. aureus* infections, in particular, where rapid protection is required, and active immunization cannot be administered rapidly enough to provide protection. The process consists of the transfusion of pre-formed antibodies from an individual or animal immunized with the *S. aureus*-specific immunogen(s) to a non-immunized recipient, thus providing immediate but temporary protection. In trials involving rabbits and the use of the multi-component 4C-Staph vaccine, passive immunity with hyperimmune serum from vaccinated rabbits directly transferred to recipient mice afforded the mice significant protection from *S. aureus* challenge. In fact, the mice that received hyperimmune serum had significantly lower bacterial loads in their kidneys and a significantly increased survival rate (58%) compared to the untreated controls (19%). The protection induced by the hyperimmune serum was considered to be due to the presence of functional antibodies capable of neutralizing toxins and opsonizing staph bacteria to facilitate opsonophagocytosis and, therefore, clearance by phagocytic immune cells (Bagnoli et al., 2015).

In subsequent studies of passive immunization, a nonhemolytic, nonlethal mutant of the *S. aureus* alpha-toxin (H35L) developed through site-directed mutagenesis was used. H35L mutant does not exhibit hemolytic activity yet retains its immunogenicity. Rabbits immunized with H35L produced antiserum that cross-reacted with wild-type alpha-toxin and neutralized the hemolytic activity of wild-type alpha-toxin in vitro. Passive immunity in mice expressed as protection against lethality from either purified wild-type alpha-toxin or high-toxin-producing strains of *S. aureus* was achieved by administering rabbit antiserum obtained from H35L immunization. The ability to generate protective antibodies using detoxified virulence factors as immunogens is incredibly useful to safely develop passive immunization strategies (Menzies & Kernodle, 1996).

Another promising strategy used polyclonal antibodies (pcAbs) generated in rabbits against multiple *S. aureus* antigens (alpha-hemolysin (Hla), staphylococcal enterotoxin B (SEB), manganese transporter component C (MntC)). In murine sepsis models, the passive transfer of these antibodies (SAvac-pcAb) provided full protective efficacy (100% survival) compared to only 10% survival in treated mice. Protection was attributed to robust opsonophagocytosis, mediated by the antibodies and neutrophils as the primary mechanism for the clearance of bacteria. Further, the antibodies cross-reacted with multiple clinical strains of *S. aureus*. This is particularly important considering the large diversity and antibiotic resistance of *S. aureus* isolates. In addition to promoting phagocytosis, the pcAbs were demonstrated to neutralize virulence factors, which diminished the pathological effects of the bacteria. These actions, combined with the ability of the pcAbs to promote opsonophagocytosis, enhanced the overall protective effect of the applied immunotherapy (Zhang et al., 2015).

15.8 CLINICAL CHALLENGES AND UNMET MEDICAL NEEDS

Developing a functional vaccine against *S. aureus* (*S. aureus*) has a number of significant challenges, as shown in Figure 15.3. First, *S. aureus* shows complicated host–pathogen interactions. There exist other bacterial pathogens that vaccines have been developed against, but *S. aureus* interacts with the human immune system in complicated ways that we don't fully understand. Developing vaccines that can reliably generate a protective immune response, therefore, becomes a problematic question. One significant challenge is the bacterium's sophisticated immune evasion tactics. *S. aureus* can invade host cells to evade immune recognition and can also exploit host metabolic processes to change the immune response being generated. This adaptation puts limitations on the number of effective vaccine strategies that can be employed. Of note, *S. aureus* also has transient genomic, transcriptional, and metabolic changes while residing in the host, a phenotype that we wouldn't see in the lab under normal experimental conditions. This variability, while inside the host, also complicates efforts to find consistent and universal vaccine targets. Also, there is the issue that this facultative pathogen is capable of exhibiting a "chameleon" phenotype of adaptation based on environmental signals, which complicates the menu of a broadly suitable vaccine (Wong Fok Lung et al., 2022).

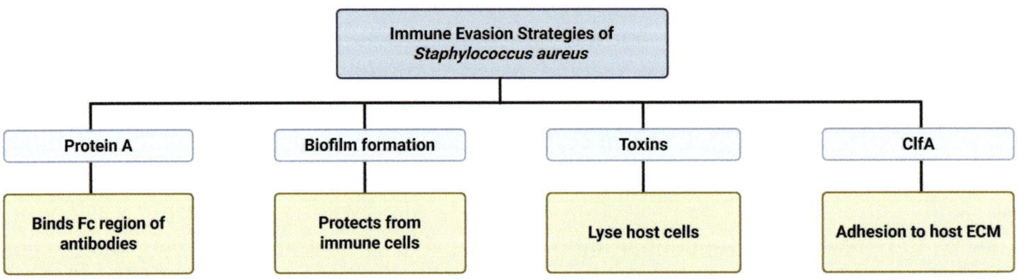

FIGURE 15.3 Immune evasion strategies of *Staphylococcus aureus*. Diagram highlighting the various immune evasion mechanisms used by *S. aureus*. These include protein A-mediated antibody neutralization, biofilm formation for immune shielding, α-toxin-mediated host cell lysis, ClfA-facilitated adhesion to host tissues, and iron scavenging through IsdA and siderophores. These strategies collectively hinder effective host immune clearance and complicate vaccine development.

The competition between *S. aureus* and the host immune cells for nutrients within infected tissues can also influence gene expression and antigen presentation, making it even more difficult to identify reliable targets to develop vaccines. In addition, the rise of precision medicine indicates vaccine development, which may integrate knowledge about human genetics and epigenetics, in an attempt to develop targeted interventions – an area still relatively nascent and presents further challenges to research and implementation (Wong Fok Lung et al., 2022)

One of the primary obstacles to the development of vaccines against *S. aureus* is the limited understanding of protective immunity in humans. Although there is a wealth of knowledge related to immune responses to a number of pathogens, *S. aureus* has a variety of virulence factors, and it remains unclear which immune components will confer protection. As such, this complication makes it very difficult to design a targeted vaccine. In fact, thus far, all clinical trials with *S. aureus* vaccines vanes own due, in large part, to the fact that these trials were driven primarily by antibody responses. These trials, as would be expected, illustrated that humoral immunity alone is insufficient for protection, and new strategies are needed. Recent work has highlighted the importance of cell-mediated immunity, particularly Th17 cells and IL-17, in protection against *S. aureus* infections. It is likely that effective vaccines will need to elicit these pathways. A significant issue, however, is the lack of reliable biomarkers for accurately predicting protective outcomes. The problem is multiplied because *S. aureus* does not have correlates of protection clearly defined by immune markers like other pathogens, and can cause difficulties in defining vaccine efficacy during development (Proctor 2012).

In addition to these obstacles, there is a wide array of diseases caused by *S. aureus*, including simple skin infections to severe illnesses such as sepsis and pneumonia. The diversity in disease states suggests that a singular, universal vaccine may not be possible; at best, targeted vaccines could be developed for individual clinical manifestations. Moreover, there is the issue of *S. aureus* colonization without symptom development, as individuals colonized in the nasal passages or on the skin may be healthy. Because colonization is often the preceding step to infection, it makes the development of a vaccine or vaccine construct, for example, to prevent infection in already colonized hosts exceedingly difficult. A further layer of complexity resides in the structure–function implications of PNAG as a possible vaccine target. One publication identified that the misunderstanding of free amines and *N*-acetylation in the number and position of amines on PNAG is limiting. Knowledge of this molecule is key to developing relevant antigenic determinants (Spellberg & Daum, 2012). Moreover, PNAG as a molecule has a complex structure, where basic use would require the synthesis of a library of PNAG Penta saccharides to test as potential vaccine epitopes – an endeavor that is also very labor-intensive and expensive (Proctor, R. A. (2012).

The challenges associated with vaccine development are worsened by the fact that many strains exist, including MRSA. A viable vaccine would require protection against multiple strains, along with the testing and validation of a large number of antigens. One possible solution involves using a conjugate vaccine containing a mutant bacteriophage Qβ as a carrier. However, ensuring proper immune presentation and response will remain a challenge. Additionally, both active and passive immunization will need to be assessed in preclinical models such as mice to ensure those approaches are effective, which is often a time-consuming and resource-intensive process. Lastly, any vaccine candidate will need to be evaluated for safety and efficacy before regulatory approval, adding another level of complexity to the development pipeline (Tan et al., 2024).

15.9 FUTURE DIRECTIONS AND RESEARCH FOCUS

One of the most exciting opportunities in *S. aureus* vaccine development is the optimization of formulations that will induce strong and durable immune responses. It is clear that although prior vaccine candidates had the goal of inducing strong humoral immunity through antibody responses, such candidates did not show protection in humans. Consequently, the current focus of researchers is to develop new strategies that will elicit both humoral and cellular immunity. In particular, T cell-mediated immunity is generally recognized as key to establishing long-term protective immunity. However, there is also an investigation into the use of advanced adjuvants that will augment the magnitude, breadth, and durability of immune response so that the pathogen can be more accurately targeted. To accomplish the goal of overcoming the translational gap between preclinical efficacy and clinical efficacy, researchers have access to more sophisticated experimental models now than they did previously. Conventional animal models are often inadequate to mimic the complexities of human immune responses to *S. aureus*, which has led to the use of next-generation systems, such as humanized mouse models and organ-on-chip platforms that more closely simulate human physiology and immune interactions, increasing predictability in vaccine efficacy and safety, and assisting in the selection and optimization of candidates before *S. aureus* vaccine clinical trials. Expanding our understanding of host–pathogen interactions is equally important. *S. aureus* has evolved elaborate methods to modulate host immune responses and cause persistent infections, and understanding these processes may expose new pathogens' weaknesses, assist in the design of targeted immunogens to neutralize immune evasion methods, and, ultimately, elicit better and more protective immunity (Clegg et al., 2021).

Antigen selection continues to be a central aspect of vaccine development. Future efforts are moving toward adding more antigens, particularly those contributing to virulence, rather than using a single antigen. A major area of change in covers vaccine design practice and emphasis will be to move toward multivalent products, including antigens targeted and involved with multiple immunological mechanisms, including but not limited to iron acquisition, immunoevasion, and bacterial survival – capability, coagulation pathways – involved in bacterial pathogenicity. This focus on developing a multivalent vaccine will increase the odds of producing a vaccine that could offer a broad-based protective cover against different *S. aureus* strains, including MRSA (Spellberg & Daum, 2012; Daum & Spellberg, 2012)

Recent immunological studies have highlighted the role of cellular immunity in *C. difficile* protection, with the conclusion that the target of a good vaccine should be the ability to elicit T helper 1 (Th1) and T helper 17 (Th17) cell responses in addition to antibodies. These T cell populations are important in the activation of phagocytes and inflammation, both of which are key to clearing *C. difficile* infections. Combinations of vaccine formats that specifically promote these specific immune responses may promote a more protective and long-lasting immune response against the bacteria. Another area of intense focus is the immune evasion mechanisms utilized by *C. difficile* and how to address these, as they are very similar to those utilized by *S. aureus*. One mechanism that is heavily studied is protein A, which is strongly virulent due to its ability to bind the Fc region of immunoglobulins. This binding interferes with opsonization, thereby blunting a necessary immune response. Future vaccine strategies have the potential to develop modifications of these proteins into non-toxic proteins to avoid their immune-modulating effects. Incorporating these modified proteins into vaccine candidates will improve immunogenicity and protective qualities (DeDent et al., 2012; Daum & Spellberg, 2012).

Additionally, vaccine trials have begun to focus on distinct high-risk populations. For example, surgical patients, those with chronic disease states, and those who have immune-compromised states (e.g., as a result of drug therapy) have been shown to be high-risk groups for *S. aureus* infections. Trials that address the safety and efficacy of vaccines in these particular population groups could assess meaningful vaccine safety or efficacy and allow for population-tailored immunization strategies to be derived for these groups. Trials targeting broader risk groups would likely also allow for easier regulatory approval and public health manoeuvrability. Combination vaccines are also attracting interest; they would allow for streamlined immunization schedules and may increase protection. By adding *S. aureus* antigens to current vaccines such as tetanus–diphtheria–pertussis (Tdap), public health could improve access and coverage and lessen the burden related to immunizations. The potential for combined vaccine strategies with therapy, such as monoclonal antibody therapy or bacteriophage therapy, is also new and could potentially offer synergism and improved overall treatment of *S. aureus* infections (Spellberg & Daum, 2012; Daum & Spellberg, 2012; DeDent et al., 2012).

Finally, new delivery methods are being invested in to promote immunogenicity and simplify delivery. Mucosal delivery to the nasal cavity, for example, may afford an opportunity to prevent *S. aureus* colonization, the first step toward reducing transmission and, ultimately, infection. As a large portion of the population is asymptomatic for *S. aureus* colonization, particularly in the nasal passage, targeting the colonization phase may reduce the likelihood of infection, which could improve public health. Given the interest in target delivery approaches, these innovations may be crucial in future vaccination strategies (DeDent et al., 2012).

In summary, the future of *S. aureus* vaccine development is focused on a multi-dimensional approach that combines immunology, new antigens, targeted clinical trials, and novel delivery platforms. To develop a safe and efficacious vaccine, we know we must overcome the complex biology of the pathogen and the pathogen's capacity to evade the immune response (Spellberg & Daum, 2012; Daum & Spellberg, 2012; DeDent et al., 2012; Clegg et al., 2021).

15.10 CONCLUSION

Developing a safe and effective vaccine against *S. aureus*, particularly MRSA, is one of the most complicated and longest-standing problems in bacterial vaccinology. Despite enormous investments and a significant level of science effort behind this challenge, numerous vaccine candidates ultimately failed or failed to demonstrate clinical efficacy in human trials and were often dismissed largely for having limited immunological breadth, ud965; for only eliciting ambiance-only immune responses, ud965; or for simply not elucidating the pathogenesis and immune evasion mechanisms used by *S. aureus* that hinder initiation and protection of the adaptive immune response. These oversights have amplified the need for a conceptual shift toward multivalent, immune-inclusive strategies that focus on developing vaccines by stimulating both arms of the adaptive immune system – immune responses combined with other factors will most like yield more substantial results over vaccine development using single-antigen approaches. Fortunately, we are seeing advances in newly developed vaccine candidates like Sta-V5, NDV-3A, and IBT-V02 that promise to address such barriers. Next-generation vaccine candidates are addressed based on broader protective immune response by utilizing mixtures of conserved protein antigens, toxoids, immune enhancers with broader and less restrictive use of immune enhancers, and so on. Notably, all of the above-named new vaccine formulations have demonstrated the ability to elicit strong antibody level responses, combined with a robust CD4+ T cell response (especially the Th1 and Th17 because of evidence of their critical role in durable protection against *S. aureus*). The combination of new adjuvants, such as nanoemulsions, alum-TLR agonist hybrids, as well as platform technologies like nucleic acid vaccines, also complements these approaches. Making equally transformative advances are novel experimental systems, such as humanized mouse models and organ-on-chip platforms, that can allow for a more accurate outcome of immune responses and disease outcomes in preclinical phase evaluation. Moreover, in conjunction with developments in systems immunology and computational vaccine design, these investigatory strategies are providing enhanced predictive power and shifting the focus to translate lab success to real-world impact.

A successful *S. aureus* vaccine will doubtless require a multi-pronged approach that includes optimization of antigen structural design, strong adjuvant systems, and insights into host–pathogen interactions. Personalization of the approaches may be necessary to maximize the impact in regard to real-world efficacy among high-risk surgical patients, those who are immunocompromised, or those with chronic illnesses. Ultimately, the confluence of immunology, bioengineering, and translational science is getting the field closer than ever to developing a long-established goal in public health: a vaccine to combat one of the most enduring and deadly bacterial infections facing modern medicine.

REFERENCES

Ansari, S., Jha, R. K., Mishra, S. K., Tiwari, B. R., & Asaad, A. M. (2019). Recent advances in *Staphylococcus aureus* infection: Focus on vaccine development. *Infection and Drug Resistance*, 1243–1255.

Bagnoli, F., Fontana, M. R., Soldaini, E., Mishra, R. P., Fiaschi, L., Cartocci, E., . . . Grandi, G. (2015). Vaccine composition formulated with a novel TLR7-dependent adjuvant induces high and broad protection against *Staphylococcus aureus*. *Proceedings of the National Academy of Sciences*, *112*(12), 3680–3685.

Chatterjee, R., Mahapatra, S. R., Dey, J., Raj Takur, K., Raina, V., Misra, N., & Suar, M. (2023). An immunoinformatics and structural vaccinology study to design a multi-epitope vaccine against *Staphylococcus aureus* infection. *Journal of Molecular Recognition*, *36*(4), e3007.

Chen, W. H., Pasetti, M. F., Adhikari, R. P., Baughman, H., Douglas, R., El-Khorazaty, J., . . . Aman, M. J. (2016). Safety and immunogenicity of a parenterally administered, structure-based rationally modified recombinant staphylococcal enterotoxin B protein vaccine, STEBVax. *Clinical and Vaccine Immunology*, *23*(12), 918–925.

Clegg, J., Soldaini, E., McLoughlin, R. M., Rittenhouse, S., Bagnoli, F., & Phogat, S. (2021). *Staphylococcus aureus* vaccine research and development: The past, present and future, including novel therapeutic strategies. *Frontiers in Immunology*, *12*, 705360.

Creech, C. B., Frenck, R. W., Fiquet, A., Feldman, R., Kankam, M. K., Pathirana, S., . . . Gurtman, A. (2020, January). Persistence of immune responses through 36 months in healthy adults after vaccination with a novel *Staphylococcus aureus* 4-antigen vaccine (SA4Ag). In *Open forum infectious diseases* (Vol. 7, No. 1, p. ofz532). US: Oxford University Press.

Daum, R. S., & Spellberg, B. (2012). Progress toward a *Staphylococcus aureus* vaccine. *Clinical Infectious Diseases*, *54*(4), 560–567.

DeDent, A., Kim, H. K., Missiakas, D., & Schneewind, O. (2012, March). Exploring *Staphylococcus aureus* pathways to disease for vaccine development. In *Seminars in immunopathology* (Vol. 34, pp. 317–333). Springer-Verlag.

Deng, J., Wang, X., Zhang, B. Z., Gao, P., Lin, Q., Kao, R. Y. T., . . . Huang, J. D. (2019). Broad and effective protection against *Staphylococcus aureus* is elicited by a multivalent vaccine formulated with novel antigens. *Msphere*, *4*(5), 10–1128.

Dwyer, M., & Gadjeva, M. (2013). Opsonophagocytic assay. In *The Complement System: Methods and Protocols* (pp. 373–379). Totowa, NJ: Humana Press.

Garrido-Palazuelos, L. I., Almanza-Orduño, A. A., Waseem, M., Basheer, A., Medrano-Félix, J. A., Mukthar, M., . . . Aguirre-Sánchez, J. R. (2024). Immunoinformatic approach for multi-epitope vaccine design against *Staphylococcus aureus* based on hemolysin proteins. *Journal of Molecular Graphics and Modelling*, *132*, 108848.

Hassanzadeh, H., Baber, J., Begier, E., Noriega, D. C., Konishi, H., Yato, Y., . . . STRIVE Investigator Team. (2023). Efficacy of a 4-Antigen *Staphylococcus aureus* vaccine in spinal surgery: The STRIVE randomized clinical trial. *Clinical Infectious Diseases*.

Hu, D. L., Omoe, K., Narita, K., Shinagawa, K., & Nakane, A. (2007). Vaccination with a nontoxic mutant of staphylococcal enterotoxin C elicits Th2 immune response for protection against *Staphylococcus aureus* infection. 弘前医学, *59*(Supplement), S210–S218.

Huesca, M., Sun, Q., Peralta, R., Shivji, G. M., Sauder, D. N., & McGavin, M. J. (2000). Synthetic peptide immunogens elicit polyclonal and monoclonal antibodies specific for linear epitopes in the D motifs of *Staphylococcus aureus* fibronectin-binding protein, which are composed of amino acids that are essential for fibronectin binding. *Infection and Immunity*, *68*(3), 1156–1163.

Inoue, M., Yonemura, T., Baber, J., Shoji, Y., Aizawa, M., Cooper, D., . . . Gurtman, A. (2018). Safety, tolerability, and immunogenicity of a novel 4-antigen *Staphylococcus aureus* vaccine (SA4Ag) in healthy Japanese adults. *Human Vaccines & Immunotherapeutics*, *14*(11), 2682–2691.

Klimka, A., Krut, O., & Krönke, M. (2023). *U.S. Patent No. 11,801,292*. Washington, DC: U.S. Patent and Trademark Office.

König, E., Gagliardi, A., Riedmiller, I., Andretta, C., Tomasi, M., Irene, C., . . . Grandi, G. (2021). Multi-antigen outer membrane vesicle engineering to develop polyvalent vaccines: The *Staphylococcus aureus* case. *Frontiers in Immunology*, *12*, 752168.

Lattar, S. M., Noto Llana, M., Denoël, P., Germain, S., Buzzola, F. R., Lee, J. C., & Sordelli, D. O. (2014). Protein antigens increase the protective efficacy of a capsule-based vaccine against *Staphylococcus aureus* in a rat model of osteomyelitis. *Infection and Immunity*, *82*(1), 83–91.

Lee, J. C. (2001). Development of antistaphylococcal vaccines. *Current Infectious Disease Reports*, *3*, 517–524.

Levy, J., Licini, L., Haelterman, E., Moris, P., Lestrate, P., Damaso, S., . . . Boutriau, D. (2015). Safety and immunogenicity of an investigational 4-component *Staphylococcus aureus* vaccine with or without AS03B adjuvant: Results of a randomized phase I trial. *Human Vaccines & Immunotherapeutics, 11*(3), 620–631.

Mahapatra, S. R., Dey, J., Raj, T. K., Misra, N., & Suar, M. (2023). Designing a next-generation multiepitope-based vaccine against *Staphylococcus aureus* using reverse vaccinology approaches. *Pathogens, 12*(3), 376.

Mancini, F. (2014). A promising novel formulation of *Staphylococcus aureus* vaccine protects mice via antibodies and cd4+ effector t cells.

Mandelli, A. P., Magri, G., Tortoli, M., Torricelli, S., Laera, D., Bagnoli, F., . . . Chiarot, E. (2024). Vaccination with staphylococcal protein A protects mice against systemic complications of skin infection recurrences. *Frontiers in Immunology, 15*, 1355764.

McNeely, T. B., Smith, S. B., & Zorman, J. K. (2013). *U.S. Patent Application No. 13/810,130.*

Menzies, B. E., & Kernodle, D. S. (1996). Passive immunization with antiserum to a nontoxic alpha-toxin mutant from *Staphylococcus aureus* is protective in a murine model. *Infection and Immunity, 64*(5), 1839–1841.

Monaci, E., Mancini, F., Lofano, G., Bacconi, M., Tavarini, S., Sammicheli, C., . . . Bertholet, S. (2015). MF59-and Al (OH) 3-adjuvanted *Staphylococcus aureus* (4C-Staph) vaccines induce sustained protective humoral and cellular immune responses, with a critical role for effector CD4 T cells at low antibody titers. *Frontiers in Immunology, 6*, 439.

Pier, G. B., & Maira-Litran, T. (2021). *U.S. Patent No. 10,919,956.* Washington, DC: U.S. Patent and Trademark Office.

Proctor, R. A. (2012). Challenges for a universal *Staphylococcus aureus* vaccine. *Clinical Infectious Diseases, 54*(8), 1179–1186.

Proctor, R. A. (2015). Recent developments for *Staphylococcus aureus* vaccines: Clinical and basic science challenges. *Eur Cell Mater, 30*(734), 315–326.

Rahman, S., Sarkar, K., & Das, A. K. (2023). Exploring staphylococcal superantigens to design a potential multi-epitope vaccine against *Staphylococcus aureus*: An in-silico reverse vaccinology approach. *Journal of Biomolecular Structure and Dynamics, 41*(22), 13098–13112.

Redi, D., Spertilli Raffaelli, C., Rossetti, B., De Luca, A., & Montagnani, F. (2018). *Staphylococcus aureus* vaccine preclinical and clinical development: Current state of the art. *New Microbiologica, 41*(3), 208–213.

Scully, I. L., Liberator, P. A., Jansen, K. U., & Anderson, A. S. (2014). Covering all the bases: Preclinical development of an effective *Staphylococcus aureus* vaccine. *Frontiers in Immunology, 5*, 109.

Skurnik, D., Merighi, M., Grout, M., Gadjeva, M., Maira-Litran, T., Ericsson, M., . . . Pier, G. B. (2010). Animal and human antibodies to distinct *Staphylococcus aureus* antigens mutually neutralize opsonic killing and protection in mice. *The Journal of clinical Investigation, 120*(9), 3220–3233.

Spellberg, B., & Daum, R. (2012, March). Development of a vaccine against *Staphylococcus aureus*. In *Seminars in immunopathology* (Vol. 34, pp. 335–348). Springer-Verlag.

Tan, Z., Yang, W., O'Brien, N. A., Pan, X., Ramadan, S., Marsh, T., . . . Huang, X. (2024). A comprehensive synthetic library of poly-N-acetyl glucosamines enabled vaccine against lethal challenges of *Staphylococcus aureus*. *Nature Communications, 15*(1), 3420.

Torre, A., Bacconi, M., Sammicheli, C., Galletti, B., Laera, D., Fontana, M. R., . . . Bensi, G. (2015). Four-component *Staphylococcus aureus* vaccine 4C-staph enhances Fcγ receptor expression in neutrophils and monocytes and mitigates *S. aureus* infection in neutropenic mice. *Infection and Immunity, 83*(8), 3157–3163.

Verkaik, N. J., Boelens, H. A., de Vogel, C. P., Tavakol, M., Bode, L. G., Verbrugh, H. A., . . . van Wamel, W. J. (2010). Heterogeneity of the humoral immune response following *Staphylococcus aureus* bacteremia. *European Journal of Clinical Microbiology & Infectious Diseases, 29*, 509–518.

Wang, Y., Mukherjee, I., Venkatasubramaniam, A., Dikeman, D., Orlando, N., Zhang, J., . . . Archer, N. K. (2024). Dry and liquid formulations of IBT-V02, a novel multi-component toxoid vaccine, are effective against *Staphylococcus aureus* isolates from low-to-middle income countries. *Frontiers in Immunology, 15*, 1373367.

Wleklinski, C. G., Neubüser, U., Ronneberger, H., Hungerer, K. D., Schaeg, W., & Blobel, H. (1982). Schutzwirkungen durch aktive Immunisierung gegen a-Hämolysin von *Staphylococcus aureus*. *Zentralblatt für Veterinärmedizin Reihe B, 29*(7), 596–603.

Wong Fok Lung, T., Chan, L. C., Prince, A., Yeaman, M. R., Archer, N. K., Aman, M. J., & Proctor, R. A. (2022). *Staphylococcus aureus* adaptive evolution: Recent insights on how immune evasion, immunometabolic subversion and host genetics impact vaccine development. *Frontiers in Cellular and Infection Microbiology*, *12*, 1060810.

Xu, S. H. I. L. U., Arbeit, R. D., & Lee, J. C. (1992). Phagocytic killing of encapsulated and microencapsulated *Staphylococcus aureus* by human polymorphonuclear leukocytes. *Infection and Immunity*, *60*(4), 1358–1362.

Yang, L. Y., Wei, C., Yang, Y., Tong, Y. N., Yang, S., Peng, L. S., . . . Sun, H. W. (2018). Immune response effects of diverse vaccine antigen attachment ways based on the self-made nanoemulsion adjuvant in systemic MRSA infection. *RSC Advances*, *8*(19), 10425–10436.

Yeaman, M. R., Filler, S. G., Chaili, S., Barr, K., Wang, H., Kupferwasser, D., . . . Ibrahim, A. S. (2014). Mechanisms of NDV-3 vaccine efficacy in MRSA skin versus invasive infection. *Proceedings of the National Academy of Sciences*, *111*(51), E5555–E5563.

Yeaman, M. R., Filler, S. G., Schmidt, C. S., Ibrahim, A. S., Edwards Jr, J. E., & Hennessey Jr, J. P. (2014). Applying convergent immunity to innovative vaccines targeting *Staphylococcus aureus*. *Frontiers in Immunology*, *5*, 463.

Yu, W., Yao, D., Yu, S., Wang, X., Li, X., Wang, M., . . . Cui, Y. (2018). Protective humoral and CD4+ T cellular immune responses of *Staphylococcus aureus* vaccine MntC in a murine peritonitis model. *Scientific Reports*, *8*(1), 3580.

Zeng, H., Zhang, J., Song, X., Zeng, J., Yuan, Y., Chen, Z., . . . Zhao, Z. (2021). An immunodominant epitope-specific monoclonal antibody cocktail improves survival in a mouse model of *Staphylococcus aureus* bacteremia. *The Journal of Infectious Diseases*, *223*(10), 1743–1752.

Zhang, J., Yang, F., Zhang, X., Jing, H., Ren, C., Cai, C., . . . Zeng, H. (2015). Protective efficacy and mechanism of passive immunization with polyclonal antibodies in a sepsis model of *Staphylococcus aureus* infection. *Scientific Reports*, *5*(1), 15553.

Zhang, Q., Jiang, T., Mao, X., Kim, J. D., Ahn, D. H., Jung, Y., . . . & Lee, B. L. (2021). Development of combination vaccine conferring optimal protection against six pore-forming toxins of *Staphylococcus aureus*. *Infection and Immunity*, *89*(10).

16 Addressing Challenges and Exploring Future Prospects in Drug Development against *Staphylococcus aureus*

Laraib Kehkashaan[1], *Asra Siddiqua*[1], *Kondakindi Venkateswar Reddy*[1], *and Naga Raju Maddela*[2]
[1]Centre for Biotechnology, University College of Engineering, Science and Technology, JNTUH, Hyderabad, India
[2]Departamento de Ciencias Biológicas, Facultad de Ciencias de la Salud, Universidad Técnica de Manabí, Portoviejo, Ecuador

ABBREVIATIONS

AMPs	antimicrobial peptides
AMR	antimicrobial resistance
ASPs	antimicrobial stewardship programs
FBDD	fragment-based drug discovery
HTS	high-throughput screening
JPIAMR	Joint Programming Initiative on Antimicrobial Resistance
MALDI-TOF MS	matrix-assisted laser desorption ionization–time-of-flight mass spectrometry
MRSA	methicillin-resistant *S. aureus*
NGS	next-generation sequencing
PCR	polymerase chain reaction
PVL	panton–valentine leukocidin
QSAR	quantitative structure–activity relationship
QSIs	quorum-sensing inhibitors
RIP	RNAIII-inhibiting peptides
SBDD	structure-based drug design
SRIA	scientific research and innovation agenda
VRSA	vancomycin-resistant *S. aureus*
WGS	whole genome sequencing

16.1 INTRODUCTION

Staphylococcus aureus remains a significant global health concern, responsible for a wide spectrum of infections ranging from mild skin conditions to life-threatening diseases such as pneumonia, endocarditis, and sepsis (Turner et al., 2019). The remarkable ability of *S. aureus* to develop antibiotic resistance adds to the clinical burden of the bacteria and methicillin-resistant *S. aureus* (MRSA) has emerged as a particularly formidable challenge in both hospital and community settings (Foster, 2017). Despite extensive research efforts, the development of effective therapeutics has been hindered by a combination of factors, including the bacterium's adaptive resistance mechanisms,

DOI: 10.1201/9781003647072-20

biofilm formation and the limited pipeline of novel antibiotics. One of the greatest difficulties in treating *S. aureus* infections is its ability to rapidly adapt to selective pressures. The widespread overuse and misuse of antibiotics have accelerated the emergence of resistant strains, leaving fewer treatment options available. Traditional antibiotic discovery efforts have struggled to keep pace with this growing resistance, thus highlighting the urgent need for innovative approaches that go beyond conventional small-molecule drugs. Additionally, the pathogen's ability to form biofilms on medical devices and tissues presents another significant challenge as these structures protect bacteria from both the host immune response and antimicrobial agents.

This review examines the critical challenges impeding drug development against *S. aureus* and explores emerging strategies that hold promise for overcoming these barriers. Beyond resistance, the complexity of *S. aureus* infections also complicates drug development efforts. The bacterium exhibits a wide range of virulence factors, allowing it to evade immune detection and persist within the host. This makes it difficult to develop a single therapeutic strategy that is universally effective. Furthermore, the economic and regulatory hurdles associated with bringing new antibiotics to market have discouraged pharmaceutical investment, further limiting the availability of novel treatment options. Despite these obstacles, promising advances in antimicrobial research are providing new opportunities for combating *S. aureus* infections. Alternative strategies, such as bacteriophage therapy, antimicrobial peptides (AMPs) and monoclonal antibodies are being explored as potential adjuncts or replacements for traditional antibiotics. Additionally, efforts to enhance host immunity and disrupt bacterial communication pathways offer new directions for therapeutic intervention. However, these novel approaches still require extensive validation in preclinical and clinical settings before they can be widely implemented. By evaluating recent scientific advancements and addressing gaps in current therapeutic approaches, this review aims to contribute to the ongoing effort to develop more effective and sustainable treatments against this persistent pathogen.

16.2 CURRENT CHALLENGES IN DRUG DEVELOPMENT AGAINST *S. AUREUS*

16.2.1 Mechanisms and Implications of Antibiotic Resistance

16.2.1.1 Genetic and Molecular Basis of MRSA and VRSA

The rise of antibiotic-resistant *S. aureus*, particularly MRSA and vancomycin-resistant *S. aureus* (VRSA), represents a major barrier in drug development. MRSA strains acquire resistance through the mecA gene, which encodes an altered penicillin-binding protein (PBP2a) with reduced affinity for β-lactam antibiotics, rendering conventional treatments ineffective (Lakhundi & Zhang, 2018). Similarly, VRSA develops resistance by acquiring the vanA gene cluster, typically from vancomycin-resistant enterococci (VRE), leading to modifications in peptidoglycan precursors that prevent vancomycin binding (Howden et al., 2010).

Beyond these well-characterized resistance mechanisms, *S. aureus* possesses an array of adaptive strategies, including efflux pumps, target site modifications and enzymatic antibiotic degradation, which collectively reduce drug efficacy. The presence of mobile genetic elements, such as plasmids and transposons, facilitates the horizontal transfer of resistance determinants, thereby accelerating the spread of multidrug-resistant (MDR) strains in both hospital and community settings (Heizmann et al., 2013). These molecular adaptations necessitate novel treatment approaches that can bypass or inhibit resistance-conferring mechanisms.

16.2.1.2 Global Epidemiology and Clinical Impact

MRSA and VRSA infections are prevalent worldwide, with significant variation in prevalence between regions due to differences in antibiotic usage, healthcare infrastructure and infection control practices. In the United States, MRSA accounts for approximately 60% of *S. aureus* infections in intensive care units, leading to increased morbidity and mortality (CDC, 2019). Similarly, in parts of Asia and Europe, community-associated MRSA (CA-MRSA) has emerged as a major public health threat, demonstrating high transmissibility and virulence (Lee et al., 2018).

The clinical impact of resistant *S. aureus* infections is profound, often resulting in prolonged hospital stays, higher medical costs and increased risk of treatment failure. Standard therapies such as vancomycin and daptomycin have shown reduced efficacy against certain strains, complicating treatment decisions and necessitating combination therapies or alternative drug strategies (Lewis et al., 2018). The continued rise of resistant strains highlights the urgency of developing new antimicrobial agents with novel mechanisms of action.

16.2.2 Biofilm-Mediated Drug Resistance

Biofilm formation is a key factor contributing to the persistence and antibiotic resistance of *S. aureus* infections. Biofilms are structured communities of bacteria encased in a self-produced extracellular matrix composed of polysaccharides, proteins, and extracellular DNA, which act as a protective barrier against immune defenses and antimicrobial agents. The biofilm lifestyle enhances bacterial survival by limiting antibiotic penetration, altering metabolic activity, and facilitating the development of persister cells, which are dormant variants that are highly tolerant to antibiotics (S. Singh et al., 2017).

Biofilm-associated resistance is particularly concerning in infections involving medical implants, such as catheters, prosthetic joints, and heart valves. The ability of *S. aureus* to transition between planktonic and biofilm states complicates treatment and highlights the need for anti-biofilm agents in future drug development efforts.

16.2.3 Diagnostic Limitations in *S. aureus* Detection and Treatment

Accurate and timely identification of *S. aureus* infections is critical for effective treatment, yet current diagnostic methods remain suboptimal. Traditional culture-based techniques require extended processing times, delaying appropriate antimicrobial therapy. Molecular assays, such as polymerase chain reaction-based tests, offer improved speed and sensitivity but are often costly and not widely available in resource-limited settings (Becker et al., 2014).

Emerging technologies, including matrix-assisted laser desorption ionization–time-of-flight mass spectrometry (MALDI-TOF MS) and CRISPR-based diagnostics, show promise in rapidly identifying *S. aureus* and its resistance genes. However, challenges remain in integrating these tools into routine clinical workflows and ensuring their accessibility across diverse healthcare settings (Cimmino et al., 2016).

16.2.4 Economic and Regulatory Barriers

Developing new antibiotics is a costly and time-intensive process, with an estimated expenditure of over $1 billion per successful drug and a development timeline spanning 10–15 years (Outterson et al., 2016). Unlike chronic disease medications, antibiotics are used for short treatment courses, making them less financially lucrative for pharmaceutical companies. Consequently, there has been a significant decline in industry investment in antibiotic research, leading to a shrinking pipeline of new drugs targeting *S. aureus.*

Regulatory approval for new antibiotics presents another hurdle, as stringent efficacy and safety requirements extend development timelines and increase costs. Additionally, concerns over antimicrobial stewardship programs limiting the widespread use of novel antibiotics have further deterred pharmaceutical investment (Outterson & Rex, 2020). To address these challenges, sustained efforts are needed to create a more sustainable model for antibiotic development that balances innovation with responsible use. Figure 16.1 illustrates the multifaceted challenges hindering the development

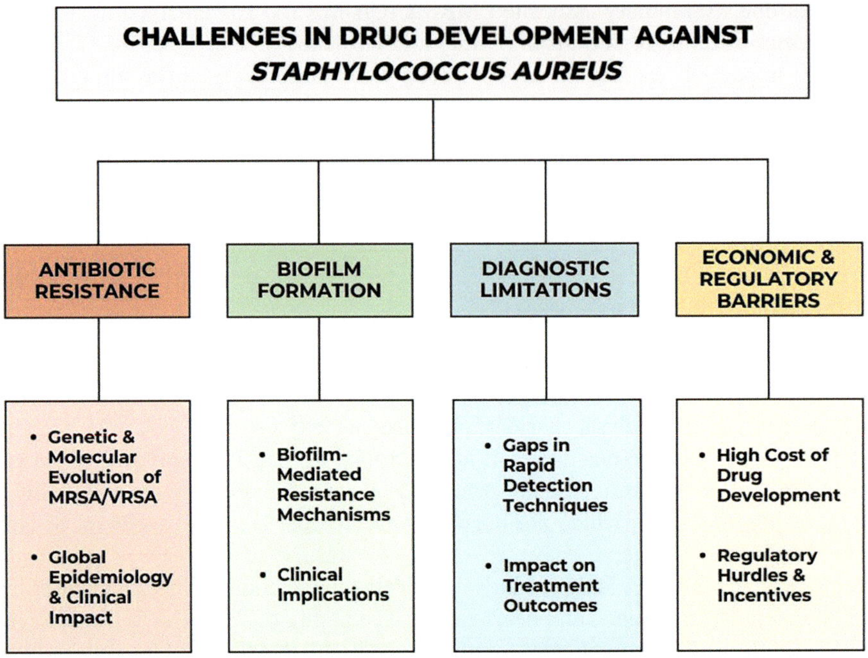

FIGURE 16.1 Overview of the key challenges in developing effective therapeutics against *Staphylococcus aureus*, including antibiotic resistance, biofilm formation, limited novel drug discovery, and regulatory barriers.

of effective therapies against *S. aureus*, including resistance mechanisms, diagnostic gaps, and regulatory barriers.

16.3 ADVANCED STRATEGIES TO COMBAT ANTIBIOTIC RESISTANCE

The increasing prevalence of antibiotic-resistant *S. aureus*, including MRSA and VRSA, necessitates novel therapeutic strategies. Traditional antibiotics are losing efficacy due to genetic adaptations, biofilm formation, and the emergence of MDR strains. To address these challenges, researchers are exploring alternative approaches, including novel antimicrobial agents, drug repurposing, antivirulence therapies, and vaccine development. These strategies aim to overcome resistance mechanisms, disrupt bacterial virulence, and enhance host immunity.

16.3.1 Development of Novel Antimicrobial Agents

16.3.1.1 Small Molecule and Peptide-Based Therapeutics

Small-molecule antibiotics remain a promising avenue in combating *S. aureus* infections. Researchers are focusing on new β-lactamase inhibitors, bacterial cell wall synthesis disruptors, and novel topoisomerase inhibitors to circumvent existing resistance mechanisms (Gopikrishnan & Haryini, 2024). Among these, oxazolidinones, diarylquinolines, and lipoglycopeptides have demonstrated efficacy against resistant strains by targeting essential bacterial processes with minimal resistance development (Kong et al., 2016).

In addition to small molecules, AMPs have gained attention due to their broad-spectrum activity and ability to bypass traditional resistance pathways. AMPs such as lantibiotics, defensins, and cathelicidins exhibit membrane-disrupting properties, making it difficult for bacteria to develop resistance (Mahlapuu et al., 2016).

16.3.1.2 Natural Product and Synthetic Derivatives

Natural products have historically served as the foundation for antibiotic discovery. Actinobacteria-derived compounds, such as teixobactin and lugdunin, have shown potent activity against MRSA by binding to precursor molecules involved in peptidoglycan synthesis (Ling et al., 2015). Additionally, plant-derived alkaloids, flavonoids, and polyphenols have exhibited antibacterial properties, often targeting *S. aureus* virulence factors rather than essential metabolic pathways (Ahmad-Mansour et al., 2021).

Synthetic derivatives of natural antibiotics, such as semi-synthetic glycopeptides and tetracyclines, have been developed to overcome existing resistance. Oritavancin and dalbavancin, for instance, offer extended half-lives and enhanced efficacy against resistant strains (Dickey et al., 2017). These compounds highlight the potential of chemical modifications in revitalizing existing antibiotic classes.

16.3.2 Drug Repurposing and Combination Therapies

16.3.2.1 Identification of Non-Traditional Antibacterial Agents

Given the high cost and slow pace of novel antibiotic development, drug repurposing presents a viable strategy for identifying new treatments against *S. aureus*. Non-antibiotic drugs, such as antipsychotics (thioridazine), antimalarials (chloroquine), and statins (simvastatin), have demonstrated antibacterial activity through mechanisms such as efflux pump inhibition, membrane disruption, and immune modulation (Thangamani et al., 2015). These drugs offer FDA-approved safety profiles, allowing for faster clinical translation than newly developed antimicrobials.

16.3.2.2 Synergistic Combinations to Overcome Resistance

Combining existing antibiotics with adjuvants has shown promise in enhancing bacterial clearance and reducing resistance selection. β-lactam antibiotics, when used in combination with β-lactamase inhibitors (e.g., clavulanic acid, avibactam), have restored activity against MRSA. Similarly, vancomycin combined with rifampin or daptomycin has demonstrated synergistic bactericidal effects against persistent *S. aureus* infections, particularly in endocarditis and osteomyelitis cases (Ye et al., 2020).

16.3.3 Anti-Virulence and Immunomodulatory Therapies

16.3.3.1 Targeting Toxins and Adhesion Mechanisms

Unlike traditional antibiotics that target bacterial survival, anti-virulence therapies focus on neutralizing pathogenic mechanisms. These approaches aim to disarm *S. aureus* rather than kill it, reducing selective pressure for resistance. Key virulence factors, such as alpha-toxin (Hla), panton–valentine leukocidin, and protein A (SpA), have been targeted using small-molecule inhibitors and monoclonal antibodies (Anderson et al., 2012). Quorum-sensing inhibitors are molecules that interfere with the bacterial quorum-sensing signaling pathways, which regulate gene expression in response to cell population density. Compounds such as RNAIII-inhibiting peptides (RIPs) have demonstrated potential in disrupting these communication systems, thereby inhibiting biofilm formation and attenuating bacterial virulence and infection severity (LaSarre & Federle, 2013).

16.3.3.2 Host Immune System Modulation Strategies

Host-directed therapies are gaining traction as a means to boost immune responses against *S. aureus* infections. Strategies include cytokine-based immunotherapies (interleukin-2, interferon-γ), neutrophil-enhancing agents, and probiotics that help reinforce innate immune defenses (van Belkum et al., 2020).

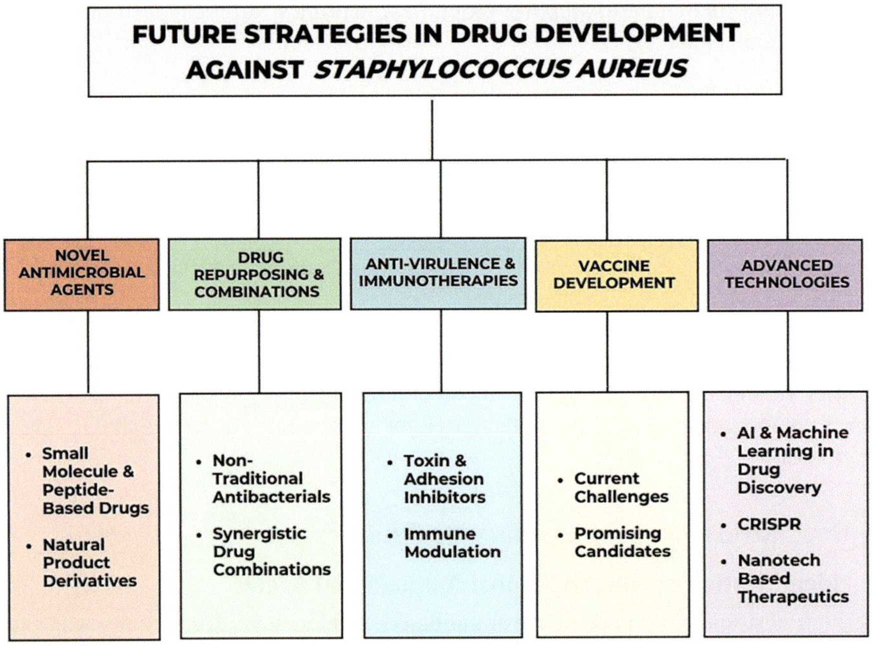

FIGURE 16.2 Future strategies for drug development against *Staphylococcus aureus*, including novel antimicrobials, repurposed drugs, anti-virulence therapies, vaccines, and advanced technologies like AI and CRISPR.

16.3.4 Vaccine Development for *S. aureus*

Despite extensive research, no licensed vaccine against *S. aureus* exists. Several vaccine candidates, such as StaphVAX (capsular polysaccharide-based) and V710 (IsdB-based), have failed in clinical trials due to limited efficacy and potential safety concerns (Fowler et al., 2013).

Recent efforts in vaccine research have focused on multivalent formulations targeting multiple virulence factors simultaneously. T cell-activating vaccines and nanoparticle (NP)-based delivery systems are being explored to enhance immune memory and long-term protection (Anderson et al., 2012). Additionally, advances in mRNA vaccine technology, as seen in COVID-19 vaccines, may offer a new platform for *S. aureus* vaccine development. As shown in Figure 16.2, multiple innovative strategies, including next-generation antimicrobials, immunotherapies, and AI-driven drug discovery, offer promising avenues for combating *S. aureus* infections.

16.4 TECHNOLOGICAL ADVANCEMENTS IN DRUG DEVELOPMENT

Building upon the challenges and advanced strategies previously discussed, recent technological breakthroughs are now reshaping the landscape of drug discovery. Advances in genomics, proteomics, high-throughput screening, CRISPR-based genetic engineering, and artificial intelligence (AI) are significantly improving our ability to identify novel targets, design new therapeutics, and optimize treatment strategies for drug discovery and development. These approaches are transforming the traditional drug development pipeline by enabling faster, more precise, and cost-effective discovery of anti-*S. aureus* agents.

16.4.1 Genomics and Proteomics for Target Discovery

16.4.1.1 Next-Generation Sequencing and Comparative Genomics

Next-generation sequencing has revolutionized bacterial genomics by providing comprehensive insights into the genetic basis of antibiotic resistance and virulence in *S. aureus*. Comparative genomic studies between antibiotic-resistant and susceptible strains have identified novel resistance determinants, regulatory elements, and metabolic pathways that could serve as drug targets. Whole genome sequencing is now being routinely used in clinical microbiology to track the evolution of resistant strains, allowing for the development of precision antimicrobial therapies (Sabat et al., 2017).

16.4.1.2 Proteomic Approaches to Identify Novel Drug Targets

While genomics provides a blueprint of bacterial physiology, proteomics offers functional insights by identifying protein expression patterns, post-translational modifications, and protein–protein interactions critical for bacterial survival. Mass spectrometry-based proteomics has helped map the *S. aureus* resistome, uncovering key membrane transporters, virulence factors, and metabolic enzymes as potential drug targets. Additionally, structural proteomics has facilitated the design of inhibitors targeting essential bacterial enzymes, such as penicillin-binding proteins and DNA gyrases (Ma et al., 2017).

16.4.2 High-Throughput Screening and Drug Design

16.4.2.1 Automated Screening Technologies

High-throughput screening allows researchers to test thousands of compounds rapidly against *S. aureus* to identify potential antimicrobial candidates. Automated platforms using robotic systems, fluorescence-based assays, and microfluidic devices have accelerated the identification of new inhibitors, particularly for biofilm formation and toxin production (Ayon, 2023).

16.4.2.2 Computational and Structure-Based Drug Design

Computational drug design has transformed the development of anti-*S. aureus* agents by enabling in silico screening, molecular docking, and quantitative structure–activity relationship modeling. Structure-based drug design leverages crystal structures of bacterial proteins to develop high-affinity inhibitors with improved pharmacokinetics. Fragment-based drug discovery has also facilitated the design of small-molecule inhibitors targeting essential bacterial enzymes, such as FtsZ (cell division protein) and ClpP (protease) (Kaczor et al., 2018).

16.4.3 CRISPR and Genetic Engineering

16.4.3.1 CRISPR-Cas Systems in Functional Genomics

The advent of CRISPR-Cas technology has revolutionized functional genomics by enabling precise gene editing, knockouts, and transcriptional regulation in *S. aureus*. CRISPR interference (CRISPRi) has been employed to silence essential genes, elucidating their roles in bacterial physiology and antibiotic resistance (Sato'o et al., 2018). This approach has significantly enhanced our understanding of genetic vulnerabilities in MRSA, paving the way for the identification of novel drug targets.

16.4.3.2 Applications in Drug Target Validation and Therapy

CRISPR-guided antimicrobials, designed to selectively eliminate antibiotic-resistant strains while sparing commensal microbiota, have shown promise in preclinical studies. Additionally,

CRISPR-Cas9 has been utilized for engineering bacteriophages with enhanced lytic activity, offering a potential alternative to antibiotics in treating drug-resistant infections (Allemailem, 2024).

16.4.4 AI and Machine Learning in Drug Discovery

AI is playing an increasingly critical role in the discovery of novel antimicrobial agents. AI-powered platforms analyze large-scale omics data, chemical libraries, and clinical datasets to identify new drug candidates with high specificity and reduced toxicity. Deep learning models have successfully predicted antibiotic activity in untested chemical compounds, leading to the discovery of halicin, a novel antibiotic with broad-spectrum activity against drug-resistant pathogens (Stokes et al., 2020).

Machine learning (ML) algorithms are being leveraged to predict bacterial resistance patterns, optimize drug combinations, and accelerate drug repurposing. By integrating genomic, proteomic, and metabolomic data, ML models can identify resistance trends in clinical isolates, enabling personalized antibiotic therapy (Ekins et al., 2019).

16.5 FUTURE PROSPECTS IN DRUG DEVELOPMENT

There is ongoing innovation in the creation of therapeutic approaches for diseases mediated by *S. aureus*. There is great potential for new treatment options due to the recent redesign of previously established therapeutic methods and the development of entirely new approaches. Here, we go over some of the most forefront and potential therapeutic approaches for treating *S. aureus* infections.

16.5.1 Innovative Therapeutic Approaches

16.5.1.1 Host-Directed Therapies and Immunotherapies

S. aureus is a leading cause of bloodstream infections, resulting in significant morbidity and mortality globally (Keihanian et al., 2018). Antibiotic resistance is a serious challenge in treating infections caused by multidrug-resistant bacteria. Vaccination may offer protection against both resistant and susceptible species. The primary goal of vaccine development is to stimulate an antibody response that can prevent poisons that kill immune cells and opsonize microorganisms.

Anti-staphylococcal monoclonal antibodies are regarded as a unique strategy when used as preventive measures for patients who are at a high risk of contracting severe *S. aureus* infections. Most strains contain alpha-toxin; a monoclonal antibody against it may be useful in preventing infections brought on by *S. aureus*, including MRSA. Anti-alpha-toxin antibodies have been shown in numerous investigations to protect against *S. aureus* infections (Tkaczyk et al., 2012).

16.5.1.2 Phage Therapy and Bacteriophage-Derived Enzymes

While bacteriophage therapies are not a new approach to treating bacterial infections, they have recently seen a resurgence with a number of positive preclinical and clinical treatments for *S. aureus* infections and other significant antibiotic-resistant bacteria. Anti-staphylococcal viruses can be used in two different ways: (1) by injecting whole viral particles and (2) by using virus-derived antimicrobial enzymes like endolysins. Two methods exist for using anti-staphylococcal viruses: (1) injecting entire viral particles and (2) employing antimicrobial enzymes derived from viruses, such as endolysins.

An example of a phage endolysin is SAL-200, which was developed by Intron Biotechnology. This *S. aureus*-specific enzyme was shown to have bactericidal activity against more than 400 strains of *S. aureus* and to improve the course of murine bacteremia. It has been demonstrated that this *S. aureus*-specific enzyme has bactericidal activity against over 400 strains and can prolong the course of murine bacteremia. Additionally, when combined with antibiotics, it improved the efficacy of treating systemic infections in mice and moth larvae (Kim et al., 2018).

16.5.2 Personalized Medicine in Antibacterial Therapy

16.5.2.1 Pharmacogenomics and Personalized Treatment Strategies

Pharmacogenomics studies how a person's genetic makeup affects how they react to drugs. It combines genomics with pharmacology. The use of pharmacogenomics in routine medical care is anticipated to increase as research advances and technology becomes more accessible, promoting a more individualized approach to healthcare. By tailoring medication regimens to each patient's unique genetic profile, pharmacogenomics seeks to develop customized medicine strategies. This method reduces the possibility of negative medication reactions while improving treatment effectiveness.

Finding resistance genes in pathogens by genomic fingerprinting can help choose antibiotics that are most likely to work against particular strains. By knowing a disease's resistance profile, medical professionals can choose therapies that the organism is vulnerable to rather than administering ineffective drugs. This strategy reduces needless drug expenses, enhances treatment results, and slows the proliferation of resistant strains (Molla & Bitew, 2024).

16.5.2.2 Biomarkers for Predictive Diagnostics and Therapeutic Efficacy

Biomarkers are typically employed for diagnosis, classification of risk by evaluating severity, clinical decision-making, therapy, and response. Antibiotic stewardship programs have also evaluated other biomarkers, such as serum presepsin, serum C-reactive protein, and interleukin (IL)-1β/IL-18 in bronchoalveolar lavage (Kyriazopoulou & Giamarellos-Bourboulis, 2022).

16.5.3 Nanotechnology-Based Antimicrobial Strategies

16.5.3.1 Nanoparticles and Nanocarriers for Drug Delivery

Widely employed as therapeutic antibacterial agents (Table 16.1), noble metal NPs include silver (Ag), gold, and platinum (Şen Karaman et al., 2018). For therapeutic drug delivery and diagnostic applications, gold NPs may be especially well-suited due to their unique properties, which include easy preparation, surface functionalities, tunable core size, photothermal and surface plasmon-related optoelectronic properties, inertness, and biocompatibility (Jagtap et al., 2017).

Because of their size- and shape-dependent optoelectronic characteristics and high surface area to volume ratios, semiconductor NPs, also known as quantum dots, or QDs, such as zinc sulfide, cadmium sulfide, and zinc oxide, have emerged as essential instruments in therapeutic drug delivery and diagnostic bioimaging. Due to their ability to load up to ten times more medicine than non-porous silica NPs, mesoporous silica NPs significantly advance the field of antibacterial drug delivery (Wagner et al., 2019).

16.5.3.2 Nanoformulations for Enhanced Antibacterial Activity

Studies on the creation of highly targeted nanomaterials to combat bacteria resistant to antibiotics have been conducted in the past decade. These NPs can boost the pharmacological action of the payload by including one or more medications without changing the cargo's integrity (Gholipourmalekabadi et al., 2017).

16.5.4 Global Collaboration and Policy Development in Antimicrobial Resistance

The excessive use and misuse of antibiotics have become a serious and growing threat to global health, particularly in the case of *S. aureus*. This overuse weakens the effectiveness of antibiotics in treating *S. aureus* infections and accelerates the rise of antimicrobial resistance (AMR), making infections more difficult to manage. The complexity of AMR in *S. aureus* requires a comprehensive, multifaceted approach that addresses its root causes in both clinical and agricultural settings while also promoting the development of innovative strategies to curb its spread. Beyond hospitals, agricultural practices contribute significantly to AMR, as the widespread use of antibiotics in livestock

TABLE 16.1
Overview of the Antibacterial Properties and Underlying Mechanisms of Action of Nanoparticles Against *Staphylococcus aureus*

Activity of Nanoparticle	Mode of Action	References
Increased antibacterial action against resistant *S. aureus*	The bacterial cell membrane is destroyed as a result of the nanoparticles' interaction with it; they also produce reactive oxygen species, deactivate proteins, inactivate enzymes, alter gene expression, and boost both innate and adaptive immunity. The nanoparticles are able to overcome the antibiotic-resistant mechanism with the aid of these bactericidal pathways. Several studies showed that by attaching to vancomycin and increasing bacterial cell death, silver nanoparticles might defeat the vancomycin-resistant strain of *S. aureus*.	Esmaeillou et al. (2017)
Inhibit the activity of biofilm formation	The glycocalyx, which has a negative charge, interacts with nanoparticles with a cationic charge, allowing them to penetrate thick biofilms. Treating triclosan, an antibacterial medication, in a micellar nanocarrier allows for deeper penetration and death of *Staphylococcus aureus* within the biofilm.	B. N. Singh et al. (2017)
Amplified intracellular drug delivery	Nanoparticles have the ability to increase the permeability and concentration of the medicine they carry inside cells. Because these particles directly affect *S. aureus* when they come into contact with it and because the released drugs diffuse to *S. aureus*, the increased cellular uptake and subsequent controlled release of the antibiotics that are entrapped or adsorbed in the nanoparticles can effectively enhance their antibacterial effects, making them more effective in treating intracellular infections.	Zhou et al. (2018)

for disease prevention, growth promotion, and treatment creates a highly selective environment that fosters the emergence and dissemination of antibiotic-resistant *S. aureus* strains.

16.5.4.1 International Initiatives and Collaborative Networks

To effectively tackle AMR, a One Health strategy that recognizes the interdependence of environmental, animal, and human health is necessary. Overcrowding, poor sanitation, and global movement all contribute to the fast spread of resistant infections, underscoring the critical need for cross-sector cooperation to contain and manage the epidemic. Antimicrobial stewardship programs are essential for optimizing the use of antibiotics, enhancing the ability to diagnose, executing treatment standards, and fortifying infection control measures in healthcare settings.

There are still a lot of unresolved underlying scientific concerns about the evolution, selection, transmission, and persistence of antibiotic resistance genes. An international collaborative platform for coordinating and funding AMR research, the Joint Programming Initiative on Antimicrobial Resistance (JPIAMR) offers a framework for research and innovation for collaborative actions, identifying critical areas that need to be addressed and offering recommendations for nations to align their national and international AMR research agendas.

The One Health viewpoint was embraced by the JPIAMR Scientific Research and Innovation Agenda, and it is recommended for national research strategies in the WHO Global Action Plan on AMR (Coque et al., 2023).

16.5.4.2 Policy Frameworks for Sustainable Drug Development

The growing threat of AMR in *S. aureus* presents a major obstacle to sustainable drug development, as it undermines the effectiveness of existing treatments and poses significant challenges to public health and healthcare systems. The United Nations' Sustainable Development Goals highlight AMR not only as a medical crisis but also as a broader socioeconomic and global public health issue.

Addressing *S. aureus* resistance requires coordinated efforts from governments, pharmaceutical companies, healthcare providers, and the public to ensure responsible antibiotic use and sustainable drug management. Governments can establish international guidelines to promote the responsible disposal and potential reuse of antibiotics, reducing environmental contamination that contributes to resistance. Meanwhile, the pharmaceutical industry can adopt sustainable practices, such as recyclable drug delivery systems and take-back programs, to minimize waste and limit the environmental spread of resistant *S. aureus* strains (Khan & Ali, 2022).

16.6 CONCLUSION

The rise of MDR *S. aureus*, including MRSA and VRSA, presents a major global health threat. Traditional antibiotics are becoming increasingly ineffective due to genetic mutations, biofilm formation, and other resistance mechanisms. The challenges in drug development range from molecular complexities to economic and regulatory barriers, making it imperative to explore innovative therapeutic strategies. This review has examined the current limitations in antibiotic treatment, novel antimicrobial agents, drug repurposing efforts, anti-virulence therapies, vaccine development, and emerging technological advancements in drug discovery.

Addressing *S. aureus* resistance requires a multifaceted approach integrating cutting-edge scientific advancements. Novel therapies, including small-molecule inhibitors, AMPs, and host-directed immunotherapies, offer potential solutions. Additionally, bacteriophage therapy and nanotechnology-based drug delivery systems are being explored as alternatives to conventional antibiotics. Personalized medicine, supported by pharmacogenomics and biomarker-driven diagnostics, may further enhance treatment precision and efficacy. However, while several promising therapies are in development, stability, cost, cytotoxicity, and large-scale production remain key challenges that must be addressed before clinical implementation. Technological advancements such as genomics, proteomics, high-throughput screening, CRISPR-based genome editing, and AI-driven drug discovery are revolutionizing the search for novel *S. aureus* treatments. AI and ML models are accelerating drug repurposing efforts, while CRISPR-based approaches provide new avenues for gene-targeted therapies. These innovations, combined with predictive analytics and big data integration, offer hope for more effective and targeted antibacterial strategies in the near future.

The fight against *S. aureus* drug resistance is not limited to scientific innovation; it also requires coordinated global efforts. International collaborations, interdisciplinary research networks, and policy-driven frameworks for sustainable drug development are essential to controlling the spread of AMR. Governments and regulatory bodies must implement incentive programs to support antibacterial drug development while ensuring responsible antibiotic use in healthcare and agriculture.

In conclusion, tackling *S. aureus* resistance necessitates a comprehensive approach that combines molecular insights, innovative therapeutics, technological advancements, and global policy initiatives. While significant progress has been made, continued research, investment, and interdisciplinary cooperation will be crucial in developing effective and sustainable solutions to combat *S. aureus* infections and AMR in the years to come.

REFERENCES

Ahmad-Mansour, N., Loubet, P., Pouget, C., Dunyach-Remy, C., Sotto, A., Lavigne, J. P., & Molle, V. (2021). *Staphylococcus aureus* toxins: An update on their pathogenic properties and potential treatments. *Toxins*, *13*(10), 677.

Allemailem, K. S. (2024). Recent advances in understanding the molecular mechanisms of multidrug resistance and novel approaches of CRISPR/Cas9-based genome-editing to combat this health emergency. *International Journal of Nanomedicine*, 1125–1143.

Anderson, A. S., Miller, A. A., Donald, R. G., Scully, I. L., Nanra, J. S., Cooper, D., & Jansen, K. U. (2012). Development of a multicomponent *Staphylococcus aureus* vaccine designed to counter multiple bacterial virulence factors. *Human Vaccines & Immunotherapeutics*, *8*(11), 1585–1594.

Ayon, N. J. (2023). High-throughput screening of natural product and synthetic molecule libraries for antibacterial drug discovery. *Metabolites*, *13*(5), 625.

Becker, K., Heilmann, C., & Peters, G. (2014). Coagulase-negative staphylococci. *Clinical Microbiology Reviews*, *27*(4), 870–926.

Centers for Disease Control and Prevention (CDC). (2019). *Antibiotic resistance threats in the United States, 2019. U.S.* Department of Health and Human Services.

Cimmino, T., Le Page, S., Raoult, D., & Rolain, J. M. (2016). Contemporary challenges and opportunities in the diagnosis and outbreak detection of multidrug-resistant infectious disease. *Expert Review of Molecular Diagnostics*, *16*(11), 1163–1175.

Coque, T. M., Cantón, R., Pérez-Cobas, A. E., Fernández-de-Bobadilla, M. D., & Baquero, F. (2023). Antimicrobial resistance in the global health network: Known unknowns and challenges for efficient responses in the 21st century. *Microorganisms*, *11*(4), 1050. https://doi.org/10.3390/microorganisms11041050

Dickey, S. W., Cheung, G. Y., & Otto, M. (2017). Different drugs for bad bugs: Antivirulence strategies in the age of antibiotic resistance. *Nature Reviews Drug Discovery*, *16*(7), 457–471.

Ekins, S., Puhl, A. C., Zorn, K. M., Lane, T. R., Russo, D. P., Klein, J. J., . . . Clark, A. M. (2019). Exploiting machine learning for end-to-end drug discovery and development. *Nature Materials*, *18*(5), 435–441.

Esmaeillou, M., Zarrini, G., Ahangarzadeh Rezaee, M., Shahbazi Mojarrad, J., & Bahadori, A. (2017). Vancomycin capped with silver nanoparticles as an antibacterial agent against multi-drug resistance bacteria. *Advanced Pharmaceutical Bulletin*, *7*(3), 479–483. https://doi.org/10.15171/apb.2017.058

Foster, T. J. (2017). Antibiotic resistance in *Staphylococcus aureus*. Current status and future prospects. *FEMS Microbiology Reviews*, *41*(3), 430–449.

Fowler, V. G., Allen, K. B., Moreira, E. D., Moustafa, M., Isgro, F., Boucher, H. W., . . . Sobanjo-ter Meulen, A. (2013). Effect of an investigational vaccine for preventing *Staphylococcus aureus* infections after cardiothoracic surgery: A randomized trial. *JAMA*, *309*(13), 1368–1378.

Gholipourmalekabadi, M., Mobaraki, M., Ghaffari, M., Zarebkohan, A., Omrani, V. F., Urbanska, A. M., & Seifalian, A. (2017). Targeted drug delivery based on gold nanoparticle derivatives. *Current Pharmaceutical Design*, *23*(20), 2918–2929.

Gopikrishnan, M., & Haryini, S. (2024). Emerging strategies and therapeutic innovations for combating drug resistance in *Staphylococcus aureus* strains: A comprehensive review. *Journal of Basic Microbiology*, *64*(5), 2300579.

Heizmann, W. R., Dupont, H., Montravers, P., Guirao, X., Eckmann, C., Bassetti, M., . . . Bodmann, K. F. (2013). Resistance mechanisms and epidemiology of multi-resistant pathogens in Europe and efficacy of tigecycline in observational studies. *Journal of Antimicrobial Chemotherapy*, *68*(suppl_2), ii45–ii55.

Howden, B. P., Davies, J. K., Johnson, P. D., Stinear, T. P., & Grayson, M. L. (2010). Reduced vancomycin susceptibility in *Staphylococcus aureus*, including vancomycin-intermediate and heterogeneous vancomycin-intermediate strains: Resistance mechanisms, laboratory detection, and clinical implications. *Clinical Microbiology Reviews*, *23*(1), 99–139.

Jagtap, P., Sritharan, V., & Gupta, S. (2017). Nanotheranostic approaches for management of bloodstream bacterial infections. *Nanomedicine: Nanotechnology, Biology and Medicine*, *13*(1), 329–341.

Kaczor, A. A., Medarametla, P., Bartuzi, D., Kondej, M., Matosiuk, D., & Poso, A. (2018). Molecular modelling approaches to antibacterial drug design and discovery. In *Frontiers in anti-infective drug discovery* (Vol. 7, pp. 153–222). Bentham Science Publishers.

Keihanian, F., Saeidinia, A., Abbasi, K., & Keihanian, F. (2018). Epidemiology of antibiotic resistance of blood culture in educational hospitals in Rasht, North of Iran. *Infection and Drug Resistance*, 1723–1728.

Khan, F., & Ali, Y. (2022). Implementation of the circular supply chain management in the pharmaceutical industry. *Environment, Development and Sustainability*, *24*(12), 13705–13731.

Kim, N. H., Park, W. B., Cho, J. E., Choi, Y. J., Choi, S. J., Jun, S. Y., . . . Kim, H. B. (2018). Effects of phage endolysin SAL200 combined with antibiotics on *Staphylococcus aureus* infection. *Antimicrobial Agents and Chemotherapy*, *62*(10), 10–1128.

Kong, C., Neoh, H. M., & Nathan, S. (2016). Targeting *Staphylococcus aureus* toxins: A potential form of anti-virulence therapy. *Toxins*, *8*(3), 72.

Kyriazopoulou, E., & Giamarellos-Bourboulis, E. J. (2022). Antimicrobial stewardship using biomarkers: Accumulating evidence for the critically Ill. *Antibiotics*, *11*(3), 367.

Lakhundi, S., & Zhang, K. (2018). Methicillin-resistant *Staphylococcus aureus*: Molecular characterization, evolution, and epidemiology. *Clinical Microbiology Reviews*, *31*(4), 10–1128.

LaSarre, B., & Federle, M. J. (2013). Exploiting quorum sensing to confuse bacterial pathogens. *Microbiology and Molecular Biology Reviews*, *77*(1), 73–111.

Lee, A. S., De Lencastre, H., Garau, J., Kluytmans, J., Malhotra-Kumar, S., Peschel, A., & Harbarth, S. (2018). Methicillin-resistant *Staphylococcus aureus*. *Nature Reviews Disease Primers*, *4*(1), 1–23.

Lewis, P. O., Heil, E. L., Covert, K. L., & Cluck, D. B. (2018). Treatment strategies for persistent methicillin-resistant *Staphylococcus aureus* bacteraemia. *Journal of Clinical Pharmacy and Therapeutics*, *43*(5), 614–625.

Ling, L. L., Schneider, T., Peoples, A. J., Spoering, A. L., Engels, I., Conlon, B. P., . . . Lewis, K. (2015). A new antibiotic kills pathogens without detectable resistance. *Nature*, *517*(7535), 455–459. https://doi.org/10.1038/nature14098

Ma, W., Zhang, D., Li, G., Liu, J., He, G., Zhang, P., . . . Liang, S. (2017). Antibacterial mechanism of daptomycin antibiotic against *Staphylococcus aureus* based on a quantitative bacterial proteome analysis. *Journal of Proteomics*, *150*, 242–251.

Mahlapuu, M., Håkansson, J., Ringstad, L., & Björn, C. (2016). Antimicrobial peptides: An emerging category of therapeutic agents. *Frontiers in Cellular and Infection Microbiology*, *6*, 194.

Molla, G., & Bitew, M. (2024). Revolutionizing personalized medicine: Synergy with multi-omics data generation, main hurdles, and future perspectives. *Biomedicines*, *12*(12), 2750.

Outterson, K., & Rex, J. H. (2020). Evaluating for-profit public benefit corporations as an additional structure for antibiotic development and commercialization. *Translational Research*, *220*, 182–190.

Outterson, K., Rex, J. H., Jinks, T., Jackson, P., Hallinan, J., Karp, S., . . . Larsen, J. (2016). Accelerating global innovation to address antibacterial resistance: Introducing CARB-X. *Nature Reviews Drug Discovery*, *15*(9), 589–590.

Sabat, A. J., Hermelijn, S. M., Akkerboom, V., Juliana, A., Degener, J. E., Grundmann, H., & Friedrich, A. W. (2017). Complete-genome sequencing elucidates outbreak dynamics of CA-MRSA USA300 (ST8-spa t008) in an academic hospital of Paramaribo, Republic of Suriname. *Scientific Reports*, *7*(1), 41050.

Sato'o, Y., Hisatsune, J., Yu, L., Sakuma, T., Yamamoto, T., & Sugai, M. (2018). Tailor-made gene silencing of *Staphylococcus aureus* clinical isolates by CRISPR interference. *PLoS One*, *13*(1), e0185987.

Şen Karaman, D., Manner, S., & Rosenholm, J. M. (2018). Mesoporous silica nanoparticles as diagnostic and therapeutic tools: How can they combat bacterial infection? *Therapeutic Delivery*, *9*(4), 241–244.

Singh, B. N., Prateeksha, N., Upreti, D. K., Singh, B. R., Defoirdt, T., Gupta, V. K., . . . Vahabi, K. (2017). Bactericidal, quorum quenching and anti-biofilm nanofactories: A new niche for nanotechnologists. *Critical Reviews in Biotechnology*, *37*(4), 525–540.

Singh, S., Singh, S. K., Chowdhury, I., & Singh, R. (2017). Understanding the mechanism of bacterial biofilms resistance to antimicrobial agents. *The Open Microbiology Journal*, *11*, 53.

Stokes, J. M., Yang, K., Swanson, K., Jin, W., Cubillos-Ruiz, A., Donghia, N. M., . . . Collins, J. J. (2020). A deep learning approach to antibiotic discovery. *Cell*, *180*(4), 688–702.

Thangamani, S., Mohammad, H., Younis, W., & Seleem, M. N. (2015). Drug repurposing for the treatment of staphylococcal infections. *Current Pharmaceutical Design*, *21*(16), 2089–2100.

Tkaczyk, C., Hua, L., Varkey, R., Shi, Y., Dettinger, L., Woods, R., . . . Sellman, B. R. (2012). Identification of anti-alpha-toxin monoclonal antibodies that reduce the severity of *Staphylococcus aureus* dermonecrosis and exhibit a correlation between affinity and potency. *Clinical and Vaccine Immunology*, *19*(3), 377–385.

Turner, N. A., Sharma-Kuinkel, B. K., Maskarinec, S. A., Eichenberger, E. M., Shah, P. P., Carugati, M., . . . Fowler, V. G., Jr. (2019). Methicillin-resistant *Staphylococcus aureus*: An overview of basic and clinical research. *Nature Reviews Microbiology*, *17*(4), 203–218.

van Belkum, A., Burnham, C. A. D., Rossen, J. W., Mallard, F., Rochas, O., & Dunne Jr, W. M. (2020). Innovative and rapid antimicrobial susceptibility testing systems. *Nature Reviews Microbiology*, *18*(5), 299–311.

Wagner, A. M., Knipe, J. M., Orive, G., & Peppas, N. A. (2019). Quantum dots in biomedical applications. *Acta Biomaterialia*, *94*, 44–63.

Ye, C., Wang, Z., Hu, Y., Deng, C., Liao, L., Sun, L., & Wang, C. (2020). Systematic review and meta-analysis of the efficacy and safety of vancomycin combined with β-lactam antibiotics in the treatment of methicillin-resistant *Staphylococcus aureus* bloodstream infections. *Journal of Global Antimicrobial Resistance*, *23*, 303–310.

Zhou, K., Li, C., Chen, D., Pan, Y., Tao, Y., Qu, W., . . . Xie, S. (2018). A review on nanosystems as an effective approach against infections of *Staphylococcus aureus*. *International Journal of Nanomedicine*, 7333–7347.

Index

L

M

N

O

P

Q

R

S

T

V

For Product Safety Concerns and Information please contact our
EU representative GPSR@taylorandfrancis.com Taylor & Francis
Verlag GmbH, Kaufingerstraße 24, 80331 München, Germany